Annual Editions: Global Issues, 34/e

Robert Weiner

http://create.mheducation.com

ISBN-10: 1260494160 ISBN-13: 9781260494167

1 2 3 4 5 6 QVS 23 22 21 20 19

Contents

Detailed Table of Contents

Unit 1: The Liberal International Order

Unit 2: Population, Natural Resources, and Climate Change

Unit 3: The Global Political Economy

Unit 4: Women and Gender Equality

Unit 5: War and Peace

Unit 6: Ethics and Values

Preface

This book engages in an analysis of contemporary global issues. A careful reading of the major elite newspapers and magazines as well as the issues that have been emphasized at recent meetings of such organizations as the International Studies Association. An effort to identify important issues and themes has also been based on an analysis of major US reports such as the National Security Strategy, the National Defense Strategy, the US Nuclear posture review, and the State Department's weekly roundup of important events. The identification of current global issues was also facilitated by the editor's attendance of many events and conferences dealing with current global issues held by the think tanks and universities in Washington, DC.

The ability of the international community to deal with global issues takes place within the framework of the forces of globalization in an international system whose structure consists both of state and non-state actors. Globalization has also been occurring in a multipolar system that is characterized by a diffusion of power. States are still the primary actors in the international system, but there has been a phenomenal growth of such non-state actors international and nongovernmental organizations, multinational corporations, and terrorist organizations. The 21st century, however, has also witnessed a process of de-globalization, with the growth of economic nationalism in advanced industrial democracies. The various articles throughout this book discuss the disruption of the international liberal order by these various forces. The six units of the book deal with the challenges which the liberal international order faces in the 21st century.

In publishing Annual Editions, we recognize the enormous role played by magazines, newspapers, and journals of the public press in a broad spectrum of areas. A number of articles are drawn from influential journals such as *Foreign Affairs*, *Foreign Policy*, and *The National Interest* as well, which deal with the most important global issues of the day. Many of these articles are appropriate for students, researchers, and professionals seeking accurate, current information to help bridge the gap between theories and the real world. These articles, however, become more useful for study when those of lasting value are carefully collected, organized, indexed, and reproduced in a low-cost format that provides easy and permanent access when the material is needed. That is the role of Annual Editions.

A number of learning tools are also included in the book. Each article is followed by a set of Critical Thinking questions designed to allow the student to engage in further research and to stimulate classroom discussion, and valuable Internet References that provide the reader with more information about the themes addressed in each article. Each article is also preceded by Learning Outcomes which helps the student to focus on the major themes of each article.

I would like to express my thanks to Mary Foust, McGraw-Hill Product Developer, without whose guidance this project would not have been completed. Special thanks are also due to Dan Torres, whose research assistance was invaluable in selecting articles that appear in this book.

Editor of This Volume

Robert Weiner is an associate lecturer in the Global Affairs Program at the University of Massachusetts/Boston and a Center Associate at the Davis Center for Russian and Eurasian Studies, Harvard University. He has worked as a consultant for Global Integrity, a Washington-based nongovernmental organization that investigates corruption in countries around the world. He is the author or editor of eight books that deal with such topics as Romanian foreign policy, the politics of East Europe, and global issues. He has authored more 20 book chapters, articles, and book reviews. Between 2001 and 2011, he was the Graduate Program Director of the Master's Program in International Relations of the McCormack Graduate School of Policy and Global Studies at the University of Massachusetts/Boston.

Academic Advisory Board Members

Members of the Academic Advisory Board are instrumental in the final selection of articles for each edition of Annual Editions. Their review of the articles for content, level, and appropriateness provides critical direction to the editors and staff. We think that you will find their careful consideration well reflected in this volume.

Unit 1

UNIT

Prepared by: Robert Weiner, *University of Massachusetts, Boston*

The Liberal International Order

This unit focuses on the major debate which foreign policy experts are engaged in about the liberal international order. President Trump has been criticized by the mainstream foreign policy establishment for following a policy of "principled realism" and disrupting the liberal international order. But as the articles in Unit 1 stress, there may be some myths associated with the notion of the liberal international order, since the conventional narrative is that it was established by the United States at the end of the Second World War to protect the interests of the international community. Since then, according to the narrative, the United States has functioned as a global hegemon in order to maintain a rules-based order. The rules-based international order was predicated on free trade, multilateral institutions such as the United Nations, international financial institutions such as the International Monetary Fund and the World Bank Group, and military alliances like The North Atlantic Treaty Organization. President Trump's critics argue that he has pursued a grand strategy which has disrupted all of the major elements of the liberal world order and has established an illiberal order, with the United States as illiberal hegemon. Other foreign policy experts in this unit argue that the United States has always pursued its national interest and put American interests first. Some analysts in this unit point out that the international security environment of the United States has changed into a more multipolar distribution of power. The focus is on the emerging powers of the BRICS (Brazil, Russia, India, China), especially China. The distribution of power in the international system has moved in a more multipolar direction, as the National Security and National Defense Strategies of the Trump administration have emphasized. With increased multipolarity, Great Power competition is back, and China and Russia are considered to be peer competitors of the United States. North Korea and Iran continue to pose a threat to the security of the United States. As the grand strategy of the National Security Strategy of the Trump administration stresses, China is a peer competitor of the United States from a military and economic point of view. The Chinese have engaged in a military and especially naval buildup, in what the United States has now dubbed the Indo-Pacific region. China is projecting its power into the South China and East China Seas. Geography still matters, as the goal from a geopolitical point of view, is to get past the "first island chain of defense" which is designed to contain China. The first island chain of defense stretches from Japan through the Ryukyus and Okinawa through Taiwan and down to the Philippines. Taiwan is a central choke-point preventing China from breaking past the first island chain. China has been building up its navy, claiming sovereignty over disputed island in the South and East China Seas, and building and militarizing artificial islands in the region.

The United States is challenging Chinese claims to its sovereignty over disputed and artificial islands, claiming the right of freedom of navigation through the disputed territories.

China is also competing with the United States in the area of science and technological innovation with the goal of becoming a "cyber superpower." China is working in such areas as robotics, artificial intelligence, and quantum computing. A good example of this is the focus of China on producing graduates in the STE (Science, Technology, Engineering, Mathematics) fields. An example of Chinese competition with the United States in the field of science and technology is the investment which the Chinese are making in the development of quantum computers, as opposed to digital computers. A quantum computer can produce vast amounts of data more quickly than a digital computer. Quantum computers also have national security implications since they will facilitate the breaking of an adversary's cryptographic codes.

Geopolitics also explains the Russian challenge to the US-led international liberal order. For example, the Russian annexation of the Crimean Peninsula in 2014 improved Moscow's geopolitical position in the Black Sea. Moscow's ability to project its power into Eastern Europe and the Middle East was significantly enhanced. Russia has also increased its pressure on the Baltic States, the Nordic States, and Poland.

One of the pillars of the liberal international order is the United Nations Thomas Weiss' article in this unit focuses on the importance of the United Nations as a multilateral organization.

The United Nations is a diplomatic site, at which multilateral diplomacy and bloc politics are practiced, as argued in several studies by the political scientist Thomas Hovet Jr. The United Nations is the result of the evolution of diplomacy as an institution. The UN represents a form of conference diplomacy which also constitutes an important ingredient of the liberal international order. On the other hand, the Trump administration has been very critical of multilateralism. The foreign policy of President Trump has placed more emphasis on the concept of national sovereignty. The members of the United Nations consist of sovereign states, where the legal fiction of state equality and sovereignty is maintained. The Trump administration has focused on US national sovereignty and urged other states to follow their national interests as well.

Article

Prepared by: Robert Weiner, *University of Massachusetts, Boston*

Will the Liberal Order Survive?

The History of an Idea

JOSEPH S. NYE, JR.

Learning Outcomes

After reading this article, you will be able to:

- Learn what the liberal international order means.
- Learn about the role of the United States in maintaining and sustaining the international liberal order.

During the nineteenth century, the United States played a minor role in the global balance of power. The country did not maintain a large standing army, and as late as the 1870s, the U.S. Navy was smaller than the navy of Chile. Americans had no problems using force to acquire land or resources (as Mexico and the Native American nations could attest), but for the most part, both the U.S. government and the American public opposed significant involvement in international affairs outside the Western Hemisphere.

A flirtation with imperialism at the end of the century drew U.S. attention outward, as did the growing U.S. role in the world economy, paving the way for President Woodrow Wilson to take the United States into World War I. But the costs of the war and the failure of Wilson's ambitious attempt to reform international politics afterward turned U.S. attention inward once again during the 1920s and 1930s, leading to the strange situation of an increasingly great power holding itself aloof from an increasingly turbulent world.

Like their counterparts elsewhere, U.S. policy makers sought to advance their country's national interests, usually in straightforward, narrowly defined ways. They saw international politics and economics as an intense competition among states constantly jockeying for position and advantage. When the Great Depression hit, therefore, U.S. officials, like others, raced to protect their domestic economy as quickly and fully as possible, adopting beggar-thy-neighbor tariffs and deepening the crisis in the process. And a few years later, when aggressive dictatorships emerged and threatened peace, they and their counterparts in Europe and elsewhere did something similar in the security sphere, trying to ignore the growing dangers, pass the buck, or defer conflict through appeasement.

By this point, the United States had become the world's strongest power, but it saw no value in devoting resources or attention to providing global public goods such as an open economy or international security. There was no U.S.-led liberal order in the 1930s, and the result was a "low dishonest decade," in the words of W. H. Auden, of depression, tyranny, war, and genocide.

With their countries drawn into the conflagration despite their efforts to avoid it, Western officials spent the first half of the 1940s trying to defeat the Axis powers while working to construct a different and better world for afterward. Rather than continue to see economic and security issues as solely national concerns, they now sought to cooperate with one another, devising a rules-based system that in theory would allow like-minded nations to enjoy peace and prosperity in common.

The liberal international order that emerged after 1945 was a loose array of multilateral institutions in which the United States provided global public goods such as freer trade and freedom of the seas and weaker states were given institutional access to the exercise of U.S. power. The Bretton Woods institutions were set up while the war was still in progress. When other countries proved too poor or weak to fend for themselves afterward, the Truman administration decided to break with U.S. tradition and make open-ended alliances, provide substantial aid to other countries, and deploy U.S. military forces abroad. Washington gave the United Kingdom a major loan in 1946, took responsibility for supporting pro-Western governments in Greece and Turkey in 1947, invested heavily in European recovery with the Marshall Plan in 1948, created NATO in 1949, led a military coalition to protect South Korea from invasion in 1950, and signed a new security treaty with Japan in 1960.

These and other actions both bolstered the order and contained Soviet power. As the American diplomat George Kennan and others noted, there were five crucial areas of industrial productivity and strength in the postwar world: the United States, the Soviet Union, the United Kingdom, continental Europe, and Northeast Asia. To protect itself and prevent a third world war, Washington chose to isolate the Soviet Union and bind itself tightly to the other three, and U.S. troops remain in Europe, Asia, and elsewhere to this day. And within this framework, global economic, social, and ecological interdependence grew. By 1970, economic globalization had recovered to the level it had reached before being disrupted by World War I in 1914.

The mythology that has grown up around the order can be exaggerated. Washington may have displayed a general preference for democracy and openness, but it frequently supported dictators or made cynical self-interested moves along the way. In its first decades, the postwar system was largely limited to a group of like-minded states centered on the Atlantic littoral; it did not include many large countries such as China, India, and the Soviet bloc states, and it did not always have benign effects on nonmembers. In global military terms, the United States was not hegemonic, because the Soviet Union balanced U.S. power. And even when its power was greatest, Washington could not prevent the "loss" of China, the partition of Germany and Berlin, a draw in Korea, Soviet suppression of insurrections within its own bloc, the creation and survival of a communist regime in Cuba, and failure in Vietnam.

Americans have had bitter debates and partisan differences over military interventions and other foreign policy issues over the years, and they have often grumbled about paying for the defense of other rich countries. Still, the demonstrable success of the order in helping secure and stabilize the world over the past seven decades has led to a strong consensus that defending, deepening, and extending this system has been and continues to be the central task of U.S. foreign policy.

Until now, that is—for recently, the desirability and sustainability of the order have been called into question as never before. Some critics, such as U.S. President-elect Donald Trump, have argued that the costs of maintaining the order outweigh its benefits and that Washington would be better off handling its interactions with other countries on a case-by-case transactional basis, making sure it "wins" rather than "loses" on each deal or commitment. Others claim that the foundations of the order are eroding because of a long-term global power transition involving the dramatic rise of Asian economies such as China and India. And still others see it as threatened by a broader diffusion of power from governments to nonstate actors thanks to ongoing changes in politics, society, and technology. The order, in short, is facing its greatest challenges in generations. Can it survive, and will it?

Power Challenged and Diffused

Public goods are benefits that apply to everyone and are denied to no one. At the national level, governments provide many of these to their citizens: safety for people and property, economic infrastructure, a clean environment. In the absence of international government, global public goods—a clean climate or financial stability or freedom of the seas—have sometimes been provided by coalitions led by the largest power, which benefits the most from these goods and can afford to pay for them. When the strongest powers fail to appreciate this dynamic, global public goods are under-produced and everybody suffers.

Some observers see the main threat to the current liberal order coming from the rapid rise of a China that does not always appear to appreciate that great power carries with it great responsibilities. They worry that China is about to pass the United States in power and that when it does, it will not uphold the current order because it views it as an external imposition reflecting others' interests more than its own. This concern is misguided, however, for two reasons: because China is unlikely to surpass the United States in power anytime soon and because it understands and appreciates the order more than is commonly realized.

Contrary to the current conventional wisdom, China is not about to replace the United States as the world's dominant country. Power involves the ability to get what you want from others, and it can involve payment, coercion, or attraction. China's economy has grown dramatically in recent decades, but it is still only 61 percent of the size of the U.S. economy, and its rate of growth is slowing. And even if China does surpass the United States in total economic size some decades from now, economic might is just part of the geopolitical equation. According to the International Institute for Strategic Studies, the United States spends four times as much on its military as does China, and although Chinese capabilities have been increasing in recent years, serious observers think that China will not be able to exclude the United States from the western Pacific, much less exercise global military hegemony. And as for soft power, the ability to attract others, a recent index published by Portland, a London consultancy, ranks the United States first and China 28th. And as China tries to catch up, the United States will not be standing still. It has favorable demographics, increasingly cheap energy, and the world's leading universities and technology companies.

Moreover, China benefits from and appreciates the existing international order more than it sometimes acknowledges. It is one of only five countries with a veto in the UN Security Council and has gained from liberal economic institutions, such as the World Trade Organization (where it accepts dispute-settlement judgments that go against it) and the International Monetary Fund (where its voting rights have increased and it fills an important deputy director position). China is now the second-largest funder of UN peacekeeping forces and has participated in UN programs related to Ebola and climate change. In 2015, Beijing joined with Washington in developing new norms for dealing with climate change and conflicts in cyberspace. On balance, China has tried not to overthrow the current order but rather to increase its influence within it.

The order will inevitably look somewhat different as the twenty-first century progresses. China, India, and other economies will continue to grow, and the U.S. share of the world economy will drop. But no other country, including China, is poised to displace the United States from its dominant position. Even so, the order may still be threatened by a general diffusion of power away from governments toward nonstate actors. The information revolution is putting a number of transnational issues, such as financial stability, climate change, terrorism, pandemics, and cybersecurity, on the global agenda at the same time as it is weakening the ability of all governments to respond.

Complexity is growing, and world politics will soon not be the sole province of governments. Individuals and private organizations—from corporations and nongovernmental organizations to terrorists and social movements—are being empowered, and informal networks will undercut the monopoly on power of traditional bureaucracies. Governments will continue to possess power and resources, but the stage on which they play will become ever more crowded, and they will have less ability to direct the action.

Even if the United States remains the largest power, accordingly, it will not be able to achieve many of its international goals acting alone. For example, international financial stability is vital to the prosperity of Americans, but the United States needs the cooperation of others to ensure it. Global climate change and rising sea levels will affect the quality of life, but Americans cannot manage these problems by themselves. And in a world where borders are becoming more porous, letting in everything from drugs to infectious diseases to terrorism, nations must use soft power to develop networks and build institutions to address shared threats and challenges.

Washington can provide some important global public goods largely by itself. The U.S. Navy is crucial when it comes to policing the law of the seas and defending freedom of navigation, and the U.S. Federal Reserve undergirds international financial stability by serving as a lender of last resort. On the new transnational issues, however, success will require the cooperation of others—and thus empowering others can help the United States accomplish its own goals. In this sense, power becomes a positive-sum game: one needs to think of not just the United States' power over others but also the power to solve problems that the United States can acquire by working with others. In such a world, the ability to connect with others becomes a major source of power, and here, too, the United States leads the pack. The United States comes first in the Lowy Institute's ranking of nations by number of embassies, consulates, and missions. It has some 60 treaty allies, and *The Economist* estimates that nearly 100 of the 150 largest countries lean toward it, while only 21 lean against it.

Increasingly, however, the openness that enables the United States to build networks, maintain institutions, and sustain alliances is itself under siege. This is why the most important challenge to the provision of world order in the twenty-first century comes not from without but from within.

Populism vs. Globalization

Even if the United States continues to possess more military, economic, and soft-power resources than any other country, it may choose not to use those resources to provide public goods for the international system at large. It did so during the interwar years, after all, and in the wake of the conflicts in Afghanistan and Iraq, a 2013 poll found that 52 percent of Americans believed that "the U.S. should mind its own business internationally and let other countries get along the best they can on their own."

The 2016 presidential election was marked by populist reactions to globalization and trade agreements in both major parties, and the liberal international order is a project of just the sort of cosmopolitan elites whom populists see as the enemy. The roots of populist reactions are both economic and cultural. Areas that have lost jobs to foreign competition appear to have tended to support Trump, but so did older white males who have lost status with the rise in power of other demographic groups. The U.S. Census Bureau projects that in less than three decades, whites will no longer be a racial majority in the United States, precipitating the anxiety and fear that contributed to Trump's appeal, and such trends suggest that populist passions will outlast Trump's campaign.

It has become almost conventional wisdom to argue that the populist surge in the United States, Europe, and elsewhere marks the beginning of the end of the contemporary era of globalization and that turbulence may follow in its wake, as happened after the end of an earlier period of globalization a century ago. But circumstances are so different today that the analogy doesn't hold up. There are so many buffers against turbulence now, at both the domestic and the international level, that a descent into economic and geopolitical chaos, as in the 1930s, is not in the cards. Discontent and frustration are likely to continue, and the election of Trump and the British vote to leave the EU demonstrate that populist reactions are common to many Western democracies. Policy elites who want to support globalization and an open economy will clearly need to pay more attention to economic inequality, help those disrupted by change, and stimulate broad-based economic growth.

It would be a mistake to read too much about long-term trends in U.S. public opinion from the heated rhetoric of the recent election. The prospects for elaborate trade agreements such as the Trans-Pacific Partnership and the Transatlantic Trade and Investment Partnership have suffered, but there is not likely to be a reversion to protectionism on the scale of the 1930s. A June 2016 poll by the Chicago Council on Global Affairs, for example, found that 65 percent of Americans thought that globalization was mostly good for the United States, despite concerns about a loss of jobs. And campaign rhetoric notwithstanding, in a 2015 Pew survey, 51 percent of respondents said that immigrants strengthened the country.

Nor will the United States lose the ability to afford to sustain the order. Washington currently spends less than four percent of its GDP on defense and foreign affairs. That is less than half the share that it spent at the height of the Cold War. Alliances are not significant economic burdens, and in some cases, such as that of Japan, it is cheaper to station troops overseas than at home. The problem is not guns versus butter but guns versus butter versus taxes. Because of a desire to avoid raising taxes or further increasing the national debt, the U.S. national security budget is currently locked in a zero-sum trade-off with domestic expenditures on education, infrastructure, and research and development. Politics, not absolute economic constraints, will determine how much is spent on what.

The disappointing track record of recent U.S. military interventions has also undermined domestic support for an engaged global role. In an age of transnational terrorism and refugee crises, keeping aloof from all intervention in the domestic affairs of other countries is neither possible nor desirable. But regions such as the Middle East are likely to experience turmoil for decades, and Washington will need to be more careful about the tasks it takes on. Invasion and occupation breed resentment and opposition, which in turn raise the costs of intervention while lowering the odds of success, further undermining public support for an engaged foreign policy.

Political fragmentation and demagoguery, finally, pose yet another challenge to the United States' ability to provide responsible international leadership, and the 2016 election revealed just how fragmented the American electorate is. The U.S. Senate, for example, has failed to ratify the UN Convention on the Law of the Sea, despite the fact that the country is relying on it to help protect freedom of navigation in the South China Sea against Chinese provocations. Congress failed for five years to fulfill an important U.S. commitment to support the reallocation of International Monetary Fund quotas from Europe to China, even though it would have cost almost nothing to do so. Congress has passed laws violating the international legal principle of sovereign immunity, a principle that protects not just foreign governments but also American diplomatic and military personnel abroad. And domestic resistance to putting a price on carbon emissions makes it hard for the United States to lead the fight against climate change.

The United States will remain the world's leading military power for decades to come, and military force will remain an important component of U.S. power. A rising China and a declining Russia frighten their neighbors, and U.S. security guarantees in Asia and Europe provide critical reassurance for the stability that underlies the prosperity of the liberal order. Markets depend on a framework of security, and maintaining alliances is an important source of influence for the United States.

At the same time, military force is a blunt instrument unsuited to dealing with many situations. Trying to control the domestic politics of nationalist foreign populations is a recipe for failure, and force has little to offer in addressing issues such as climate change, financial stability, or Internet governance. Maintaining networks, working with other countries and international institutions, and helping establish norms to deal with new transnational issues are crucial. It is a mistake to equate globalization with trade agreements. Even if economic globalization were to slow, technology is creating ecological, political, and social globalization that will all require cooperative responses.

Leadership is not the same as domination, and Washington's role in helping stabilize the world and underwrite its continued progress may be even more important now than ever. Americans and others may not notice the security and prosperity that the liberal order provides until they are gone—but by then, it may be too late.

Critical Thinking

1. Why doesn't China pose a threat to the liberal international order?
2. How is the liberal international order expected to change in the 21st century?
3. Why will the United States continue to be the dominant power in the international liberal order?

Internet References

Summary of the 2018 National Defense Strategy

https://dod.defense..gov/Portals/1/.../2018-National-DefenseStrategy-Summary-pdf

Testing the Value of the Postwar International Order—Rand Corp.

https://www.rand.org/pubs/research_reports/RR2226.html

The National Security Strategy of the United States

https://www.whitehouse.gov/articles/new-national-securitystrategy-new-era/

Joseph S. Nye, Jr., is University Distinguished Service Professor at the Harvard Kennedy School of Government and the author of *Is the American Century Over?*

Article Prepared by: Robert Weiner, *University of Massachusetts, Boston*

The Geopolitics of Cyberspace after Snowden

Ron Deibert

Learning Outcomes

After reading this article, you will be able to:

- Discuss the effects of Snowden's revelations on the freedom of the Internet.
- Explain the major changes that have occurred recently in the use of the Internet.

For several years now, it seems that not a day has gone by without a new revelation about the perils of cyberspace: the networks of Fortune 500 companies breached; cyberespionage campaigns uncovered; shadowy hacker groups infiltrating prominent websites and posting extremist propaganda. But the biggest shock came in June 2013 with the first of an apparently endless stream of riveting disclosures from former US National Security Agency (NSA) contractor Edward Snowden. These alarming revelations have served to refocus the world's attention, aiming the spotlight not at cunning cyber activists or sinister data thieves, but rather at the world's most powerful signals intelligence agencies: the NSA, Britain's Government Communications Headquarters (GCHQ), and their allies.

The public is captivated by these disclosures, partly because of the way in which they have been released, but mostly because cyberspace is so essential to all of us. We are in the midst of what might be the most profound communications evolution in all of human history. Within the span of a few decades, society has become completely dependent on the digital information and communication technologies (ICTs) that infuse our lives. Our homes, our jobs, our social networks—the fundamental pillars of our existence—now demand immediate access to these technologies.

With so much at stake, it should not be surprising that cyberspace has become heavily contested. What was originally designed as a small-scale but robust information-sharing network for advanced university research has exploded into the information infrastructure for the entire planet. Its emergence has unsettled institutions and upset the traditional order of things, while simultaneously contributing to a revolution in economics, a path to extraordinary wealth for Internet entrepreneurs, and new forms of social mobilization. These contrasting outcomes have set off a desperate scramble, as stakeholders with competing interests attempt to shape cyberspace to their advantage. There is a geopolitical battle taking place over the future of cyberspace, similar to those previously fought over land, sea, air, and space.

Three major trends have been increasingly shaping cyberspace: the big data explosion, the growing power and influence of the state, and the demographic shift to the global South. While these trends preceded the Snowden disclosures, his leaks have served to alter them somewhat, by intensifying and in some cases redirecting the focus of the conflicts over the Internet. This essay will identify several focal points where the outcomes of these contests are likely [to] be most critical to the future of cyberspace.

Big Data

Before discussing the implications of cyberspace, we need to first understand its characteristics: What is unique about the ICT environment that surrounds us? There have been many extraordinary inventions that revolutionized communications throughout human history: the alphabet, the printing press, the telegraph, radio, and television all come to mind. But arguably the most far-reaching in its effects is the creation and development of social media, mobile connectivity, and cloud computing—referred to in shorthand as "big data." Although these three technological systems are different in many ways, they share one very important characteristic: a vast and rapidly growing volume of personal information, shared (usually voluntarily) with entities separate from the individuals to whom the information applies. Most of those entities are privately owned companies, often headquartered in political jurisdictions other than the one in which the individual providing the information lives (a critical point that will be further examined below).

We are, in essence, turning our lives inside out. Data that used to be stored in our filing cabinets, on our desktop computers, or even in our minds, are now routinely stored on equipment maintained by private companies spread across the globe. This data we entrust to them includes that which we are conscious of and deliberate about—websites visited, emails sent, texts received, images posted—but a lot of which we are unaware.

For example, a typical mobile phone, even when not in use, emits a pulse every few seconds as a beacon to the nearest WiFi router or cellphone tower. Within that beacon is an extraordinary amount of information about the phone and its owner (known as "metadata"), including make and model, the user's name, and geographic location. And that is just the mobile device itself. Most users have within their devices several dozen applications (more than 50 billion apps have been downloaded from Apple's iTunes store for social networking, fitness, health, games, music, shopping, banking, travel, even tracking sleep patterns), each of which typically gives itself permission to extract data about the user and the device. Some applications take the practice of data extraction several bold steps further, by requesting access to geolocation information, photo albums, contacts, or even the ability to turn on the device's camera and microphone.

We leave behind a trail of digital "exhaust" wherever we go. Data related to our personal lives are compounded by the numerous and growing Internet-connected sensors that permeate our technological environment. The term "Internet of Things" refers to the approximately 15 billion devices (phones, computers, cars, refrigerators, dishwashers, watches, even eyeglasses) that now connect to the Internet and to each other, producing trillions of ever-expanding data points. These data points create an ethereal layer of digital exhaust that circles the globe, forming, in essence, a digital stratosphere.

Given the virtual characteristics of the digital experience, it may be easy to overlook the material properties of communication technologies. But physical geography is an essential component of cyberspace: *Where* technology is located is as important as *what* it is. While our Internet activities may seem a kind of ephemeral and private adventure, they are in fact embedded in a complex infrastructure (material, logistical, and regulatory) that in many cases crosses several borders. We assume that the data we create, manipulate, and distribute are in our possession. But in actuality, they are transported to us via signals and waves, through cables and wires, from distant servers that may or may not be housed in our own political jurisdiction. It is actual matter we are dealing with when we go online, and that matters—a lot. The data that follow us around, that track our lives and habits, do not disappear; they live in the servers of the companies that own and operate the infrastructure. What is done with this information is a decision for those companies to make. The details are buried in their rarely read terms of service, or, increasingly, in special laws, requirements, or policies laid down by the governments in whose jurisdictions they operate.

The vast majority of Internet users now live in the global South.

Big State

The Internet started out as an isolated experiment largely separate from government. In the early days, most governments had no Internet policy, and those that did took a deliberately laissez-faire approach. Early Internet enthusiasts mistakenly understood this lack of policy engagement as a property unique to the technology. Some even went so far as to predict that the Internet would bring about the end of organized government altogether. Over time, however, state involvement has expanded, resulting in an increasing number of Internet-related laws, regulations, standards, and practices. In hindsight, this was inevitable. Anything that permeates our lives so thoroughly naturally introduces externalities—side effects of industrial or commercial activity—that then require the establishment of government policy. But as history demonstrates, linear progress is always punctuated by specific events—and for cyberspace, that event was 9/11.

We continue to live in the wake of 9/11. The events of that day in 2001 profoundly shaped many aspects of society. But no greater impact can be found than the changes it brought to cyberspace governance and security, specifically with respect to the role and influence of governments. One immediate impact was the acceleration of a change in threat perception that had been building for years.

During the Cold War, and largely throughout the modern period (roughly the eighteenth century onward), the primary threat for most governments was "interstate" based. In this paradigm, the state's foremost concern is a cross-border invasion or attack—the idea that another country's military could use force and violence in order to gain control. After the Cold War, and especially since 9/11, the concern has shifted to a different threat paradigm: that a violent attack could be executed by a small extremist group, or even a single human being who could blow himself or herself up in a crowded mall, hijack an airliner, or hack into critical infrastructure. Threats are now dispersed across all of society, regardless of national borders. As a result, the focus of the state's security gaze has become omnidirectional.

Accompanying this altered threat perception are legal and cultural changes, particularly in reaction to what was widely perceived as the reason for the 9/11 catastrophe in the first place: a "failure to connect the dots." The imperative shifted from the micro to the macro. Now, it is not enough to simply look for a needle in the haystack. As General Keith Alexander (former head of the NSA and the US Cyber Command) said, it is now necessary to collect "the entire haystack." Rapidly, new laws have been introduced that substantially broaden the reach of law enforcement and intelligence agencies, the most notable of them being the Patriot Act in the United States—although many other countries have followed suit.

This imperative to "collect it all" has focused government attention squarely on the private sector, which owns and operates most of cyberspace. States began to apply pressure on companies to act as a proxy for government controls—policing their own networks for content deemed illegal, suspicious, or a threat to national security. Thanks to the Snowden disclosures, we now have a much clearer picture of how this pressure manifests itself. Some companies have been paid fees to collude, such as Cable and Wireless (now owned by Vodafone), which was paid tens of millions of pounds by the GCHQ to install surveillance equipment on its networks. Other companies have been subjected to formal or informal pressures, such as court orders, national security letters, the with-holding of operating licenses, or even appeals to patriotism. Still others became the targets of computer exploitation, such as US-based Google, whose back-end data infrastructure was secretly hacked into by the NSA.

This manner of government pressure on the private sector illustrates the importance of the physical geography of cyberspace. Of course, many of the corporations that own and operate the infrastructure—companies like Facebook, Microsoft, Twitter, Apple, and Google—are headquartered in the United States. They are subject to US national security law and, as a consequence, allow the government to benefit from a distinct home-field advantage in its attempt to "collect it all." And that it does—a staggering volume, as it turns out. One top-secret NSA slide from the Snowden disclosures reveals that by 2011, the United States (with the cooperation of the private sector) was collecting and archiving about 15 billion Internet metadata records *every single day.* Contrary to the expectations of early Internet enthusiasts, the US government's approach to cyberspace—and by extension that of many other governments as well—has been anything but laissez-faire in the post-9/11 era. While cyberspace may have been born largely in the absence of states, as it has matured states have become an inescapable and dominant presence.

Domain Domination

After 9/11, there was also a shift in US military thinking that profoundly affected cyberspace. The definition of cyberspace as a single "domain"— equal to land, sea, air, and space—was formalized in the early 2000s, leading to the imperative to dominate and rule this domain; to develop offensive capabilities to fight and win wars within cyberspace. A Rubicon was crossed with the Stuxnet virus, which sabotaged Iranian nuclear enrichment facilities. Reportedly engineered jointly by the United States and Israel, the Stuxnet attack was the first de facto act of war carried out entirely through cyberspace. As is often the case in international security dynamics, as one country reframes its objectives and builds up its capabilities, other countries follow suit. Dozens of governments now have within their armed forces dedicated "cyber commands" or their equivalents.

The race to build capabilities also has a ripple effect on industry, as the private sector positions itself to reap the rewards of major cyber-related defense contracts. The imperatives of mass surveillance and preparations for cyberwarfare across the globe have reoriented the defense industrial base. It is noteworthy in this regard how the big data explosion and the growing power and influence of the state are together generating a political-economic dynamic. The aims of the Internet economy and those of state security converge around the same functional needs: collecting, monitoring, and analyzing as much data as possible. Not surprisingly, many of the same firms service both segments. For example, companies that market facial recognition systems find their products being employed by Facebook on the one hand and the Central Intelligence Agency on the other.

As private individuals who live, work, and play in the cyber realm, we provide the seeds that are then cultivated, harvested, and delivered to market by a massive machine, fueled by the twin engines of corporate and national security needs. The confluence of these two major trends is creating extraordinary tensions in state-society relations, particularly around privacy. But perhaps the most important implications relate to the fact that the market for the cybersecurity industrial complex knows no boundaries—an ominous reality in light of the shifting demographics of cyberspace.

Southern Shift

While the "what" of cyberspace is critical, the "who" is equally important. There is a major demographic shift happening today that is easily overlooked, especially by users in the West, where the technology originates. The vast majority of Internet users now live in the global South. Of the 6 billion mobile devices in circulation, over 4 billion are located in the developing world. In 2001, 8 of every 100 citizens in developing nations owned a mobile subscription. That number has now jumped to 80. In Indonesia, the number of Internet users increases each month by a stunning 800,000. Nigeria had 200,000 Internet users in 2000; today, it has 68 million.

Remarkably, some of the fastest growing online populations are emerging in countries with weak governmental structures or corrupt, autocratic, or authoritarian regimes. Others are developing in zones of conflict, or in countries that have only recently gone through difficult transitions to democracy. Some of the fastest growth rates are in "failed" states, or in countries riven by ethnic rivalries or challenged by religious differences and sensitivities, such as Nigeria, India, Pakistan, Indonesia, and Thailand. Many of these countries do not have long-standing democratic traditions, and therefore lack proper systems of accountability to guard against abuses of power. In some, corruption is rampant, or the military has disproportionate influence.

Consider the relationship between cyberspace and authoritarian rule. We used to mock authoritarian regimes as slow-footed, technologically challenged dinosaurs that would be inevitably weeded out by the information age. The reality has proved more nuanced and complex. These regimes are proving much more adaptable than expected. National-level Internet controls on content and access to information in these countries are now a growing norm. Indeed, some are beginning to affect the very technology itself, rather than vice versa.

In China (the country with the world's most Internet users), "foreign" social media like Facebook, Google, and Twitter are banned in favor of nationally based, more easily controlled alternatives. For example, We Chat—owned by China-based parent company Tencent—is presently the fifth-largest Internet company in the world after Google, Amazon, Alibaba, and eBay, and as of August 2014 it had 438 million active users (70 million outside China) and a public valuation of over $400 billion. China's popular chat applications and social media are required to police the country's networks with regard to politically sensitive content, and some even have hidden censorship and surveillance functionality "baked" into their software. Interestingly, some of We Chat's users outside China began experiencing the same type of content filtering as users inside China, an issue that Tencent claimed was due to a software bug (which it promptly fixed). But the implication of such extraterritorial applications of national-level controls is certainly worth further scrutiny, particularly as China-based companies begin to expand their service offerings in other countries and regions.

It is important to understand the historical context in which this rapid growth is occurring. Unlike the early adopters of the Internet in the West, citizens in the developing world are plugging in and connecting after the Snowden disclosures, and with the model of the NSA in the public domain. They are coming online with cybersecurity at the top of the international agenda, and fierce international competition emerging throughout cyberspace, from the submarine cables to social media. Political leaders in these countries have at their disposal a vast arsenal of products, services, and tools that provide their regimes with highly sophisticated forms of information control. At the same time, their populations are becoming more savvy about using digital media for political mobilization and protest.

While the digital innovations that we take advantage of daily have their origins in high-tech libertarian and free-market hubs like Silicon Valley, the future of cyberspace innovation will be in the global South. Inevitably, the assumptions, preferences, cultures, and controls that characterize that part of the world will come to define cyberspace as much as those of the early entrepreneurs of the information age did in its first two decades.

Who Rules?

Cyberspace is a complex technological environment that spans numerous industries, governments, and regions. As a consequence, there is no one single forum or international organization for cyberspace. Instead, governance is spread throughout numerous small regimes, standard-setting forums, and technical organizations from the regional to the global. In the early days, Internet governance was largely informal and led by non-state actors, especially engineers. But over time, governments have become heavily involved, leading to more politicized struggles at international meetings.

Although there is no simple division of camps, observers tend to group countries into those that prefer a more open Internet and a tightly restricted role for governments versus those that prefer a more centralized and state-led form of governance, preferably through the auspices of the United Nations. The United States, the United Kingdom, other European nations, and Asian democracies are typically grouped in the former, with China, Russia, Iran, Saudi Arabia, and other nondemocratic countries grouped in the latter. A large number of emerging market economies, led by Brazil, India, and Indonesia, are seen as "swing states" that could go either way.

Prior to the Snowden disclosures, the battle lines between these opposing views were becoming quite acute—especially around the December 2012 World Congress on Information Technology (WCIT), where many feared Internet governance would fall into UN (and thus more state-controlled) hands. But the WCIT process stalled, and those fears never materialized, in part because of successful lobbying by the United States and its allies, and by Internet companies like Google. After the

Snowden disclosures, however, the legitimacy and credibility of the "Internet freedom" camp have been considerably weakened, and there are renewed concerns about the future of cyberspace governance.

The original promise of the Internet as a forum for free exchange of information is at risk.

Meanwhile, less noticed but arguably more effective have been lower-level forms of Internet governance, particularly in regional security forums and standards-setting organizations. For example, Russia, China, and numerous Central Asian states, as well as observer countries like Iran, have been coordinating their Internet security policies through the Shanghai Cooperation Organization (SCO). Recently, the SCO held military exercises designed to counter Internet-enabled opposition of the sort that participated in the "color revolutions" in former Soviet states. Governments that prefer a tightly controlled Internet are engaging in partnerships, sharing best practices, and jointly developing information control platforms through forums like the SCO. While many casual Internet observers ruminate over the prospect of a UN takeover of the Internet that may never materialize, the most important norms around cyberspace controls could be taking hold beneath the spotlight and at the regional level.

Technological Sovereignty

Closely related to the questions surrounding cyberspace governance at the international level are issues of domestic-level Internet controls, and concerns over "technological sovereignty." This area is one where the reactions to the Snowden disclosures have been most palpably felt in the short term, as countries react to what they see as the US "home-field advantage" (though not always in ways that are straightforward). Included among the leaked details of US- and GCHQ-led operations to exploit the global communications infrastructure are numerous accounts of specific actions to compromise state networks, or even the handheld devices of government officials—most notoriously, the hacking of German Chancellor Angela Merkel's personal cellphone and the targeting of Brazilian government officials' classified communications. But the vast scope of US-led exploitation of global cyberspace, from the code to the undersea cables and everything in between, has set off shockwaves of indignation and loud calls to take immediate responses to restore "technological sovereignty."

For example, Brazil has spearheaded a project to lay a new submarine cable linking South America directly to Europe, thus bypassing the United States. Meanwhile, many European politicians have argued that contracts with US-based companies that may be secretly colluding with the NSA should be cancelled and replaced with contracts for domestic industry to implement regional and/or nationally autonomous data-routing policies—arguments that European industry has excitedly supported. It is sometimes difficult to unravel whether such measures are genuinely designed to protect citizens, or are really just another form of national industrial protectionism, or both. Largely obscured beneath the heated rhetoric and underlying self-interest, however, are serious questions about whether any of the measures proposed would have any more than a negligible impact when it comes to actually protecting the confidentiality and integrity of communications. As the Snowden disclosures reveal, the NSA and GCHQ have proved to be remarkably adept at exploiting traffic, no matter where it is based, by a variety of means.

A more troubling concern is that such measures may end up unintentionally legitimizing national cyberspace controls, particularly for developing countries, "swing states," and emerging markets. Pointing to the Snowden disclosures and the fear of NSA-led surveillance can be useful for regimes looking to subject companies and citizens to a variety of information controls, from censorship to surveillance. Whereas policy makers previously might have had concerns about being cast as pariahs or infringers on human rights, they now have a convenient excuse supported by European and other governments' reactions.

Spyware Bazaar

One by-product of the huge growth in military and intelligence spending on cybersecurity has been the fueling of a global market for sophisticated surveillance and other security tools. States that do not have an in-house operation on the level of the NSA can now buy advanced capabilities directly from private contractors. These tools are proving particularly attractive to many regimes that face ongoing insurgencies and other security challenges, as well as persistent popular protests. Since the advertised end uses of these products and services include many legitimate needs, such as network traffic management or the lawful interception of data, it is difficult to prevent abuses, and hard even for the companies themselves to know to what ends their products and services might ultimately be directed. Many therefore employ the term "dual-use" to describe such tools.

We leave behind a trail of digital "exhaust" wherever we go.

Research by the University of Toronto's Citizen Lab from 2012 to 2014 has uncovered numerous cases of human rights activists targeted by advanced digital spyware manufactured by Western companies. Once implanted on a target's device, this spyware can extract files and contacts, send emails and text messages, turn on the microphone and camera, and track the location of the user. If these were isolated incidences, perhaps we could write them off as anomalies. But the Citizen Lab's international scan of the command and control servers of these products—the computers used to send instructions to infected devices—has produced disturbing evidence of a global market that knows no boundaries. Citizen Lab researchers found one product, Finspy, marketed by a UK company, Gamma Group, in a total of 25 countries— some with dubious human rights records, such as Bahrain, Bangladesh, Ethiopia, Qatar, and Turkmenistan. A subsequent Citizen Lab report found that 21 governments are current or former users of a spyware product sold by an Italian company called Hacking Team, including 9 that received the lowest ranking, "authoritarian," in the *Economist's* 2012 Democracy Index.

Meanwhile, a 2014 Privacy International report on surveillance in Central Asia says many of the countries in the region have implemented far-reaching surveillance systems at the base of their telecommunications networks, using advanced US and Israeli equipment, and supported by Russian intelligence training. Products that provide advanced deep packet inspection (the capability to inspect data packets in detail as they flow through networks), content filtering, social network mining, cellphone tracking, and even computer attack targeting are being developed by Western firms and marketed worldwide to regimes seeking to limit democratic participation, isolate and identify opposition, and infiltrate meddlesome adversaries abroad.

Pushing Back

The picture of the cyberspace landscape painted above is admittedly quite bleak, and therefore one-sided. The contests over cyberspace are multidimensional and include many groups and individuals pushing for technologies, laws, and norms that support free speech, privacy, and access to information. Here, too, the Snowden disclosures have had an animating effect, raising awareness of risks and spurring on change. Whereas vague concerns about widespread digital spying were voiced by a minority and sometimes trivialized before Snowden's disclosures, now those fears have been given real substance and credibility, and surveillance is increasingly seen as a practical risk that requires some kind of remediation.

The Snowden disclosures have had a particularly salient impact on the private sector, the Internet engineering community, and civil society. The revelations have left many US companies in a public relations nightmare, with their trust weakened and lucrative contracts in jeopardy. In response, companies are pushing back. It is now standard for many telecommunications and social media companies to issue transparency reports about government requests to remove information from websites or share user data with authorities. US-based Internet companies even sued the government over gag orders that bar them from disclosing information on the nature and number of requests for user information. Others, including Google, Microsoft, Apple, Facebook, and WhatsApp, have implemented end-to-end encryption.

Internet engineers have reacted strongly to revelations showing that the NSA and its allies have subverted their security standards-setting processes. They are redoubling efforts to secure communications networks wholesale as a way to shield all users from mass surveillance, regardless of who is doing the spying. Among civil society groups that depend on an open cyberspace, the Snowden disclosures have helped trigger a burgeoning social movement around digital-security tool development and training, as well as more advanced research on the nature and impacts of information controls.

Wild Card

The cyberspace environment in which we live and on which we depend has never been more in flux. Tensions are mounting in several key areas, including Internet governance, mass and targeted surveillance, and military rivalry. The original promise of the Internet as a forum for free exchange of information is at risk. We are at a historical fork in the road: Decisions could take us down one path where cyberspace continues to evolve into a global commons, empowering individuals through access to information and freedom of speech and association, or down another path where this ideal meets its eventual demise. Securing cyberspace in ways that encourage freedom, while limiting controls and surveillance, is going to be a serious challenge.

Trends toward militarization and greater state control were already accelerating before the Snowden disclosures, and seem unlikely to abate in the near future. However, the leaks have thrown a wild card into the mix, creating opportunities for alternative approaches emphasizing human rights, corporate social responsibility, norms of mutual restraint, cyberspace arms control, and the rule of law. Whether such measures will be enough to stem the tide of territorialized controls remains to be seen. What is certain, however, is that a debate over the future of cyberspace will be a prominent feature of world politics for many years to come.

Critical Thinking

1. What is the role of the UN in the governance of the Internet?
2. What countries support a free Internet and what countries support state control of the Internet?
3. What countries are considered "swing" states in the use of the Internet?

Internet References

Berkman Center for Internet and Society
https://cyber.law.harvard.edu

Internet Governance Forum
intgovforum.org

Internet Society
internetsociety.org

National Security Agency
https://www.nsa.gov

Ron Deibert is a professor of political science and director of the Canada Center for Global Security Studies and the Citizen Lab at the University of Toronto. His latest book is *Black Code: Inside the Battle for Cyberspace* (Signal, 2013).

Article

Prepared by: Robert Weiner, *University of Massachusetts, Boston*

The Myth of the Liberal Order: From Historical Accident to Conventional Wisdom

GRAHAM ALLISON

Learning Outcomes

After reading this article, you will be to:

- Learn what caused the long peace after World War II.
- Learn why the spread of democracy was not a vital US interest.

Among the debates that have swept the U.S. foreign policy community since the beginning of the Trump administration, alarm about the fate of the liberal international rules-based order has emerged as one of the few fixed points. From the international relations scholar G. John Ikenberry's claim that "for seven decades the world has been dominated by a western liberal order" to U.S. Vice President Joe Biden's call in the final days of the Obama administration to "act urgently to defend the liberal international order," this banner waves atop most discussions of the United States' role in the world.

About this order, the reigning consensus makes three core claims. First, that the liberal order has been the principal cause of the so-called long peace among great powers for the past seven decades. Second, that constructing this order has been the main driver of U.S. engagement in the world over that period. And third, that U.S. President Donald Trump is the primary threat to the liberal order—and thus to world peace. The political scientist Joseph Nye, for example, has written, "The demonstrable success of the order in helping secure and stabilize the world over the past seven decades has led to a strong consensus that defending, deepening, and extending this system has been and continues to be the central task of U.S. foreign policy." Nye has gone so far as to assert: "I am not worried by the rise of China. I am more worried by the rise of Trump."

Although all these propositions contain some truth, each is more wrong than right. The "long peace" was the not the result of a liberal order but the byproduct of the dangerous balance of power between the Soviet Union and the United States during the four and a half decades of the Cold War and then of a brief period of U.S. dominance. U.S. engagement in the world has been driven not by the desire to advance liberalism abroad or to build an international order but by the need to do what was necessary to preserve liberal democracy at home. And although Trump is undermining key elements of the current order, he is far from the biggest threat to global stability.

These misconceptions about the liberal order's causes and consequences lead its advocates to call for the United States to strengthen the order by clinging to pillars from the past and rolling back authoritarianism around the globe. Yet rather than seek to return to an imagined past in which the United States molded the world in its image, Washington should limit its efforts to ensuring sufficient order abroad to allow it to concentrate on reconstructing a viable liberal democracy at home.

Conceptual Jell-O

The ambiguity of each of the terms in the phrase "liberal international rules-based order" creates a slipperiness that allows the concept to be applied to almost any situation. When, in 2017, members of the World Economic Forum in Davos crowned Chinese President Xi Jinping the leader of the liberal economic order—even though he heads the most protectionist, mercantilist, and predatory major economy in the world—they revealed that, at least in this context, the word "liberal" has come unhinged.

What is more, "rules-based order" is redundant. Order is a condition created by rules and regularity. What proponents of the liberal international rules-based order really mean is an order that embodies good rules, ones that are equal or fair. The United States is said to have designed an order that others willingly embrace and sustain.

Many forget, however, that even the UN Charter, which prohibits nations from using military force against other nations or intervening in their internal affairs, privileges the strong over the weak. Enforcement of the charter's prohibitions is the preserve of the UN Security Council, on which each of the five great powers has a permanent seat—and a veto. As the Indian strategist C. Raja Mohan has observed, superpowers are "exceptional"; that is, when they decide it suits their purpose, they make exceptions for themselves. The fact that in the first 17 years of this century, the self-proclaimed leader of the liberal order invaded two countries, conducted air strikes and Special Forces raids to kill hundreds of people it unilaterally deemed to be terrorists, and subjected scores of others to "extraordinary rendition," often without any international legal authority (and sometimes without even national legal authority), speaks for itself.

Cold War Order

The claim that the liberal order produced the last seven decades of peace overlooks a major fact: the first four of those decades were defined not by a liberal order but by a cold war between two polar opposites. As the historian who named this "long peace" has explained, the international system that prevented great-power war during that time was the unintended consequence of the struggle between the Soviet Union and the United States. In John Lewis Gaddis' words, "Without anyone's having designed it, and without any attempt whatever to consider the requirements of justice, the nations of the postwar era lucked into a system of international relations that, because it has been based upon realities of power, has served the cause of order—if not justice—better than one might have expected."

During the Cold War, both superpowers enlisted allies and clients around the globe, creating what came to be known as a bipolar world. Within each alliance or bloc, order was enforced by the superpower (as Hungarians and Czechs discovered when they tried to defect in 1956 and 1968, respectively, and as the British and French learned when they defied U.S. wishes in 1956, during the Suez crisis). Order emerged from a balance of power, which allowed the two superpowers to develop the constraints that preserved what U.S. President John F. Kennedy called, in the aftermath of the Cuban missile crisis of 1962, the "precarious status quo."

What moved a country that had for almost two centuries assiduously avoided entangling military alliances, refused to maintain a large standing military during peacetime, left international economics to others, and rejected the League of Nations to use its soldiers, diplomats, and money to reshape half the world? In a word, fear. The strategists revered by modern U.S. scholars as "the wise men" believed that the Soviet Union posed a greater threat to the United States than Nazism had. As the diplomat George Kennan wrote in his legendary "Long Telegram," the Soviet Union was "a political force committed fanatically to the belief that with US there can be no permanent modus vivendi." Soviet Communists, Kennan wrote, believed it was necessary that "our society be disrupted, our traditional way of life be destroyed, the international authority of our state be broken, if Soviet power [was] to be secure."

Before the nuclear age, such a threat would have required a hot war as intense as the one the United States and its allies had just fought against Nazi Germany. But after the Soviet Union tested its first atomic bomb, in 1949, American statesmen began wrestling with the thought that total war as they had known it was becoming obsolete. In the greatest leap of strategic imagination in the history of U.S. foreign policy, they developed a strategy for a form of combat never previously seen, the conduct of war by every means short of physical conflict between the principal combatants.

To prevent a cold conflict from turning hot, they accepted—for the time being—many otherwise unacceptable facts, such as the Soviet domination of Eastern Europe. They modulated their competition with mutual constraints that included three noes: no use of nuclear weapons, no overt killing of each other's soldiers, and no military intervention in the other's recognized sphere of influence.

American strategists incorporated Western Europe and Japan into this war effort because they saw them as centers of economic and strategic gravity. To this end, the United States launched the Marshall Plan to rebuild Western Europe, founded the International Monetary Fund and the World Bank, and negotiated the General Agreement on Tariffs and Trade to promote global prosperity. And to ensure that Western Europe and Japan remained in active cooperation with the United States, it established NATO and the U.S.-Japanese alliance.

Each initiative served as a building block in an order designed first and foremost to defeat the Soviet adversary. Had there been no Soviet threat, there would have been no Marshall Plan and no NATO. The United States has never promoted liberalism abroad when it believed that doing so would pose a significant threat to its vital interests at home. Nor has it ever refrained from using military force to protect its interests when the use of force violated international rules.

Nonetheless, when the United States has had the opportunity to advance freedom for others—again, with the important caveat that doing so would involve little risk to itself—it has acted. From the founding of the republic, the nation has embraced radical, universalistic ideals. In proclaiming that "all" people "are created equal," the Declaration of Independence did not mean just those living in the 13 colonies.

It was no accident that in reconstructing its defeated adversaries Germany and Japan and shoring up its allies in Western Europe, the United States sought to build liberal democracies that would embrace shared values as well as shared interests. The ideological campaign against the Soviet Union hammered home fundamental, if exaggerated, differences between "the free world" and "the evil empire." Moreover, American policy makers knew that in mobilizing and sustaining support in Congress and among the public, appeals to values are as persuasive as arguments about interests.

In his memoir, *Present at the Creation*, former U.S. Secretary of State Dean Acheson, an architect of the postwar effort, explained the thinking that motivated U.S. foreign policy. The prospect of Europe falling under Soviet control through a series of "'settlements by default' to Soviet pressure" required the "creation of strength throughout the free world" that would "show the Soviet leaders by successful containment that they could not hope to expand their influence throughout the world." Persuading Congress and the American public to support this undertaking, Acheson acknowledged, sometimes required making the case "clearer than truth."

Unipolar Order

In the aftermath of the disintegration of the Soviet Union and Russian President Boris Yeltsin's campaign to "bury communism," Americans were understandably caught up in a surge of triumphalism. The adversary on which they had focused for over 40 years stood by as the Berlin Wall came tumbling down and Germany reunified. It then joined with the United States in a unanimous UN Security Council resolution authorizing the use of force to throw the Iraqi military out of Kuwait. As the iron fist of Soviet oppression withdrew, free people in Eastern Europe embraced market economies and democracy. U.S. President George H. W. Bush declared a "new world order." Hereafter, under a banner of "engage and enlarge," the United States would welcome a world clamoring to join a growing liberal order.

Writing about the power of ideas, the economist John Maynard Keynes noted, "Madmen in authority, who hear voices in the air, are distilling their frenzy from some academic scribbler of a few years back." In this case, American politicians were following a script offered by the political scientist Francis Fukuyama in his best-selling 1992 book, *The End of History and the Last Man*. Fukuyama argued that millennia of conflict among ideologies were over. From this point on, all nations would embrace free-market economics to make their citizens rich and democratic governments to make them free. "What we may be witnessing," he wrote, "is not just the end of the Cold War, or the passing of a particular period of postwar history, but the end of history as such: that is, the end point of mankind's ideological evolution and the universalization of Western liberal democracy as the final form of human government." In 1996, *The New York Times* columnist Thomas Friedman went even further by proclaiming the "Golden Arches Theory of Conflict Prevention": "When a country reaches a certain level of economic development, when it has a middle class big enough to support a McDonald's, it becomes a McDonald's country, and people in McDonald's countries don't like to fight wars; they like to wait in line for burgers."

This vision led to an odd coupling of neoconservative crusaders on the right and liberal interventionists on the left. Together, they persuaded a succession of U.S. presidents to try to advance the spread of capitalism and liberal democracy through the barrel of a gun. In 1999, Bill Clinton bombed Belgrade to force it to free Kosovo. In 2003, George W. Bush invaded Iraq to topple its president, Saddam Hussein. When his stated rationale for the invasion collapsed after U.S. forces were unable to find weapons of mass destruction, Bush declared a new mission: "to build a lasting democracy that is peaceful and prosperous." In the words of Condoleezza Rice, his national security adviser at the time, "Iraq and Afghanistan are vanguards of this effort to spread democracy and tolerance and freedom throughout the Greater Middle East." And in 2011, Barack Obama embraced the Arab Spring's promise to bring democracy to the nations of the Middle East and sought to advance it by bombing Libya and deposing its brutal leader, Muammar al-Qaddafi. Few in Washington paused to note that in each case, the unipolar power was using military force to impose liberalism on countries whose governments could not strike back. Since the world had entered a new chapter of history, lessons from the past about the likely consequences of such behavior were ignored.

As is now clear, the end of the Cold War produced a unipolar moment, not a unipolar era. Today, foreign policy elites have woken up to the meteoric rise of an authoritarian China, which now rivals or even surpasses the United States in many domains, and the resurgence of an assertive, illiberal Russian nuclear superpower, which is willing to use its military to change both borders in Europe and the balance of power in the Middle East. More slowly and more painfully, they are discovering that the United States' share of global power has shrunk. When measured by the yardstick of purchasing power parity, the U.S. economy, which accounted for half of the world's GDP after World War II, had fallen to less than a quarter of global GDP by the end of the Cold War and stands at just one-seventh today. For a nation whose core strategy has been to overwhelm challenges with resources, this decline calls into question the terms of U.S. leadership.

This rude awakening to the return of history jumps out in the Trump administration's National Security Strategy and National Defense Strategy, released at the end of last year and the beginning of this year, respectively. The NDS notes that in the unipolar decades, "the United States has enjoyed uncontested or dominant superiority in every operating domain." As a consequence, "we could generally deploy our forces when we wanted, assemble them where we wanted, and operate how we wanted." But today, as the NSS observes, China and Russia "are fielding military capabilities designed to deny America access in times of crisis and to contest our ability to operate freely." Revisionist powers, it concludes, are "trying to change the international order in their favor."

The American Experiment

During most of the nation's 242 years, Americans have recognized the necessity to give priority to ensuring freedom at home over advancing aspirations abroad. The Founding Fathers were acutely aware that constructing a government in which free citizens would govern themselves was an uncertain, hazardous undertaking. Among the hardest questions they confronted was how to create a government powerful enough to ensure Americans' rights at home and protect them from enemies abroad without making it so powerful that it would abuse its strength.

Their solution, as the presidential scholar Richard Neustadt wrote, was not just a "separation of powers" among the executive, legislative, and judicial branches but "separated institutions sharing power." The Constitution was an "invitation to struggle." And presidents, members of Congress, judges, and even journalists have been struggling ever since. The process was not meant to be pretty. As Supreme Court Justice Louis Brandeis explained to those frustrated by the delays, gridlock, and even idiocy these checks and balances sometimes produce, the founders' purpose was "not to promote efficiency but to preclude the exercise of arbitrary power."

From this beginning, the American experiment in self-government has always been a work in progress. It has lurched toward failure on more than one occasion. When Abraham Lincoln asked "whether that nation, or any nation so conceived, . . . can long endure," it was not a rhetorical question. But repeatedly and almost miraculously, it has demonstrated a capacity for renewal and reinvention. Throughout this ordeal, the recurring imperative for American leaders has been to show that liberalism can survive in at least one country.

For nearly two centuries, that meant warding off foreign intervention and leaving others to their fates. Individual Americans may have sympathized with French revolutionary cries of "Liberty, equality, fraternity!"; American traders may have spanned the globe; and American missionaries may have sought to win converts on all continents. But in choosing when and where to spend its blood and treasure, the U.S. government focused on the United States.

Only in the aftermath of the Great Depression and World War II did American strategists conclude that the United States' survival required greater entanglement abroad. Only when they perceived a Soviet attempt to create an empire that would pose an unacceptable threat did they develop and sustain the alliances and institutions that fought the Cold War. Throughout that effort, as NSC-68, a Truman administration national security policy paper that summarized U.S. Cold War strategy, stated, the mission was "to preserve the United States as a free nation with our fundamental institutions and values intact."

Sufficient unto the Day

Among the current, potentially mortal threats to the global order, Trump is one, but not the most important. His withdrawal from initiatives championed by earlier administrations aimed at constraining greenhouse gas emissions and promoting trade has been unsettling, and his misunderstanding of the strength that comes from unity with allies is troubling. Yet the rise of China, the resurgence of Russia, and the decline of the United States' share of global power each present much larger challenges than Trump. Moreover, it is impossible to duck the question: Is Trump more a symptom or a cause?

While I was on a recent trip to Beijing, a high-level Chinese official posed an uncomfortable question to me. Imagine, he said, that as much of the American elite believes, Trump's character and experience make him unfit to serve as the leader of a great nation. Who would be to blame for his being president? Trump, for his opportunism in seizing victory, or the political system that allowed him to do so?

No one denies that in its current form, the U.S. government is failing. Long before Trump, the political class that brought unending, unsuccessful wars in Afghanistan, Iraq, and Libya, as well as the financial crisis and Great Recession, had discredited itself. These disasters have done more to diminish confidence in liberal self-government than Trump could do in his critics' wildest imaginings, short of a mistake that leads to a catastrophic war. The overriding challenge for American believers in democratic governance is thus nothing less than to reconstruct a working democracy at home.

Fortunately, that does not require converting the Chinese, the Russians, or anyone else to American beliefs about liberty. Nor does it necessitate changing foreign regimes into democracies. Instead, as Kennedy put it in his American University commencement speech, in 1963, it will be enough to sustain a world order "safe for diversity"—liberal and illiberal alike.

That will mean adapting U.S. efforts abroad to the reality that other countries have contrary views about governance and seek to establish their own international orders governed by their own rules. Achieving even a minimal order that can accommodate that diversity will take a surge of strategic imagination as far beyond the current conventional wisdom as the Cold War strategy that emerged over the four years after Kennan's Long Telegram was from the Washington consensus in 1946.

Critical Thinking

1. What does Graham Allison think of Fukuyama's notion of the end of history?
2. What is meant by a rules-based order?
3. Why should the United States accept an international order based on liberal regimes?

Internet References

A World Imagined, Nostalgia and World Order
https://www.-cato.org/publications/policy...world-imagined-nostalgia-libera;-order

Summary of the 2018 National Defense Strategy
https://dod.defense..gov/Portals/1/.../2018-National-DefenseStrategy-Summary-pdf

Testing the Value of the Postwar International Order—Rand Corp.
https://www.rand.org/pubs/research_reports/RR2226.html

The National Security Strategy of the United States
https://www.whitehouse.gov/articles/new-national-securitystrategy-new-era/

Graham Allison is Douglas Dillon Professor of Government at the Harvard Kennedy School.

Article

Prepared by: Robert Weiner, *University of Massachusetts, Boston*

The Liberal International Economic Order on the Brink

KRISTEN HOPEWELL

Learning Outcomes

After reading this article, you will be able to:

- Learn why Trump's Trade policy is out of step with the reality of the world today.
- Learn why the International economic order has always served US economic interests.

The liberal international economic order was constructed during the era of American hegemony and heavily shaped by US power. From its dominant position in the post–World War II international system, the United States engaged in an unprecedented building of multilateral institutions and rules to govern an increasingly integrated global economy, based broadly on the principles of open markets and trade. This integrated system of global economic governance—centered on the World Trade Organization (WTO), the International Monetary Fund (IMF), and the World Bank—has been a historically distinct and defining feature of American hegemony. US power has been exercised in and through international rules and institutions. Yet while these institutions served as a channel for the projection of American power, their rules and norms also served, at least to some extent, to rein in the arbitrary and indiscriminate use of power by the hegemon.

The election of Donald Trump as president, propelled in part by a surge of sentiment that blames "unfair trade" for the current economic and social ills of the United States, has put the future of the US-led economic order in doubt. Trump has promised to arbitrarily restrict access to the US market and take unilateral retaliatory trade actions against other nations. In one of his first acts as president, Trump pulled the United States out of the Trans-Pacific Partnership after the trade pact with 11 other nations was painstakingly negotiated by his predecessor's administration. He has threatened to withdraw from the North American Free Trade Agreement (NAFTA) and the Korea-US Free Trade Agreement (KORUS), and to exit the WTO or simply ignore any of its rulings that do not go in Washington's favor. Under a president hostile to free trade and international cooperation, the United States appears to be abandoning its commitment to the liberal international economic order and abdicating its leadership role.

Privileges of Power

When asked what he thought of Western civilization, Mahatma Gandhi is reputed to have replied, "I think it would be a good idea." Precisely the same could be said of the liberal international economic order. In rightly seeking to criticize Trump's agenda and the danger it represents, there has been a tendency to fall back on a largely fictitious vision of the past—a romanticized image of the pre-Trump liberal order and America's role within it. But the reality is not so simple. Liberal ideas and institutions have provided a powerful source of ideological legitimacy and support to the world order constructed under American hegemony. However, the American commitment to liberal principles has always been partial, selective, and self-serving. US leadership has been experienced by many other countries as coercive rather than benevolent.

In the trading system, realism and liberalism coexist in tension. While it is formally based on the liberal principles of free trade and the sovereign equality of states, in practice power politics prevail. Power asymmetries deeply shape the content of trade agreements: the powerful are able to impose liberalization on the weak while maintaining scope for their own protectionist policies.

Although the United States has been the primary driver of liberalization in the global economy—pushing other countries to open their markets to its goods, services, and capital—its own policies have often deviated from the principles of free trade. Washington has pursued a two-pronged strategy: in areas of vulnerability, it has used trade protections to shield sensitive sectors from foreign competition, while in areas where American companies are, or have the potential to become, world leaders, it has actively used various forms of state support, combined with its influence in shaping international trade rules, to give its domestic industries a competitive edge and promote their global market dominance.

Trump's notion that America is disadvantaged by global trade agreements, or that it has been unfairly taken advantage of, is thus deeply perplexing to other countries. As the most powerful player in the international system, the United States wrote the rules of international trade. The multilateral trading system was designed by Washington to serve its own interests, frequently at the expense of other, weaker nations.

This was strikingly illustrated in the last successful round of multilateral trade negotiations, the Uruguay Round (1986–94). The Uruguay Round was part of a larger neoliberal turn in the global economic order driven by the United States. Grounded in a deep antipathy toward the state and a zeal for the "free market," neoliberalism gained ascendance in the United States in the 1980s and was spread globally by Washington through the multilateral economic institutions, pushing a policy program of fiscal austerity, privatization, deregulation, and trade and capital liberalization. Driven by more than mere ideology, the "Washington Consensus" was a strategy aimed at restoring and reasserting US power, after its relative decline in the 1970s, by opening foreign markets to its exports, investment, and multinationals. Washington aggressively pushed this agenda on other states through the structural adjustment programs of the IMF and the World Bank, as well as the Uruguay Round.

Facing changes in the nature of its competitive advantage and seeking to address its growing trade deficit, the United States used the Uruguay Round to drive a dramatic expansion of global trade rules into new areas of US competitiveness, including services, investment, and intellectual property. Washington extracted deep concessions from developing countries, while the gains they were promised in return failed to materialize. Ultimately, the Uruguay Round was pushed through by a raw use of power on the part of the United States (including threats of unilateral trade sanctions and withdrawing access to the world's largest market). Robert Wade of the London School of Economics has likened it to "a slow-motion Great Train Robbery" since the agreement produced a significant transfer of resources from poor to rich countries. It constrained the space for development policy, prohibiting many of the same industrial-policy tools, such as subsidies and controls on foreign investment, that advanced economies like the United States had themselves used to develop.

For many developing countries, their experience of the coercive power wielded by Washington in trade negotiations and through the international financial institutions left a deep legacy of distrust that still profoundly shapes how they view American hegemony—and their willingness to welcome the emergence of new challengers. For most developing countries, the prescriptions of the Washington Consensus resulted in one or two "lost decades" of economic stagnation. Meanwhile, the countries that achieved the greatest development success—such as China and other major emerging economies—departed widely from the neoliberal orthodoxy and instead pursued state-led development policies.

The failure of the Washington Consensus profoundly undermined the credibility of US leadership in the global order. This was only compounded by the global financial crisis of 2008. Inadequately regulated financial markets in the United States imperiled the entire global economy and led to sharp drops in output, trade, and employment, as well as an increase in poverty felt around the world.

We should thus be careful not to idealize the world economic order, or US leadership, before Trump. There has always been a gap between how the United States views its role in the world and how it is seen by others. While paying lip service to the principles of free trade, international law, and multilateralism, Washington routinely violated those same principles when it served its interests, blatantly abusing the privileges of its power and disregarding the rules of the system it created. The hypocrisy of the US-led international economic order—Washington's failure to live up to its liberal principles—damaged its legitimacy and undercut support for US leadership.

Tables Turned

Even before Trump, the core pillar of the US-led economic order—the WTO—had already been deeply disrupted by contemporary power shifts. The rise of new powers such as China, India, and Brazil as major players in the global economy has proved profoundly destabilizing for the multilateral trading system. While the emerging powers are supporters of the system, they challenged the longstanding dominance of the United States and other advanced industrial nations, demanding a greater role in global trade governance. The resulting confrontation led to the collapse of the Doha Round of trade negotiations.

The Doha Round was a US project launched in 2001 over strong opposition from developing countries, which were corralled into participating through a combination of inducements and coercion. The rise of Brazil, India, and China, however, transformed the negotiations into a battle between the Global North and the Global South. In all eight previous trade rounds, the United States was the primary aggressor, pushing other countries to liberalize their markets. But in the Doha Round, for the first time, the emerging powers turned the tables and made the protectionist trade policies of the US and other advanced industrialized nations—particularly their massive agriculture subsidies, one of the most flagrant examples of hypocrisy in the trading system—a central target at the WTO. The rebalancing of power fundamentally shifted the terms of the prospective Doha agreement and the concessions among states that would be necessary to secure a deal.

In 2008, the WTO looked close to concluding the Doha Round. Most nations were broadly satisfied with the draft agreement on the table. In a marked change from the past, the prospective deal was seen as generally favorable for developing countries: they would benefit from liberalization by developed countries, while being granted "special and differential treatment," allowing them to undertake less liberalization and providing them with substantial flexibility. At that point, however, it was Washington that balked. Members of Congress and business and farm lobbying groups complained that the draft agreement did not require enough of the emerging powers. The United States aggressively worked to "rebalance" the deal by pressing the large emerging economies to make concessions on manufactured goods and agriculture beyond the terms that had already been negotiated.

From the perspective of the emerging powers, these demands were out of proportion to the reforms that the United States itself was willing to undertake. Washington was minimalist in making concessions but maximalist in making demands of others. It was trying to change the terms of the deal, and its new demands—which had escalated in response to the complaints of its domestic lobby groups—were unfair and unjustified. Brazil, India, and China now had sufficient power to stand up to America and refused to be pushed into what they perceived as an unbalanced agreement. With these two sides evenly matched, neither was able to impose its will on the other. The result has been a permanent impasse.

The collapse of the Doha Round represents the disruption of the US-led liberal international economic order. The WTO's existing rules remain in force and subject to its dispute settlement mechanism. However, amid a more equitable distribution of power among states, the core function of the WTO—the negotiation of successive multilateral trade agreements to drive forward the opening and liberalization of global markets—has broken down. The Doha Round impasse is an important signal of the weakening of Washington's ability to impose its will globally. The WTO has spun beyond the control of the United States—or any other single state or group of states. Even if America maintains a preponderance of power in the international system, its capacity to direct and steer global economic governance—until now a distinct and defining feature of its hegemony—has been severely diminished.

Confronted by a strong counterweight to its power at the WTO, the United States abandoned multilateralism in trade and abdicated its traditional leadership role within the institution, beginning under President George W. Bush and continuing under the Obama administration. Instead, Washington turned to creating new bilateral and regional free trade agreements (FTAs), in which it had greater leverage to extract concessions and secure more favorable deals. This was part of an intensified strategy of "competitive liberalization" that sought to reward cooperative countries and punish uncooperative ones. However, this strategy backfired as other nations, including China, turned to negotiating their own competing trade deals, leading to a proliferation of FTAs and further undermining the multilateral trading system.

The US abandonment of the Doha Round was seen as particularly egregious because Doha had been sold to the developing world as a "Development Round," with the promise that it would deliver real benefits to the Global South and redress the historical imbalances built into previous multilateral trade agreements. A key aspect of the Doha mandate was the principle that developing countries would not be required to engage in an equal exchange of concessions with the advanced industrialized nations, but rather that the final bargain would be struck on the basis of "less than full reciprocity," favoring developing countries. But Washington had shown once again that it would accept less than full reciprocity only in its own favor.

For American negotiators, the vague development promises attached to the Doha Round were primarily a political maneuver to ensure its launch. The United States had never intended to fulfill such promises if doing so would require a sacrifice of its own commercial interests. For developing countries, the realist logic of the trading system—its inability to serve the needs of weaker states—was laid bare. Washington was unwilling to live up to the development mandate of the Doha Round and refused to agree to a trade pact that would genuinely advance the interests of developing countries.

The End of Hypocrisy

Trump's approach to international economic relations is typically treated as a radical break with previous US trade policy. But "America First" is nothing new. The United States has always put its own interests first and often used bullying to get its way. Trump's capricious behavior makes it impossible to predict with any certainty what he will do in the future. But so far, despite his bellicose rhetoric on trade, the actual tactics he is employing are not at all unfamiliar.

Trump's trade agenda increasingly seems to boil down to trying to renegotiate trade agreements by using "all possible sources of leverage" to extract greater concessions from trading partners (as his 2017 Trade Policy Agenda states), while simultaneously reducing their access to the US market. Trump may be taking this approach to an extreme, but Washington has always used its economic and political might to tilt the terms of trade agreements in its favor and leave room for its own protectionist policies.

The United States has frequently taken unilateral steps to restrict access to its market. Many of the specific protectionist measures that Trump is deploying—such as antidumping and countervailing duties, safeguards, and Section 301 investigations—are drawn from a preexisting toolkit that has long been used to restrict imports. In 2016, before Trump's election, the United States was the world's biggest user of antidumping actions (protectionist tariffs imposed on imports alleged to be priced below fair value) and other trade defense measures. The United States also accounts for half of all countervailing duty actions (tariffs imposed on imports to counter alleged subsidies) by G20 nations.

Rather than using the WTO's mechanism for resolving trade disputes, which is intended to ensure they are resolved in a fair and orderly manner, the Trump administration has launched an investigation of China's intellectual property practices under a contentious section of US trade law. Section 301 enables the government to unilaterally take retaliatory action against foreign countries that maintain any law or practice that it deems to "unjustifiably" or "unreasonably" restrict or burden US commerce. The sweeping nature of Section 301 makes it a blunt instrument with which the United States can leverage its economic might to force other countries to open their markets to its exports. While Section 301 has rarely been used in recent years, it has a long and bitter history. It was deployed over 100 times against US trading partners after it became law in the 1970s. Most strikingly, Washington used the threat of aggressive unilateral action under Section 301 to force countries to swallow their opposition and accept the Uruguay Round agreements in 1994.

While Trump's overt hostility toward the WTO is a marked departure from previous American leaders, the United States had already soured on the WTO before him. Ultimately, it was America that killed the Doha Round by walking away from the deal on the table. In doing so, it dealt a major blow to the multilateral trading system. When thwarted from securing its preferred outcomes in Doha, Washington turned away from the WTO, disengaging from the institution and focusing instead on bilateral and regional trade agreements.

Trump has now threatened to ignore WTO rulings he does not like, but that is not entirely new either. The United States has often intensely resisted changing its laws and policies to comply with WTO rulings (in disputes over aircraft, cotton subsidies, softwood lumber, country-of-origin labeling, and the calculation of antidumping duties, to name just a few), turning many of these cases into sagas that persist for years and even decades. Beyond his rhetoric, the substance of what Trump has done so far represents an intensification and deepening of existing patterns of US behavior in the trading system rather than a fundamental change.

If, as the political scientist Mlada Bukovansky has argued, hypocrisy put "a velvet glove on the iron fist" of US power, that glove has now come off. Under Trump, the US has dispensed with pretense, engaging in blatantly self-serving behavior and the raw use of power without the guise of liberalism. But that iron fist has also gotten weaker.

Trump's belligerence—his efforts to strong-arm trading partners and threats of aggressive unilateral trade actions—comes at a time when America's relative economic power is declining. It has less control over global economic governance than it used to, and less scope for unilateral action. While the United States remains the world's largest economy, its relative weight has diminished significantly, constituting just over 20 percent of the global economy today, compared with 40 percent in 1960. Its position at the epicenter of global capitalism has been eroded by growing multipolarity, and particularly by the rise of China, which is now the world's second-largest economy, its leading goods trader, the biggest market for many inputs and consumer goods, and a booming source of outward foreign investment.

If Trump were to proceed with slapping huge tariffs on imports from China, Beijing would not just sit idly by. It would retaliate against US exports. Trump risks triggering a global trade war. Many American firms and industries rely on China as their largest export market. Reports indicate that US companies have been reluctant to participate in the Section 301 investigation against China for fear of reprisals that would damage their sales and ability to operate in the Chinese market.

And it is not just China that has greater capacity to resist US pressure. In its efforts to renegotiate the terms of KORUS, the Trump administration issued a list of unilateral concessions it wanted South Korea to make, including immediately eliminating tariffs on key US exports, while the United States would stall its own promised tariff reductions on Korean goods. This carried echoes of a past era when Washington forced "voluntary export restraints" on South Korea, Japan, and other competitors in the 1970s and 1980s. This time, however, Seoul has refused to capitulate to the American demands.

Hegemonic Hubris

Trump's plan to "Make America Great Again" will most likely do precisely the opposite by accelerating American decline.

Trump has taken hegemonic hubris to unprecedented heights. He assumes that the United States can simply impose its will on other countries, but Washington has less weight to throw around than ever before and it is being met with stronger opposition and competition for global leadership.

The failure of the Doha Round at the WTO revealed the diminished ability of the United States to simply overpower other countries to get its way in trade negotiations. More recently, when it tried to pressure its allies to stay out of China's new Asian Infrastructure Investment Bank, it failed utterly; China's financial power simply exerted too strong a gravitational pull on other nations. Trump's worldview is out of step with the reality of the world today and in denial about the new balance of power.

Trump's trade proposals would only serve to further weaken US economic and political power. If he were to follow through with his most extreme threats, such as withdrawing from existing trade agreements, the consequences would be profoundly damaging to the United States itself. US withdrawal from the WTO would be economic suicide: it would profoundly disrupt global trade flows, ushering in unprecedented economic chaos. The United States would immediately face high tariffs and other trade barriers, causing its exports to plummet. If Trump followed through on his threat to withdraw from NAFTA, he would undermine the competitiveness of US manufacturing, which is heavily based on integrated North American supply chains, just when US firms and industries face increased competition from China. Going ahead with Trump's threat to terminate KORUS would harm Washington's relationship with its closest ally against North Korea.

It can only be hoped that Trump's toughest talk is merely that—empty threats intended to strong arm US trading partners into making greater concessions—but given the rash and reckless nature of Trump's approach to policy making, that may be wishful thinking. And even if these threats ultimately prove empty, Trump's overt hostility toward major trading partners is doing irreparable damage to vital American alliances.

With his erratic and bullying behavior, Trump risks destroying the credibility, legitimacy, and authority the United States still has as a leader on the international stage—and driving its allies into the arms of China, its chief rival and challenger to its hegemony. China has made no secret of its global ambitions, aggressively courting the world with trade, investment, and aid. And many countries, especially in the developing world, already see Beijing as a more helpful and reliable partner than Washington. As America shows itself to be an increasingly volatile and unreliable trading partner, countries have enormous incentive to cement closer ties with China. It would be unwise to underestimate China's pull. Its remarkable economic growth and success in fostering development are an immense source of soft power to attract other nations. So is China's considerable financial power, including both its largesse in dispensing investment and aid, and its massive and still rapidly growing market.

Dangerous Distraction

Trump's anti-trade campaign has been founded on the notion that the American economy is in decline. This is a fallacy. The United States is richer today than ever before. Contrary to Trump's claims that America has been unfairly disadvantaged by existing global trade arrangements, the reality is that it has benefited hugely from the liberal international order it constructed. Globalization has fueled a vast expansion of the American economy. Since 1980, real average per capita income has grown by 80 percent. But while trade has benefited the US economy overall, many workers—especially lower-skilled workers—have lost out. While this is partly due to trade, the far more significant factor has been technological change and automation, compounded by the internal movement of jobs within the United States to economic sectors and regions with weaker unionization.

There is now a massive disconnect between the aggregate wealth of the US economy and the profound sense of economic insecurity felt by much of its population, amid rising inequality, stagnant real wages, and the increasing precariousness of employment. But trade is not the problem. It is merely a scapegoat for a president who feeds on racism and xenophobia. Pitting American workers against their counterparts in China, Mexico, or other US trading partners only serves to misdirect attention from the true source of America's economic and social ills—a set of neoliberal economic policies that have benefited a small fraction of the population at the expense of the rest over the past thirty years. These policies have provided tax cuts for the wealthy, reduced the power of unions, decimated social safety nets, reduced investment in human and physical capital, and increased instability in financial markets.

America's economic woes will not be solved through protectionism or attacks on its trading partners. Not only is Trump's anti-trade rhetoric a distraction from the real and pressing problems facing the United States; his economic policies—tax cuts for the rich, reducing health-care coverage by repealing the Affordable Care Act, deregulating Wall Street—would severely exacerbate these problems. Under Trump, Washington is doubling down on a market-fundamentalist agenda that has been discredited and abandoned virtually everywhere else in the world.

At a time when US economic and political dominance is increasingly challenged, particularly in the face of a rising China, Trump's kamikaze approach to managing the economy and foreign relations appears bent on self-destruction. We can only hope that he does not destroy the liberal international economic order in the process. While the existing order has too often failed to live up to its liberal ideals, the fallout from its collapse—a descent into autarky, chaos, and conflict—would be disastrous.

For too long, the United States has taken support for its global leadership for granted and abused its power with little concern for the consequences. While it was able to get away with this in a unipolar world, that is no longer the case when real challengers to its hegemony are emerging. If Washington wants to maintain its hegemony, it should be recommitting to the liberal global order and working to strengthen that order by bringing it closer to realizing its liberal ideals of fairness, rule of law, and mutual benefit. By forsaking those principles altogether, Trump risks leaving the United States a pariah and weaker than ever before.

Critical Thinking

1. What economic tools has the United States used to coerce other countries in trade negotiations?
2. Why would US withdrawal from the World Trade Organization create economic chaos?
3. Why would countries cement economic ties with China?

Internet References

A World Imagined, Nostalgia and World Order
https://www.-cato.org/publications/policy...world-imagined-nostalgia-libera;-order

Summary of the 2018 National Defense Strategy
https://dod.defense..gov/Portals/1/.../2018-National-DefenseStrategy-Summary-pdf

Testing the value of the postwar International Order—Rand Corp.
https://www.rand.org/pubs/research_reports/RR2226.html

The National Security Strategy of the United States
https://www.whitehouse.gov/articles/new-national-securitystrategy-new-era/

KRISTEN HOPEWELL is a senior lecturer in international political economy at the University of Edinburgh. She is the author of *Breaking the WTO: How Emerging Powers Disrupted the Neoliberal Project* (Stanford University Press, 2016).

Article Prepared by: Robert Weiner, *University of Massachusetts, Boston*

The United Nations and Sovereignty in the Age of Trump

THOMAS G. WEISS

Learning Outcomes

After reading this article, you will be able to:

- Learn why no country can unilaterally solve global problems.
- Learn how the UN can become a more effective mechanism for cooperation.

After serving as a British intelligence officer during World War II, Brian Urquhart was the second official recruited for the United Nations Secretariat in 1946. Looking back decades later on a distinguished career as an international civil servant, he quipped, "The UN is the last bastion of national sovereignty." It was not a compliment. He was lamenting the world organization's inability to come to the rescue of desperate people subjected to violent attacks on their human rights. Presidents, princes, and prime ministers claimed that what they did at home was exclusively their business. Other UN member states agreed.

More recently, the international community of states occasionally has applied the doctrine of the "responsibility to protect" to revoke the license for mass murder claimed by sovereign thugs who abuse their citizens. Of course, states have also agreed to limit their sovereign prerogatives through international treaties of various sorts, and globalization means that they are largely powerless to halt some other intrusions—for instance through financial and technology transfers, and the free flow of information. In short, sovereignty ain't quite what it used to be.

Nonetheless, the UN and other intergovernmental organizations—even the supposedly supranational European Union—remain firmly grounded in sovereignty, which US President Donald Trump made even clearer when he uttered the s-word 21 times in his September 2017 address to the UN General Assembly. His mantra was well received by such major powers as Russia, China, and India as well as minor ones including Zimbabwe and Cuba. These countries customarily emphasized sovereignty to ward off criticism from Washington. No longer.

The contrast with Barack Obama was stark. In his first address to the General Assembly in 2009, Obama referred to sovereignty just once while reaffirming the US commitment to international cooperation and multilateralism.

Although Trump and Vice President Mike Pence have tried unpersuasively to square the circle, "America First" actually means "America Alone." The sting of Trump's strident use of the phrase on the world's biggest stage may have eased momentarily when he told other leaders that they were right to put their own countries first. However, his address was primarily a declaration of war on international obligations and cooperation, and an unconvincing effort to reassert the power of a single state to address global problems.

We should recall that Trump's slogan has its isolationist origins in the America First Committee, which was the largest and best-organized antiwar group ever. It was founded in 1940 by the likes of proto-fascists Charles Lindbergh and Father Coughlin to keep the United States out of World War II. The America First Committee collapsed after Pearl Harbor. Trump's contemporary adaptation has reached the height of power.

Among the "alternative facts" being peddled by the Trump administration is the claim that unilateralism is the way to address pressing problems—from terrorism to North Korea's nuclear weapons to the planet's changing climate. Yet no country is powerful enough to impose its will and solve global problems on its own. Tending one's own garden is not a realistic strategy in 2018. Human welfare depends increasingly on effective collaboration across national boundaries. This obvious reality, like many others, is denied by the evidence-averse White House.

The United Nations is still the logical place to convene discussions and orchestrate action to address global problems. This preeminent universal-membership institution provides the means to confront a multitude of life-threatening problems that national actions alone cannot address effectively. At the same time, the organization's current limitations—not only its sovereignty-bound foundations but also its atomized and wasteful operations—are obvious to anyone but the most blinkered UN cheerleaders. Multilateralism all too often appears unable to deliver.

How, in the era of Trump, can the UN become a more effective mechanism for cooperation in a world of sovereign states? Can the UN of 2018 provide a dose of sanity and marshal global efforts for the collective pursuit of survival with dignity for the human race?

The ninth secretary-general, António Guterres, took over from his lackluster predecessor, Ban Kimoon, on January 1, 2017. Hopes for a new beginning remain unrealistically high among civil society, UN staff, and many member states. We should temper such expectations, especially since Trump's inauguration followed shortly after that of Guterres. After all, we are speaking about transforming a seven-decade-old international bureaucracy—a herculean task that will be even harder given an American administration that seems uncooperative, to say the least.

Enlightened Realism

US presidents have often pursued what they perceive as vital national interests through the United Nations and other international bodies. In 2002, George W. Bush's administration "unsigned" the Rome Statute establishing the International Criminal Court on the grounds that it was against US interests—except when it was not and Washington wanted to use the court to prosecute Sudanese and Libyan war criminals. The Security Council issued Bush a blank check in Afghanistan, though it declined to authorize the war in Iraq.

Obama relied on the Security Council for a green light to intervene in Libya but got none for Syria because of Moscow and Beijing's vetoes. However, that same paralyzed council suddenly came to life when Washington and other powers sought in 2013 to dismantle Syria's chemical weapons capacity and turned to the International Atomic Energy Agency to verify the 2015 agreement barring Iran from developing nuclear weapons.

Such opportunism is commonplace. But most people overlook the fact that the purpose of the United Nations at its birth was to defeat fascism. Its creation reflected a radically different US attitude toward consistent multilateral cooperation in the face of a truly existential threat. It began with the "Declaration by United Nations," signed by 26 countries in January 1942, which built on the Atlantic Charter of August 1941. That declaration committed the Allies to multilateralism—not only to crush Nazi Germany and Imperial Japan in the short term, but also to maintain international peace and security as well as to foster postwar economic prosperity and social stability over the longer term.

Observers customarily trace the collapse of this early idealism to the end of the war, when the combination of the atomic bombings of Hiroshima and Nagasaki and the growing tensions between the West and the Soviet Union instilled a new cynicism about the prospects for peace. Few recall the powerful mixture of realism and idealism that emerged earlier in the Lend-Lease program and the wartime United Nations. Alongside US military muscle, multilateralism was integral to US decision making during World War II.

A wide array of Allied wartime initiatives—on issues including international criminal justice, postwar reconstruction, refugee assistance, international development, economic regulations, public diplomacy, and agricultural and educational policy—not only sustained the military enterprise but also were intended to lay the foundations for future stability. Wartime planners rejected unilateral military might and lawlessness as policy options for the postwar order.

The United Nations Conference on International Organization, convened in San Francisco in 1945, and the subsequent creation of the UN system were not peripheral but rather central to US decision making. At a moment when one might have expected the fallout from the failed League of Nations to favor Hobbesian unilateralism, those overseeing the Allied war machine and thinking about the future were resolute: multilateralism and the rule of law, not going it alone and the law of the jungle, should underpin the postwar order. The bleak alternative was on display in the Third Reich and the Japanese Empire, which epitomized the right of might.

Decisions to collaborate in the construction of international organizations for peace and prosperity were central to the mobilization against fascism and reflected enlightened realism about the merits of multilateralism. A genuine cooperative strategy motivated peoples and kept states allied. The postwar vision was more than just propaganda, though "business as usual" once again became the default option with the Cold War's onset. It was not the weakness of international cooperation but the intensity of the Cold War that replaced multilateralism with a diminished vision based on the lowest common denominator of narrowly defined national interests.

The bottom line at the time of the UN's creation was straightforward: neither governments nor analysts calculated that a return to the world of 1913—that is, before World War I and without even a toothless League of Nations—was desirable. Unfortunately, the Trump administration has forgotten this lesson. The received wisdom is that middle and smaller powers prefer multilateralism while major powers favor unilateralism. However, the UN's wartime origins suggest that collaboration is also useful for the most powerful nations when the political conditions are right.

Going It Alone

Of course, plenty of contemporary leaders traffic in nationalism, nativism, and populism. But Trump's impact will be the biggest for the UN, since he heads what has long been its most influential member state and largest funder.

Bowing to the anti-abortion views of his base, one of Trump's first moves was a nonnegotiable elimination of US funding for the UN Population Fund, the organization devoted to the reproductive health of girls and women. The administration also urged drastic cuts to peacekeeping funds, setting an initial target of $1 billion in cuts from an $8 billion budget, but ultimately settling for a $600 million overall cut, resulting in a $170 million reduction in the US contribution. The savings amount to a rounding error in the US federal budget, but the cuts will seriously impede UN operations.

The administration's insistence that the United States should abandon its traditional role as the leading proponent of free trade was an even bigger indication of narrow nationalism run amok. Trump began by ripping up the Trans-Pacific Partnership, thereby forgoing the potential benefits of expanded access to Asian markets and the chance to advance human rights and environmental protection standards as conditions for participation. Trump has continued in this vein by threatening to scrap the North American Free Trade Agreement, while calling for a border tax and tariffs on steel and other imports.

China has made the most of this opportunity delivered to it on a silver platter. It now can dictate the standards for international commerce in Asia; it has picked up new trading partners, in Asia and worldwide; it even presents itself as the new champion of free trade. Even those who viewed the rise of China and other emerging powers as inevitable underestimated how quickly the stature of the United States would diminish and its credibility evaporate. A strident and unpredictable Washington seems determined to start a world trade war; meanwhile, Beijing is the calm and predictable voice for free trade and stability.

Of still greater significance was the May announcement that the United States would abandon the Paris Agreement on climate change and make no effort to meet voluntary targets to curb planet-warming emissions. Once again, China is the direct beneficiary of Trump's myopia. Beijing is happy to fill an unexpected new role as the leading advocate for action on climate change, just as it has become the world's largest producer of greenhouse gases. Green-technology producers in China are forging ahead toward a dominant global market position while the Trump administration vows to create jobs in coal mines—never mind that ten times as many US workers are currently employed in renewable energy and other green technology jobs as in the coal industry.

The 2020 presidential election will be held just before the US withdrawal takes legal effect. By then, perhaps, more voters will have come to their senses. In any case, the fact that many US cities, states, and corporations are mobilizing in support of the Paris Agreement and have vowed to adhere to its goals means that a multilateral approach will still be a viable option for the United States. The final communiqué from the July 2017 Group of 20 meeting in Hamburg—which noted that the climate agreement was irreversible and nonnegotiable, notwithstanding Trump's demands for a new deal—showed that the other 19 members of the G-20 remain committed while Washington pouts.

Trump's isolationism starkly contrasts with the views and approaches of virtually all of Washington's friends. Rubbing salt in their wounds, the administration announced in December that it was pulling out of UN negotiations toward a global compact on migration, shortly after having set its withdrawal from the UN Educational, Scientific, and Cultural Organization (UNESCO). Of course, even as committed an internationalist as Obama occasionally reverted to heavy-handed pressure on the UN: the US had already stopped paying its bills to UNESCO in 2010 to protest its admission of Palestine.

Political Theater

Trump's UN debut resembled many of his domestic speeches in its incoherence and hostility toward liberal values and institutions. He paid lip service to the UN's values, but went on to suggest that he remains committed to dismantling the web of intergovernmental organizations and the rule of law created and nurtured by the United States since World War II.

The General Assembly has long been the setting for memorable performances by bombastic and narcissistic leaders. In 1960, Soviet leader Nikita Khrushchev banged his shoe on a table to demonstrate his diplomatic dyspepsia. In 1974, the Palestine Liberation Organization's Yasser Arafat ceremoniously checked his revolver at the door and then brandished an olive branch. Venezuelan President Hugo Chávez began his 2006 speech by sniffing the chamber, indicating that it still smelled of sulfur—because "the devil," George W. Bush, had addressed the assembly the previous day.

During the "general debate" that opens each year's session of the General Assembly, Brazil by tradition speaks first, followed by the host nation, the United States. Thus, the start of the seventy-second session of the General Assembly on September

19, 2017, provided a stage for Trump's reality show at UN headquarters. Speaking for almost three times the suggested 15-minute limit, he hardly set a new standard for succinctness.

After a tepid opening, he reverted to complaints about the US paying an unfair share of the UN's bills, self-promotional chest-thumping, and rants about North Korea and Iran. While expected, Trump's warning that he was prepared to annihilate North Korea was a first—threatening nuclear war in the assembly hall supposedly devoted to the peaceful resolution of disputes. Overall, the speech was received as a declaration of unrepentant nationalism in the very home of internationalism.

Of course, great powers have never been reluctant to throw their weight around, which the UN Charter recognizes by putting five veto-wielding permanent members in the Security Council. In one way, oddly enough, Trump's approach resembles that of a far more respected, cultivated, and dignified world leader of an earlier generation. Charles de Gaulle was temporarily successful in attacking multilateral institutions while at the same time expressing an atavistic Gallic nationalism.

General de Gaulle dismissed the UN as "lemachin" (the thing) in order to create space for France to maneuver outside the US-Soviet confrontation. A few years later, as president of the Fifth Republic, he withdrew France from the integrated military command structure of the North Atlantic Treaty Organization (NATO), seeking autonomy within the Western alliance. He also temporarily left vacant the French seat in the European Economic Community (which would become the European Union) in order to ensure that members retained full sovereignty and that Britain was excluded.

Although steeped in history, de Gaulle seemingly forgot that the predecessor of "the thing" had liberated occupied France. He also was unable to recognize the essential contribution of NATO and expanded European integration in maintaining peace and ensuring growth and prosperity in France and on the continent.

Trump's unilateral and nativist perspective—he has heaped scorn on "obsolete" NATO as well as the UN—will be revealed as just as shortsighted as de Gaulle's over half a century ago. It is important to hold the fort in the interim. Just as the multilateral institutions discounted by de Gaulle were resilient enough and ready to expand operations and membership after his departure, the UN and NATO—and proponents of international cooperation more broadly—should be prepared for a resurgence after Trump disappears from the international scene.

Guterres to the Rescue?

Trying to reform the UN has been virtually a perpetual task since the ink dried on the signatures to the Charter in 1945. Efforts never cease to improve the UN's effectiveness; to make it more inclusive, transparent, and accountable; and to pull together its autonomous parts. The results have been modest and uneven at best. Those who see the institution as essential interpret these efforts optimistically as an encouraging sign of life. Critics see the sclerosis of an aging institution that is not worth saving. Meaningful change and renovation, if not structural transformation, are indisputably necessary.

The decibel level of the criticism is rising, including but certainly not only in Washington. What Richard Haass, president of the Council on Foreign Relations, calls "a world in disarray" has become more complex and uncertain precisely when predictable collective action is so desperately required. Powerful and less powerful countries and their publics appear skeptical about intergovernmental organizations; they are likely to move toward more of a cost-benefit, transactional approach to multilateralism of all stripes, including under UN auspices. The multilateral narrative has less visceral appeal than in 1945, or even a few years ago.

This context was already obvious during the 2016 US presidential race. Simultaneously, António Guterres was running his successful campaign to become the ninth occupant of the UN's top floor. The stage was set for only the second time—the first was in 1996—when the campaigns for US president and UN secretary-general ran in parallel.

The 2016 UN contest produced a slate of thirteen nominees—seven of whom were women. (Over the previous seven decades, through all eight prior campaigns for the UN's top post, only three women were actively considered.) All the candidates pursued their campaigns both in person and through lobbyists. Member states were also lobbied by 1 for 7 Billion, a campaign endorsed by 750 civil society organizations from around the world "committed to getting the best UN Secretary-General."

In a refreshing break from past practice, the General Assembly gathered for two-hour hearings with each candidate from April to September 2016. It also organized an open public event for all of them, while civil society debates in New York and London augmented these governmental gatherings. The candidates' resumes were available for public and private scrutiny. Each of them circulated a "vision statement" that contained proposals for reshaping the unwieldy UN and making better use of its 80,000 international civil servants and the 120,000 soldiers and civilians engaged in peacekeeping operations. As a result, the selection process was somewhat open and transparent. It no longer resembled a papal conclave, though only an inveterate Pollyanna would have hoped to eliminate backroom horse-trading in the Security Council's small electoral college of the five permanent members.

Guterres, the frontrunner for several months and the winner of five straw polls in the usually divided Security Council, secured the recommendation of that body. The General Assembly then appointed him by acclamation in early October 2016. Having previously served as a two-term prime minister of Portugal and then for over a decade as the UN High Commissioner for Refugees, he is the first socialist former head of government to hold the top UN post. While thus not *The Wall Street Journal*'s preferred candidate, his distinguished government and UN management experience, together with his evident energy and diplomatic finesse, made him the best of the declared candidates. Other leading contenders included UNESCO Director-General Irina Bokova, UN Development Program Administrator Helen Clark, and former UN climate chief Christiana Figueres. Many inside and outside governments and the Secretariat insisted the post should finally go to a woman, but the declared female candidates never fared well in the straw polls.

Guterres might not have prevailed under the old rule, which called for a geographical rotation and would have dictated that it was an Eastern European's turn. The outgoing secretary-general, Ban Ki-moon, arguably would not have been selected in 2006 if the new procedures had been in effect.

Perhaps these welcome, albeit modest, steps toward a more merit-based and transparent process will become part of the standard for filling other senior UN positions. This appears to have happened in the May 2017 election of Tedros Adhanom Ghebreyesus, a former Ethiopian foreign minister and health minister, as director-general of the World Health Organization, and the November 2017 election of French culture minister Audrey Azoulay as UNESCO director-general. Greater meritocracy at the UN may well also have an impact at the World Bank and the International Monetary Fund, where the top jobs have always been reserved for US and European nationals, respectively.

Impossible Job?

In his first remarks after taking the oath of office in January 2017, Guterres signaled that the prevention of armed conflict would be a priority, a familiar plea from secretaries-general since the 1990s. His two other stated emphases are management reform of the bureaucracy and getting the UN's pillars—peace and security; human rights and humanitarian action; and sustainable development—to work together instead of separately.

Guterres's reform agenda needs to reflect the emerging, long-term opportunities signaled by global unpredictability and instability. The UN is the logical forum for organizing for cooperative responses to global problems. The secretary-general is the most visible advocate for and manager of an essential global institution. It is more crucial than many believe.

The waste and overlap in the UN and its network of organizations are hardly news. High-level panels, international commissions, academic treatises, and media analyses have all highlighted the fragmentation of UN activities and the resulting turf wars over scarce resources. Can Guterres replicate the administrative slimming-down and decentralization that he implemented while he was the high commissioner for refugees? If not, the world body risks continuing along the path to obsolescence.

A year into his five-year term, the secretary-general's honeymoon is over. His position remains what the first incumbent, Trygve Lie, glumly called "the most impossible job in the world." Based on his experience, Guterres is fully aware of the UN's political flaws and structural and staffing shortcomings. We must hope that he has the fortitude not to shy away from the Sisyphean task of transforming the way that the UN does business.

On the eve of opening his first General Assembly, the secretary-general laid out numerous proposals for overhauling the UN's operational machinery, and the US Permanent Mission to the UN along with twelve other countries co-hosted a high-level event on UN reform. Given the number of previous proposals for such efforts that ended up on the cutting-room floor, it would be unrealistic to expect major changes anytime soon.

Nonetheless, could Guterres use the Trump administration's tightening of financial screws to do what long has needed to be done? In an interview with PBS following his late-October meeting with Trump at the White House, the secretary-general said: "The UN can also be very useful to the United States, especially if we will be able, as I strongly intend, to have a more dynamic, more reformed, more nimble UN." If Guterres fails in this task, a self-reinforcing dynamic will result: the UN's inability to manage global problems will mean additional blowback for multilateralism.

The world certainly would not be better off without the UN—both the "first UN" of member states and the "second UN" of staff. To deny this proposition is to assert that we would be better off without the cooperative international campaigns to eradicate smallpox, to promote women's rights, to orchestrate efforts to counter climate change, and to aid and protect victims of war. Yet it is equally difficult to maintain that the world would not have been far better off if those same two UNs had improved their performance in numerous ways—for example, if member states had acted earlier to halt Rwanda's genocide in 1994; or if peacekeepers had been prevented from raping children in the Central African Republic or spreading cholera in Haiti; or if UN personnel had performed better in implementing development projects and setting global norms and standards.

State sovereignty is a convenient crutch on which everyone leans to explain international inaction—undoubtedly the UN's worst ailment. Article 2 incorporates sovereignty as the point of departure in the UN Charter. However, sovereignty can be and has been interpreted more inclusively—including during World War II—to justify cooperation in the face of common, existential threats. Sovereigns can calculate and define their interests to help or hinder efforts to improve the quality of human life and address those transboundary threats that former Secretary-General Kofi Annan aptly called "problems without passports."

In the age of Trump, sanity and survival are at stake. "We are calling for a great reawakening of nations," Trump told the General Assembly, thereby ignoring history: the United States actually helped to create the world organization to curb the horrors of nationalism. Instead, he should be calling for a great reawakening of the United Nations.

Critical Thinking

1. Where does President Trump's concept of America first come from?
2. What was the effect of the Cold War on the vision of multilateralism?
3. Why has the multilateral narrative lost its appeal?

Internet References

The United Nations
www.un.org/en/

US Permanent Mission to the UN
https://usun.state.gov/

Thomas G. Weiss is a professor of political science at the Graduate Center, City University of New York. His latest book, *Would the World Be Better Without the UN?* will be published in March by Polity Press, 2018.

Article

Prepared by: Robert Weiner, *University of Massachusetts, Boston*

The Players Change, but the Game Remains

STEPHEN KOTKIN

Learning Outcomes

After reading this article, you will be able to:

- Why geopolitics can explain the relationship between Asia and the United States?
- Learn about the Chinese expansion of its sphere of influence.

Geopolitics didn't return; it never went away. The arc of history bends toward delusion. Every hegemon thinks it is the last; all ages believe they will endure forever. In reality, of course, states rise, fall, and compete with one another along the way. And how they do so determines the world's fate.

Now as ever, great-power politics will drive events, and international rivalries will be decided by the relative capacities of the competitors—their material and human capital and their ability to govern themselves and their foreign affairs effectively. That means the course of the coming century will largely be determined by how China and the United States manage their power resources and their relationship.

Just as the free-trading United Kingdom allowed its rival, imperial Germany, to grow strong, so the free-trading United States has done the same with China. It was not dangerous for the liberal hegemon to let authoritarian competitors gain ground, the logic ran, because challengers would necessarily face a stark choice: remain authoritarian and stagnate or liberalize to continue to grow. Either way, the hegemon would be fine. It didn't end well the first time and is looking questionable this time, too.

China will soon have an economy substantially larger than that of the United States. It has not democratized yet, nor will it anytime soon, because communism's institutional setup does not allow for successful democratization. But authoritarianism has not meant stagnation, because Chinese institutions have managed to mix meritocracy and corruption, competence and incompetence, and they have somehow kept the country moving onward and upward. It might slow down soon, and even implode from its myriad contradictions. But analysts have been predicting exactly that for decades, and they've been consistently wrong so far.

Meanwhile, as China has been powering forward largely against expectations, the United States and other advanced democracies have fallen into domestic dysfunction, calling their future power into question. Their elites steered generations of globalization successfully enough to enable vast social mobility and human progress around the world, and they did quite well along the way. But as they gorged themselves at the trough, they overlooked the negative economic and social effects of all of this on citizens in their internal peripheries. That created an opening for demagogues to exploit, which they have done with a vengeance.

The Great Depression ended an earlier age of globalization, one that began in the late nineteenth century. Some thought the global financial crisis of 2008 might do the same for the current wave. The system survived, but the emergency measures implemented to save it—including bailouts for banks, but not for ordinary people—revealed and heightened its internal contradictions. And in the decade following, antiestablishment movements have grown like Topsy.

Today's competition between China and the United States is a new twist on an old story. Until the onset of the nineteenth century, China was by far the world's largest economy and most powerful country, with an estimated 40 percent share of global GDP. Then it entered a long decline, ravaged from without and within—around the same time the United States was born and began its long ascent to global dominance. The United States' rise could not have occurred without China's weakness, given how important U.S. dominance of Asia has been to American primacy. But nor could China's revival have occurred without the United States' provision of security and open markets.

So both countries have dominated the world, each has its own strengths and weaknesses, and for the first time, each confronts the other as a peer. It is too soon to tell how the innings ahead will play out. But we can be confident that the game will continue.

Beware of What You Wish for

To understand the world of tomorrow, look back to yesterday. In the 1970s, the United States and its allies were rich but disordered and stagnant; the Soviet Union had achieved military parity and was continuing to arm; China was convulsed by internal turmoil and poverty; India was poorer than China; Brazil, ruled by a military junta, had an economy barely larger than India's; and South Africa was divided into homelands under a regime of institutionalized racism.

Four decades later, the Soviet Union has dissolved, and its successor states have embraced capitalism and private property. China, still politically communist, chose markets over planning and has grown to have the world's second-largest economy. Once-destitute India now has the sixth-largest economy. Brazil became a democracy, experienced an economic takeoff, and now has the eighth-largest economy. South Africa overturned apartheid and became a multiracial democracy.

The direction of these changes was no accident. After World War II, the United States and its allies worked hard to create an open world with ever-freer trade and ever-greater global integration. Policymakers bet that if they built it, people would come. And they were right. Taken together, the results have been extraordinary. But those same policymakers and their descendants weren't prepared for success when it happened.

Globalization creates wealth by enticing dynamic urban centers in richer countries to invest abroad rather than in hinterlands at home. This increases economic efficiency and absolute returns, more or less as conventional economic theory suggests. And it has reduced inequality at the global level, by enabling hundreds of millions of people to rise out of grinding poverty.

But at the same time, such redirected economic activity increases domestic inequality of opportunity and feelings of political betrayal inside rich countries. And for some of the losers, the injury is compounded by what feels like cultural insult, as their societies become less familiar. Western elites concentrated on harvesting globalization's benefits rather than minimizing its costs, and as a result, they turbocharged the process and exacerbated its divisive consequences.

Too many convinced themselves that global integration was fundamentally about economics and sameness and would roll forward inexorably. Only a few Cassandras, such as the political scientist Samuel Huntington, pointed out that culture was more powerful and that integration would accentuate differences rather than dissolve them, both at home and abroad. In 2004, he noted that

> in today's America, a major gap exists between the nation's elites and the general public over the salience of national identity compared to other identities and over the appropriate role for America in the world. Substantial elite elements are increasingly divorced from their country, and the American public, in turn, is increasingly disillusioned with its government. Soon enough, "outsider" political entrepreneurs seized the moment.

Having embraced an ideology of globalism, Western elites left themselves vulnerable to a mass political challenge based on the majoritarian nationalism they had abandoned. The tribunes of the popular insurgencies may traffic in fakery, but the sentiments of their voters are real and reflect major problems that the supposed experts ignored or dismissed.

That Was Then

For all the profound changes that have occurred over the past century, the geopolitical picture today resembles that of the 1970s, and even the 1920s, albeit with one crucial exception. Diminished but enduring Russian power in Eurasia? Check. Germany at the core of a strong but feckless Europe? Check. A distracted U.S. giant, powerful enough to lead but wavering about doing so? Check. Brazil and South Africa dominating their regions? Check. Apart from the stirrings of older Indian, Ottoman, and Persian power centers, the most important difference today is the displacement of Japan by China as the central player in the Asian balance of power.

China's industriousness has been phenomenal, and the country has certainly earned its new position. But it could never have achieved what it has over the last two generations without the economic openness and global security provided by the United States as a liberal hegemon. From the late nineteenth and into the twentieth century, the United States—unlike the Europeans and the Japanese—spent relatively little effort trying to establish direct colonial rule over foreign territory. It chose instead to advance its interests more through voluntary alliances, multilateral institutions, and free trade. That choice was driven largely by enlightened self-interest rather than altruism, and it was backed up by global military domination. And so the various multinational bodies and processes of the postwar system are actually best understood not as some fundamentally new chimera called "the liberal international order" but as mechanisms for organizing and extending the United States' vast new sphere of influence.

Strong countries with distinctive ideologies generally try to proselytize, and converts generally flock to a winner. So it should hardly be surprising that democracy, the rule of law, and other American values became globally popular during the postwar years, given the power of the U.S. example (even in spite of the fact that U.S. ideals were often more honored in the breach than the observance). But now, as U.S. relative power has diminished and the U.S. brand has run into trouble, the fragility of a system dependent on the might, competency, and image of the United States has been exposed.

Will the two new superpowers find a way to manage their contest without stumbling into war? If not, it may well be because of Taiwan. The thriving Asian tiger is yet another tribute to the wonders of globalization, having become rich, strong, and democratic since its unprepossessing start seven decades ago. But Beijing has been resolute in insisting on reclaiming all territories it regards as its historical possessions, and Chinese President Xi Jinping has personally reaffirmed that Taiwan is Chinese territory and a "core interest." And the People's Liberation Army, for its part, has gradually amassed the capability to seize the island by force.

Such a radical move might seem crazy, given how much chaos it could provoke and how deeply China's continued internal success depends on external stability. But opinion polls of the island's inhabitants have recorded a decisive trend toward a separate Taiwanese identity, the opposite of what Beijing had expected from economic integration. (Western elites aren't the only ones who harbor delusions.) Will an increasingly powerful Beijing stand by and watch its long-sought prize slip away?

This Is Now

Over the last decade, Russia has confounded expectations by managing to weather cratering oil prices and Western sanctions. Vladimir Putin's regime may be a gangster kleptocracy, but it is not only that. Even corrupt authoritarian regimes can exhibit sustained good governance in some key areas, and smart macroeconomic policy has kept Russia afloat.

China, too, has a thuggish and corrupt authoritarian regime, and it, too, has proved far more adaptable than most observers imagined possible. Its elites have managed the development of a continent-sized country at an unprecedented speed and scale, to the point where many are wondering if China will dominate the world. In 1800, one would have expected China to dominate a century later—and instead, Chinese power collapsed and American power skyrocketed. So straightline projections are perilous. But what if that early-nineteenth-century forecast was not wrong but early?

Authoritarianism is all-powerful yet brittle, while democracy is pathetic but resilient. China is coming off a long run of stable success, but things could change quickly. After all, Mao Zedong led the exact same regime and was one of the most barbaric and self-destructive leaders in history. Just as many people once assumed that China could never rise so far, so fast, now some assume that its rise must inevitably continue—with as little justification.

Xi's decision to centralize power has multiple sources, but one of them is surely an appreciation of just how formidable the problems China faces are. The natural response of authoritarian regimes to crises is to tighten their grip at the top. This allows greater manipulation of events in the short term, and sometimes impressive short-term results. But it has never yet been a recipe for genuine long-term success.

Still, for now, China, backed by its massive economy, is projecting power in all directions, from the East China and South China Seas, to the Indian Ocean, to Central Asia, and even to Africa and Latin America. Wealth and consistency have combined to yield an increasingly impressive soft-power portfolio along with the hard-power one, enabling China to make inroads into its opponent's turf.

Australia, for example, is a rich and robust liberal democracy with a high degree of social solidarity and a crucial pillar of the American order—and it happens to be smack in the path of China's expansion. Beijing's influence and interference there have been growing steadily over the last generation, both as a natural consequence of economic interdependence and thanks to a deliberate long-term campaign on the part of China to lure Australia into a twenty-first-century version of Finlandization. Similar processes are playing out across Asia and Europe, as China embarks on building a Grand Eurasia centered on Beijing, perhaps even turning Europe away from the Atlantic.

Right now, the United States' debasement is giving China a boost. But as Adam Smith noted, there is indeed "a great deal of ruin in a nation," and the United States remains the strongest power in the world by far. Furthermore, this will not be a purely bilateral game. Yes, the United Kingdom allowed Germany to rise and lead a hegemonic challenge against it—twice. But it also allowed the United States to rise, and so when those challenges came, it was possible, as Winston Churchill understood, for the New World, with all its power and might, to come to the aid of the Old.

In the same way, the United States has allowed China to rise but has also facilitated the growth of Europe, Japan, India, Brazil, and many others. And however much those actors might continue to chafe at aspects of American leadership or chase Chinese investment, they would prefer the continuation of the current arrangements to being forced to kowtow to the Middle Kingdom.

The issue of the day might seem to be whether a Chinese sphere of influence can spread without overturning the existing U.S.-created and U.S.-dominated international order. But that ship has sailed: China's sphere has expanded prodigiously and will continue to do so. At the same time, China's revival has earned it the right to be a rule-maker. The real questions, therefore, are whether China will run roughshod over other countries, because it can—and whether the United States will share global leadership, because it must.

Are a hegemon's commitments co-dependent, so that giving up some undermines the rest? Can alliances and guarantees in one place unwind while those in another remain strong? In short, is retrenchment possible, or does even a hint of retreat have to turn into a rout? A well-executed U.S. transition from hegemonic hyperactivity to more selective global engagement on core interests might be welcome both at home and abroad, however much politicians and pundits would squeal. But cases of successful peaceful retrenchment are rare, and none has started from such an apex.

History tells us nothing about the future except that it will surprise us. Three-D printing, artificial intelligence, and the onrushing digital and genetics revolutions may upend global trade and destabilize the world radically. But in geopolitics, good outcomes are possible, too—realism is not a counsel of despair. For today's gladiators to buck the odds and avoid falling at each other's throats like most of their predecessors did, however, four things will be necessary. Western policy makers have to find ways to make large majorities of their populations benefit from and embrace an open, integrated world. Chinese policy makers have to continue their country's rise peacefully, through compromise, rather than turning to coercion abroad, as well. The United States needs to hew to an exactly right balance of strong deterrence and strong reassurance vis-a-vis China and get its house in order domestically. And finally, some sort of miracle will have to take care of Taiwan.

Critical Thinking

1. What is the relationship between globalization and populism and authoritarianism in the international system?
2. What is the geopolitical importance of Australia to China?
3. What changes have occurred in the international order over the past 70 years?

Internet References

Maritime Strategy in a New Era of Great Power Competition

https://www.hudson.org/research/4070-maritime-strategy-in-a-newera–of-great-power.../

The National Security Strategy of the US

https://www.whitehouse.gov/articles/new-national-security-strategy-new-era/

Stephen Kotkin is John P. Birkelund '52 Professor in History and International Affairs at Princeton University and a Senior Fellow at Stanford's Hoover Institution.

Article Prepared by: Robert Weiner, *University of Massachusetts, Boston*

Trump and Russia

The Right Way to Manage Relations

EUGENE RUMER, RICHARD SOKOLSKY, AND ANDREW S. WEISS

Learning Outcomes

After reading this article, you will be able to:

- Learn about the policy the Trump administration should follow in dealing with Russia.
- Learn about Russian foreign policy under Putin.

Relations between the United States and Russia are broken, and each side has a vastly different assessment of what went wrong. U.S. officials point to the Kremlin's annexation of Crimea and the bloody covert war Russian forces are waging in eastern Ukraine. They note the Kremlin's suppression of civil society at home, its reckless brandishing of nuclear weapons, and its military provocations toward U.S. allies and partners in Europe. They highlight Russia's military intervention in Syria aimed at propping up Bashar al-Assad's brutal dictatorship. And they call attention to an unprecedented attempt through a Kremlin-backed hacking and disinformation campaign to interfere with the U.S. presidential election last November.

Russian President Vladimir Putin and his circle view things differently. In Ukraine, Moscow sees itself as merely pushing back against the relentless geopolitical expansion of the United States, NATO, and the EU. They point out that Washington and its allies have deployed troops right up to the Russian border. They claim that the United States has repeatedly intervened in Russian domestic politics and contend, falsely, that former U.S. Secretary of State Hillary Clinton even incited antigovernment protests in Moscow in December 2011. And they maintain that the United States is meddling in Syria to overthrow a legitimate government, in just the latest example of its unilateral attempts to topple regimes it doesn't like.

The gap between these two narratives is dangerous. Not only do heightened tensions raise the risk of a military accident or confrontation in Europe and beyond; they are also largely a reflection of deeply entrenched resentments within the Russian national security establishment that are likely to persist well beyond the Putin era. The differences between the United States and Russia run deep, and they are not amenable to easy solutions.

The challenge facing the Trump administration is to skillfully manage, rather than permanently resolve, these tensions with Moscow. Trying to appease Putin, perhaps by making unilateral concessions, would only convince him that he is winning and encourage him to continue wrong-footing the United States and the West. But a more confrontational approach would risk generating a provocative and dangerous response from Russia. So Washington will need to chart a middle path. That means both seeking ways to cooperate with Moscow and pushing back against it without sleepwalking into a collision.

Of course, that advice presupposes a U.S. administration that views Russia the same way previous ones have: as a problematic yet important partner on discrete issues that also poses a significant national security threat. U.S. President Donald Trump, however, appears eager to jettison established bipartisan approaches to dealing with Moscow. As he wrote on Twitter in January, "Having a good relationship with Russia is a good thing, not a bad thing. Only 'stupid' people, or fools, would think that it is bad!" And for months, he mocked the U.S. intelligence community's warnings about Russian cyberattacks aimed at interfering with the U.S. democratic process and repeatedly praised Putin's leadership.

Such antics suggest that Trump may attempt an abrupt reconciliation with Russia that would dramatically reverse the policies of President Barack Obama. It is hard to overstate the lasting damage that such a move would do to the U.S. relationship with Europe, to the security of the continent, and to an already fraying international order.

Putin's Game

Any consideration of U.S. policy toward Russia must start with a recognition of that country's manifold weaknesses. The Russian economy may not be "in tatters," as Obama once remarked, but the boom that allowed Putin, during his first two terms in office, to deliver steady increases in prosperity in exchange for political passivity is a distant memory. Absent major structural reforms, which Putin has refused to undertake for fear of losing control, the economy is doomed to "eternal stagnation," as Ksenia Yudaeva, a senior Russian central bank official, put it last year.

Following Putin's return to the presidency in 2012, the regime has retooled the sources of its legitimacy. It has fostered a fortress mentality, mobilizing the public to defend Russia against foreign adversaries and mounting an unrelenting search for Western-backed fifth columnists. The apparent spur-of-the-moment decision to annex Crimea transformed the Russian domestic political landscape overnight, propelling Putin to unprecedented levels of popularity. And in Syria, the Kremlin has capitalized on its intervention to highlight Russia's return to global prominence.

Unfortunately, tighter economic constraints are not likely to dissuade Putin from engaging in future foreign policy adventures. The collapse of oil prices that began in 2014 hit the Russian economy hard, as did the sanctions the West applied in response to Russian aggression in Ukraine that same year. Yet Putin has shown little restraint in the international arena since. His defiant approach appears to have strong support from the Russian elite, which faithfully rallies to the cause of standing up to the United States and reasserting Russia's great-power status.

Indeed, Russia has always been much more than a mere "regional power," as Obama once dismissed it; the country figures prominently in important issues across the globe, from the Iran nuclear program to the security of the entire transatlantic community. That will not change. But even if one accepts that Russia is a declining power, history shows that such states can cause considerable damage on their way down. And if there is one thing that can be said for certain about Putin, it is that he is a skilled and opportunistic risk taker capable of forcing others to deal with him on his own terms.

The United States must also reckon with another fundamental characteristic of Russia's foreign policy: its desire for de facto control over its neighbors' security, economic, and political orientation. Both Democratic and Republican administrations have long considered this unacceptable. Yet it constitutes one of the Russian regime's core requirements for security.

Absent an abrupt change in these fundamental realities, it will be hard to significantly improve U.S. relations with Russia. The country's intervention in Ukraine has demolished much of the post–Cold War security order and, along with it, any semblance of trust on either side. And it would be irresponsible for Washington to turn a blind eye to the Kremlin's reliance on hacking, disinformation, and Cold War–style subversion in its efforts to undermine the United States.' international reputation and to meddle in democratic processes in Europe and beyond. The best course of action is for the United States to stand firm when its vital interests are threatened, to expose and counter Moscow's penchant for irregular tactics, and to carefully manage the rivalry that lies at the heart of the bilateral relationship.

The Big Picture

In recent years, Russia and the West have been heading toward something that looks a lot like a second Cold War. This confrontation may lack the geopolitical and ideological scope of the first, but it still carries a high risk of actual conflict. The close encounters that NATO aircraft and warships have had with Russian jets are no accident; they are part of a deliberate Kremlin strategy to intimidate Moscow's adversaries.

For now, the Kremlin is likely to try to downplay sources of tension, setting the stage for friendly initial encounters with the new U.S. president and his team. Assuming Moscow follows that course, Washington will have to proceed with caution as Putin, the consummate dealmaker, seeks to shape the terms of a new relationship. In negotiating those terms, the Trump administration should adhere to five overarching principles.

First, it must make clear that the United States' commitment to defend its NATO allies is absolute and unconditional. To do so, the United States should bolster deterrence through an ongoing series of defense improvements and increased military deployments on the alliance's eastern flank. It should also ramp up the pressure on fellow NATO members to spend more on defense.

Second, the United States needs to steadfastly uphold the principles enshrined in the 1975 Helsinki Final Act and the 1990 Charter of Paris for a New Europe—both of which commit Moscow to recognize existing borders and the right of all countries to choose their own allies. It may be hard to imagine a feasible scenario for returning Crimea to Ukraine, but the annexation remains a flagrant violation of international law that no country should recognize or reward. That means keeping in place the U.S. and EU sanctions that ban transactions and economic cooperation with Russian-occupied Crimea.

Third, as Washington reengages with Moscow, it must not run roughshod over Russia's neighbors. Appeasing Russia on

Ukraine or caving in to its demand for a sphere of influence in its neighborhood would set a terrible precedent and undermine U.S. standing in the world. The inherent fragility of Russia's neighbors will create many openings for future Russian meddling, so the United States and its allies will need to remain vigilant and become more deeply engaged in such a complex region.

Fourth, Washington and its partners in the EU should commit themselves to supporting Ukrainian political and economic reform through skillful diplomacy and a generous flow of resources. It will probably take a generation or longer to turn this pivotal country into a prosperous, European-style state, not least because of Russia's undisguised desire for Ukraine's reformist experiment to fail. If Ukraine receives steady Western support based on clear and achievable conditions, its success will have a lasting positive impact on Russia's trajectory by demonstrating a viable alternative to the Kremlin's top-down approach to governance.

Fifth, as the United States attempts to support democracy in Russia and other former Soviet states, it should make a sober-minded assessment of local demand for it and the best use of limited resources. Russia's democratic deficit will hinder better relations with the West for as long as it persists. The same problem will continue to complicate U.S. ties with many of Russia's neighbors. But too often, Washington has overestimated its ability to transform these societies into functioning democracies.

In applying these principles, the United States needs to remain mindful of the risks of overreaching. That will mean making sharp distinctions between what is essential, what is desirable, and what is realistic.

Needs and Wants

Improved communication belongs in the first category. In response to Russia's moves in Ukraine, the Obama administration suspended most routine channels of communication and cooperation with the Russian government and encouraged U.S. allies to follow suit. As the crisis has dragged on, it has become harder to address differences, avoid misunderstandings, and identify points of cooperation in the absence of regular interactions at various levels. The Trump administration should entertain the possibility of resuming a wide-ranging dialogue, even though the Russians may well prove as unwilling to engage in a serious give-and-take as they did during the George W. Bush and Obama administrations, or may choose to use the talks solely to score political points. But even if the Kremlin isn't ready to engage forthrightly, the Trump administration should put four essential priorities above all else in its early discussions with the Russian government.

First, the Trump administration should respond to Russian meddling in the U.S. presidential election in ways that get the Russians' attention. As a parting shot, Obama imposed sanctions on Russian entities involved in the hacking and ejected 35 Russian diplomats from the United States. Yet much more needs to be done. A carefully calibrated covert response in cyberspace would send the message that the United States is prepared to pay back the Kremlin and its proxies for their unacceptable actions. Trump should also work to protect the large swaths of government and private-sector networks and infrastructure in the United States that remain highly vulnerable to cyberattacks. The lack of a concerted response to Russia's meddling would send precisely the wrong signal, inviting further Kremlin exploits in France and Germany, which are holding their own elections this year. In the meantime, the U.S. government should explore whether it can work with major actors in the cyber-realm, such as China and Russia, to develop new rules of the road that might limit some of the most destabilizing kinds of offensive operations.

Second, the Trump administration should ensure that military-to-military channels are open and productive. Russia's provocations carry the very real risk of a military confrontation arising from a miscalculation. Washington should prioritize getting Russia to respect previously agreed-on codes of conduct for peacetime military operations, however difficult that might be. The situation is especially dangerous in the skies over Syria, where Russian pilots frequently flout a set of procedures agreed to in 2015 to avoid in-air collisions with U.S. and other jets.

Third, in Ukraine, Trump should focus on using diplomatic tools to de-escalate the military side of the conflict and breathe new life into the Minsk accords, a loose framework of security and political steps that both sides have refused to fully embrace. The existing package of U.S. and EU sanctions represents an important source of leverage over Moscow, and so it should not be reversed or scaled back in the absence of a major change in Russian behavior in Ukraine. At the same time, the United States and its EU allies must work to keep Ukraine on a reformist path by imposing strict conditions on future aid disbursements to encourage its government to fight high-level corruption and respond to the needs of the Ukrainian people.

The fourth and final priority for the Trump administration is to remain realistic about the prospects of promoting transformational change in Russia. As the last 25 years have shown again and again, Russia resists outside efforts at modernization. In other words, the United States should not treat Russia as a project for political, social, or economic engineering.

Then there are goals that, although not essential, remain desirable. In this category should go issues on which Washington and Moscow have a good track record of cooperation thanks to overlapping, if not identical, interests. These include cooperation on preventing nuclear proliferation, reducing the threat of nuclear terrorism, and protecting the fragile environment in the Arctic. Because these issues are largely technical in nature, they do not require the time and attention of senior officials. A great deal of progress can be made at lower levels.

On more ambitious arms control efforts, however, progress will require high-level decisions that neither side is eager to make. Such is the case with resolving the impasse over the Intermediate-Range Nuclear Forces Treaty, which the United States claims Russia has violated, and securing further reductions in the size of both countries' strategic and tactical nuclear arsenals.

Even so, the Trump administration should keep the door open to further progress on arms control. The U.S.-Russian arms control edifice is in danger of collapsing: the Anti-Ballistic Missile Treaty and the Treaty on Conventional Armed Forces in Europe are no longer in force, the Intermediate-Range Nuclear Forces Treaty may soon fall apart, and the New START treaty is due to expire in 2021. Neither Russia nor the United States is ready for a new arms control agreement, primarily because of conflicting agendas. Moscow wants to constrain U.S. deployments of missile defense systems and high-tech conventional weapons, while Washington wants to further reduce the number of Russian strategic and tactical nuclear weapons. But neither would be served by abandoning arms control completely. At a minimum, both would benefit from more conversations about their force structures and nuclear doctrines, with an eye toward ensuring stability, especially in crises.

Fact and Fantasy

Of course, Washington's ability to achieve what is essential and what is desirable will be limited by what is realistic. In a perfect world, Trump would focus on keeping relations from deteriorating further. Instead, he and his team appear to be fanning expectations of a big breakthrough and a grand bargain.

Indeed, much of what Trump says he believes about Russia appears unrealistic, to put it mildly. For starters, he has made the mystifying choice to ridicule the U.S. intelligence community's finding that it was Russia that was behind the hacking of e-mails from the Democratic National Committee and the Clinton campaign. If Trump's and his advisers' statements are to be believed, even a brazen attempt originating at the highest levels of the Russian government to undermine Americans' confidence in their country's democratic process is less important than the poor cybersecurity practices of the Democratic National Committee and Clinton's inner circle.

Trump appears to hold an equally unrealistic view of the Ukrainian crisis, saying of Putin during the campaign, "He's not going to go into Ukraine, all right?"—even as thousands of Russian troops were already there. When asked by *The New York Times* on the eve of the election about Putin's behavior in Ukraine and Syria and the ongoing crackdown against Putin's political opponents, Michael Flynn, Trump's pick for national security adviser, called these issues "besides the point." He added, "We can't do what we want to do unless we work with Russia, period."

But as Trump will likely discover, reality has a way of interfering with attempts to transform relations with Moscow. Every U.S. president from Bill Clinton on has entered office attempting to do precisely that, and each has seen his effort fail. Clinton's endeavor to ease tensions fell apart over NATO expansion, the Balkan wars, and Russian intervention in Chechnya; George W. Bush's collapsed after the 2008 Russian-Georgian war; and Obama's ran aground in Ukraine. Each administration encountered the same obstacles: Russia's transactional approach to foreign policy, its claim to a sphere of influence, its deep insecurities about a yawning power gap between it and the United States, and its opposition to what it saw as Western encroachment. Finding common ground on these issues will be difficult.

It appears that at the core of Trump's vision for improved relations is a coalition with Russia against the Islamic State—to, in his words, "knock the hell out of ISIS." Yet such cooperation is unlikely to materialize. The Russians have shown no interest in beating back ISIS in Syria, choosing instead to attack the main opposition forces arrayed against the Assad regime. Russia's and Iran's support for Assad may have fundamentally changed the course of the civil war in Syria, but their crude methods and disregard for civilian casualties have probably only emboldened the radical jihadists. Help from the Russian military would be a mixed blessing, at best, for the U.S.-led coalition against ISIS, given the pervasive lack of trust on both sides and the very real risk that sensitive intelligence and targeting information would find its way into the hands of Moscow's allies in Damascus and Tehran.

Trump has also expressed interest in developing stronger economic ties with Russia as a foundation for improved diplomatic relations, at least according to the Kremlin's summary of Putin's congratulatory call to Trump after the election. Here, too, he is likely to be disappointed. Clinton, Bush, and Obama all placed high hopes on trade as an engine of better relations with Russia. All were frustrated by the fact that the two countries are, for the most part, not natural trading partners, to say nothing of the effects of Russia's crony capitalism, weak rule of law, and predatory investment climate.

Proceed with Caution

Trump inherited a ruptured U.S.-Russian relationship, the culmination of more than 25 years of alternating hopes and disappointments. As both a candidate and president-elect, he repeatedly called for a new approach. "Why not get along with

Russia?" he has asked. The answer is that at the heart of the breakdown lie disagreements over issues that each country views as fundamental to its interests. They cannot be easily overcome with the passage of time or a summit meeting or two. Thus, the challenge for the new administration is to manage this relationship skillfully and to keep it from getting worse.

Should Trump instead attempt to cozy up to Moscow, the most likely outcome would be that Putin would pocket Washington's unilateral concessions and pursue new adventures or make demands in other areas. The resulting damage to U.S. influence and credibility in Europe and beyond would prove considerable. Already, the rules-based international order that the United States has upheld since the end of World War II is in danger of unraveling, and there is mounting concern throughout Europe, Asia, and beyond that Trump does not consider it worth preserving. What's more, there's no telling how Trump will respond if and when he has his first showdown with Putin, although his behavior toward those who cross him suggests that things would not end well.

Reduced tensions with Russia would no doubt help further many of the United States' political and security priorities. But policy makers must keep in mind that the abiding goal should be to advance U.S. interests, support U.S. allies across the world, and uphold U.S. principles—not to improve relations with Russia for their own sake. Indeed, it's possible to stand up for American interests and principles while pursuing a less volatile relationship with Russia. The Nixon administration sowed mines in a harbor in North Vietnam, a Soviet ally, while seeking détente with Moscow. The Reagan administration aggressively challenged Soviet-backed regimes and groups in Eastern Europe, Asia, Africa, and Latin America at the same time as it signed arms control agreements with Moscow.

Likewise, the Trump administration can, for example, counter Russian aggression in Ukraine while looking for ways to cooperate on efforts to keep weapons of mass destruction out of the wrong hands. Such an approach has a far greater chance of success than pure confrontation or pure concession. Russian leaders have long expressed their preference for realpolitik; they will respect a country that stays true to its principles, knows its interests, and understands power.

Critical Thinking

1. Why is Russia more than a regional power?
2. Is Russia an enemy or competitor of the United States? Why or why not?
3. Is there a new Cold War between Russia and the United States? Why or why not?

Internet References

Embassy of the Russian Federation in the US
www.russianembassy.org

Russian Ministry of Foreign Affairs
government.ru

The National Security Strategy of the United States
https://www.whitehouse.gov/articles/new-national-securitystrategy-new-era/

Eugene Rumer is a senior fellow in and Director of the Russia and Eurasia Program at the Carnegie Endowment for International Peace.

Richard Sokolsky is a senior fellow in the Carnegie Endowment's Russia and Eurasia Program.

Andrew S. Weiss is vice president for Studies at the Carnegie Endowment.

This article draws from a longer study they undertook for a joint task force of the Carnegie Endowment for International Peace and the Chicago Council on Global Affairs.

Article

Prepared by: Robert Weiner, *University of Massachusetts, Boston*

Russia as it Is: A Grand Strategy for Confronting Putin

MICHAEL MCFAUL

Learning Outcomes

After reading this article, you will be able to:

- Learn about the grand strategy of containment that the United States should follow in dealing with Putin's Russia.
- Learn why Putin considers the United States to be an enemy.

Relations between Russia and the United States have deteriorated to their most dangerous point in decades. The current situation is not, as many have dubbed it, a new Cold War. But no one should draw much comfort from the ways in which today's standoff differs from the earlier one. The quantitative nuclear arms race is over, but Russia and the United States have begun a new qualitative arms race in nuclear delivery vehicles, missile defenses, and digital weapons. The two countries are no longer engulfed in proxy wars, but over the last decade, Russia has demonstrated less and less restraint in its use of military power. The worldwide ideological struggle between capitalism and communism is history, but Russian President Vladimir Putin has anointed himself the leader of a renewed nationalist, conservative movement fighting a decadent West. To spread these ideas, the Russian government has made huge investments in television and radio stations, social media networks, and Internet "troll farms," and it has spent lavishly in support of like-minded politicians abroad. The best description of the current hostilities is not cold war but hot peace.

Washington must accept that Putin is here to stay and won't end his assault on Western democracy and multilateral institutions anytime soon. To deal with the threat, the United States desperately needs a new bipartisan grand strategy. It must find ways to contain the Kremlin's economic, military, and political influence and to strengthen democratic allies, and it must work with the Kremlin when doing so is truly necessary and freeze it out when it is not. But above all, Washington must be patient. As long as Putin remains in power, changing Russia will be close to impossible. The best Washington can hope for in most cases is to successfully restrain Moscow's actions abroad while waiting for Russia to change from within.

Ups and Downs

At the end of the Cold War, both U.S. and Russian leaders embraced the promise of closer relations. So what went wrong? Russia's renewed international power provides part of the explanation. If Russia were too weak to annex Crimea, intervene in Syria, or interfere in U.S. elections, Moscow and Washington would not be clashing today. But not all rising powers have threatened the United States. Germany and Japan are much stronger than they were 50 years ago, yet no one is concerned about a return to World War II rivalries. What is more, Russia's relations with the United States were much more cooperative just a few years back, well after Russia had returned to the world stage as a great power.

In Russian eyes, much of the blame falls on U.S. foreign policy. According to this argument, the United States took advantage of Russia when it was weak by expanding NATO and bombing Serbia in 1999, invading Iraq in 2003, and allegedly helping overthrow pro-Russian governments in Georgia in 2003 and Ukraine in 2004. Once Russia was off its knees, it had to push back against U.S. hegemony. At the 2007 Munich Security Conference, Putin championed this line of analysis: "We are seeing a greater and greater disdain for the basic principles of international law. . . . One state, and, of course, first and foremost the United States, has overstepped its national borders in every way."

There is some truth to this story. The expansion of NATO did exacerbate tensions with Moscow, as did Western military interventions in Serbia and Iraq. Democratic upheavals in Georgia and Ukraine threatened Putin's ability to preserve autocracy at home, even if Putin grossly exaggerated the U.S. role in those so-called color revolutions.

Yet this account omits a lot of history. After the end of the Cold War, U.S. presidents were truly committed to, in Bill Clinton's words, "a strategic alliance with Russian reform" and Russia's integration into the international system. Just as the United States and its allies helped rebuild, democratize, and integrate Germany and Japan after World War II, the thinking went, so it would rebuild Russia after the Cold War. It is true that the United States and Europe did not devote enough resources or attention to this task, leaving many Russians feeling betrayed. But it is revisionism to argue that they did not embrace Moscow's new leaders, support democratic and market reforms, and offer Russia a prominent place in Western clubs such as the G-8.

The most powerful counterargument to the idea that U.S. foreign policy poisoned the well with Russia is that the two countries managed to work together for many years. The cooperative dynamic of U.S.-Russian relations established after the fall of the Soviet Union survived not only U.S. provocations but also two Russian military operations in Chechnya and the 1998 Russian financial crisis, after which foreign governments accused the Kremlin of wasting Western aid. And even the U.S. withdrawal from the Anti-Ballistic Missile Treaty, in 2002, and another, larger round of NATO expansion, in 2004, did not end the cooperative dynamic that U.S. President George W. Bush and Putin had forged after the 9/11 attacks. Russia's invasion of Georgia in 2008 pushed U.S.-Russian relations to a low point in the post-Cold War era. But even this tragedy did not permanently derail cooperation.

How It All Went Wrong

Even after all these ups and downs, U.S.-Russian relations experienced one last spike in cooperation, which lasted from 2009 to 2011. In 2009, when U.S. President Barack Obama met for the first time with Russian President Dmitry Medvedev and Putin, who was then serving as Russia's prime minister, the U.S. president tried to convince the two Russians that he was a new kind of American leader. He had opposed the Iraq war long before it was popular to do so, he explained, and had always rejected the idea of regime change. At least at first, Medvedev seemed convinced. Even Putin showed signs of softening. Over the next few years, Russia and the United States signed the New Strategic Arms Reduction Treaty (or New START), worked through the UN to impose tough new sanctions on Iran, managed Russia's entry into the World Trade Organization, coordinated to defuse violence in Kyrgyzstan after the collapse of the government there, and arranged a vast expansion of the network used to transport U.S. soldiers and supplies to Afghanistan through Russia. In 2011, in perhaps the most impressive display of renewed cooperation, Russia acquiesced in the Western intervention in Libya. At the height of the so-called reset, in 2010, polls showed that around 50 percent of Americans saw Russia as a friendly country and that some 60 percent of Russians viewed the United States the same way.

This period of relative harmony began to break down in 2011, owing primarily to the way that Putin reacted to popular democratic mobilizations against autocracies in Egypt, Libya, Syria—and Russia itself. The Libyan uprising in 2011 marked the beginning of the end of the reset; the 2014 revolution in Ukraine marked the start of the hot peace.

Popular mobilization inside Russia was especially unnerving to Putin. He had enjoyed solid public support during most of his first eight years as president, thanks primarily to Russia's economic performance. By 2011, however, when he launched a campaign for a third term as president (after having spent three years as prime minister), his popularity had fallen significantly. The implicit bargain that Putin had struck with Russian society during his first two terms—high economic growth in return for political passivity—was unraveling. Massive demonstrations flooded the streets of Moscow, St. Petersburg, and other large cities after the parliamentary election in December 2011. At first, the protesters focused on electoral irregularities, but then they pivoted to a grander indictment of the Russian political system and Putin personally.

In response, Putin revived a Soviet-era source of legitimacy: defense of the motherland against the evil West. Putin accused the leaders of the demonstrations of being American agents. Obama tried to explain that the United States had not prompted the Russian demonstrations. Putin was unconvinced. After his reelection in the spring of 2012, Putin stepped up his attacks on protesters, opposition parties, the media, and civil society and placed under house arrest the opposition leader he feared the most, the anticorruption blogger Alexei Navalny. The Kremlin further restricted the activity of nongovernmental organizations and independent media outlets and imposed significant fines on those who participated in protests that the authorities deemed illegal. Putin and his surrogates continued to label Russian opposition leaders as traitors supported by the United States.

Putin's anti-American campaign was not just political theater intended for a domestic audience: Putin genuinely believed that the United States represented a threat to his regime. Some pockets of U.S.-Russian cooperation persisted, including a joint venture between the Russian state-owned oil giant Rosneft and ExxonMobil, an agreement brokered by Obama and Putin in which Syria pledged to eliminate its chemical weapons, and Russian support for the international negotiations that produced the Iran nuclear deal. But most of these ended in 2014, after the fall of the pro-Russian Ukrainian government and the subsequent Russian invasion of Ukraine. Once again, Putin blamed the Obama administration, this time for supporting the revolutionaries who toppled Ukrainian President Viktor Yanukovych.

Putin was never inclined to believe in Washington's good faith. His training as a KGB agent had led him to distrust the United States along with all democratic movements. But in the early years of his presidency, he had held open the possibility of close cooperation with the West. In 2000, he even suggested that Russia might someday join NATO. After the 9/11 attacks, Putin firmly believed that Russia could work with the United States in a global war on terrorism. In 2008, after he stepped aside as president, he allowed Medvedev to pursue closer ties with Washington. But the Western intervention in Libya confirmed Putin's old suspicions about U.S. intentions. Putin believed that the United States and its allies had exploited a UN resolution that authorized only limited military action in order to overthrow the Libyan dictator Muammar al-Qaddafi. In Putin's view, Obama had turned out to be a regime changer, no different from Bush.

Confronting the Kremlin

Four years after Russia annexed Crimea, the United States has still not articulated a bipartisan grand strategy for dealing with Russia. Such a strategy is necessary because Washington's conflict with the Kremlin doesn't revolve around mere policy disagreements: rather, it is a contest between Putinism and democracy. No tweaking of U.S. policy on Syria or NATO will influence Putin's thinking. He has been in power for too long– and he is not likely to leave in the foreseeable future. U.S. policy makers must dispense with the fantasy that Putin's regime will collapse and democracy will emerge in Russia in the near term. The United States and its allies must continue to support human rights and democracy and embrace people inside Russia fighting for those values. But real political change will likely begin only after Putin steps down.

The United States also has to give up on the idea that Russia can or should be integrated into multilateral institutions. The theory that integration would moderate Russian behavior has not been borne out by events. The United States must dig in for a long and difficult confrontation with Putin and his regime. On most issues, the aim should be to produce a stalemate, as preserving the status quo will often be the best the United States can hope for.

Containment must start at home. Limiting Putin's ability to influence U.S. elections should be priority number one. The Trump administration should mandate enhanced cybersecurity resilience. If the federal government can require all cars to have seat belts, then federal authorities can require elementary cybersecurity protections such as dual authentication for all processes related to voting during a presidential election. Those who operate the systems that maintain voter registries must be required to receive training about how to spot common hacking techniques and an even more rigorous set of standards must be adopted for the vote count. In a dozen states, including large battlegrounds such as Florida and Pennsylvania, at least some precincts lack paper trails for each ballot cast. These sloppy practices have to end. Every precinct must be able to produce a paper record for every vote.

Congress should also pass laws to provide greater transparency about Russian media activities inside the United States, including a requirement for social media companies to expose fake accounts and disinformation. Foreign governments should not be allowed to buy ads anywhere to influence voter preferences. Beyond elections, the federal government must devote more time and money to blocking Russian threats to all national electronic infrastructure.

To further counter Putin's ideological campaign, the United States should organize democracies around the world to develop a common set of laws and protocols regulating government-controlled media. Through regulation, Washington should encourage social media platforms to grant less exposure to Kremlin-created content. Algorithms organizing search results on Google or YouTube should not overrepresent information distributed by the Russian government. When such material does appear in searches, social media companies should make its origins clear. Readers must know who created and paid for the articles they read and the videos they watch.

On their own, without government intervention, social media platforms should provide sources from more reliable news organizations; every time an article or video from the Kremlin-backed news channel RT appears, a BBC piece should pop up next to it. Social media companies have long resisted editorial responsibilities; that era must end.

In Europe, Putin's success in courting Hungarian President Viktor Orban and nurturing several like-minded political parties and movements within NATO countries underscores the need for a deeper commitment to ideological containment on the part of Washington's European allies. Those allies must pay greater attention to combating Russian disinformation and devote more time and resources to promoting their own values. NATO members must also meet their defense spending pledges, deploy more soldiers to the alliance's frontline states, and reaffirm their commitment to collective security.

No theater in the fight to contain Russia is more important than Ukraine. Building a secure, wealthy, democratic Ukraine, even if parts of the country remain under Russian occupation for a long time, is the best way to restrain Russian ideological and military aggression in Europe. A failed state in Ukraine will confirm Putin's flawed hypothesis about the shortcomings of U.S.-sponsored democratic revolutions. A successful democracy in Ukraine is also the best means for inspiring democratic reformers inside Russia and other former Soviet republics. The United States must increase its military, political, and economic support for Ukraine. Washington should also impose new sanctions on Russians involved in violating Ukraine's sovereignty and ratchet them up until Putin begins to withdraw.

In the Middle East, the United States needs a more aggressive strategy to contain Russia's most important regional ally, Iran. It should continue to arm and support Syrian militias fighting Iranian soldiers and their allies in Syria and should promote anti-theocratic and prodemocratic ideas in the region, including inside Iran. Abandoning the fight in Syria would deliver a tremendous victory to Moscow and Tehran. The goals of U.S. policy toward Iran must remain denying Tehran a nuclear weapon, containing its destabilizing actions abroad, and encouraging democratic forces inside the country, but not coercive regime change from the outside.

The United States must contain the Kremlin's ambitions in Asia, as well. Strengthening existing alliances is the obvious first step. Putin has sought to weaken U.S. ties with Japan and South Korea. To push back, the United States should make its commitment to defend its allies more credible, starting by abandoning threats to withdraw its soldiers from South Korea. It should also begin negotiations to rejoin the Trans-Pacific Partnership. A harder but still important task will be to divide China from Russia. In 2014, Putin suffered a major setback when China did not support his annexation of Crimea at the UN. But today, putting daylight between the two countries will not be easy, as Putin and Chinese President Xi Jinping have forged a united front on many issues. When opportunities do arise, such as working with Beijing toward North Korean denuclearization, Washington must act.

Western countries must also develop a coherent strategy to contain the Russian government's economic activities. Europe must reduce its dependence on Russian energy exports. Projects such as the planned Nord Stream 2 natural gas pipeline from Russia to Germany are no longer appropriate and should be discontinued. Putin uses government-owned and supposedly private companies to advance his foreign policy interests; the United States and Europe must impose greater financial sanctions on the activities those firms undertake in the service of Kremlin interests abroad if Russia continues to occupy Ukraine or assault the integrity of democratic elections. At a minimum, the West must adopt new laws and regulations to require greater transparency around Russian investments in the United States, Europe, and, as far as possible, the rest of the world. Russian officials and businesspeople tied to the Kremlin cannot be allowed to hide their wealth in the West. Genuine private-sector companies inside Russia should be encouraged to engage with Western markets, but authorities must expose the ill-gotten financial assets that Putin and his cronies have parked abroad. The goal should be to underscore the economic benefits of free markets and access to the West while highlighting the economic costs of state ownership and mercantilist behavior.

On the other side of the equation, Western foundations and philanthropists must provide more support for independent journalism, including Russian-language services both inside and outside Russia. They should fund news organizations that need to locate their servers outside Russia to avoid censorship and help journalists and their sources protect their identities.

More generally, the United States and its democratic allies must understand the scope of their ideological clash with the Kremlin. Putin believes he is fighting an ideological war with the West, and he has devoted tremendous resources to expanding the reach of his propaganda platforms in order to win. The West must catch up.

How Do You Solve a Problem Like Putin's Russia?

Containing Russia does not mean rejecting cooperation in every area. The United States selectively cooperated with the Soviet Union during the Cold War; it should do so with Russia now. First on the list must be striking new arms control deals or at least extending existing ones, most urgently New START, which is set to expire in 2021 and contains crucial verification measures. Combating terrorism is another area for potential partnership, as many terrorist organizations consider both Russia and the United States to be their enemies. But such cooperation will have to remain limited since the two countries have different ideas about what groups and individuals qualify as terrorists, and some of Russia's allies in the fight against terrorism, such as Iran, Syria, and Hezbollah, are at odds with the United States. U.S. and Russian officials might also seek to negotiate an agreement limiting mutual cyberattacks. Yet Washington should not pursue engagement as an end in itself. Good relations with Russia or a friendly summit with Putin should be not the goal of U.S. diplomacy but the means to achieve concrete national security ends.

Some might argue that the United States cannot pursue containment and selective cooperation at the same time. The history of the Cold War suggests otherwise. President Ronald Reagan, for example, pursued a policy of regime change against Soviet-backed communist dictatorships in Afghanistan, Angola, Cambodia, and Nicaragua while negotiating arms control deals with Soviet leaders.

On global issues in which Russia does not need to be involved, the United States should isolate it. Since the end of the

Cold War, U.S. presidents have been eager to give their counterparts in the Kremlin symbolic leadership roles as a way to signal respect. Those days are over. Conversations about Russia rejoining the G-8 must end. Western governments should boycott sporting events held in Russia. Let the athletes play, but without government officials in the stands. Given Moscow's politicization of Interpol arrest requests, Interpol must suspend Russian participation. Even Russia's presence at NATO headquarters must be rethought. The more the United States can do without Russia, the better.

Even as the United States isolates the Russian government, it must continue to develop ties with Russian society. By canceling exchange programs, banning U.S. civil society organizations, and limiting Western media access to Russian audiences, Putin has tried to cut the Russian people off from the West. The United States and Europe need to find creative ways to reverse this disturbing trend. Happily, far more opportunities exist to do so today than did during the Cold War. Washington should promote student and cultural exchanges, dialogues between U.S. and Russian nongovernmental organizations, trade, foreign investment, and tourism.

Strategic Patience

But no matter how effective a containment strategy U.S. policy makers put in place, they must be patient. They will have to endure stalemate for a long time, at least as long as Putin is in power, maybe even longer, depending on who succeeds him. In diplomacy, Americans often act like engineers; when they see a problem, they want to fix it. That mentality has not worked with Putin's Russia, and if tried again, it will fail again.

At the same time, American leaders must say clearly that they do not want endless conflict with Russia. When the current confrontation winds down, most likely because of political change inside Russia, future U.S. presidents must stand ready to seize the moment. They will have to do better at encouraging democracy within Russia and integrating Russia into the West than their predecessors have done. Past politicians and the decisions they made created today's conflict. New politicians who make different decisions can end it.

Critical Thinking

1. What strategy is Putin pursuing to undermine Western liberal democracy?
2. Are relations between Russia and the United States marked by a new Cold War? Why or why not?
3. What steps can the United States take to contain Russia?

Internet References

Embassy of the Russian Federation in the US
www.russianembassy.org

Permanent Mission of Russia to the UN
Russianun.nu/en

Russian Ministry of Foreign Affairs
Government.ru

Michael McFaul is director of the Freeman Spogli Institute for International Studies at Stanford University and the author of *From Cold War to Hot Peace: An American Ambassador in Putin's Russia.* From 2012 to 2014, he served as U.S. Ambassador to Russia.

Article Prepared by: Robert Weiner, *University of Massachusetts, Boston*

China's South Asian Miscalculation

RAFFAELLO PANTUCCI

Learning Outcomes

After reading this article, you will be able to:

- Learn about the relationship between Chinese foreign policy and the Indian perception that it is being encircled by China.
- Learn about the miscalculations that China has made in its relations with India.

At a conference in China a few years ago, I watched as a Chinese expert gave a presentation laying out Beijing's view of the military conflict that it faced in nearby seas. It was largely a story about the United States and East Asian competitors, and China's aggressive assertions of ownership of islands in the South China Sea. At the end of the presentation, a former Indian officer raised his hand and indignantly asked why India had not been mentioned as a competitor.

In a moment of surprising candor, the Chinese expert responded that he did not include India because, from his perspective, it did not pose much of a threat to China. The answer riled the Indian participant, but it reflected a fundamental calculation that exists in Beijing about India. It is a calculation that could cause serious complications for China's broader South Asian vision, and ultimately provoke a clash between the two Asian giants.

At stake is China's Belt and Road Initiative (BRI), a much-discussed and puzzled-over concept. It has been variously described as a Chinese power grab; an attempt by China to promote its companies' overseas interests and build infrastructure to suit its own interests; an effort by Beijing to claim leadership of the international order; or, by Beijing's own account, a project to bind together a "community of common destiny." But it is really best understood as an umbrella concept that acts as a central organizing principle for China's foreign policy.

The core of this scheme—building trade and economic corridors that emanate from China in every direction—strengthens China's position in the global order and across the Eurasian landmass. The aim of these corridors is not only to help Chinese firms go out into the world and increase China's trade connections. Most importantly, they will help China develop domestically.

Ostensibly, this is a benign concept. By improving trade and transportation links through investments in infrastructure, China is enhancing the global commons. Few would say that more economic connectivity and prosperity is a bad thing. But the reality is of course very different. China is advancing its own national interests, and is doing so by offering a one-size-fits-all policy—which means that it can appear to be proffering the same opportunity to European powers and Southeast Asian neighbors alike. While this is a perfectly understandable self-interested approach, Beijing has been blind to geopolitical problems that it is exacerbating and which may in the long term disrupt its entire strategy.

Fears of Encirclement

Nowhere is this more evident than in South Asia, where Beijing's miscalculations regarding India have created conflict with a regional power that has the capability and desire to disrupt China's outward push. Chinese strategists see South Asia as a region of great potential opportunity where China can expand its influence. It is a region full of poor countries with large and growing young populations and governments that want access to Chinese investment—an arena where Beijing can expect to reap great rewards. The Chinese see few strategic competitors on the immediate horizon and worry more about nontraditional security threats like terrorism, insurgent groups, and criminal networks than they do about state-based ones.

In stark contrast, Indian strategists see an increasingly assertive China steadily encircling their country. China has developed important strategic investment relationships with all the countries that share a land border with India, while a growing Chinese presence in the Maldives and Sri Lanka has given it a string of island harbors connected to footholds that China has established on the mainland through port investments in places like Gwadar in Pakistan and Kyaukpyu in Myanmar. India looks and feels surrounded by countries that are increasingly either in Beijing's economic thrall or hosting its bases.

For New Delhi this territorial encirclement is coupled with regular border disputes of varying substance with China, either directly on their still-contentious borders, or in proxy locations like Bhutan where both have interests—China in terms of territorial claims and India through longstanding treaty obligations. But the question is not simply a territorial one.

The same pattern can be found on the global stage, marked by persistent Chinese efforts to stymie India's advancement and interests. For example, Beijing has blocked India's bid for membership in the Nuclear Suppliers Group (comprising countries that seek to control the proliferation of nuclear weapons) and used its veto power in the United Nations Security Council to keep Masood Azhar, the leader of Pakistan-based Jaish-e-Mohammed, which is already designated by the UN as a terrorist organization, from being added to a UN sanctions list.

Indian analysts see these moves as a matter of Beijing poking New Delhi in the eye while steadily encircling it. They also feel that the world is failing to lend India the support it deserves as the world's largest democracy, in contrast to China's one-party system. The story is one of growing confrontation, as hawkish national security establishments on both sides increasingly outflank economic pragmatists who want to take advantage of the potential benefits of a more cooperative relationship between the two Asian giants.

Auspicious Beginnings

This was not how China meant for things to go. When Xi Jinping first ascended to power five years ago, Beijing was clearly focused on moving in a positive direction with New Delhi. Prime Minister Li Keqiang's maiden foreign visit in May 2013 was to India—the first stop on a tour that took him on to Pakistan and then to Europe (presaging in many ways one of the routes of the Belt and Road Initiative).

During his stop in New Delhi, Li reawakened a long-dormant plan for a Bangladesh-China-India-Myanmar Economic Corridor (BCIM-EC)—first broached in 1999 when Jiang Zemin, China's president at the time, was pushing his own western development strategy within China—by establishing a joint working group to move the project forward. Moving on to Pakistan, Li signed a memorandum of understanding to get started on a China-Pakistan Economic Corridor (CPEC) project. Both corridors were the early paving stones for the Belt and Road Initiative, which Xi christened in Astana, Kazakhstan, in September 2013. The BRI went on to subsume both corridors and more under its vast reach.

The positive tone of Li's visit was reciprocated that October, when Indian Prime Minister Manmohan Singh paid a return visit to China. Since it was the first time in 60 years that Chinese and Indian premiers had exchanged visits in the same year, the event was marked with some fanfare, including a speech by Singh at the Central Party School in Beijing. While highlighting mutual interests and welcoming China's rise, Singh stated, "Our strategic partnerships with other countries are defined by our own economic interests, needs, and aspirations. They are not directed against China or anyone else. We expect a similar approach from China." This call for reciprocity was a clear signal from New Delhi that the opening would work only if Beijing treated India as a peer.

Simmering Tensions

In retrospect, Singh's visit seems to have marked the apex of recent relations between the two countries. He lost power in national elections in May 2014, leading to Narendra Modi's ascension to the premiership. By the time Xi finally paid his own visit to India that September, tensions had started to simmer. A standoff between Indian and Chinese troops in a disputed border region in Ladakh marred the visit, though that and other contemporaneous border confrontations were widely dismissed as routine skirmishes that received extra attention only because they occurred during visits by senior officials.

Tensions had already surfaced when Singh visited Beijing in 2013. One scathing opinion piece in *The New India Express* by former Indian government official and veteran China watcher Jayadeva Ranade declared that the visit had yielded "no tangibles" and cast doubt on the notion that the BCIM-EC was making "incremental progress" as China's "Southern Silk Route." But India would have to carefully study the plan in case it might disrupt its neighborhood, Ranade warned. Such voices are common in New Delhi; they reflect the view of a hawkish faction of the establishment that has gained growing power under Modi.

Xi set the tone for his September 2014 visit to India before he left Beijing with a declaration that he hoped Chinese investments in India would rise as high as $100 billion. Xi began his visit in Modi's home state of Gujarat. In between public walks together and other displays of bonhomie, the two leaders signed deals amounting to around $20 billion in Chinese investments. They also presided over agreements to establish a Chinese-funded business park in Gujarat and a "twinning" relationship between Gujarat and the southern coastal Chinese province of Guangdong.

It seemed to be an auspicious start to a new era of friendly and lucrative ties between the Elephant and the Dragon. Soon afterward in Beijing, Xi hosted a foreign-policy work conference that focused on border diplomacy and relations with neighboring powers as a priority for his administration. But four years later the agreements struck during his visit seem empty. Chinese foreign direct investment in India has stalled at around $4 billion, according to comments by Vice Minister for Finance Shi Yaobin in August 2016. Indian figures for the period from April 2000 to December 2016 show total Chinese inflows of only $1.6 billion. (The discrepancy is likely a product of Chinese investments flowing through secondary locations like Hong Kong.) Neither is anywhere near the $100 billion Xi had touted.

The business park in Gujarat has failed to materialize. The financier on the Chinese side, the China Development Bank, has announced several times that the project will start soon, but the Gujarat Industrial Development Corporation's website shows little evidence of progress. The two Chinese firms that set up shop in the state's industrial parks (TBEA, a manufacturer of electrical transmission equipment, and Baosteel, an iron and steel company) were vastly outnumbered by the many international firms already present.

The pairing between Guangdong and Gujarat also seems to have taken a turn for the worse. In early 2017, the Gujarat Chamber of Commerce and Industry called for a boycott of Chinese products, citing dumping of ceramics, plastics, and other Chinese exports, China's close ties with Pakistan, and other ways that China was "working against our interests." By 2018, the chamber was actively courting other foreign investors specifically to crowd out Chinese investments.

These moves were coupled with a national push to investigate Chinese pharmaceutical suppliers in India over quality concerns. That industry is particularly important in Gujarat, and Chinese firms in the sector had been at the forefront of the businesses that took advantage of Xi's visit to increase their presence in India.

Pakistani Frictions

The exact reasons for this chill in India-China business ties are open to interpretation, but they are most likely linked to a deterioration in India's relations with Pakistan. The trigger was an escalation in violence between India and Pakistan, particularly two attacks by non-state armed groups on Indian targets, launched from bases in Pakistan, that took place in 2016. Indian-Pakistani relations, but this set of events sent them into a downward spiral.

The first was a January 2016 strike on an Indian airbase at Pathankot in Punjab state, which led to the deaths of seven Indian servicemen and six attackers. A subsequent attack in September that year hit an Indian Army base in Uri, in the state of Jammu and Kashmir, resulting in the deaths of 19 soldiers. Both locations are situated close to the India-Pakistan border. That second attack led to retaliatory "surgical strikes" by India deep into Pakistani-controlled territory in Kashmir. Both attacks were allegedly linked to Jaish-e-Mohammed, the terrorist group led by Masood Azhar—the very man China has kept off the UN blacklist.

Despite these Pakistani provocations on Indian territory, China continued to push ahead with its China-Pakistan Economic Corridor. Reportedly making investments of over $50 billion (a number that fluctuates depending on what is included), China has clearly thrown its economic weight behind Pakistan, and has shown no sign of slowing down. Some routes cut through disputed territory in Gilgit-Baltistan—the northernmost tip of Pakistani-controlled Kashmir, which is still claimed by India. China is disregarding Indian concerns and treating the situation as de facto resolved. The actual projects underway in Gilgit-Baltistan are very limited, but the broader corridor has provided Pakistan with a bulwark against Indian and other external pressure. This has weakened India's ability to respond, angering New Delhi and placing a strain on China-India relations.

It has also had an impact in other respects. The energy that had been injected into the BCIM corridor in the wake of Li and Xi's visits to India seemed to wane. When I talked to Chinese experts and officials in mid-2017, they placed the blame for the stalled corridor firmly in Modi's court. They said they had tried to engage with their Indian counterparts on projects under the auspices of the BCIM plan, but their efforts led nowhere. They saw little evidence that India was moving forward with its side of the corridor by doing feasibility studies or taking other steps necessary to bring the concept to fruition.

Checkbook Diplomacy

While the Indian end of the corridor appeared to stall, China pushed ahead in forging closer ties with the two nations in between, Myanmar and Bangladesh. In Myanmar, China faced pushback over environmental concerns and other scandals surrounding the proposed Chinese-financed Myitsone Dam in 2011, which came to a head amid democratic reforms that seemed to push the country toward closer relations with the West. Since then, however, growing disillusionment in Western capitals about the ability of the new civilian leader, Aung San Suu Kyi, to rein in the military has pushed the country away from the West and made it amenable once again to China's embrace.

Meanwhile, a game of one-upmanship started to play out between China and India in Bangladesh. Modi extended a $2 billion line of credit to Bangladesh in 2015—on top of a previous $1 billion facility and a joint Indian effort with Japan to provide Bangladesh with a much-needed deep-sea port in Mataburi that beat out the Chinese-financed alternative 25 kilometers away in Sonadia. But this gambit was trumped in October 2016 when Xi visited Bangladesh and oversaw the signing of trade and investment deals worth around $13.6 billion, as well as $20 billion in loan agreements.

That, in turn, prompted a riposte from India in the form of $10 billion in investment and $5 billion in loans (including $500 million worth of defense assistance) after Bangladeshi Prime Minister Sheikh Hasina visited New Delhi in April 2017. The actual number was slimmed down in October when Indian Finance Minister Arun Jaitley paid a return visit and confirmed a $4.5 billion loan facility. But India's show of economic might was a clear signal to Beijing that it, too, could play checkbook diplomacy.

To India's north, the encirclement appears to continue with a growing Chinese footprint in Nepal. A $2.5 billion hydropower project that had previously been shelved was restarted when a new pro-China government took power after national elections in early 2018. An $8 billion railway between Lhasa and Kathmandu has been proposed under the Belt and Road Initiative.

India's response has been more limited, with proposed investments totaling in the hundreds of millions of dollars. According to data from the Nepalese Department of Industry, China has been the country's leading source of foreign direct investment since 2015. While Nepal's principal foreign partner is still India, there has been a gradual shift over time toward Beijing. Attempts by New Delhi to assert its influence through a 2015–16 blockade in response to alleged mistreatment of minority communities have seemingly backfired and simply provided greater access to Beijing.

In nearby Bhutan, a small Himalayan kingdom that has historical treaty agreements with India but no formal relations with Beijing, the rivalry between its big neighbors came into sharp focus last year when China started to build roads and bases in disputed territory on the Doklam Plateau. The Bhutanese government complained and India sent in soldiers to back up its ally, confronting Chinese troops. The tense standoff lasted for weeks before both sides sought to diplomatically de-escalate.

With no claim over the territory but a strong alliance with Bhutan, India believed it held the upper hand. Chinese experts and officials I spoke to at the time rejected New Delhi's declarations of diplomatic victory. They asserted that the episode grew out of a ridiculous intervention over territory that had little to do with India. One senior Chinese security official went even further and dismissed India's capabilities to back up its position, asserting that once winter came the less hardy Indian forces would be forced to stand down.

The Chinese narrative made it clear that Beijing did not take India seriously, regarding it as a power that was unable to compete with China economically, militarily, or strategically. The de-escalation seemed to be a simple recognition that China had pushed India as far as it could in this particular context and had achieved what it wanted for now. Analysis of satellite imagery subsequently showed how little the Chinese position had actually changed. Indian strategists are already bracing for a repeat performance of the standoff later this year, once winter ends and the area becomes more accessible.

Finally, in the seas around South Asia, China has continued to develop a footprint and establish what appear to be strategic stakes in island nations including Sri Lanka, the Maldives, and Mauritius, peppering the Indian Ocean with its presence and heightening India's feeling of encirclement. China's navy has already established a military base at the other end of the Indian Ocean in Djibouti. It remains a matter of time before it links up the dots to complete what is often called the "string of pearls." Yet China refuses to acknowledge such aims. In China's portrayal, its relationship with each nation in the region is unique, with specific interests and dynamics. Only time will tell whether this is a coherent and cohesive strategy, as India fears, or a set of disparate bilateral relationships. In many ways, the entire distinction is moot; if they all come to fruition, China will have created a network of strategic alliances entirely encircling India.

Chinese media, meanwhile, scoff at Indian capabilities. India's test of its Agni-V nuclear-capable intercontinental ballistic missile in April 2017 was mocked in a *Global Times* editorial that stated, "Chinese don't feel India's development has posed any big threat. . . . And India wouldn't be considered as China's main rival in the long run."

Hubris

There is both arrogance and hubris in China's response to the challenges India poses—arrogance in the form of a belief that there is little chance that India can be a genuine competitor, and hubris based on the fact that India, like much of the world, is eager to get its share of the economic opportunity that China offers. The reality is that this is the paradox at the heart of India's engagement with China. While New Delhi sees China as a threat, it is still keen for Chinese investment. The economic relationship between the two powers has enormous potential, and expanding it is something that both can see as a long-term goal they would like to achieve.

New Delhi declined to send senior representatives to a May 2017 summit in Beijing for the Belt and Road Initiative, but it welcomed the opportunity to participate in the Asian Infrastructure Investment Bank (AIIB), a new multilateral development institution led by China. Indeed, India has become home to the new bank's largest projects. In 2017 alone, the AIIB approved $1.5 billion in loans for projects in India, with another $3 billion expected this year. Earlier, Modi actively sought Chinese investment when he was chief minister in his home state of Gujarat, visiting Beijing four times and ensuring that the state was the largest recipient of Chinese investment.

Currently the trade imbalance is grossly in China's favor and a source of some contention, but the potential for Chinese companies in India is also manifest. Smartphone maker Xiaomi reported in early 2017 that its sales revenue in India had surpassed $1 billion; by September, it had joined South Korea's Samsung as the top phone sellers in the country. At the start of 2018, a competing manufacturer, Huawei, highlighted the opportunity it saw in the Indian market, announcing that it would use an "India first" policy to develop its presence in the market and achieve its goal of becoming the world's third-largest smartphone brand in the next five years. In the short term, it sought to capture at least 10 percent of the Indian market by March 2018 and was producing an ever-increasing volume of its phones in India. More recently, the Chinese Internet giant Alibaba announced an investment of $150 million in Zomato, an Indian online food ordering service.

Chinese investors are deeply interested in the opportunities presented by India's mostly young population of 1.4 billion. This is just one of the many reasons why Beijing has to find a way to fix its troubled relationship with New Delhi. As Foreign Minister Wang Yi put it recently at a press conference, "The Chinese 'dragon' and the Indian 'elephant' must not fight each other, but dance with each other. In that case, one plus one will equal not only two, but also eleven."

Some similarities emerge when we compare China's relations with India and South Asia with China's relations with Russia and the tussle for influence between them that is playing out in parts of the former Soviet Union. The trend in Central Asia is a gradual expansion of Beijing's power and influence, to Moscow's detriment. This is something that Russia can do little about, in part because it lacks the means, but also because it needs Chinese investment and economic relations to offset tightening Western sanctions.

In contrast, India would like Chinese investment, but it is not shut out of the international system as Russia is, nor is it in nearly as desperate an economic situation. Rather, India is an ascendant power with growing wealth and influence and a keen desire to highlight its place on the international stage.

India can act as a competitor to China and an obstacle to the Belt and Road Initiative in South Asia. India's anger with Pakistan is shared elsewhere around the world (and in hushed voices, even in Beijing). Nepal and Bangladesh will always find themselves umbilically tied to India thanks to cultural, ethnic, and historical affinities. Sri Lanka may have accepted Chinese investment, but (along with the Maldives and Mauritius) it continues to see India as an important partner. In all of these contexts, China may be able to increase its influence, but the nations of South Asia, unlike their Central Asian counterparts, have a clear alternative on offer in India. This is the trump card that New Delhi could play against China. It is one that Beijing has failed to consider adequately.

Critical Thinking

1. What is the Chinese one belt one road project?
2. What is meant by the "string of pearls"?
3. What is Chinese strategic maritime policy?

Internet References

China's Belt and Road is Full of Holes

https://www.csis.org/analysis/chinas=belt-and-road-full-of-holes

The Indo-Pacific Region Takes Center Stage at Shangri-la

https://www.csis.org/analysis/indo-pacific-region-takes-center-stage-shangri-la

Raffaello Pantucci is the director of international security studies at the Royal United Services Institute for Defence and Security Studies in London.

Article

Prepared by: Robert Weiner, *University of Massachusetts, Boston*

The Liberal World Order Loses Its Leader

[T]he United States is voluntarily abdicating much of its global leadership role. And it is doing so in what appears to be a fit of petulance, as though it has been aggrieved and made into a victim by the very international order it helped to build.

MLADA BUKOVANSKY

Learning Outcomes

After reading this article, you will be to:

- Learn why the international order was created.
- Learn about the different elements that make up the international liberal order.

All political orders, domestic and international, mix elements of oppression and legitimacy, of violence and benevolence. Leaders also embody a mix of qualities, some admirable and some reprehensible; no leader is a pure paragon of virtue. Any logical inconsistency deriving from the fact that the United States has been seen simultaneously as an imperial oppressor and subjugator of nonwhite races, and as a virtuous human rights and democracy–promoting liberal leader of the "free world," need not disconcert us. The United States historically has not always been a unitary actor in world affairs, and its leaders have many disparate traditions and narratives of national identity on which to draw when articulating its ideals, interests, and role in the world.

That the liberal international order led by the United States has provided disproportionate advantages to some while relegating others to the margins of survivability is hardly news. But insofar as this order has promised to progressively extend the benefits of economic openness and human rights to those on the margins, its virtues and its legitimacy have counterbalanced its vices and its oppressiveness. The ambitious scope of its broadly progressive aims renders the order something of an historical anomaly, and this makes the US abdication of leadership especially sad.

It takes work to sustain the legitimacy of any political order, and to prevent its corrosion into a tool of a privileged few. Institutions purporting to serve a higher good are vulnerable to co-optation by self-serving interests; when such interests appear to have the upper hand, an institution risks losing broader legitimacy. The US executive branch has been responsible for presenting a coherent vision and setting priorities for America's role in the world, and mustering resources to support that vision and those priorities. It presents to the world a set of values and virtues capable of attracting cooperation, drawing others to causes worthy of respect and collaboration: causes such as promoting economic growth and development, financial stability, and ecological sustainability; advancing human rights; defending allies against aggression; and curbing the proliferation of biological, chemical, and nuclear weapons. If these goals appear to be mere cover for naked self-interest, if the vices and the corruption are more easily discernible than the virtues evoked by its avowed public purposes, America's "soft power" is undermined. Without the legitimacy that comes with soft power—the power of attraction—the other options open to a leader are to threaten reluctant followers with force or buy them off; in other words, coercion or bribery.

The cacophony of disturbing news since the election of Donald Trump as US president suggests that coercion and bribery, rather than soft power, will be the modus operandi of his administration. We have seen populist authoritarian tendencies, increased visibility of pseudo-nativist racism, snubbing of allies, threats of nuclear war, aggressive reversal of both domestic and international efforts to address climate change, a revival of protectionism, fawning over Vladimir Putin and friendliness toward other authoritarian leaders (at least the ones who might support a Trump-franchised construction project), understaffing of the State Department, demonization of the free press, steamrolling of ethics rules, and general disrespect for the rule of law. None of this is likely to burnish the image of the United States.

Trump, a consummate brand-peddler, is tossing the post–World War II American brand of leadership out the window, replacing it with chants of "America First" and the highly leveraged glitz of the Trump logo. The vices characteristic of this administration have probably always been present in American politics, but they now seem so much more visible than the virtues the United States has at times presented to the world.

Rather than being openly challenged by a rival great power, as students of history might expect, the United States is voluntarily abdicating much of its global leadership role. And it is doing so in what appears to be a fit of petulance, as though it has been aggrieved and made into a victim by the very international order it helped to build. As though countries such as China, India, Brazil, and Mexico have somehow received unwarranted advantages from trade and capital-account liberalization—advantages that were, after all, what helped sell the process of globalization and persuaded these and other countries to open their markets in the first place. As though our European and Asian allies were wrong to exploit the security guarantees through which we bound them into our global network of military bases.

Does this abdication spell the death of liberal internationalism—of an order which, whatever its oppressions, omissions, and hypocrisies, emphasizes the values of multilateral cooperation, peaceful settlement of disputes, open economic exchange, respect for the rule of law, and progressive promotion of individual rights and liberties?

Steps Toward Liberation

Although the situation appears dire, it bears remembering that international order does not rise and fall with a single US president. International order is built over time. The United Nations did not spring forth from the San Francisco Conference of 1945 fully formed like Athena from the head of Zeus (or Roosevelt). Its political foundations were laid in wartime Allied conferences and its legal framework drew extensively on its predecessor, the League of Nations, which itself incorporated and adapted practices of conference diplomacy begun under the Concert of Europe (1815–48).

Global leadership is also built over time; it consolidates when traditions and institutions of the leading power or powers are interwoven with the fabric of international society. Thus the Concert of Europe reflected principles central to the organization and ideologies of its leading members, such as a balance of power and dynastic legitimacy, as well as the idea that great powers had special responsibilities for the system's stability—an idea at once self-serving and crucial to the maintenance of such order as there was. And that order had broader value insofar as limited wars were better than the alternative.

By the time the League of Nations came into existence after World War I, democratic institutions had been consolidated in at least some leading European states in addition to the morally and materially influential United States. The language of the League's Covenant hinted at, though it did not fully articulate, the idea of national self-determination. Its structure featured an Assembly where each member state had a voice and a vote. But since the legitimacy of imperialism had yet to be systematically called into question, and racism was still endemic in the Euro-Atlantic world, not all were deemed worthy of self-government.

Both imperialism and racial discrimination were still considered morally defensible in the nineteenth and early twentieth centuries, even by many liberals. The writings of John Stuart Mill are unambiguous on this point, though his assessment did not go unchallenged by his contemporaries. The League's architecture included plenty of patronizing language and provisions regarding the status of the "less civilized" peoples purportedly in need of administrative oversight and possibly outright colonial control by the "more civilized" great powers. And yet, despite the grip of imperial powers on the core decision-making institutions as well as their substantial military superiority, others found in the language of self-determination the basis for making claims against those very imperial powers.

After World War II this trend accelerated under the United Nations, where the push to delegitimize colonialism reached its crescendo in the 1960s. This agenda was not driven—and indeed was often resisted—by the major world powers. Whatever the anti-imperial rhetoric of the United States, it only supported anticolonial movements that explicitly aligned with its geostrategic interests. When leading liberation figures and writers like Congo's Patrice Lumumba dared to articulate their grievances in a Marxist idiom, they cast themselves in the role of enemy. Connections between pan-Africanists, nationalist decolonizers, and domestic agitators for reform made it possible for the US government to rationalize the application of oppressive power simultaneously on enemies at home and abroad. Yet decolonization happened anyway, as did the civil rights movement. The rhetoric of liberation from time to time eludes the control of those who seek to use it only for their own narrow purposes.

Thus the shape of the international order does not always perfectly align with the ideologies and interests, nor even with the military might, of its leading powers; global order and global leadership do not rise and fall in perfect synchrony. The international order may be transformed by the agitation of those who seek to extend to themselves the liberation promised by its ideals but denied by its practices at a given point in time. That is, after all, what decolonization within the framework of the United Nations was at least partly about. Nevertheless, without an ideological affinity between the governing ideals of a leading state or states and the broader international institutional framework, the resources available to those on the margins who seek to make their own claims based on the proclaimed ideals are very limited. However, selective the attention given to human rights by the US State Department has been over the years, for example, things would probably have been worse had it not engaged in such advocacy at all.

Order versus Justice?

Hedley Bull, a prominent Australian theorist in the mid-twentieth century, noted a tension between the maintenance of order and the quest for justice in international relations. What international order there was, he argued, was based on a configuration of power wherein stronger, more developed Western "great powers" came to form an international society of states, embodying some basic rules of mutual recognition and coexistence. The international society of states was not necessarily peaceful, but its armed conflicts were normally limited (if large-scale war broke out, this signified a breakdown of the order). To the extent that demands for justice called for eradicating the privileges and dominant positions of the powerful states that served as custodians of international society, such demands were in tension with the maintenance of order.

Earlier in the twentieth century, the British scholar E.H. Carr argued that the core problem of international relations was how to enact peaceful change. Buried in the back of his book *The Twenty Years' Crisis, 1919–1939* is the observation that the problem of peaceful change might require that the strong negotiate away some of their privileges to the weak, on the model of trade union bargaining. To do that, the strong had to acknowledge their dominant position, and also that those in a weaker position had legitimate claims to some sort of redistribution of power or resources.

Carr's cynical assessment of liberal internationalism was based on the observation that liberal states tended to cast their own particular interests, preferences, and privileges as being in the interest of all: the idealistic logic of liberalism blinded its proponents to the fact that their liberal institutions did not actually serve all of humanity but rather helped to sustain their own dominant position in the international system. Carr saw the British and the Americans as exemplars of this sort of imperial liberal hubris and self-deception. Anticipating a reprise of the historical pattern of the rise and fall of imperial powers, he appealed for a greater realism that would see more clearly the dilemma later posed by Bull as the order-versus-justice problem: those disadvantaged by the system will seek its revision, through violence if they deem it necessary. Therefore, those in a position of privilege might want to forestall this by trying to address the claims of the disadvantaged.

At the close of World War II, it appeared as though the United States and its allies had learned this lesson and found a formula through which to enact at least partly the sort of peaceful change Carr had in mind, and in the process to resolve at least some of the tension between order and justice in international relations. The postwar international order was anchored by the multilateral institutions of the United Nations, the International Monetary Fund (IMF), the International Bank for Reconstruction and Development (also known as the World Bank), and the General Agreement on Tariffs and Trade (GATT), now the World Trade Organization. These were mechanisms designed to promote the economic, social, and political development of the so-called Third World—and also (at least initially) of the Second World, the nations of the Eastern bloc.

The overall social purpose of the economic growth to be facilitated by progressive reduction of trade barriers was to lessen the disparities between rich and poor nations, and also to ease class struggles between labor and capital within the developed capitalist countries. The new institutions embodied not just an orthodox liberalism of minimal state intervention but, in the term of the American international affairs scholar John Ruggie, an "embedded" liberalism that accepted the role of states and multilateral institutions in cushioning societies vulnerable to the dislocations of raw market forces. In his 2014 book *Forgotten Foundations of Bretton Woods*, Eric Helleiner shows how international development was also on the agenda at the negotiations that shaped the international economic institutions. Representatives from the Global South and from Eastern Europe actively contributed to this focus, aligning with supporters within the United States even as other American and British negotiators resisted the push to make development a central goal of the postwar order.

Although development concerns were ultimately marginalized in the final drafts creating the institutions, the IMF and the World Bank went on to recognize the special needs of developing countries as they established rules and mechanisms for financial stability. The GATT allowed countries to protect certain industries deemed strategically important and acknowledged that developing nations needed to proceed more slowly with trade liberalization despite the overall goal of progressive reduction of tariff barriers. The Charter of the United Nations also articulates goals of economic, social, and political development for all countries and peoples. In supporting these institutions, the United States initially showed itself willing to bear disproportionately the costs of international order, not only by making the largest budgetary contributions, but also by taking the lead on liberalization and not expecting immediate reciprocity from countries still too weak to open their markets to US industrial competition.

Even on the later-emerging issue of the global spread of nuclear weapons, the Nuclear Nonproliferation Treaty, which took effect in 1970, promised concessions to the nonnuclear nations in the form of peaceful nuclear technology as well as eventual disarmament. The reality that these promises remain largely unfulfilled does not render irrelevant the fact that they were made in the first place; the treaty is there as a resource for those who wish to highlight its promises. Thus in the early days of the post–World War II order it appeared as though the United States was following Carr's advice: making concessions to the weak and disadvantaged so as to ease wealth and power disparities and thereby forestall more violent attempts at transformation.

Beyond Containment

It is easy to claim that the broader geostrategic purpose of containing the Soviet Union was the reason for, and limit to, this supposed generosity and benign leadership on the part of the United States. But the institutions of the post–World War II order were not simply the tools of US containment policy; they had higher purposes and a more extensive legitimacy, and they neither fully excluded nor successfully contained the Soviet bloc. Socialism and communism were perhaps better integrated into the postwar order than is normally admitted. The Soviets and Eastern Europeans sent representatives to the Bretton Woods negotiations, though ultimately they declined to join the IMF and World Bank and formed their own alternative arrangements for economic cooperation. And of course the United Nations never excluded but was in fact strongly influenced by the concerns and priorities of the socialist and communist powers. The Soviets had their seat on the Security Council and influence over the Non-Aligned Movement. However cynical their claims may have been, the communist states did not openly reject the idea of democracy the way they rejected capitalism, but rather claimed to be providing alternative forms of economic development and popular sovereignty.

One might even argue that socialist ideas about state-led planning and development, as well as concerns for the status of workers, were a central feature of postwar "embedded liberalism," at least in the way that some political parties in the Western bloc interpreted it. Socialists influenced politics and policy not just in Eastern Europe but throughout the continent and in the United States as well, whatever damage the anticommunist campaigns of Senator Joseph McCarthy and his colleagues may have done to that cause (though perhaps no one did as thorough a job discrediting communism as the communists themselves). Antitrust legislation and protection of collective bargaining rights remained features of the political landscape in even the most resolutely capitalist of countries, though of course the advent of neoliberalism under Margaret Thatcher and Ronald Reagan chipped away at them—precisely because they were associated with socialism.

Modest Proposal

Ambitious and encompassing as they were, the goals and principles of the post–World War II order meshed well with certain trends and traditions in American politics, though they ran against others. In his Pulitzer Prize–winning 1997 book *Promised Land, Crusader State*, the historian Walter A. McDougall argued that over the course of the nation's history, US foreign policy has been informed by a number of disparate traditions: exceptionalism, isolationism, the hemispheric regionalism of the Monroe Doctrine, expansionism, progressive imperialism, liberal internationalism, containment, and something he called "global meliorism," by which he meant the impulse to engage in humanitarian intervention, democracy promotion, and state-building around the world.

McDougall claimed that although in the early decades of the republic there was some coherence in how the first four traditions on his list were put into practice, over time it has become clear that these different strands, each still mustering a political constituency to support it, pull policy in different and often contradictory directions. He suggested that in the post–Cold War world the United States should lay aside the crusading instinct and be guided instead by a modest realism aligning aspirations with resources, which, while staying true to US traditions and institutions, would avoid arrogance and overreach, respect international institutions without denying the importance of national interests and capabilities, and above all cherish and actively preserve the liberties enshrined in the Constitution. What he seemed to be calling for at that opportune moment, when the security dilemmas of the Cold War had receded, was a focus on the domestic agenda and the revitalization of US institutions. In retrospect that was probably useful counsel, given the growing political polarization and erosion of trust that has gripped the country in recent decades.

McDougall's modest and cautionary prescriptions have not been followed. The terrorist attacks of September 11, 2001, seem to have flipped a switch, forestalling that option. But even before this, the United States did not curtail its "global meliorism." It chose NATO expansion. Its leaders still spoke of making the world safe for democracy. And, of course, it was an aggressive advocate of further globalization, especially of finance, and expansion of US-based multinationals into new markets.

After 9/11, the United States became mired in the "forever wars" in Afghanistan and Iraq, even as the associated emphasis on democracy promotion and humanitarianism facilitated a renewed interweaving of US strategic interests with broader public goods. Consider the vast expansion of the US Agency for International Development's work in the context of the Afghanistan and Iraq wars, as well as that agency's cooperation with and funding of nongovernmental aid organizations and private foundations. This was an attempt to lend broader legitimacy to—and tap more resources for—US foreign policy priorities by drawing on democratic and humanitarian rationales.

Fighting terrorism and supporting humanitarianism and democracy seemed to align for a time, but the tensions between these goals have become undeniable. Costly development projects in Afghanistan and Iraq have not yielded broader benefits; they have become nodes of corruption. Civilian casualties of US-backed antiterrorist missions are often painfully visible in the age of social media. Both at home and abroad, the United States has done things in the war on terror that undercut its claims to be a bastion of liberty and a promoter of global public goods such as development and freedom. From the revelations of torture at Abu Ghraib to the exposure of continued racial inequality and bias in America, the image of the "shining city on a hill" is deeply tarnished. No longer attracted by the soft power of the US example and the international ordering principles that align with it, revisionist powers such as China unsurprisingly choose to go their own way—but increasingly, so do allies such as France and Germany.

A modest and restrained realism concentrating on domestic renewal and curtailing "global meliorism" may have been an option in the early post-Cold War days. But at this point foreign entanglements are not easy to unravel, as the Obama administration discovered when trying to live up to its promise to pull US troops out of Afghanistan and concentrate more resources on its ambitious and progressive domestic agenda. US military personnel, corporations, and development and humanitarian relief workers are still spread across the globe.

A Narrower Vision

The Trump administration did not create this context, but it has given little indication that it respects and understands the history and contours of US global leadership. It has utterly failed to articulate any sense of broader responsibility for the maintenance of international order. The chaotic and contradictory signals emanating from the White House have caused even America's closest allies to look inward, or elsewhere, for reliable partners in tackling global problems. For the deeply cynical, it might just come as a relief that the US executive branch no longer pretends to believe in its own myths.

Within the global economy, rising powers have already begun the work of reconfiguring existing institutions so as to increase their voice and influence. This might be construed as a mark of success rather than failure of the post–World War II economic order, or at least of its ability to adjust to a shift in the global distribution of economic power.

China has begun to articulate its own overarching aims in a bid for global economic leadership, as the Asian Infrastructure Investment Bank and the "One Belt, One Road" initiative for expanding infrastructure capacity across some 60 countries indicate. Chinese leadership does not come with a vision of promoting liberty and democracy, which may heighten its appeal to others jealous of their sovereignty. But it is not likely to secure a deeper popular legitimacy, especially in those regions witnessing the expansion of Chinese military power.

Perhaps, the European Union can recover from its most recent institutional crises and continue to carry the torch of economic interdependence, human rights and freedoms, rule of law, political cooperation, and the pooling of sovereignty to address common problems. To do so it will have to refurbish its image and sell itself as something more than a tool of elite interests. The United States, for its part, still has substantial military and economic power, so we now face not so much a power vacuum as a leadership vacuum. But this matters. Failure to mobilize support with soft power means hard power resources will be expended more rapidly, possibly accelerating geopolitical decline.

If the promises of liberal internationalism no longer have the attractive pull they once had, it is possible to imagine a rolling back of some commitments and aims with a narrower but renewed focus on critical issues. Rather than leading on every front with a sweeping vision that meshes with a narrative of national identity, the United States might now opt to concentrate its resources on the global policy areas that matter most—climate change and nuclear proliferation come to mind. But this narrowing of vision would still need to attend to the roots of what makes political orders legitimate: they must be seen as working for a broader constituency and not as serving the interests of a privileged few. They must offer the hope of justice in addition to the comforts of order.

Some will argue that it is appropriate to scale back the liberal internationalist vision even further and concentrate diplomatic energies on shoring up basic rules of coexistence, leaving sovereign prerogatives intact. Why worry about inspiring and mobilizing virtuous action on common problems in a world dominated by narrow interests? Why not just accept a transactional model of politics, grant the primacy of self-interest, and leave it at that? This could mean allowing authoritarians to operate unchecked; it could mean scaling back the aspirational liberal vision and accepting that the requirements of order will entail the suppression of demands for justice. The problem with such a multipolar international society of sovereigns is that it is hard to see how it would mobilize enough popular, corporate, and sovereign resources to tackle collectively the very real problems facing the planet in the twenty-first century. The embattled liberal order may have deep imperfections, but it still comes closer than the alternatives to resolving the tensions between order and justice in the international system.

Critical Thinking

1. Discus E.H. Carr's concept of international order.
2. Why is President Trump disrupting the liberal international order?
3. Does the liberal international order really exist? Why or why not?

Internet References

A World Imagined, Nostalgia and World Order

https://www.-cato.org/publications/policy...world-imagined-nostalgia-libera;-order

Summary of the 2018 National Defense Strategy

https://dod.defense..gov/Portals/1/.../2018-National-DefenseStrategy-Summary-pdf

Testing the Value of the Postwar International Order—Rand Corp.

https://www.rand.org/pubs/research_reports/RR2226.html

The National Security Strategy of the United States

https://www.whitehouse.gov/articles/new-national-securitystrategy-new-era/

Mlada Bukovansky is a professor of government at Smith College.

Article

Prepared by: Robert Weiner, *University of Massachusetts, Boston*

The Rise of Illiberal Hegemony: Trump's Surprising Grand Strategy

Barry R. Posen

Learning Outcomes

After reading this article, you will be to:

- Learn what President Trump's Grand Strategy is or is not.
- Learn how President Trump is changing the liberal international order.

On the campaign trail, Donald Trump vowed to put an end to nation building abroad and mocked U.S. allies as free riders. "'America first' will be the major and overriding theme of my administration," he declared in a foreign policy speech in April 2016, echoing the language of pre-World War II isolationists. "The countries we are defending must pay for the cost of this defense, and if not, the U.S. must be prepared to let these countries defend themselves," he said—an apparent reference to his earlier suggestion that U.S. allies without nuclear weapons be allowed to acquire them.

Such statements, coupled with his mistrust of free trade and the treaties and institutions that facilitate it, prompted worries from across the political spectrum that under Trump, the United States would turn inward and abandon the leadership role it has played since the end of World War II. "The US is, for now, out of the world order business," the columnist Robert Kagan wrote days after the election. Since Trump took office, lis critics have appeared to feel vindicated. They have seized on his continued complaints about allies and skepticism of unfettered trade to claim that the administration has effectively withdrawn from the world and even adopted a grand strategy of restraint. Some have gone so far as to apply to Trump the most feared epithet in the U.S. foreign policy establishment: "isolationist."

In fact, Trump is anything but. Although he has indeed laced his speeches with skepticism about Washington's global role, worries that Trump is an isolationist are out of place against the backdrop of the administration's accelerating drumbeat for war with North Korea, its growing confrontation with Iran, and its uptick in combat operations worldwide. Indeed, across the portfolio of hard power, the Trump administration's policies seem, if anything, more ambitious than those of Barack Obama.

Yet Trump has deviated from traditional U.S. grand strategy in one important respect. Since at least the end of the Cold War, Democratic and Republican administrations alike have pursued a grand strategy that scholars have called "liberal hegemony." It was hegemonic in that the United States aimed to be the most powerful state in the world by a wide margin, and it was liberal in that the United States sought to transform the international system into a rules-based order regulated by multilateral institutions and transform other states into market-oriented democracies freely trading with one another. Breaking with lis predecessors, Trump has taken much of the "liberal" out of "liberal hegemony." He still seeks to retain the United States' superior economic and military capability and role as security arbiter for most regions of the world, but he has chosen to forgo the export of democracy and abstain from many multilateral trade agreements. In other words, Trump has ushered in an entirely new U.S. grand strategy: illiberal hegemony.

No Dove

Grand strategy is a slippery concept, and for those attempting to divine the Trump administration's, its National Security Strategy—a word salad of a document—yields little insight. The better way to understand Trump's approach to the world is to look at a year's worth of actual policies. For all the talk of avoiding foreign adventurism and entanglements, in practice, his administration has remained committed to geopolitical competition with the world's greatest military powers and to the formal and informal alliances it inherited. It has threatened new wars to hinder the emergence of new nuclear weapons states, as did its predecessors; it has pursued ongoing wars against the Taliban in Afghanistan and the Islamic State (or ISIS) in Iraq and Syria with more resources and more violence than its predecessors. It has also announced plans to invest even more money in the Department of Defense, the budget of which still outstrips that of all of the United States' competitors' militaries combined.

When it comes to alliances, it may at first glance seem as if Trump has deviated from tradition. As a candidate, he regularly complained about the failure of U.S. allies, especially those in NATO, to share the burden of collective defense. However uninformed these objections were, they were entirely fair; for two decades, the defense contributions of the European states in NATO have fallen short of the alliance's own guidelines. Alliance partisans on both sides of the Atlantic find complaints about burden sharing irksome not only because they ring true but also because they secretly find them unimportant. The actual production of combat power pales in comparison to the political goal of gluing the United States to Europe, no matter what. Thus the handwringing when Trump attended the May 2017 NATO summit and pointedly failed to mention Article 5, the treaty's mutual-defense provision, an omission that suggested that the United States might not remain the final arbiter of all strategic disputes across Europe.

But Trump backtracked within weeks, and all the while, the United States has continued to go about its ally-reassurance business as if nothing has changed. Few Americans have heard of the European Reassurance Initiative. One would be forgiven for thinking that the nearly 100,000 U.S. troops that remained deployed in Europe after the end of the Cold War would have provided enough reassurance, but after the Russian invasion of Ukraine in 2014, the allies clamored for still more reassurance, and so was born this new initiative. The ERI is funded not in the regular U.S. defense budget but in the Overseas Contingency Operations appropriation—the "spend whatever it takes without much oversight" fund originally approved by Congress for the global war on terrorism. The ERI has paid for increased U.S. military exercises in Eastern Europe, improved military infrastructure across that region, outright gifts of equipment to Ukraine, and new stockpiles of U.S. equipment in Europe adequate to equip a U.S. armored division in case of emergency. At the end of 2017, Washington announced that for the first time, it would sell particularly lethal antitank guided missiles to Ukraine. So far, the U.S. government has spent or planned to spend $10 billion on the ERI, and in its budget for the 2018 fiscal year, the Trump administration increased the funding by nearly $1.5 billion. Meanwhile, all the planned new exercises and deployments in eastern Europe are proceeding apace. The U.S. military commitment to NATO remains strong, and the allies are adding just enough new money to their own defense plans to placate the president. In other words, it's business as usual.

In Asia, the United States appears, if anything, to be more militarily active than it was during the Obama administration, which announced a "pivot" to the region. Trump's main preoccupation is with the maturation of North Korea's nuclear weapons program—a focus at odds with his campaign musings about independent nuclear forces for Japan and South Korea. In an effort to freeze and ultimately reverse North Korea's program, he has threatened the use of military force, saying last September, for example, "The United States has great strength and patience, but if it is forced to defend itself or its allies, we will have no choice but to totally destroy North Korea." Although it is difficult to tell if Pyongyang takes such threats seriously, Washington's foreign policy elite certainly does, and many fear that war by accident or design is now much more likely. The Pentagon has backed up these threats with more frequent military maneuvers, including sending long-range strategic bombers on sorties over the Korean Peninsula. At the same time, the administration has tried to put economic pressure on North Korea, attempting to convince China to cut off the flow of critical materials to the country, especially oil.

Across the Pacific, the U.S. Navy continues to sustain a frenetic pace of operations—about 160 bilateral and multilateral exercises per year. In July, the United States conducted the annual Malabar exercise with India and Japan, bringing together aircraft carriers from all three countries for the first time. In November, it assembled an unusual flotilla of three aircraft carriers off the Korean Peninsula during Trump's visit to Asia. Beginning in May 2017, the navy increased the frequency of its freedom-of-navigation operations, or FONOPS, in which its ships patrol parts of the South China Sea claimed by China. So busy is the U.S. Navy, in fact, that in 2017 alone, its Seventh Fleet, based in Japan, experienced an unprecedented four ship collisions, one grounding, and one airplane crash.

During his trip to Asia in November, Trump dutifully renewed U.S. security commitments, and Prime Minister Shinzo Abe of Japan seems to have decided to allow no daylight between him and the president, including on North Korea. Given Trump's litany of complaints about the unfairness of U.S. trade relationships in Asia and his effective ceding of the economic ground rules to China, one might be surprised that U.S. allies in the region are hugging this president so closely. But free security provided by a military superpower is a difficult thing to replace, and managing relations with one that sees the world in more zero-sum economic terms than usual is a small price to pay.

The Trump administration has increased its military activities across the Middle East, too, in ways that should please the critics who lambasted Obama for his arm's-length approach to the region. Trump wasted no time demonstrating his intent to reverse the mistakes of the past. In April 2017, in response to evidence that the Syrian government had used chemical weapons, the U.S. Navy launched 59 cruise missiles at the air base where the attack originated. Ironically, Trump was punishing Syria for violating a redline that Obama had drawn and a chemical weapons disarmament agreement that Obama had struck with Syria, both of which Trump pilloried his predecessor for having done. Nevertheless, the point was made: there's a new sheriff in town.

The Trump administration has also accelerated the war against ISIS. This Pentagon does not like to share information about its activities, but according to its own figures, it appears that the United States sent more troops into Iraq and Syria, and dropped more bombs on those countries, in 2017 than in 2016. In Afghanistan, Trump, despite having mused about the mistakes of nation building during the campaign, has indulged the inexplicable compulsion of U.S. military leaders ("my generals," in his words) to not only remain in the country but also escalate the war. Thousands of additional U.S. troops have been sent to the country, and U.S. air strikes there have increased to a level not seen since 2012.

Finally, the administration has signaled that it plans to confront Iran more aggressively across the Middle East. Trump himself opposed the 2015 nuclear deal with Iran, and his advisers appear eager to push back against the country, as well. In December, for example, Nikki Haley, the U.S. ambassador to the UN, stood in front of debris from what she claimed was an Iranian missile and alleged that Tehran was arming rebels in Yemen, where Iran and Saudi Arabia are engaged in a proxy war. Behind the scenes, the Trump administration seems to have been at least as supportive of the Saudi intervention in Yemen as was its predecessor. The Obama administration lent its support to the Saudis in order to buy their cooperation on the Iran deal, and given that Trump despises that agreement, his backing of the Saudis can be understood only as an anti-Iran effort. Barring a war with North Korea—and the vortex of policy attention and military resources that conflict would create—it seems likely that more confrontation with Iran is in the United States' future.

The Trump administration's defense budget also suggests a continued commitment to the idea of the United States as the world's policeman. Trump ran for office on the proposition that, as he put it on Twitter, "I will make our Military so big, powerful & strong that no one will mess with us." Once in office, he rolled out a defense budget that comes in at roughly 20 percent more than the 2017 one; about half the increase was requested by the administration, and the other half was added by Congress. (The fate of this budget is unclear: under the Budget Control Act, these increases require the support of the Democrats, which the Republicans will need to buy with increased spending on domestic programs.) To take but one small example of its appetite for new spending, the administration has ramped up the acquisition of precision-guided munitions by more than 40 percent from 2016, a move that is consistent with the president's oft-stated intention to wage current military campaigns more intensively (as well as with an expectation of imminent future wars).

Trump also remains committed to the trillion-dollar nuclear modernization program begun by the Obama administration. This program renews every leg of the nuclear triad—missiles, bombers, and submarines. It is based on the Cold War-era assumption that in order to credibly deter attacks against allies, U.S. nuclear forces must have the ability to limit the damage of a full-scale nuclear attack, meaning the United States needs to be able to shoot first and destroy an adversary's entire nuclear arsenal before its missiles launch. Although efforts at damage limitation are seductive, against peer nuclear powers, they are futile, since only a few of an enemy's nuclear weapons need to survive in order to do egregious damage to the United States in retaliation. In the best case, the modernization program is merely a waste of money, since all it does is compel U.S. competitors to modernize their own forces to ensure their ability to retaliate; in the worst case, it causes adversaries to develop itchy trigger fingers themselves, raising the risk that a crisis will escalate to nuclear war. If Trump were truly committed to America first, he would think a bit harder about the costs and risks of this strategy.

Primacy without a Purpose

Hegemony is always difficult to achieve, because most states jealously guard their sovereignty and resist being told what to do. But since the end of the Cold War, the U.S. foreign policy elite has reached the consensus that liberal hegemony is different. This type of dominance, they argue, is, with the right combination of hard and soft power, both achievable and sustainable. International security and economic institutions, free trade, human rights, and the spread of democracy are not only values in their own right, the logic goes; they also serve to lure others to the cause. If realized, these goals would do more than legitimate the project of a U.S.-led liberal world order; they would produce a world so consonant with U.S. values and interests that the United States would not even need to work that hard to ensure its security.

Trump has abandoned this well-worn path. He has denigrated international economic institutions, such as the World Trade

Organization, which make nice scapegoats for the disruptive economic changes that have energized his political base. He has abandoned the Paris climate agreement, partly because he says it disadvantages the United States economically. Not confident that Washington can sufficiently dominate international institutions to ensure its interests, the president has withdrawn from the Trans-Pacific Partnership, launched a combative renegotiation of the North American Free Trade Agreement, and let the Transatlantic Trade and Investment Partnership wither on the vine. In lieu of such agreements, Trump has declared a preference for bilateral trade arrangements, which he contends are easier to audit and enforce.

Pointing out that recent U.S. efforts to build democracy abroad have been costly and unsuccessful, Trump has also jettisoned democracy promotion as a foreign policy goal, aside from some stray tweets in support of anti-regime protesters in Iran. So far as one can tell, he cares not one whit about the liberal transformation of other societies. In Afghanistan, for example, his strategy counts not on perfecting the Afghan government but on bludgeoning the Taliban into negotiating (leaving vague what exactly the Taliban would negotiate). More generally, Trump has often praised foreign dictators, from Vladimir Putin of Russia to Rodrigo Duterte of the Philippines. His plans for more restrictive immigration and refugee policies, motivated in part by fears about terrorism, have skated uncomfortably close to outright bigotry. His grand strategy is primacy without a purpose.

Such lack of concern for the kinder, gentler part of the American hegemonic project infuriates its latter-day defenders. Commenting on the absence of liberal elements in Trump's National Security Strategy, Susan Rice, who was a national security adviser in the Obama administration, wrote in December, "These omissions undercut global perceptions of American leadership; worse, they hinder our ability to rally the world to our cause when we blithely dismiss the aspirations of others."

But whether that view is correct or not should be a matter of debate, not a matter of faith. States have long sought to legitimate their foreign policies, because even grudging cooperation from others is less costly than mild resistance. But in the case of the United States, the liberal gloss does not appear to have made hegemony all that easy to achieve or sustain. For nearly 30 years, the United States tested the hypothesis that the liberal character of its hegemonic project made it uniquely achievable. The results suggest that the experiment failed.

Neither China nor Russia has become a democracy, nor do they show any sign of moving in that direction. Both are building the military power necessary to compete with the United States, and both have neglected to sign up for the U.S.-led liberal world order. At great cost, Washington has failed to build stable democratic governments in Afghanistan and Iraq. Within NATO, a supposed guardian of democracy, Hungary, Poland, and Turkey are turning increasingly authoritarian. The European Union, the principal liberal institutional progeny of the U.S. victory in the Cold War, has suffered the loss of the United Kingdom, and other member states flaunt its rules, as Poland has done regarding its standards on the independence of the judiciary. A new wave of identity politics—nationalist, sectarian, racist, or otherwise—has swept not only the developing world but also the developed world, including the United States. Internationally and domestically, liberal hegemony has failed to deliver.

What Restraint Looks Like

None of this should be taken as an endorsement of Trump's national security policy. The administration is overcommitted militarily; it is cavalier about the threat of force; it has no strategic priorities whatsoever; it has no actual plan to ensure more equitable burden sharing among U.S. allies; under the guise of counterterrorism, it intends to remain deeply involved militarily in the internal affairs of other countries; and it is dropping too many bombs, in too many places, on too many people. These errors will likely produce the same pattern of poor results at home and abroad that the United States has experienced since the end of the Cold War.

If Trump really wanted to follow through on some of his campaign musings, he would pursue a much more focused engagement with the world's security problems. A grand strategy of restraint, as I and other scholars have called this approach, starts from the premise that the United States is a very secure country and asks what few things could jeopardize that security. It then recommends narrow policies to address those potential threats.

In practice, restraint would mean pursuing a cautious balance-of-power strategy in Asia to ensure that China does not find a way to dominate the region—retaining command of the sea to keep China from coercing its neighbors or preventing Washington from reinforcing them, while acknowledging China's fears and, instead of surrounding it with U.S. forces, getting U.S. allies to do more for their own defense. It would mean sharing best practices with other nuclear powers across the globe to prevent their nuclear weapons from falling into the hands of nonstate actors. And it would mean cooperating with other countries, especially in the intelligence realm, to limit the ability of nihilistic terrorists to carry out spectacular acts of destruction. The United States still faces all these threats, only with the added complication of doing so in a world in which its relative power position has slipped. Thus, it is essential that U.S. allies, especially rich ones such as those in Europe, share more of the burden, so that the United States can focus its own power on the main threats. For example, the Europeans should build most of the military power to deter Russia, so that the United States can better concentrate its resources to sustain command of the global commons—the sea, the air, and space.

Those who subscribe to restraint also believe that military power is expensive to maintain, more expensive to use, and generally delivers only crude results; thus, it should be used sparingly. They tend to favor free trade but reject the notion that U.S. trade would suffer mightily if the U.S. military were less active. They take seriously the problem of identity politics, especially nationalism, and therefore do not expect other peoples to welcome U.S. efforts to transform their societies, especially at gunpoint. Thus, other than those activities that aim to preserve the United States' command of the sea, restraint's advocates find little merit in Trump's foreign policy; it is decidedly unrestrained.

During the campaign, Trump tore into the United States' post-Cold War grand strategy. "As time went on, our foreign policy began to make less and less sense," he said. "Logic was replaced with foolishness and arrogance, which led to one foreign policy disaster after another." Many thought such criticisms might herald a new period of retrenchment. Although the Trump administration has pared or abandoned many of the pillars of liberal internationalism, its security policy has remained consistently hegemonic. Whether illiberal hegemony will prove any more or any less sustainable than its liberal cousin remains an open question. The foreign policy establishment continues to avoid the main question: Is U.S. hegemony of any kind sustainable, and if not, what policy should replace it? Trump turns out to be as good at avoiding that question as those he has condemned.

Critical Thinking

1. What is meant by liberal and illiberal hegemony?
2. How has the Trump administration maintained the formal and informal alliances of the United States?
3. How has Trump remained committed to competition with the world's greatest military powers?

Internet References

A World Imagined, Nostalgia and World Order

https://www.-cato.org/publications/policy...world-imagined-nostalgia-libera;-order

Testing the value of the postwar International Order—Rand Corp.

https://www.rand.org/pubs/research_reports/RR2226.html

The National Security Strategy of the United States

https://www.whitehouse.gov/articles/new-national-securitystrategy-new-era/

Summary of the 2018 National Defense Strategy

https://dod.defense..gov/Portals/1/.../2018-National-DefenseStrategy-Summary-pdf

Barry R. Posen is Ford International Professor of Political Science and Director of the Security Studies Program at the Massachusetts Institute of Technology.

Article

Prepared by: Robert Weiner, *University of Massachusetts, Boston*

Time to Re-engage

Whipsawed by years of foreign policy activism and then by general retreat, the United States is at risk of losing an opportunity to cement hard-won gains in Central Asia/Afghanistan.

S. FREDERICK STARR

Learning Outcomes

After reading this article, you will be able to:

- Learn to view central Asia as a region.
- Learn why the United States should support secular governments in central Asia.

For some time now many in official Washington have viewed Central Asia and Afghanistan as Eurasia's flyover zone, to be quickly traversed as they "pivot" to East Asia, Russia, or back to the Middle East. Only in its final year did the Obama Administration pay the region much attention, by accepting Kazakhstan's proposal to set up an annual group consultation with Washington. Even then, the State Department, in a move directly at odds with U.S. interests and the Central Asians' own wishes, left Afghanistan out of the new grouping. As Talleyrand allegedly said of Louis XVIII, we seem to have learned nothing and forgotten nothing.

A host of clichés and semi-truths have fed the dismissive attitude harbored by many Americans toward the region. Many see "the Stans" as a group of former Soviet countries in Russia's "backyard" that failed to adopt the Washington Consensus on privatization through shock therapy, and now plod along with dysfunctional economies and authoritarian governments. Civil society languishes, beckoning the State Department to wag its finger at what it brands religious repression. True, Chevron and other American firms are making money pumping oil there, but who needs that oil, with all the shale oil and gas in the United States? As to Afghanistan, is it not a quagmire that sucks in and destroys all who enter? These and similar notions are advanced by those calling for further disengagement.

Those favoring reengagement cite a completely different body of evidence. They note that most of the regional economies are making steady progress, and so may offer economic opportunities for U.S. business. GDP growth there last year ranged from 1.2 percent (Kazakhstan), 2.2 percent (Afghanistan), 3.47 (Kyrgyzstan), 4.2 percent (Tajikistan), 6.5 percent (Turkmenistan), and 7.1 percent (Uzbekistan). This may lag behind China's 6.9 percent or India's 7.6 percent—assuming for a moment that any of these numbers are accurate—but it soundly beats Russia's –3.73 percent and compares favorably with Pakistan and Iran. A group of former World Bank economists predicts that all Central Asian economies have a realistic chance of reaching middle-class status by 2050, with Kazakhstan already there.[1] Boosting their chances is the fact that all of them, including Afghanistan, boast rising generations of talented and well-educated young men and women whose values place them squarely in what the French sociologist Claude Levi-Straus once called the "global monoculture."

On the geopolitical level, champions of reengagement note that the region as the only one on earth surrounded by nuclear powers and vulnerable to destabilizing competition or, worse, external control, which is Putin's clear intent. But for thousands of years domination by an external power has been a formula for instability there and remains so today. The U.S. government, with aid from allies, should back Central Asian sovereignties and help them build security "from within." The engagers also stress that Central Asia is one of the historic seats of Islam, yet its governments are secular, with secular laws and courts. Their systems still suffer the effects of Soviet statism and repression. But for all their flaws, they offer a better model for Muslim societies than most countries in the Middle East and are far more open to modern learning.

As the debate between disengagers and re-engagers rages on, dramatic but little-noticed changes are occurring within Central Asia itself. First, the transition in Uzbekistan following the death of Islam Karimov this past September went smoothly, with even the OSCE acknowledging that the election, while imperfect, marked a step forward. This effectively killed prospects for a dynastic succession in Kazakhstan, which President Nursultan Nazarbayev had ruled out anyway. During the election campaign in Uzbekistan, the acting President, Shavkat Mirziyoyev, launched a movement to resolve outstanding problems with all neighbors, reopened air flights to neighboring Tajikistan, which had been grounded for more than two decades, called for visa-free access from 15 countries, including the United States, sent his Foreign Minister to Kabul to open negotiations on expanding transport and trade, and promised for the first time to make the Uzbek currency convertible. In his inaugural speech he spoke boldly of making the offices of Governors and Mayors elective, and of setting up electronic complaint bureaus in every government office.

If President Mirziyoyev follows through on even half of these proposals, Central Asia and Afghanistan will be set on a new and more promising course. Meanwhile, Kazakhstan, in sharp contrast to Russia, is using the oil crisis to diversify its economy, mount a world exposition on renewable energy, and set up a regional financial center in Astana where disputes can be resolved through British law. Over the next two years, Kazakhstan will also sit on the UN Security Council.

Meanwhile, gas-rich Turkmenistan has also felt the pinch of lower energy prices but is nonetheless forging ahead with construction of the TAPI gas pipeline across Afghanistan to Pakistan and India. America tried to develop this megaproject but failed, yet TAPI directly serves U.S. interests. TAPI will produce large royalties for Afghanistan and will provide thousands of jobs for a country in which the U.S. government has invested so much in lives and treasure. Finally, the two poorest countries of the region, Kyrgyzstan and Tajikistan, will soon benefit from the new CASA 1000 electric transmission grid, which will enable them to reap much-needed income from sending some of their huge potential supply of hydroelectric energy to Afghanistan and Pakistan.

In short, there are powerful currents of change and development afoot in Central Asia today. These present the U.S. government with a choice: It can either remain disengaged and continue to try to leave Afghanistan, or it can commit to finishing the job in Afghanistan and facilitating the security and revival of Central Asia as a whole, on the grounds that an insecure Afghanistan and Central Asia will continue to impinge on U.S. security concerns, as it did so tragically on 9/11. Exhausted from years of unrequited activism in the Middle East, Americans have difficulty even framing this choice. Instead, they debate the past. Those who would minimize America's future engagement with Greater Central Asia view American actions to date as largely fruitless and, in Afghanistan, tied to expenditures so vast and ill-advised that no possible gains can offset them. Both judgments are wrong.

This gloomy view misses more than half of the story. The legacy of U.S. policy towards the post-Soviet states of Central Asia since 1992 may be mixed, but it is largely positive. U.S. policy helped solidify the sovereignty of these states, helped them develop sound institutions to a reasonable extent, and helped them grow economically. It didn't turn them into liberal democracies, nor did it instill human rights as Eleanor Roosevelt would have understood them, although undeniable progress has been achieved in both areas.

However, Hillary Clinton tried but failed to help turn the region into a transportation hub, ultimately ceding the initiative to China. Nor was there any real coherence, let alone mutual reinforcement, among U.S. strategies in the areas of economics, security, and governance. A prime reason for this is that little or no coordination exists among the main executive agencies responsible for Afghanistan and Central Asia—namely, the State Department, the Pentagon, and the Department of Commerce—and within particular agencies. Security, trade, reform, and rights are pursued as separate goals, with USAID and the Pentagon each optimizing the efforts of its particular "stovepipe" with little concern for the single overarching strategy that both claim to be advancing. Obviously, such an approach precludes achieving synergies among them, let alone the trade-offs and, yes, *deals* that would make such synergies possible.

This problem begins at the top, and can only be solved at the top: Until the Secretary of State engages more actively with his counterparts in the Pentagon, Treasury, and Commerce, no progress will be possible and the total value of U.S. programs in Central Asia will continue to be less than the sum of their parts.

But what about Afghanistan? All five of the former Soviet states of Central Asia view Afghanistan as their greatest immediate source of danger and, if it is stabilized, as one of their major future partners. Skeptics in the U.S. are right to note the dangers that have increased steadily over the past two years. How could it have been otherwise, since in the same announcement that President Obama announced his "surge" he declared that he would shortly begin withdrawing U.S. forces? The Taliban and dozens of other extremist groups that had filtered into the country from the Middle East and Pakistan simply set their watches. But some of the best minds in the U.S. military believe stability and development are still attainable if Washington would somewhat increase its force level, declare its intentions for the longer term, and, equally important, remove some of the constraints imposed by Obama's self-defeating rules of engagement.

Meanwhile, in several regions of Afghanistan new power plants are opening and legal businesses are expanding. In spite of everything, Afghanistan today has far more human and natural resources than South Korea at end of the Korean War. The most urgent step toward winding down the conflict and reaping this potential is for the U.S. to regularize its support for Afghan security and the Afghan economy and to pursue the two in an integrated fashion, rather than separately, as has been the case up to now. Instead of focusing on what the U.S. government will not do, as was the unmistakable message sent during the Obama Administration, it is time to define what it *will* do.

This is essential both for Afghanistan and the wider region. The alternative would be to create a power vacuum in a region where some of the world's largest geopolitical tectonic plates crunch against each other. Other powers are watching for a signal from Washington and will adjust their actions depending on what we do. If the Trump Administration continues Obama's plan to walk away from America's trillion dollar investment in Afghanistan, others will fill the vacuum thus created. Worse, America's friends and foes alike will question the seriousness of America's commitments everywhere, and its tenacity in advancing them.

What, then, should be the pillars of a strategy of dynamic, sustained, but limited engagement with Central Asia and Afghanistan? The key elements of such a strategy can be stated in ten simple propositions.

First, the U.S. government should support the sovereignty of all the countries of the region and self-determination for the region as a whole. Central Asia must not become a zone of conflict among its big neighbors. Its security will arise either from its internal strengths or from an order imposed from without. History suggests that external control by anyone, whether China, Russia, or another power, generates instability. U.S. policy should therefore work to enhance security from within, by strengthening economies and self-defense forces and by fostering open and effective government.

Second, as part of the proposed "security from within rather than from without" strategy, the U.S. government must declare its support for the "balanced" approach to relations with major powers that all regional governments have adopted. On this point, the interests of the United States and Central Asian countries and Afghanistan are identical.

Third, the U.S. government should preserve and significantly deepen its new annual consultative mechanism with the combined countries of the region ("C5 Plus 1") and add Afghanistan to this regional entity. It should accord it much greater importance, welcoming inputs from regional partners and bringing to it proposals that flow from America's strategic goals.

Fourth, the U.S. government should welcome any effort by regional governments to form a consultative and cooperative organization for themselves only, without the involvement of any outside powers. Such an initiative would be for Central Asia as a region, and should not be seen as being directed against anyone. In light of the fact that it was Putin who closed down an earlier regional body along these lines, the Central Asia Economic Union, the U.S. government should be prepared to indicate early its support of any new entity the regional countries might choose to create.

Fifth, the U.S. government should not welcome further expansion of the Eurasian Economic Union because it limits sovereignty without bringing commensurate benefits. In spite of its claims to the contrary, it promotes political integration, Russia-based security, economic control from the north, Russian legal and economic practices, Russian sources of information, and Russian foreign policy objectives. Further, it constrains relations between its members and their other powerful neighbors, including China, India, Pakistan, and Iran, as well as with the United States and Europe. Moreover, such an expansion of the EEU promises eventually to curtail Russia's own evolution as an open society, which will directly and negatively affect American and Western interests elsewhere.

Sixth, working mainly through government-to-government contacts but also through NGOs that are open to working with official bodies, the U.S. government should promote the development of responsive and effective governments that over time can evolve toward more open and democratic practices. In doing so, it must work with all countries of the region and avoid picking favorites. In advancing this program, the U.S. government must work *with* regional governments, not *on* them.

Seventh, U.S. policy should once again embrace the plan to promote the entire region as an open road, railroad, and energy transit zone. The goal is to shape existing transport plans so as to bring maximum benefits to the region itself. An important first step would be to open direct and frequent flights between all six capitals. A second step would require direct access by road and railroad to the southern ports of Gwadar in Pakistan (via Kandahar in Afghanistan) and Chabahar in Iran. Both ports are needed in order to foster competition and drive down transport costs for Central Asians. It also means opening the route from Pakistan via Afghanistan to the new Caspian port of Turkmenbashi and keeping the Caspian Sea open to unimpeded transit between Aktau, Turkmenbashi, and Alat. In embracing regional transport and trade as a goal, the U.S. government should not and cannot become a major funder of infrastructure. Instead, it should help regional states develop their own priorities rather than respond passively to plans devised by major powers or international financial institutions. U.S. policy can advance this role by supporting country-based and region-wide planning efforts and by using its convening power to promote the priorities thus identified.

Eighth, the U.S. government should declare as a strategic interest the establishment and improvement of secular governments, legal systems, courts, and public education, while protecting the rights of both believers and nonbelievers. The defense of secular institutions in Muslim societies will call for diverse initiatives to roll back the worst practices of the Soviet era and to replace them with "best practices" gleaned from other countries. U.S. officials should present the American government as a partner rather than a teacher. By advancing this goal in and with Muslim majority countries with historic claims to being a heartland of the faith, the U.S. government will promote a model that can be applied elsewhere. Through C5 Plus 1 and other channels, the U.S. government should participate as an equal in discussions of this issue, which happens also to be a challenge to the West itself.

Ninth, the U.S. government, working with regional governments, European partners, and the private sector, should ensure that all people in the region gain direct access to Western news media and information. This "information agenda" is significantly an American issue. The closing of USIA and the erosion of the U.S.-sponsored radio stations are only the most visible part of the problem. More serious is the failure to reframe the issue in terms of all the new media and technologies that have mushroomed onto the scene in the past few years. Progress in this area will require the State Department, as a first step, to convene qualified experts from both the private and public sectors to review the entire situation and make economically feasible proposals.

Tenth, in all of the above, U.S. policy should facilitate the involvement of U.S.-based firms by working to remove impediments to investments there and improving business contacts and two-way trade.

It is one thing to announce strategic objectives and another to achieve them. Effective implementation will require that U.S. policy consistently treat Greater Central Asia as a region as well as an agglomeration of six sovereign states. It will also require that Washington bring together its security, political, and economic concerns into a single negotiation with each country. This task falls under the purview of the Assistant Secretary of State for Central and South Asia, with support on important relationships from the Secretary of State. This must result in agreements that embody trade-offs and synergies between U.S. interests and those of its partners, and must be made in the context of advancing clearly stated U.S. goals in Central Asia as a whole. The absence of such deal-making is one of the principal failures of current U.S. strategy. The achievement of agreements such as those called for here, as well as their implementation, will require a much more effective interagency process at all levels than exists at present.

In developing and implementing a strategy based on the region as a whole, the U.S. government must recognize the leadership role of Kazakhstan and Uzbekistan and do everything possible to facilitate cordial and equitable relations between the two largest economies of the region, and their collaboration in areas of common interest. Far from diminishing the standing of the other states in U.S. policy, such an approach recognizes that positive and productive relations between these regional drivers, and between them and Washington, will facilitate the development of strong and beneficial links among all the other countries in the region and between them and the United States.

This strategy must be explained to Russia, China, India, Pakistan, Turkey, and other regional powers as an approach based on self-restraint, and that it is not aimed against the legitimate interests of anyone. U.S. officials can explain this in bilateral meetings, but in general they must avoid negotiating with Russia or China over the heads of the Central Asian governments.

Finally, effective implementation of the regional strategy proposed here will require high-level leadership. For example, when it comes to connecting the capitals of all six countries with regularly scheduled and frequent airline flights, leadership from the Secretary of State and Secretary of Commerce will likely be required. Thanks to the consultative formats they maintain with Russia, China, Japan, and the European Union, governments of the region have come to expect such high-level involvement, and will judge a hands-off approach from top American officials as a sign of disinterest. A sure way to counter this perception would be for the U.S. President to make a whirlwind tour of all the capitals of the region. This is the more important since the U.S. President is the only head of state of a major power never to have visited any of the five former Soviet republics.

It would be a grave mistake for the United States to disengage from Central Asia and Afghanistan. Discussion of this issue, whether with respect to Afghanistan or its five northern neighbors, is often couched in terms of dollars and cents. But rarely do such arguments address the cost of disengagement. The present "exit strategy" in Afghanistan, which one pundit called "all exit and no strategy," will negatively and dramatically affect all Central Asia. It will invite Russia's destabilizing meddling in the five former Soviet republics and leave the way open for Putin to launch further efforts to establish himself as an arbiter of Afghan affairs. By undercutting the "balanced" strategies of all these states and of Afghanistan itself, America's disengagement will sharply reduce the extent and depth of their interactions with the West and drastically diminish the likelihood of their becoming more open and democratic societies. As this happens, the entire region, which holds such promise as a model of Muslim societies with secular governments and laws, will feel the increased impact of radical Muslim extremism and terrorism. This in turn will cause China to deepen its security reach into the region, which will inevitably trigger yet another round of instability and conflict.

During the 1990s, under both Democrats and Republicans, the U.S. government took a proactive approach to the new states of Central Asia, even as it ignored growing problems in Afghanistan. After 9/11 America sacrificed thousands of lives and untold treasure in Afghanistan while demoting its northern neighbors to minor roles. This leaves America today with a stark choice: On the one hand it can abandon Afghanistan to Pakistan and Iran, with supporting roles for Russia and China, leaving the rest of Central Asia to Russia and China, with supporting roles for growing number of extremists from a decaying Afghanistan. Or, on the other hand, it can embrace a limited but proactive and integrated policy toward the region as a whole in the realistic expectation that the U.S. government can help it emerge from a difficult and painful period intact and with viable economies and institutions.

Yes, there are costs of a re-engagement with Afghanistan and Central Asia, but they are bound to be not only far less than the cost of disengagement and inaction but also more enduring. In the decades since the Korean War the United States made this calculation in South Korea and resolved to persist, with impressive results. We should follow a similarly strategic course today with respect to Afghanistan and Central Asia.

The problem is that as of this writing the U.S. government has neither an integrated strategy towards the region as a whole nor effective coordination between the various agencies that carry out what passes for American policy there. Glaring flaws in the conception, negotiating practices, and implementation of American strategy must be fixed if any enduring progress is to be achieved.

But simply by acknowledging them, and accepting that they are fully under our own control, we will set ourselves on a more positive course. The derangements in U.S. strategy and its execution can be fixed. If we choose to do so, we might surprise ourselves at how relatively inexpensive success can be.

Note

1. Harinder Kohli, Johannes Linn, eds., Central Asia 2050 (Center on Global Interests, 2016).

Critical Thinking

2. What should US strategy toward Central Asia be?
3. What is the relationship between Afghanistan and all of the other "stans"?
4. Why is the author optimistic about the future of the "stans"? Is his optimism justified?

Internet References

Central Asia Program, George Washington University
Centralasiaprogram.org

Central Asia Security Program
https://centralasiasecurity.org/en

Sigur Center for Asian Studies
https://sigur.elliott.gwu.edu/

Article

Prepared by: Robert Weiner, *University of Massachusetts, Boston*

The Chinese Century

HAL BRANDS

Learning Outcomes

After reading this article, you will be able to:

- Learn about the five economic theories, which the United States held about China.
- Learn about why the United States should be concerned about China's rise.

No one can say we didn't see it coming. Since the end of the Cold War, and even before, it has been obvious that a rapidly rising China could eventually menace America's position and influence in East Asia—and, perhaps, globally as well. Since the Taiwan Strait Crisis in 1995–96, moreover, there have been accumulating signs that Beijing is not a status quo power, but rather one determined to reshape the East Asian order. For decades, then, there has been no shortage of warnings about the emerging China challenge.

As early as 1975, Henry Kissinger predicted, "If they keep developing the way they have, they could be a pretty scary outfit." In the 1980s, Pentagon analysts worried that the long-run "possibility of Chinese conflicts with U.S. allies and friends and the United States itself cannot be excluded." After the Cold War ended, the George H. W. Bush administration speculated that Washington might have to "contain, or balance" Beijing. By 1997–98, the CIA was publicly reporting China's determination "to assert itself as the paramount East Asian power," and even to become a global power "on a par with the United States by the middle of the 21st century." And in the early 2000s, Andrew Marshall—the legendary director of the Office of Net Assessment in the Department of Defense—argued that America must gear up "for a long-term competition between the US and China for influence and position within the Eurasian continent and the Pacific Rimland." All of this, besides warnings by prominent intellectuals such as Robert Kagan, Aaron Friedberg and John Mearsheimer, who predicted—as early as the mid-1990s—that China's rise was likely to be disruptive.

Today, these warnings increasingly seem vindicated. From China's maritime expansionism and illegal island building in the South China Sea to its efforts to coerce and intimidate neighbors from India to Japan, from ambitious geo-economic projects meant to draw surrounding countries more deeply into its grasp to a multidecade military buildup shrouded in opacity and deception, there is abundant evidence that Beijing seeks to dominate its periphery. As Adm. Harry Harris, the commander of U.S. Pacific Command, remarked in 2016, one would "have to believe in a flat earth" not to perceive China's agenda. Indeed, Chinese leaders themselves have become quite candid about the country's vaulting geopolitical ambitions. In October 2017, at the Nineteenth National Congress of the Chinese Communist Party, President Xi Jinping proclaimed that China had entered a "new era" and could now "take center stage in the world."

The point at which the Chinese threat goes from a distant prospect to an urgent near-term reality is thus rapidly approaching, if it has not already arrived. For years, American strategists have known this moment was coming—yet Washington consistently has been slow to react. The rapidity of China's rise has been matched by the lethargy of America's response. Understanding why is critical to addressing that challenge in the years ahead.

To be clear, America has not simply stood aside as China reaches for regional hegemony. After the Cold War, Washington maintained a powerful military presence in East Asia, partially as a check on a potentially aggressive China. Since then, America has periodically modernized its bilateral alliances, developed deeper partnerships with non-ally countries such as Singapore and Vietnam, built and deployed new and advanced military capabilities to the region, and designed creative operational concepts. From the Bush-era "transformation agenda" to Air-Sea Battle and the Third Offset Strategy, the China threat has never been far from Pentagon planners' minds. From the mid-2000s onward, in fact, the Pentagon gradually shifted a larger proportion of its air and naval forces into the Asia-Pacific region, as part of a quiet rebalance under George W. Bush and a more ostentatious version under Obama. Meanwhile, the United States has sought to prevent China from bullying its neighbors—by sending two carrier strike groups to deter potential action against Taiwan in 1996, for instance—and it has worked to strengthen regional diplomatic resistance to Beijing's assertiveness in the South China Sea and elsewhere. None of this has been lost on Chinese observers, many of whom accuse America of seeking to thwart Beijing's ascent.

What is also clear, however, is that these efforts have not kept pace with the challenge. The regional military balance has shifted sharply in China's favor: the RAND Corporation reported in 2015 that the Asia-Pacific was "approaching a series of tipping points" at which Beijing might believe it could successfully use force in crises involving Taiwan or perhaps even the South China Sea. "By next decade," another close observer predicts, "China's military buildup will give it the ability to dominate the air and sea lines of communication in the western Pacific." The United States has also struggled to respond to Chinese coercion short of war; it has largely failed, as former Obama administration officials have acknowledged, to prevent Beijing from dramatically strengthening its position in the South China Sea through island building and intimidation of neighboring states. China's economic tentacles have spread across the region, giving Beijing increased leverage in countries as far afield as Australia. The president of the Philippines, Rodrigo Duterte, exaggerated when he said in 2016 that "America has lost" in the Asia-Pacific, but the situation has grown steadily grimmer.

And far from systematically containing China, the United States has powerfully assisted its rise. By opening U.S. markets to China, bringing Beijing into the World Trade Organization, and promoting technology transfer and foreign investment in China, America has contributed enormously to China's astonishing economic growth. Moreover, America has encouraged China to increase its international reach by pushing Beijing to become more involved in addressing challenges from climate change to nuclear proliferation. It is hard to square the accusation that America "is bent on containing China," as Hillary Clinton pointed out in 2010, with the reality that "China has experienced breathtaking growth and development" since reestablishing relations with the United States.

For historians looking back on this period, America's behavior will present a puzzle. Hegemonic powers are not supposed to tolerate, much less assist, the rise of challengers; they are supposed to fiercely and even violently resist. Was it not "the growth of Athenian power and the fear which this caused in Sparta" that led to the Peloponnesian War? So why has the United States responded so counterintuitively to the rise of such a formidable rival?

Part of the answer undoubtedly lies in the particular strategies China has pursued to dull America's response. From the late 1980s through the late 2000s, China adhered to Deng Xiaoping's dictum: "Hide our capabilities and bide our time." Accordingly, Beijing sought to project an unthreatening image and stressed its commitment to rising peacefully, so as to avoid conflict even as it assiduously developed "comprehensive national power." After 2008, Chinese behavior became more truculent, as the global financial crisis and perceptions of American retrenchment convinced China's leaders that the country's geopolitical window was opening. But even as China started pushing harder throughout the region, it did so via incremental tactics—island building, using paramilitary and Coast Guard forces to control disputed areas in the South China Sea, and taking other measures short of war—that were deliberately designed not to provoke a U.S. military response. Give credit where due: China has proven adept at strengthening its position while limiting resistance to its rise.

Yet if these strategies have proven effective, it is also because America has constrained itself. Over the past quarter century, U.S. policy has been shaped by a set of core ideas about how best to respond to China's emergence. All of these ideas were sincere and well intentioned, all contained elements of real insight, and in many cases they were reasonable and even arguably appropriate, given the context in which they took root. But all these ideas pointed in the same direction—toward policies that now appear to represent more of an underreaction than an overreaction to China's rise—and they seem increasingly suspect today. Good policy begins with getting the underlying ideas right. When it comes to China, America is overdue for an intellectual reset.

That reset should begin with the idea that China will inevitably—if perhaps only eventually—become a satisfied democracy at peace with its neighbors and the world. This belief took root at the outset of the post–Cold War era, when authoritarian regimes were tumbling everywhere, and the Chinese government, which had been badly shaken during the Tiananmen Square protests, looked like the next to fall. By 1993, predicted Winston Lord, the former U.S. ambassador to China, "There will be a more moderate, humane government in Beijing." A more democratic government, in turn, would presumably be more peaceful in its foreign relations and more positively disposed toward the United States. The reconsolidation of the Communist regime in the early 1990s was thus an ideological cold shower of sorts, for it indicated that history had not yet ended in Beijing. In response, however, U.S. officials simply tweaked the argument.

The Chinese dictatorship might be here to stay in the short term, they acknowledged, but surely it was unsustainable over the long run. In the meantime, promoting greater trade ties with Beijing would hurry the inevitable along by hastening the development of a middle class that would demand greater political rights. Indeed, promoting globalization and promoting democracy went hand in hand, because greater commerce would expose Chinese citizens to liberalizing currents from abroad, and because only truly free societies could compete in an integrated global market. "The more China liberalizes its economy, the more fully it will liberate the potential of its people," Bill Clinton argued. "And when individuals have the power, not just to dream, but to realize their dreams, they will demand a greater say." The crucial implication was that the United States could trade freely with China—and thereby contribute to Beijing's skyrocketing national wealth and power—without fearing the geopolitical consequences, because the forces of economic modernization and integration would ultimately transform the regime. "Trade freely with China," George W. Bush remarked, "and time is on our side."

This idea, dubbed the "Soothing Scenario" by James Mann, had the virtue of appealing to Americans' deep-seared belief that their form of government is both morally desirable and universally desired. It reflected the immense ideological optimism of the post-Cold War moment, and was well attuned to the best social-science literature on the relationship between prosperity and democracy, and between democracy and peace. Not least, it was useful in providing assurance that Americans could deepen their commercial engagement with China—and enjoy the vast economic rewards that engagement brought—without forsaking either their moral values or their national-security interests. For if it was true, as Bill Clinton argued, that "the choice between economic rights and human rights, between economic security and national security, is a false one," then Americans need not confront the choice at all.

As reasonable as this idea may initially have seemed, however, over time it has become steadily harder to defend. China not moved toward democracy over the past quarter century, even as its national wealth, per capita wealth and integration into the global economy have shot upward. Chinese leaders, rather, have used prosperity to buy legitimacy while also ruthlessly but skillfully repressing dissent. According to the Polity IV data set, China is just as authoritarian as it has been for decades—and the human-rights crackdowns, repression of civil society and centralization of power under Xi Jinping indicate that the regime is actually becoming less liberal. China may still eventually become a democracy, and it is conceivable, as Sinologist David Shambaugh argues, that increasing authoritarianism is actually an anxious response to pressures from below. Yet it is doubtful that China will become a democracy before it grows powerful enough to severely disrupt the international order. The United States has long felt it had the luxury of effectively underwriting the growth of a potentially dangerous authoritarian power, because that power would not remain authoritarian indefinitely. That wager no longer seems so promising today.

Neither does a second wager—that Beijing will become a "responsible stakeholder." Since the late 1980s, U.S. officials have believed—correctly—that China's assistance is critical to addressing an array of global problems, from trade disputes and piracy to terrorism and climate change. Likewise, there has been a bipartisan consensus that the United States can best obtain China's help on these issues, and moderate Chinese behavior more broadly, by drawing Beijing into the international system and demonstrating that it can gain wealth, power and respect by accepting its rules.

This idea gained its name in 2005, when Deputy Secretary of State Robert Zoellick announced that America ultimately wanted a China that would "work with us to sustain the international system that has enabled its success." Yet its origins date to the Clinton and even George H. W. Bush years. In 1989, Deputy Secretary of State Lawrence Eagleburger argued that the United States could gain China's cooperation on weapons proliferation and other important issues only if it refrained from isolating that country after the Tiananmen massacre. It was already clear, Madeleine Albright agreed in 1997, that

> China will be a rising force in Asian and world affairs. The history of this century teaches us the wisdom of trying to bring such a power into the fold as a responsible participant in the international system, rather than driving it out into the wilderness of isolation.

As a result, U.S. officials have worked energetically over a quarter century to enable China's integration into the global economy, on the theory that a rich China would be a satisfied China and that globalization would foster lucrative relationships that Beijing would be loath to disrupt. At the same time, both Democratic and Republican administrations have consistently sought to give Beijing a greater voice in world affairs, by bringing it into international diplomatic forums and encouraging it to play a larger role in security and political issues in the Asia-Pacific and globally.

The responsible-stakeholder policy was not naive in the context in which it developed. There was little international or domestic support for a policy of isolating Beijing, except perhaps in the immediate aftermath of Tiananmen, and it was inconceivable that the United States would not at least try to integrate the world's most populous country into the broader global order. In some respects, that integration has undoubtedly occurred: China's economy is far more globalized, and its participation in international diplomacy is far more extensive than at the end of the Cold War. Likewise, there are ways in which the policy has arguably succeeded. China has, on several important occasions, cooperated with U.S. policy objectives. It supported the indefinite extension of the nuclear nonproliferation treaty in the 1990s; it assisted, albeit reluctantly, in the economic isolation of Iran during the dispute over that country's nuclear program; and during the late Obama years it participated in groundbreaking international diplomacy to curb climate change. More recently, China has even styled itself as the defender of multilateralism and globalization in response to the parochial nationalism of Donald Trump.

After more than two decades of experience, however, the limitations of the responsible stakeholder have also come into view. In particular, it is evident that China is at best a selective stakeholder, and that the core goal of this policy—persuading Beijing to define its national interests as America might like them to be defined—is probably unachievable.

Beijing has never been willing to alter what it perceives to be its most crucial national-security interests to suit Washington's concept of global order; witness the unending, and continually unavailing, U.S. efforts to obtain the desired level of Chinese cooperation in pressuring North Korea over its nuclear weapons and missile programs. Worse still, China does not seem to have moderated its behavior, or fundamentally bought into the U.S.-led international system, as it has grown more powerful. If anything, its expansionist tendencies in the South China Sea and East China Sea; its efforts to bully neighbors along its maritime and territorial peripheries; its increasingly frequent resorts to diplomatic, economic and paramilitary coercion; its harassment of U.S. military aircraft and vessels in international waters; its ongoing military buildup; and many other actions tell a different story. Such behavior, Aaron Friedberg observes, compels us "to re-examine the pleasing assumption that the country is fast on its way to becoming a status quo power." Even where China has benefited from the existing system, in fact, it has frequently declined to play by the rules. Xi Jinping may be a rhetorical champion of free trade and globalization, but Chinese economic policies often tend toward the protectionist and mercantilist.

When Chinese commentators speak of "Asia for Asians," when Chinese leaders demand that its neighbors show greater deference to Chinese prerogatives, when China continually seeks to undermine U.S. alliances and partnerships in the Asia-Pacific, one thinks not of a "responsible stakeholder" but of a proud and ambitious nation determined to bend the system to its liking. Nothing could be more normal; this is how rising powers usually behave. And so, as China's power continues to grow relative to America's, one should only expect Beijing to become less, rather than more, accommodating. U.S. leaders should certainly continue to engage China on issues where cooperation is possible. But the idea that China will simply accept the international order that America designed is an illusion that must be punctured.

Washington would also do well to part with a third shibboleth: that hostility toward China will be met in kind. Since the 1990s, U.S. officials have ritualistically averred that China is at an inflection point in its relationship with the outside world, and that America must refrain from behavior that will incline Beijing to strident nationalism and hostility. Treat China as a friend, the thinking goes, and it may become a friend. Treat it as a threat or rival, and it will surely reciprocate. "If you treat China as an enemy, China will become an enemy," remarked Joseph Nye, who served as assistant secretary of defense, in the 1990s. "It will become a self-fulfilling prophecy."

This concept, too, has endured across administrations: one can find numerous examples of officials from the George W. Bush and Obama eras voicing the same sentiment. "We don't want to fence them in," said Adm. Timothy Keating, then the commander of U.S. Pacific Command, in 2008. "We want to draw them out . . . and assure them we mean them no ill will."

This is a noble sentiment, and one that was never entirely misplaced. It was originally meant to prevent a security dilemma in which actions taken by one country to protect its own interests seem threatening to another country, triggering a spiral of hostility and leading ultimately to conflict. It reflected an accurate perception that the prospects for international peace and stability would improve dramatically if China and the United States could avoid antagonism. And in the 1990s, at a time when China was still relatively weak and American dominance was unchallenged, it made sense to go the extra mile in trying to reassure Beijing that Washington did not think the relationship was destined for conflict. In recent years, however, the problems and limitations of this concept have become painfully apparent.

For one thing, the "self-fulfilling prophecy" warning has sometimes encouraged an unwillingness even to discuss honestly the problems that China poses. In 2016, for instance, the White House reportedly ordered the Pentagon to stop using the term "great-power competition" to characterize the relationship, despite the obvious fact that such a competition was well underway. At other points, this concept has thwarted sharper and potentially more effective measures to curb Chinese expansionism. As Ely Ratner, who served as deputy national security adviser to Vice President Biden, has written, the United States repeatedly responded to Chinese island building and coercion of neighbors in the South China Sea by denouncing that behavior, while simultaneously looking "for ways to reduce tensions and avoid conflict whenever possible." That stance succeeded in avoiding an unwanted military conflict. It also allowed China "to reach the brink of total control" over that crucial waterway.

Finally, this concept has had the perverse outcome of effectively denying agency to the Chinese themselves, suggesting that the most important factor driving Chinese behavior is not what China wants but what America does, and therefore assuming that if Beijing is treated as a friend, it will not view Washington as an enemy. Neither of these things is necessarily true; both, in fact, are doubtful. "Because China and the United States have longstanding conflicts over their different ideologies, social systems, and foreign policies," the Chinese military reportedly argued as early as 1993, "it will prove impossible to fundamentally improve Sino-U.S. relations." And given how determined China has appeared in recent years to assert its own power and weaken the U.S. position in the Asia-Pacific, it is hard to argue that the chief problem in the relationship is a lack of American reassurance. The idea of a self-fulfilling prophecy may have been worth testing in the 1990s, but it has passed its expiration date by now.

So has a fourth idea that has dampened the U.S. response to the rising threat in the East: the notion that America should be more concerned about a weak China than a strong China. International-relations realists would find this idea quite odd—wouldn't America want its principal geopolitical rival to be as weak and inhibited as possible? Yet even today, many informed observers insist that Chinese debility is more dangerous than Chinese strength. "We have more to fear from a weakened, threatened China than a successful rising China," Barack Obama explained in 2016.

> If China fails; if it is not able to maintain a trajectory that satisfies its population . . . then not only do we see the potential for conflict with China, but we will find ourselves having more difficulty dealing with these other challenges that are going to come.

The logic of this idea, as Thomas Christensen has written, is that China has become "too big to fail." If China experienced economic collapse or even prolonged economic stagnation, the world would struggle to maintain global prosperity and growth. If China fell into political turmoil, it could cause a humanitarian catastrophe within that country and destabilize large swaths of Asia. A faltering China would also, presumably, be less capable of helping to solve critical global problems. And in a worst-case scenario, a China plagued by economic and political upheaval might channel its frustrations outward in a fit of international aggression. "One of the worst possible Chinese futures from a U.S. perspective would be not China's continued rise but its stagnation or even internal collapse," Christensen writes. U.S. policy must therefore focus on encouraging continued Chinese growth, self-confidence and global influence, rather than seeking to thwart China's rise or, worse still, to undermine its economy and polity.

This concept fits nicely with America's preference for a positive-sum global order—one in which U.S. interests and the system as a whole benefit from the strength and prosperity of its members. Moreover, no one can dispute that a failed or failing China would indeed create grave problems for its neighbors, the United States, and the entire global community. Yet this idea, too, is becoming less and less adequate to addressing a U.S.-China relationship that appears more competitive every day.

If taken literally, of course, this precept would cause U.S. officials to focus more on empowering China than on restraining its more dangerous impulses—even as those impulses become more and more disruptive to peace and stability in the Asia-Pacific. Indeed, this idea has frequently led officials, such as President Obama, to argue that the international system cannot function without a strong and assertive China—just as China develops the coercive capabilities and refines the assertive practices that are increasingly allowing it to challenge that system. And as grave as the difficulties caused by a weak China would be, those caused by a strong China, a potential peer rival that is already seeking primacy along its periphery and may well harbor long-term ambitions that are greater still, could be even graver. After all, could there be any threat starker than a revisionist authoritarian power with roughly four times the population of the United States?

Chinese debility may indeed have been more threatening than Chinese strength in the 1990s, or even a decade ago. Today, however, a weaker China no longer looks so bad.

The final illusion that needs dispelling is that China is the adversary of tomorrow, not today. For many years, U.S. observers have seen the China challenge coming. Yet they have persistently argued that the challenge remains on the fairly distant horizon. "China is like that long book you've always been meaning to read," a U.S. official once told me, "but you always end up waiting until next summer."

The reasons for this tendency have been multiple. Some American observers have mentally pushed the Chinese threat into the distance because they doubt that Beijing will ultimately be able to maintain rapid economic growth or hold its authoritarian political system together. In other cases, U.S. observers have simply underestimated how quickly China would advance in developing high-end power-projection and antiaccess/area-denial capabilities. "In the past decade or so," Adm. Robert Willard, the commander of U.S. Pacific Command, acknowledged in 2009, "China has exceeded most of our intelligence estimates of their military capability and capacity, every year." In part, this is just because China is a devilishly difficult intelligence target; in part, it is because some U.S. analysts could hardly imagine that a country that was so recently poor and underdeveloped could threaten America's dominance. "Officers and analysts reared during the Cold War," observe James Holmes and Toshi Yoshihara, "found it hard to shed the image of China's People's Liberation Army Navy (PLAN) as a backward force."

The United States has also found it difficult to confront the China challenge because other threats keep getting in the way. Prior to 9/11, the Bush administration was gearing up for a concerted effort to maintain U.S. military and geopolitical advantages against a rapidly modernizing China; the Pentagon's 2001 Quadrennial Defense Review depicted the emerging antiaccess challenge as a critical problem to be overcome. After 9/11, however, America's strategic focus understandably shifted as the nation began what has now amounted to nearly two decades of military conflict in the greater Middle East. To be fair, the Bush administration still took modest steps to strengthen U.S. capabilities in the Asia-Pacific, but as time went on, the diversion of attention and resources began to tell. By 2009, Secretary of Defense Robert Gates was reducing or canceling investments in critical high-end capabilities such as the F-22 fighter and a new stealth bomber in favor of lower-end capabilities crucial to ongoing operations in Iraq and Afghanistan. "We must not be so preoccupied with preparing for future . . . conflicts," he wrote, "that we neglect to provide all the capabilities necessary to fight and win conflicts such as those the United States is in today."

Yet if China was a problem for the future in 2001 or even 2009, yesterday's future has now arrived. China may or may not be able to sustain its economic and political trajectory in the coming decades; its growth rate has already declined substantially from the double-digit norm of the thirty years years before 2010. But it would be Pollyannaish in the extreme to count on the wheels coming off soon enough, and completely enough, to avoid a serious geopolitical challenge—not least because that challenge is already here. China's gray-zone coercion is not some hypothetical problem of tomorrow; it is reshaping the geopolitics of the Asia-Pacific in real time. China's "incremental salami slicing tactics," writes Patrick Cronin, are progressively adding up to "major changes in the status quo." And while China is still reluctant to hazard military confrontation with Washington, U.S. advantages in that realm have also eroded alarmingly.

According to the RAND Corporation, the United States may already face grave challenges in defending Taiwan at an acceptable price. Those challenges are growing more acute in other contingencies, as well. In 2014, Frank Kendall, the undersecretary of defense for acquisition, technology, and logistics, warned that U.S. superiority was being "challenged in ways that I have not seen for decades." He added, "This is not a future problem. This is a here-now problem." And in 2017, the normally circumspect chairman of the Joint Chiefs of Staff, Gen. Joseph Dunford, told Congress that America might lose its ability to project power into contested regions within five years, absent corrective measures. The worst of the China challenge may or may not still lie in the future, but the challenge is plenty severe today.

The fact that the five ideas discussed here face mounting doubts does not mean that U.S. officials were knaves ever to have entertained them. It was a bit whiggish—but not absurd—to think that China would evolve toward democracy at a time when so many authoritarian regimes were making that transition; it was certainly worth testing whether a nation of more than one billion people could be made a satisfied member of the international system. The United States could afford to emphasize reassurance and engagement in dealing with Beijing in the 1990s or even the early 2000s; there was a time when the China challenge was indeed far less immediate than other pressing dangers.

The trouble with the ideas that have guided America's China policy, then, is not that they were clearly wrong to begin with. It is that they have now outlived their utility. As China's revisionist ambitions and growing assertiveness become more pronounced, these concepts serve principally to obscure the nature and dimensions of the challenge, and to weaken the impetus to a sharper response. Ideas that seemed reasonable enough in their time have become increasingly dangerous today.

None of this is to say that America should precipitously shift from a strategy featuring strong elements of engagement to one that represents a caricature of Cold War-style containment—an approach that seeks not simply to stymie China's geopolitical ascendancy, but to halt its economic growth, destabilize its political system and isolate it diplomatically. Such a strategy is unworkable, given China's economic heft, significant diplomatic influence and status as primary trading partner to many U.S. partners and allies. Indeed, it is an ironic effect—call it a catastrophic success—of decades of U.S. and international engagement of China that containment of that country is now impossible.

Fortunately, such a policy is also unnecessary, because there are myriad intermediate steps Washington can take to compete more effectively: increasing military spending and accelerating development of capabilities that can puncture the Chinese A2/AD bubble; helping U.S. allies acquire cheap and plentiful A2/AD capabilities of their own to constrain Chinese maritime advances; imposing greater—if still selective—diplomatic and economic costs in response to gray-zone expansionism; limiting Beijing's investment in sectors critical to national security and otherwise reducing the vulnerability of America and its friends to Chinese economic coercion; and investing more in the geo-economic elements of statecraft. These steps all entail a willingness to court increased tensions with Beijing, but they are not inherently incompatible with continued commerce or cooperation on issues of mutual concern. If the United States and the Soviet Union could collaborate on nuclear nonproliferation, arms control, and the eradication of smallpox amid the Cold War, surely a more competitive strategy toward China would not preclude all forms of engagement today.

Regardless of how America responds to the Chinese challenge, however, its policy must be rooted in reality. Preventing an increasingly confident great power from remaking the East Asian order, and perhaps challenging U.S. interests globally, will be the defining challenge of American statecraft in the twenty-first century. Meeting that challenge will be hard enough even if America dispenses with its China illusions; doing so will probably be impossible if it does not.

Critical Thinking

1. Why was China's rise underestimated by the United States?
2. Why did the United States think that the economic development of China would result in a democratic transformation?

Internet References

China's Belt and Road Is Full of Holes
https://www.csis.org/analysis/chinas=belt-and-road-full-of-holes

Fairbank Center for Chinese Studies
https://fairbank.fas.harvard.edu/

Kissinger Institute on China and the US
https://www.wilsoncenter.org/program/kissinger-institute-china-and-the-united-states/

The Indo-Pacific Region Takes Center Stage at Shangri-La
https://www.csis.org/analysis/indo-pacific-region-takes-center-stage-shangri-la

HAL BRANDS is the Henry A. Kissinger Distinguished Professor of Global Affairs at the Johns Hopkins School of Advanced International Studies (SAIS), senior fellow at the Center for Strategic and Budgetary Assessments, and author of *American Grand Strategy in the Age of Trump.*

Article

Prepared by: Robert Weiner, *University of Massachusetts, Boston*

Life in China's Asia: What Regional Hegemony Would Look Like

JENNIFER LIND

Learning Outcomes

After reading this article, you will be able to:

- What is meant by a regional hegemon?
- The relationship between economic and regional hegemony.

For now, the United States remains the dominant power in East Asia, but China is quickly closing the gap. Although an economic crisis or domestic political turmoil could derail China's rise, if current trends continue, China will before long supplant the United States as the region's economic, military, and political hegemon.

As that day approaches, U.S. allies and partners in the region, such as Australia, Japan, the Philippines, and South Korea, will start to face some difficult questions. Namely, should they step up their individual defense efforts and increase their cooperation with other countries in the region, or can they safely decide to accept Chinese dominance, looking to Beijing as they have looked to Washington for the past half century?

It may be tempting to believe that China will be a relatively benign regional hegemon. Economic interdependence, one argument goes, should restrain Chinese aggression: because the legitimacy of the Chinese Communist Party (CCP) rests on economic growth, which depends on trade, Beijing would maintain peaceful relations with its neighbors. Moreover, China claims to be a different sort of great power. Chinese officials and scholars regularly decry interventionism and reject the notion of "spheres of influence" as a Cold War relic. Chinese President Xi Jinping has said that his country has "never engaged in colonialism or aggression" thanks to its "peace-loving cultural tradition." In this view, life in China's Asia would not be so different from what it is today.

But this is not how regional hegemons behave. Great powers typically dominate their regions in their quest for security. They develop and wield tremendous economic power. They build massive militaries, expel external rivals, and use regional institutions and cultural programs to entrench their influence. Because hegemons fear that neighboring countries will allow external rivals to establish a military foothold, they develop a profound interest in the domestic politics of their neighborhood, and even seek to spread their culture to draw other countries closer.

China is already following the strategies of previous regional hegemons. It is using economic coercion to bend other countries to its will. It is building up its military to ward off challengers. It is intervening in other countries' domestic politics to get friendlier policies. And it is investing massively in educational and cultural programs to enhance its soft power. As Chinese power and ambition grow, such efforts will only increase. China's neighbors must start debating how comfortable they are with this future, and what costs they are willing to pay to shape or forestall it.

Economic Centrality

Over the past few decades, China has become the number one trading partner and principal export destination for most countries in East Asia. Beijing has struck a number of regional economic deals, including free-trade agreements with Australia, Singapore, South Korea, the Association of Southeast Asian Nations, and others. Through such arrangements, which exclude the United States, Beijing seeks to create a Chinese-dominated East Asian community. Beijing is also building an institutional infrastructure to increase its influence at the expense of U.S.-led institutions, such as the International Monetary

Fund (IMF) and the World Bank, and Japanese-led ones, such as the Asian Development Bank. In 2014, China, along with Brazil, Russia, and India, set up the $100 billion New Development Bank, which is headquartered in Shanghai. In 2015, China founded the $100 billion Asian Infrastructure Investment Bank, which 80 countries have now joined. Furthermore, Xi's much-heralded Belt and Road initiative will promote Chinese trade and financial cooperation throughout the region and provide massive Chinese investment in regional infrastructure and natural resources. The China Development Bank has already committed $250 billion in loans to the project.

Such policies mimic the economic strategies of previous regional hegemons. China was the predominant economic and military power in East Asia until the nineteenth century. It granted or withheld trade privileges according to an elaborate system of tribute, in which other countries had to send diplomatic missions, bestow gifts, and kowtow to the Chinese emperor. The Chinese then determined the prices and quantities of all goods traded. Imperial China consolidated its economic power by investing in agriculture and railroads, extracting minerals, and encouraging close commercial integration throughout the region.

In Latin America, the United States followed the same playbook to establish itself as the region's central economic player. In the nineteenth century, American firms flocked to the region in search of fruit, minerals, sugar, and tobacco. The U.S. company United Fruit managed to gain control of the entire fruit export trade in Central America. Finance was another powerful tool; as the Uruguayan journalist Eduardo Galeano has argued, a U.S. "banking invasion" diverted local capital to U.S. firms. Washington encouraged American banks to assume the debts of European creditors to minimize the influence of European rivals. For almost 100 years, Washington used diplomacy to advance its economic interests through initiatives promoting U.S. regional trade and investment, such as the Big Brother policy in the 1880s, "dollar diplomacy" in the early 1900s, and the Alliance for Progress in the 1960s.

The United States also built a regional institutional architecture to advance its agenda. In 1948, it created the Organization of American States (headquartered in Washington, D.C.) to promote regional security and cooperation. American influence ensured that the OAs remained silent on, or even legitimized, various U.S. military and political interventions in Latin America. Other development institutions, including the IMF, the World Bank, the Inter-American Development Bank, the U.S. Agency for International Development, and the Export-Import Bank of the United States, also advanced U.S. interests. Through "tied aid," such organizations required sponsored projects to hire U.S. vendors. The IMF, as Galeano has argued, was "born in the United States, headquartered in the United States, and at the service of the United States."

Another regional hegemon, Japan, pursued similar strategies in its empire that dominated the region in the early twentieth century. Vowing to eject the Western colonial powers, Tokyo declared itself the head of a "Greater East Asia Co-Prosperity Sphere." To feed its industrial economy and military, Tokyo extracted raw materials from countries it conquered. To promote Japan's centrality, and to prevent economic activities by rival countries, it reformed and managed local economies in a regional network, standardizing the region's currency in a "yen bloc" and dispatching Japanese banks throughout the area so that they controlled the majority of the region's bank deposits. Tokyo also created the Southern Development Bank, which provided financial services and printed currency in occupied territories.

Similarly, in Eastern Europe after World War II, the Soviet Union relied on economic and financial statecraft to dominate the region. Moscow blocked all trade with Western Europe and forbade Eastern European states from accepting aid under the 1948 Marshall Plan. Instead, it created the Council for Mutual Economic Assistance to manage and integrate the regional economy. Soviet investment, trade agreements, and trade credits made Eastern European countries economically dependent on Moscow, both as their primary export market and as their supplier of raw materials and energy. And by selling raw materials at below-market prices, Moscow encouraged local political leaders to become dependent on its subsidies.

Economic dominance lets regional hegemons use economic coercion to advance their agendas. In Latin America, the United States has long sought to coerce countries through sanctions. In addition to the long-standing (and failed) U.S. embargo of Cuba, Washington used financial pressure to weaken President Salvador Allende in Chile in the 1970s and embargoed Nicaragua to undermine the Sandinista government in 1985. Similarly, in Eastern Europe, Moscow sought to control independent-minded leaders, imposing sanctions against Yugoslavia in 1948, Albania in 1961, and Romania in 1964.

Beijing has already begun to employ such economic coercion. In 2017, China punished South Korea and the Japanese-South Korean business conglomerate Lotte for cooperating with the U.S.-built THAAD missile defense program. (Lotte had sold the land on which THAAD was deployed to the South Korean government.) Beijing banned Chinese tour groups from visiting South Korea, Chinese regulators closed 80 percent of Lotte supermarkets and other Korean-owned businesses (ostensibly for fire-code violations), and state-run media urged boycotts of Korean products. Beijing has also used economic coercion against Japan (banning the export of Chinese rare-earth metals to the country after a 2009 ship collision) and Norway (embargoing Norwegian fish exports after the Chinese dissident Liu Xiaobo won the Nobel Peace Prize in 2010). And in 2016, when Mongolia hosted the Dalai Lama, Beijing imposed extra fees on commodities moving through the country and froze all diplomatic activity—including negotiations about a $4 billion Chinese loan. "We hope that Mongolia has taken this lesson to heart," the Chinese foreign ministry said in a statement. Apparently it has: the Mongolian government has announced that the spiritual leader will not be invited back.

Such coercion will be less necessary in the future as leaders preemptively adjust their policies with Beijing in mind. Consider the Philippines: in the past, the country has stood up to China—for example, filing a complaint about Chinese territorial assertiveness with an international tribunal at The Hague in 2013. But more recently, Philippine President Rodrigo Duterte, who has received $24 billion in investment pledges from Beijing, has warmed relations with China and distanced his country from the United States.

The Pursuit of Military Hegemony

Following the example of previous hegemons, China is also expanding its regional military reach. Since the 1990s, Chinese military spending has soared, and the CCP is modernizing weaponry and reforming its military organizations and doctrine. The People's Liberation Army (PLA) has adopted the doctrine of "anti-access, area denial" to push the U.S. military away from its shores and airspace. China has also built the region's largest coast guard and controls a vast militia of civilian fishing vessels. In 2017, the PLA opened its first overseas military base in Djibouti; it will likely build more bases along the African east coast and the Indian Ocean in coming years. Meanwhile, in the South China Sea, China has built six large islands that house air force bases, missile shelters, and radar and communications facilities. Already, the U.S. military finds itself constrained by the expanding bubble of Chinese air defenses, by China's growing ability to find and strike U.S. naval vessels, and by an increased missile threat to U.S. air bases and ports.

Beijing is using these capabilities to more forcefully assert its territorial claims. By transiting disputed waters and massing ships there, Beijing is pressuring Japan militarily over a cluster of small islands called the Diaoyu by China and the Senkaku by Japan. Elsewhere, to deny access to disputed areas, the PLA swarms fishing and coast guard vessels, and fires water cannons at other countries' ships. Last summer, after asserting ownership of an oil-rich area in Vietnam's exclusive economic zone, Beijing threatened to use military force if Vietnam did not stop drilling. Vietnam stopped drilling.

Contemporary China's quest for regional military dominance follows the behavior of previous regional hegemons, including China itself. As the historian Peter Perdue has argued, modern China is a product of invasions that subdued all of modern Xinjiang and Mongolia, and reached Tibet, as well. Chinese dynasties, he has written, "never shrank from the use of force," including the "righteous extermination" of rival states and rebels. Throughout Asia, Chinese military garrisons subdued invaders and pirates.

Subsequent hegemons dominated their regions through military force, too. Starting in the late nineteenth century, the United States began to build what would become the Western Hemisphere's preeminent military. In that period, the United States acquired territory through numerous wars against Mexico and Spain. Over the next few decades (often to advance the United States' commercial interests), U.S. forces invaded Latin American countries more than 20 times, most often the Dominican Republic, Haiti, Mexico, and Nicaragua. During the Cold War, the United States repeatedly used military force to counter leftist movements in Latin America: it blockaded Cuba in 1962, sent troops to the Dominican Republic in 1965, mined Nicaraguan harbors in the 1980s, and invaded Grenada in 1983 and Panama in 1989.

Japan also built and maintained its empire through military force. Its nineteenth-century military modernization yielded stunning victories over China and Russia. Through these and other military campaigns, Japan seized territories such as Korea and Taiwan and wrested colonial possessions from France, Germany, the United Kingdom, and the United States. The Japanese military then administered the empire, fighting counterinsurgencies and suppressing independence movements.

In Europe after World War II, the Soviet Union dominated its sphere of influence with the region's most powerful army. It stationed troops in Czechoslovakia, East Germany, Hungary, and Poland. To shape the region to its liking, the Kremlin was willing to use force. It dispatched Soviet troops to quell uprisings in Hungary in 1956 and Czechoslovakia in 1968.

These hegemons did not tolerate the presence of rival great powers in their regions. Likewise, China today is chafing against the U.S. presence in Asia and actively working to undermine it. Chinese officials and defense white papers criticize U.S. alliances as outdated and destabilizing. Xi himself, calling for a new "Asian security architecture," has argued that these relationships fail to address the region's complex security needs. Meanwhile, by cultivating close ties with Seoul and encouraging the Philippines' tilt toward China, Beijing has sought to draw U.S allies away.

Nosy Neighbor

Beijing is also interfering in the domestic politics of other countries. Citing China specifically, Canadian intelligence officials have warned of foreign agents who might be serving as provincial cabinet ministers and government employees. And in 2016, a scandal erupted in Australia after it was revealed that Sam Dastyari, a senator who had defended Chinese territorial claims in the South China Sea, had financial ties to a Chinese firm, prompting new laws banning foreign political donations.

Historically, regional hegemons have intervened extensively in domestic politics to support friendly governments and undermine parties and leaders perceived as hostile. Within China's tribute system, the emperor delegated the administration of subservient states to local leaders, an approach known as "using barbarians to govern barbarians." But local independence went only so far. As the sixteenth-century statesman Chang Chu-cheng said of such vassals, "Just like dogs, if they wag their tails, bones will be thrown to them; if they bark wildly, they will be beaten with sticks; after the beating, if they submit again, bones will be thrown to them again; after the bones, if they bark again, then more beating."

Japan similarly intervened in domestic politics during its imperial heyday. In the Philippines, for example, it abolished all political parties except for the pro-Japanese one. Elsewhere, it delegated control to friendly local leaders and police, and trained such leaders at institutes in Japan. If officials in China, Korea, and Manchuria did not cooperate, Tokyo relied on a Japanese paramilitary organization that intimidated, blackmailed, and assassinated local leaders.

For its part, the United States meddled in Latin American politics countless times. Through the Roosevelt Corollary to the Monroe Doctrine, Washington claimed the right to intervene in its neighbors' affairs. It relied on covert and overt, violent and nonviolent methods to support anticommunist leaders and to undermine or depose leftist ones. The U.S. diplomat Robert Olds explained the approach in blunt terms in 1927: "Central America has always understood that governments which we recognize and support stay in power, while those which we do not recognize and support fall." During the Cold War, the U.S. military and the CIA funded, armed, and trained anticommunist forces throughout Latin America at institutions such as the U.S. Army School of the Americas in Panama. U.S.-trained forces sought to depose leftist governments in Cuba, Ecuador, El Salvador, and Nicaragua. Washington also supported coups in Guatemala in 1954 and Chile in 1973.

Moscow was similarly busy in Eastern Europe. After World War II, the Soviet Union installed communist parties in its neighbors' governments, in which advancement depended on loyalty to Moscow. Under Stalin, Soviet secret police harassed, tortured, and murdered opposition leaders. After Stalin, the Soviets relied on subtler tactics, such as bringing foreign elites to train in communist party schools and to build networks with Soviet and regional politicians. Through the Brezhnev Doctrine, Moscow claimed the authority to intervene in its neighbors' politics in order to defend socialism from hostile forces.

Playing Hardball for Soft Power

China today is seeking to increase its influence in East Asia and beyond through extensive educational and cultural activities. The media is central to this effort. The state-run media organizations Xinhua and the China Global Television Network have bureaus all over the world. Hollywood studios regularly seek Chinese funding for their projects, as well as distribution rights in China's vast market. Wary of offending the CCP, studios have started preemptively censoring their content. Censorship has also begun to infect the publishing industry. To gain access to China's vast market, publishers are increasingly required to censor books and articles containing specific words or phrases (for example, "Taiwan," "Tibet," and "Cultural Revolution"). Prominent publishing houses, including Springer Nature—the world's largest academic book publisher—have succumbed to Beijing's demands and are increasingly self-censoring.

Beijing also promotes Chinese influence in education. China has become the world's third most popular destination for foreign study, welcoming more than 440,000 students from over 200 countries in 2016. Many students receive support from the Chinese government. Overseas, in 142 countries, Beijing has created more than 500 Confucius Institutes to promote Chinese language and culture. A study by the U.S.-based National Association of Scholars argues that Confucius Institutes are decidedly nontransparent about their connections to the CCP. Their teachers must observe CCP restrictions on free speech and are pressured to "avoid sensitive topics," such as human rights, Tibet, and Taiwan.

The CCP also infiltrates college campuses abroad. Beijing enlists members of the 60-million-strong Chinese diaspora: at universities around the world, Chinese Students and Scholars Associations demonstrate in support of visiting Chinese leaders and protest the Dalai Lama and other speakers the CCP deems hostile. Beijing also monitors and silences Chinese critics abroad by mobilizing harassment on social media and by threatening their families back home. In Australia, concerns about Chinese interference and espionage at universities led intelligence officials to issue warnings about an "insidious threat" from foreign governments seeking to shape local public opinion.

Past regional hegemons similarly promoted their influence through culture and education, and by co-opting leaders of civil society. As the China expert Suisheng Zhao writes, "Chinese culture was seen as a great lasting power to bridge periods of disunity and to infuse new governments . . . with values supportive of the traditional Chinese order." China spread its language, literature, Confucian philosophy, and bureaucratic traditions to Japan, Korea, Vietnam, and other countries. Chinese emperors also followed the advice of one minister in the Han dynasty who proposed subduing barbarians with "five baits": silk clothing and carriages, sumptuous food, entertainments and female attendants, mansions with slaves, and imperial favors such as banquets and awards.

U.S. hegemony in Latin America also relied heavily on soft power. In 1953, the U.S. government created the U.S. Information Agency, which, according to President Dwight Eisenhower, would show countries that U.S. objectives "are in harmony with and will advance their legitimate aspirations for freedom, progress, and peace." U.S. TV stations started Latin American channels that broadcast American films and programs. The U.S. government built news agencies and radio stations and infiltrated or intimidated opposition media outlets. In Chile and the Dominican Republic, for example, the CIA and the USIA engaged in an intense propaganda effort against undesirable political candidates, spreading misinformation and silencing opposition media.

Likewise, imperial Japan created the East Asia Development League to shape regional perceptions and guide the activities of Japanese people living in the empire. Tokyo controlled civil society by creating and infiltrating organizations such as youth groups, martial arts clubs, student unions, secret societies, and religious organizations. Its Greater East Asia Cultural Policy sought to eradicate Western culture. For example, Tokyo banned Coca-Cola on the grounds that it had been invented "to bring the people under the soul- and mind-shattering influence of the insidious drug, and so to make them more apt for Anglo-American exploitation." Tokyo prohibited the use of European languages and established Japanese as the area's official language, dispatching hundreds of teachers throughout Asia. Japan transmitted its culture through radio programs, newspapers, and comic books, as did cultural institutes that sponsored exhibitions, lectures, and films.

The Soviet Union secured its influence in Eastern Europe through extensive cultural activities. As the writer Anne Applebaum details in her book Iron Curtain, Soviet-backed communist parties took over radio stations and newspapers and intimidated or shut down independent media. The Soviets created influential youth organizations and co-opted writers, artists, and other intellectual leaders by offering well-paid jobs, lavish houses with servants, and free education for their children.

Moscow also created a vast organization known as VOKS (a Russian acronym for All-Union Society for Cultural Relations With Foreign Countries) to disseminate Soviet ideas and culture and bring Western intellectuals under communist influence. VOKS brought thousands of visitors to the Soviet Union and sponsored scientific research, filmmaking, athletics, ballet, music, and publishing. It also spent lavishly at international fairs and expositions—such as the Brussels World's Fair of 1958—to showcase Soviet technology and culture.

Contemplating Life in China's Asia

When examining China's current behavior in the context of previous regional hegemonies, some common themes stand out. First, economic interdependence has a dark side. Although interdependence raises the cost of conflict, it also creates leverage. China's centrality in regional trade and finance increases its coercive power, which Beijing has already begun to exercise. Second, history shows that regional hegemons meddle extensively in their neighbors' domestic politics. Indeed, Beijing has already begun to reverse its much-touted policy of nonintervention. As China grows stronger, its neighbors can expect Beijing to increasingly interfere in their domestic politics.

East Asian countries need to decide whether this is something they are willing to accept. In particular, Japan, the only country with the potential power to balance China, faces an important choice. Since World War II, Japan has adhered to a highly restrained national security policy, spending just one percent of its GDP on defense. For obvious historical reasons, the Japanese people are suspicious of military statecraft, and they worry about a lagging economy and the expense of caring for an aging population. They may decide to continue devoting their wealth to butter rather than guns.

This would be a perfectly valid choice, but before making it, the Japanese people should contemplate their life in China's Asia. Beijing and Tokyo are already embroiled in a bitter territorial dispute over the Diaoyu/Senkaku Islands. To gain control of the islands, weaken the U.S.-Japanese relationship, and advance other interests, Beijing can be expected to use greater military and economic coercion and to meddle in Japanese politics. Beyond a hegemon's normal reasons to intervene, China harbors deep historical resentment toward Japan. Imagine if the United States had actually hated Cuba.

If Japan decided that Chinese hegemony would be unacceptable, its national security policy would need to change. The United States' global interests and commitments allow Washington to devote only some of its resources to Asia. It would not have the capability, let alone the will, to balance Beijing alone. Japan would need to become more like West Germany: a U.S. ally that, although outgunned and directly threatened by a hostile great power, mobilized substantial military might and was a true partner with the United States in securing its national defense.

Tokyo and Washington could use diplomacy to offer countries an alternative to Chinese regional dominance. To do so, they should look to a core group of maritime countries with similar values and overlapping interests—namely, Australia, India, New Zealand, and the Philippines. Other potentially interested actors, such as Indonesia, Malaysia, Singapore, and Thailand, should be welcomed, too. But the first step on this path is a Japanese—and broader East Asian—debate about the prospect of living in China's Asia.

Critical Thinking

1. What can the United States do to counterbalance China's regional hegemony in Asia?
2. What is the relationship between Great Powers and regional hegemons?
3. Compare and contrast regional hegemony in Latin America and Eastern Europe.

Internet References

Fairbank Center for Chinese Studies
https://fairbank.fas.harvard.edu/

Kissinger Institute on China and the US
https://www.wilsoncenter.org/program/kissinger-institute-china-and-the-united-states/

The Asian Infrastructure Investment Bank
Aiib.org/

The Regional Economic Comprehensive Partnership
https://en.wikipedia.org/wiki/regional-comprehensive-economic-partnership

Jennifer Lind is an associate professor of Government at Dartmouth College.

Unit 2

UNIT

Prepared by: Robert Weiner, *University of Massachusetts, Boston*

Population, Natural Resources, and Climate Change

The articles in this unit engage in an analysis of the material base of global affairs, which still matter in a liberal international order. A revolution has occurred in energy technology which is having a profound impact on the traditional sources of energy derived from the traditional fossil fuels. A critical question also is whether or not the planet has the carrying capacity to sustain a population which is projected to grow to 10 billion.

One of the most important questions is the extent to which such developments will have an effect on climate change and warming. A major development in connection with this has been the decision of the Trump administration to withdraw from the Paris Climate Agreement effective in 2020.The agreement reached in Paris in 2015 has the goal of preventing the average global temperature from rising 2 degrees Celsius beyond postindustrial levels (levels which existed before the industrial revolution). Some experts are convinced that 2 degrees is not low enough. For the agreement to become adopted, there was a "double trigger" of 55 countries which were responsible for 55 percent of the emissions of the greenhouse gases had to approve the agreement. The "double trigger" was met, and the Paris Agreement subsequently went into effect. However, the participation of the United States in the Paris Agreement became problematical with the election of Donald Trump as President. During the presidential campaign, Trump had stated that he believed that the idea of climate change linked to the warming of the planet caused by human activity was a hoax. One of the first actions that President Trump took was to reduce the extent of the powers of the Environmental Protection Agency. One of the key components of President Trump's agenda is to increase the production of natural gas and oil from shale, as well as increase the production of coal. On June 1, 2017, President Trump announced that the United States was withdrawing from the Paris Agreement. A follow-up conference to the Paris Agreement met in Morocco on November 7–18. However, a shadow was cast over the conference by the election of Donald Trump. The US delegation to the conference decided to go ahead with the conference in spite of the lack of support from the Trump administration. On an optimistic note the Conference opened with the news that the Paris Agreement had received the necessary number of ratifications on November 4,2016 to go into effect. That is, 55 countries that were responsible for 55 percent of the carbon emissions had ratified the Paris Agreement. The conference adopted the "Marrakech Act" or Proclamation which also linked climate change to Sustainable Development. The proclamation emphasized that the climate was warming at "an alarming rate and it was necessary to move forward to reduce greenhouse gas emissions. "The purpose of the Marrakech Conference was to work on the implementation of the Paris Agreement through the development of "rules of the road. "The question of providing adequate funding for the UN's Green Climate Fund also became problematic with the decision by the Trump administration not to contribute to the fund, subject to Congressional approval. The purpose of US participation in the Conference was to continue to look after US interests. Consequently, the United States participated in the 23rd meeting of the parties to the Framework Convention on Climate Change. The 23rd Conference of Parties (COP) took place in Bonn, Germany, November 6–17, 2017. The conference was huge, with over 30,000 people participating. The reason for the large number of participants was because besides the 9200 government representatives, there also were representatives from many different civil associations, businesses, and various subnational entities such as cities from around the world, states, and provinces from various countries. There also were representatives from local communities and indigenous peoples. Much of the work of the conference took place within groups and blocs of states characterized by varying degrees of internal cohesion. The groups included the Umbrella Group which represented non-EU developing countries, the Environmental Integrity Group, the Alliance of Small Island Developing

States, the Coalition for Rainforest Nations, the "Like-Minded Group of Developing Nations," the Independent Association of Latin America and the Caribbean, the Bolivarian Alliance for the Peoples of Our America, and Brazil, India, and China. Most of the above-mentioned groups have emerged at various stages of the negotiations and conventions dealing with climate change. There also was a lot of activity on the sidelines of the conference An American bipartisan group consisting of representatives of cities, states, governors, county executives, businesses, universities and colleges, cultural institutions, tribal groups, and think tanks, called "We Are Still In," pledged their support to the international community for the implementation of the agreement. COP 24 is scheduled to meet in Katowice, Poland in November 2018.

Article Prepared by: Robert Weiner, *University of Massachusetts, Boston*

Here's Looking At You, 2050

How a less Christian Europe, an aging population in the West, and the empowerment of women are going to shape the future.

PAUL TAYLOR

Learning Outcomes

After reading this article, you will be able to:

- Learn about the relationship between demographic trends.
- Learn about the sharp demographic variances across regions and within countries.

The 21st century is still just a teenager, but we can already forecast with a fair degree of confidence what its demographic profile will look like by 2050.

Population growth will have slowed down. Global aging will have risen to unprecedented levels. Birthrates will drop. The working-age share of the world's population will shrink. Poverty will ameliorate in poor countries; income inequality will worsen in wealthy ones. And for the first time ever, Islam will challenge Christianity as the world's largest religion.

What's notable about these disparate trends is how much they are interrelated. They're driven not just by the traditional demographic triad of births-deaths-and-migration, but by myriad powerful new forces that define modernity—from the empowerment of women, to improvements in health care, to the information and technology revolutions, to the concurrent rise of secularization and religious fundamentalism.

However, the fact that they are connected does not mean they are universal. Beneath the broad umbrella of global demographic change, there will be sharp variances across regions (and sometimes within countries).

Consider the most basic demographic metric of all—population size. By midcentury, the world's fastest-growing region, Africa, is projected to see its population more than double, while the slowest-growing region, Europe, is expected to see its population decline by about 4 percent.

This means that in 2050 there will be around 3.5 times more Africans (2.5 billion) than Europeans (707 million). In 1950, there were nearly twice as many Europeans as Africans. Demography is a drama in slow motion. But tick by tock, it transforms the world.

The staggering reversal of population fortunes is largely the result of the huge continental differences in birthrates—1.6 children per woman in Europe today versus 4.7 children per woman in Africa. By midcentury, however, those rates are expected to increase in Europe and decrease in Africa, as both continents converge toward the projected global rate of roughly 2.25 down from 5 in 1950 and 2.5 in 2015.

Declining global birthrates mostly stem from women's empowerment. As more girls and women have acquired more education, economic independence, and control over their reproductive decisions, they have had fewer babies. In the 35 years from 2015 to 2050, the world's population is expected to rise by only 32 percent. During the 20th century, it nearly quadrupled.

As population growth slows, median ages will rise—the result not just of fewer children but also of steady increases in human longevity. By 2050, the share of the global population that is 60 or older will nearly double to 21.5 percent, from today's 12.3 percent. Aging will be most pronounced in economic powerhouses like Japan, where the median age by midcentury will be 53, South Korea (54), Germany (51), China (50), and the United States (42). The global median age will be 36, up from today's 30.

These aging societies will be hard-pressed to maintain their economic vitality as the working-age shares of their populations decline and the fiscal pressures on their health care systems and old-age social insurance programs grow.

Meanwhile, the less-developed countries throughout Africa, the Middle East, Latin America, and parts of Asia are still experiencing a youth bulge (albeit one with less girth than in the past). Countries like India, Nigeria, Egypt, South Africa, and Kenya will see the working-age shares of their population grow between now and midcentury. Their challenge will be to make the investments in human and physical capital needed to take advantage of this demographic dividend.

Since the turn of the millennium, these disparate age structures, along with the incessant march of technology, have already yielded different economic outcomes around the world. According to the World Bank, 1 billion people have climbed out of extreme poverty since 2000, the vast majority of them in poor countries, where inexpensive mobile technology has unlocked new economic opportunity in rural villages and urban slums.

Economics isn't the only human realm in the throes of demographic change. The world's religious profile is also undergoing a major shift, driven mostly by differences in fertility rates and age structures around the globe, as well as by faith-switching.

Islam is the fastest-growing major religion in the world. According to projections by the Pew Research Center, by midcentury, the number of Muslims (2.8 billion, or 30 percent of the world's population) and Christians (2.9 billion, or 31 percent) will for the first time be at near parity. If current trends continue, Islam will surpass Christianity as the world's largest religion around 2070.

Muslim women have the highest fertility rate of any major religion (an average of 3.1 children per woman). Christian women are not far behind (2.7), which places them above the global average of 2.5 and well above the replacement rate of 2.1.

In fact for the first time, the Christian populations in France and the United Kingdom are projected to drop below half by midcentury. The growth of Christianity is expected to slow in part due to followers leaving the faith, a phenomenon most prevalent in developed countries like the United States (where the Christian population is projected to decline from more than three-quarters in 2010 to two-thirds in 2050). In comparison, throughout Europe, the Muslim population is expected to rise slightly (to 10.2 percent from today's 5.9 percent). These projections are dependent on migration patterns that could be affected by geopolitical developments.

That caveat applies to all projections: Demography is the future we already know—except when it isn't. Harold Macmillan, a 20th-century British prime minister, put it well. When asked what might blow his government off course, he replied, "Events, dear boy, events."

Critical Thinking

1. What will account for the reversal of population size between Europe and Africa by 2050?
2. What accounts for declining global birthrates?
3. What will account for the growth in the Muslim religion and the decline in the Christian religion?

Internet References

Demographic Trends Shaping the US and the World in 2017

www.pewresearch.org/.../10-demographictrends-shaping-the-u-s-and-the-world-in-2017

The UN Commission on Population and Development

www.un.org/esa/population/cpd/about.com.htm

United Nations Population Fund

https://www.unfpa.org/world-population-trends

Article

Prepared by: Robert Weiner, *University of Massachusetts, Boston*

The Timely Disappearance of Climate Change Denial in China

From Western plot to party line, how China embraced climate science to become a green-energy powerhouse.

GEOFF DEMBICKI

Learning Outcomes

After reading this article, you will be able to:

- Learn about China's policy at the Copenhagen Climate Conference.
- Why China did not originally believe in climate change?
- What changed the Chinese attitude toward global warming?

In December 2009, climate-watchers the world over were trying to make sense of how the most promising attempt to date at preventing a global climate disaster went so horribly wrong. The Copenhagen Climate Change Conference had just come to a close, and the summit, which had brought together 192 countries, was meant to create the world's first legally binding treaty on global warming. But in its final days, during negotiations between China and the United States, talks had sputtered, teetered, and ultimately collapsed. To observers eager for good news, the result came as a stunning and disheartening anticlimax.

To most of the West, it appeared that China had come intent on playing the spoiler. The country's coal consumption had been growing steadily for decades as the government pushed industrialization. In the four years preceding Copenhagen, the country added 500 new 600-megawatt coal plants; it was responsible for more than 40 percent of global coal consumption in 2009. From the outside, the rationale for China's alleged resistance was rather simple. It just wasn't in China's interest to put the brakes on its rapid growth for environmental considerations. What could the country possibly gain by capping emissions?

Back in Beijing, however, there was no doubt about the threat of climate change. Behind closed doors, officials were telling a different story about the failed negotiations in Copenhagen.

"It was unprecedented for a conference negotiating process to be so complicated, for the arguments to be so intense, for the disputes to be so wide and for progress to be so slow," observed an internal report commissioned by the Environment Ministry for the minister, vice minister, and various other subordinates in the immediate aftermath, and obtained by *The Guardian* in February 2010. The report's authors concluded that the plan pushed by the United States, which proposed cuts on all countries instead of just developed ones, had been "a conspiracy by developed nations to divide the camp of developing nations." The report also lauded China's decision to oppose a legally binding climate treaty, trumpeting, "The overall interests of developing countries have been defended." Far from being the destructors of a progressive plan for climate change policy, the view from within China was that its delegates had possibly faced down a vast Western plot.

It was a strong reaction but one mostly rooted in diplomatic objections—a rejection of a deal that could be seen as asking China and India to pay for the sins of countries that had grown rich and modern by their bad behavior. But just over a month later, the idea of the Western plot took a strange, sharp turn. While speaking at a diplomatic event in New Delhi, Xie Zhenhua, China's top climate change negotiator—as well as vice minister of China's National Development and Reform Commission, the country's economic planning agency—surprised an audience of foreign environment ministers by saying that "we need to adopt an open attitude" about whether humans or natural atmospheric changes were to blame for the climate's warming. It was a shot against the very foundation of climate science.

Though the remark flummoxed the diplomats in the crowd, it could have been written off as a negotiating ploy. Chinese leaders had been cagey about the politics of global warming and had assented to sign the 1997 Kyoto Protocol, Copenhagen's predecessor, on the condition they not be forced to limit emissions. It was known that there were still powerful forces in the government that were antagonistic toward any plan that could curtail the country's freedom to burn fossil fuels. But this was something new. Back in China, the public backlash against Copenhagen—and climate science in general—had already begun. But this was something new. Back in China, the public backlash against Copenhagen—and climate science in general—had already begun.

On Jan. 17, 2010, a highly popular—and provocative—television host named Larry Hsien Ping Lang devoted an entire episode of his current affairs talk show, *Larry's Eyes on Finance*, to the "great swindle" of global warming. Lang, a University of Pennsylvania-educated economist who was once described as China's version of Larry King, told his millions of viewers that the goal of Europe and the United States at the Copenhagen negotiations was to prevent China from being a global leader.

"The Western countries manufactured the climate myth without any scientific integrity," and they have proceeded to "demonize and constrain China in the name of climate," Lang said. Clips of the episode were viewed tens of millions of times on Youku, China's YouTube.

Lang's worldview seemed to resonate. "[The weather] is obviously getting colder and colder, but they are still lying through their teeth. These disgusting Westerners never stop trying to topple China," argued one online commenter in response to Lang's show. "These foreign bastards are so worried that China will rise and surpass the United States. Because they are jealous of China, they even made up lies about China . . . the scientists are all puppets controlled by politics," read another. The commenter continued: "Copenhagen liars! American liars!"

Over the next year, more than a half-dozen books on the West's climate conspiracy were published in China. Social media posts theorizing an American conspiracy proliferated.

Then something strange happened. After 2011, no more climate skeptic books were published. China's state leaders stopped their skeptical statements, and the intense online discussions diminished. Just as it was gaining steam, the conspiracy theory seemed to disappear. And along with it, any public mention of climate change denial. As climate skeptics were gaining a steady foothold in U.S. politics, why did China's suddenly vanish?

The origins of climate change skepticism in China can be traced to a scientist named Coching Chu. A pioneering meteorologist in the 1920s and 1930s, Chu later became vice president of the Chinese Academy of Sciences and attained national fame after authorities decided to teach his life story in school textbooks. In the early 1970s, toward the tail end of his career, he drew from historical Chinese records to hypothesize that global temperatures had risen and fallen by several degrees Celsius during the past 5,000 years—due to natural fluctuations. It was a different conclusion from that reached by researchers in the United States and Europe, some of whom speculated that the planet was cooling. Others were already finding links between human activity and the steady rise in global temperatures. And though to most, Chu's work on cooling was a footnote at the end of his career, China's climate skeptics latched on—less for the particulars of his conclusions than for the fact that he'd reached them independently of the West.

It is hard to overstate how critical that distinction would become in validating Chu's work in his native country. The conviction that Western powers are trying to control and humiliate the country is a recurring theme of China's modern political development—and closely linked to its wave of climate skepticism.

This sense of aggrieved nationalism has a legitimate historical basis. China was often treated like a lesser power by Europe and Japan in the mid-19th and early 20th centuries. Even as the Communists opened China to globalization in the 1970s and 1980s, wounded national pride remained a potent undercurrent of political life. It would eventually give rise to an intellectual movement that started in the late 1990s loosely known as the "New Left." Members of this cadre believed China had too firmly embraced Western-style capitalism and that to address widening inequality the state must take more control over economic life.

In 2006, after decades of unchecked industrial growth, China's cities were choked with smog and the country was poised to surpass the United States as the top emitter of greenhouse gases—a superlative it would claim the following year. The Communist government responded by enacting the Renewable Energy Law that same year, which ordered that 15 percent of the country's electricity needs be met by alternatives to fossil fuels. It created some paradoxical numbers. In 2009, air pollution was so bad that China spent an estimated $110 billion dealing with the health impacts, according to the World Bank. At the same time, it had quickly become the biggest global investor in clean energy, spending nearly $35 billion in 2009 alone, compared with about $19 billion in the United States. There was a strong economic rationale for doing so. Former President Hu Jintao argued that China must "seize preemptive opportunities in the new round of the global energy revolution."

These competing forces of distrust of the West, a nascent but promising commitment to clean energy, and a willful belief in the country's right to develop came to a head in Copenhagen as negotiations stalled over the disagreement about who should bear the burden for cutting emissions. Though China was at least theoretically primed to support action against climate change, the particulars of the deal—and even the negotiations—were equally set to derail an agreement. "It was a very frenetic, emotional, high-pressure time," said Mark Lynas, a U.K. writer and environmentalist who was in Copenhagen as an advisor to Mohamed Nasheed, then president of the Maldives.

At one point during the talks, Lynas found himself in a room with then-President Barack Obama, German Chancellor Angela Merkel, and two dozen other heads of state. Posted against a wall, Lynas was shocked by what he witnessed. First, China's delegation snubbed the meeting by sending a lower-level diplomat in the place of the premier Wen Jiabao. Then, it opposed targets such as a peak in worldwide emissions by 2020 and a long-term emissions drop of 50 percent. Merkel appeared to be furious, Lynas later recalled, and at one point, then-Australian Prime Minister Kevin Rudd banged his microphone in annoyance. In Lynas's opinion, China was "torpedoing" hopes of an effective climate treaty to ensure its access to cheap supplies of coal. He recounted the meeting in a *Guardian* article: "China wrecked the talks, intentionally humiliated Barack Obama, and insisted on an awful 'deal.'"

Wen said his name was never included on invitations to the meeting Lynas attended. And *China Daily*, a government-run English language paper, later argued that China "played a vital role" in salvaging the talks by convening a last-minute meeting with Brazil, India, and South Africa.

In the immediate aftermath of Copenhagen, China's foreign ministry spokeswoman Jiang Yu accused Britain of demonizing and isolating China on the world stage. "We urge them to correct mistakes, fulfill their obligations to developing countries in an earnest way, and stay away from activities that hinder the international community's cooperation in coping with climate change," she said. It reinforced the narrative that the Communist leadership had been teaching for decades in China's schools: The West was conspiring against them. Some far-left nationalists took it further. They began to argue that global warming is a hoax.

John Chung-En Liu first stumbled upon China's particular brand of climate denial in 2013, when he walked into a bookshop in Beijing and saw *Low-Carbon Plot: The Life and Death Battle Between China and the West* displayed prominently on a shelf featuring a host of similar tomes. Written in 2010 by a then-relatively unknown Chinese writer named Gou Hongyang, it argues that Europe and the United States invented the idea of climate change as a way to exercise control over China. "Behind the back of the demonizing of 'carbon,' we must recognize that it is the sinister intention of the Developed Countries to attempt to use 'carbon' to block the living space of the Developing Countries," he wrote. In another section, he argued, "We can see it clearly from the Copenhagen Summit that the struggle between [the] two camps has intensified."

Liu had never seen anything like it. He bought as many of the books as he could find.

Later, as a sociologist at Occidental College in Los Angeles, he did an exhaustive search for conversations about climate change denial on Weibo, a popular Chinese social networking site, and sifted through decades' worth of issues of *China Daily.* Liu has published his findings on climate-skeptic literature in a 2015 journal article titled: "Low-Carbon Plot: Climate Change Skepticism With Chinese Characteristics." Before long, he was considered the foremost expert on China's denier community.

When I met him earlier this spring at the university, he was dressed in a button-down shirt, slacks, and glasses—the typical outfit of academia. In his small, tidy office, he produced a stack of eight oversize paperback books that, after years of research, he has concluded are the most influential and widely read skeptic books published in China. "[They] all came out right after Copenhagen," he said.

With titles like *In the Names of* CO_2, whose cover depicts a flaming dollar sign hurtling toward Earth, and *The Global Struggle behind the Low-Carbon Hoax*, Liu believes the books are crucial to understanding the worldview of China's climate skeptics: Science isn't neutral, and whichever country produces it controls the world. "The Europeans have made great effort on climate science for so many years," In the *Names of* CO_2 argues. "They have tons of publications and an enormous amount of data to back up their claim." This, it explains, gives the West "discursive power"—a Chinese buzzword used to describe Western domination of global conversation.

Here, the skeptics Liu studied do have something of a point. When the Intergovernmental Panel on Climate Change (IPCC) produced its fourth assessment report in 2007, only 28 Chinese scientists participated in the review, representing less than 2 percent of the report's contributors. Three years later, the deputy director general of China's National Climate Center, Xuedu Lu, said: "The majority of the IPCC's references came from Europe and North America. Developing countries also want their voices to be heard in the drafting stage." Only two of the climate skeptic books that Liu studied were written by scientists. Their aim was to make technical critiques of the IPCC consensus. One argued that global "temperature change is different from what the IPCC [predicted]," Liu said.

Most of the authors in Liu's collection, though, focused on the geopolitical implications of climate science. They argued that the West leveraged its scientific authority to impose restrictive policies on China. This position is central to a 2011 book by Deng Guangchi titled *Low Carbon War: The Transformation of the 4th Industrial Revolution*. "The United States uses [climate policies] as camouflage to force developing countries, China in particular, to lower carbon emissions and halt their industrialization processes," it reads. Liu disagrees, but he can see where this line of thinking originates. "For such a long time China has had this antagonistic relationship with the U.S.," he said. "This is not about the science. It's about who can emit how much, and it's about the West trying to contain China's development."

This argument is being driven by a wider distrust of the capitalist system that China has been embracing over the past decades. One of its most outspoken critics is the populist and conspiracy theorist Song Hongbing, who wrote the 2007 best-selling (in China) *book Currency Wars*. It argues that Western financial elites, such as the Rothschild family, are trying to dominate the world under the guise of open borders and free trade. The book's 2011 sequel claims that the adoption of financial markets for greenhouse gases—in the form, say, of a cap and trade system—is part of this plot. "Who would spend so much time and money spreading the idea about carbon emissions?" Song says. "How can we believe that things like carbon currencies, carbon trading, and carbon tariffs are not driven by a strong economic incentive?"

Low Carbon War raises similar concerns about shifting to renewable energy. "Because developing countries do not have leading new energy technology, in the end, they have to spend an enormous amount of money to purchase it from the European Union," it reads.

But as Liu points out, underneath all the vitriol and paranoia, the core of this climate change movement was less about science and more about power politics. As the author of *The Empire of Carbon Brokers: Carbon Capitalism and Our Bible*—whose cover displays a big red grenade in the shape of Earth with a smoking factory on top—writes: "The key is that China should not argue whether climate change is real or not with the West, but be part of the game." In the end, "many of them are agnostic," Liu says. "It doesn't really matter if climate change happens or not. It's really about this huge power play."

And if China has to compete against the West, it might as well win.

In 2012, with assistance from Yale University, Beijing's Renmin University of China conducted a rare national climate survey that resulted in some seemingly contradictory findings. On the one hand, it suggested that 93 percent of Chinese people think climate change is happening and the majority of respondents believe it "will harm themselves and their own family." (For comparison, a Yale survey of Americans taken around the same time found that only 70 percent believe in climate change, and a far smaller portion says it will affect them.) Yet the Renmin study found that just 55 percent of Chinese people think humans are the primary cause of global warming, a percentage roughly comparable with the United States.

The numbers were surprising, says Binbin Wang, co-founder of the Beijing-based China Center for Climate Change Communication, who helped design the survey. She had anticipated high rates of belief, but the response she found was off the charts. Though the months and years following Copenhagen marked a high point for China's climate skeptics, internally the Communist leadership was coming to a consensus that global warming warranted serious action, quietly but decisively pushing out nonbelievers. Without a big announcement of change, the subtle but steady shift in messaging had gone largely unnoticed by the rest of the world. "The leaders needed to rethink where China should go," Wang said. One of their main considerations was China's slowing economic growth and whether green technology had the potential to reverse it.

The survey revealed that most of the public did not agree with the skeptics. Climate change denial was no longer an acceptable opinion. Indeed, this seems to be the moment that China's climate skeptics vanished. The authors Liu studied stopped writing books about global warming; no new titles were published after 2011. The pockets of intense online discussion they'd inspired appeared to subside. A new perspective had taken hold.

By the time China adopted its 12th Five-Year Plan in 2011, a green strategy had begun to crystalize. The plan proposed to turn low-carbon industries into a major driver of the economy. China aimed to spend $761 billion by 2020 transitioning off fossil fuels. "It is a historical moment," Beijing-based economist Hu Angang wrote at the time, "the point at which China launches—and joins—the global green revolution and adopts a concrete plan of action for responding to climate change." Leaders in China also saw global warming as a looming threat to domestic stability. "In China's thousands of years of civilization, the conflict between humanity and nature has never been as serious as it is today," Environment Minister Zhou Shengxian said. "In China's thousands of years of civilization, the conflict between humanity and nature has never been as serious as it is today," Environment Minister Zhou Shengxian said.

China's new climate policies were accompanied by extensive state outreach, Wang says. It appears many Chinese people were receptive to the message—and the messenger. Renmin's 2012 climate survey found that 86 percent of respondents trust the central government as a source of information about global warming. Similarly high percentages trust scientific institutes and China's news media, which are largely controlled by the state. This helps explain why so many of the people Wang surveyed accept global warming.

As for the 45 percent of respondents who aren't sure if humans are to blame? Wang says it's lack of education. The 4,200 Chinese adults she and other researchers contacted came from affluent urban centers along with poorer and less-educated rural regions, whose dwellers can plainly see that droughts and extreme weather are becoming more common—even if they're unsure exactly why. "Many Chinese have at least heard about climate change because most of them experience it," she said.

Liu has tried to find out what happened to his once-buzzing hive of deniers. "I tried to hire a student to look into 'Do we have any new things coming in from this camp?' and so far nothing really," he said.

There's the possibility of a de facto censorship. Although "we can assume that the Chinese government does not actively suppress such skepticism," Liu said in a 2015 academic article, it certainly does not provide the conditions that would allow for the kind of climate denial you see in Western countries to flourish: a large network of anti-government think tanks heavily supported by oil, gas, and coal companies. ExxonMobil, for one, has spent $33 million since 1998 funding organizations like the Heartland Institute, which questions the link between humans and climate change, according to research from the publication *DeSmogBlog*. And Greenpeace estimates that Koch Industries has spent $100 million over a similar period. "In the United States, casting doubt on the human cause of climate change has been one of the major strategies of industry," said Anthony Leiserowitz, the director of the Yale program on Climate Change Communication, who helped provide academic assistance for Wang's study. "There's been a very concerted disinformation campaign."

It's possible the Chinese skeptics played a similar, if truncated, role during a period of internal debate in the government, with fossil fuel-friendly interests in the government helping to get the books published.

In China, when the party line cohered around the greener path in 2012, the space for that debate disappeared. The leadership in the state-run energy companies was largely purged during a recent anti-corruption campaign and now they would have very clear incentives not to promote denial, even if state-owned fossil fuel companies like Sinopec and China National Offshore Oil Corp. wanted to question the existence of climate change. "Top [oil and gas] executives have always been very much aware of the fact that their promotion depends on the Party," a report from the Oxford Institute for Energy Studies stated.

In the years leading up to the 2015 climate change negotiations in Paris, China's government made low-carbon growth one of its top priorities. "It's a totally different situation in China than the U.S.," Wang said.

Another reason China's skepticism receded is that climate change stopped serving the same ideological purpose. In Copenhagen, China felt attacked and humiliated by the United States and Europe. But in Paris, China worked closely with the United States to negotiate the world's first comprehensive climate agreement. Obama phoned Chinese President Xi Jinping shortly after the talks "to express appreciation for the important role China played," the White House said in December 2015. Climate change no longer made China look weak. It was now a story of China's strength. "The current leadership is really setting its sight on having China be the preeminent global power of our time," said Victor Shih, an associate professor at the University of California at San Diego and a well-known commentator on China. "[Fighting climate change] does give China the opportunity to do so."

But it doesn't mean climate skepticism in China has disappeared—not completely. When Liu and his colleague Bo Zhao, an assistant professor at Oregon State University, did an extensive search of all the Weibo posts mentioning climate change in the months before and after the Paris negotiations, they found contrarian opinions. A U.S.-educated physics researcher named Wan Weigang speculated in one of the more popular climate denier posts that, "Maybe in five years, the global warming theory will be cited as a joke." They found other statements that could have been lifted right out of the skeptic books in Liu's office: "Climate replaces guns, cannons, and warships in old times to become the weapon to constrain, oppress, and exploit poor countries."

But nobody in China's government is publicly echoing those opinions. In the media and academia, they have also all but disappeared. "It's really fringe," Liu said. "It's not mainstream." It seems likely to stay that way, too.

If climate change has been a piece of the larger game that China was playing with the West, it's possible that, almost a decade after the collapse at Copenhagen, Beijing is finally—and decisively—winning.

Shortly after Donald Trump won the presidency, Xi told him in a call that China will continue fighting climate change "whatever the circumstances." Though the new U.S. president has staffed his administration with skeptics such as Scott Pruitt, the head of the Environmental Protection Agency, China released data suggesting it could meet its 2030 Paris targets a decade early. "The financial elites I talk with," Shih said, "they think that the fact that the Trump presidency has so obviously withdrawn from any global effort to try to limit greenhouse gases provides China with an opportunity to take leadership."

The paths both countries are taking couldn't be more divergent. While Trump rescinded Obama's Clean Power Plan with a promise to end America's "war on coal," China aims to close 800 million tons of coal capacity by 2020. The U.S. Office of Energy Efficiency & Renewable Energy is facing a budget cut of more than 50 percent when China is pouring over $361 billion into renewable energy. All this "is likely to widen China's global leadership in industries of the future," concluded a recent report from the Institute for Energy Economics and Financial Analysis. Yet the United States is still the most important global actor on climate change. "All the rest of the world, including China, we are looking at Trump—what will he do?" Wang said. But no matter what happens, she added, "the green transformation for China and the world is a reality."

During the U.S. presidential campaign, reporters dredged up a 2012 tweet that sounded as if it might as well have been drafted by one of Liu's writers. "The concept of global warming was created by and for the Chinese in order to make U.S. manufacturing non-competitive," Trump wrote. But perpetuating hoaxes and plots won't win the coming fight against climate change. It'll be the ability—and the willingness—to adapt. And while China has seized onto climate change as the issue on which it could be both a technological and moral leader, the United States has taken a great leap back.

In November, the world will come together again in Bonn, Germany, for the latest United Nations conference on climate change. We'll have to wait and see who the spoiler will be this time.

Critical Thinking

1. Why were the Chinese skeptical about climate change?
2. What persuaded the Chinese to believe in climate change?
3. How is President Trump's policy on climate change affecting China?

Internet References

Center for Governance and Sustainability
http://www.umb.edu/cgs

Climate Summit 2014
http://www.un.org/climatechange/summit

UN Framework Convention on Climate Change
http://unfccc.int/2860.php

Article

Prepared by: Robert Weiner, *University of Massachusetts, Boston*

Paris Isn't Burning

Why the Climate Agreement Will Survive Trump

BRIAN DEESE

Learning Outcomes

After reading this article, you will be able to:

- Learn why the Paris Climate Agreement will survive the US withdrawal.
- Learn about the decoupling of greenhouse gas emissions and economic growth.
- Learn why China's role in maintaining the Paris Agreement is so important.

For decades, the world has understood the threat of climate change. But until recently, the economic and political obstacles to tackling the problem stymied global action. Today, that calculus has changed. Technological progress has made clean energy a profitable investment, and growing popular pressure has forced politicians to respond to the threat of ecological disaster. These trends have enabled major diplomatic breakthroughs, most notably the 2015 Paris agreement. In that pact, 195 countries pledged to make significant reductions in their greenhouse gas emissions. "We've shown what's possible when the world stands as one," proclaimed U.S. President Barack Obama after the talks concluded.

Now, however, that agreement is under threat. When it comes to climate change, U.S. President Donald Trump has replaced urgency with skepticism and threatened to pull the United States out of the Paris agreement. He has spent the early months of his presidency attempting to roll back the Obama administration's environmental regulations and promising the return of the U.S. coal industry.

The Trump administration has not yet decided whether to formally leave the Paris agreement. Whatever it decides, the agreement itself will survive. Negotiators designed it to withstand political shocks. And the economic, technological, and political forces that gave rise to it are only getting stronger. U.S. policy cannot stop these trends. But inaction from Washington on climate change will cause the United States serious economic and diplomatic pain and waste precious time in the race to save the planet. Sticking with the deal would mitigate the damage and is clearly in the U.S. national interest, but Washington's failure to otherwise lead on climate change would still hurt the United States and the world. So U.S. businesses, scientists, engineers, governors, mayors, and citizens must step forward to demonstrate that the country can still make progress and that, in the end, it will return to climate leadership.

More for Less

A decade ago, the Paris agreement could never have been negotiated successfully. Effective collective action on climate change was simply too difficult to achieve because of the vast costs involved. But since then, rapid reductions in the price of renewable energy and increases in the efficiency of energy consumption have made fighting climate change easier, and often even profitable. By the time of the Paris negotiations, the world had reached a milestone that energy analysts had previously thought was decades away: in many places, generating energy from solar or wind sources was cheaper than generating it from coal. According to research from Bloomberg New Energy Finance, in 2015, clean energy attracted twice as much investment globally as fossil fuels.

As a result, the world has adopted clean energy far faster than experts expected. Consider the projections of the International Energy Agency, the world's most respected forecaster of energy-market trends. In 2002, the agency predicted that it would take 28 years for the world to generate more than 500 terawatt-hours of wind energy; instead, it took eight. And in 2010, the agency projected that it would take until 2024 to install 180 gigawatts of solar capacity; that level was reached in 2015, almost a decade ahead of schedule.

This improbable progress has upended the once dominant assumption that economic growth and rising greenhouse gas emissions must go hand in hand. Between 2008 and 2016, the U.S. economy grew by 12 percent while carbon emissions from energy generation fell by about 11 percent—the first time the link between the two had been broken for more than a year at a time. This decoupling of emissions and economic growth has begun to occur in at least 35 countries, including China, where many believe that emissions will peak and begin to decline in the next few years, more than a decade earlier than the 2030 target China has set for itself. In fact, 2016 was the third year in a row when global emissions did not rise even as the global economy grew. Before this streak, only recessions had ever brought emissions down. This quiet shift represents a seismic change in the political economy of clean energy. Once, countries had to trade faster economic growth for reducing emissions. Now, they are racing against one another to claim the economic benefits of clean energy.

The pace of change will likely continue to outstrip projections. Technological breakthroughs in energy storage will make renewable power cheap enough to use in more places and accelerate the move to electric cars and other electric transportation systems. China plans to invest $340 billion in renewable energy sources by 2020; Saudi Arabia is investing $50 billion. In the last year alone, India doubled its solar capacity. It is installing solar panels so fast that Prime Minister Narendra Modi's audacious goal of reaching 100 gigawatts of solar capacity by 2022 no longer seems like a pipe dream.

We're All Environmentalists Now

As new technologies upend the economics of climate change, the politics surrounding the environment are changing, too. In 2008, the U.S. embassy in Beijing made a routine decision to place an air-quality monitor on its roof and tweet out the readings. It began as a way to provide information to Americans and other expats living in Beijing about how safe it was to go outside at any particular moment; most Chinese were unable to see the tweets, since China's "Great Firewall" generally blocks Twitter. But as more Chinese citizens acquired smartphones, app developers created ways for them to bypass the filter and access the air-quality updates. Beijing's middle-class residents reacted with outrage at the prospect of exposing their children to dangerous air pollution. Schools built giant domes for pupils to play in, safe from the polluted air. Many children started wearing heavy-duty masks on their way to school. The furor forced the Chinese government into action. By 2013, it had installed hundreds of air-quality monitors in over 70 of China's largest cities. That same year, the government promised to spend billions of dollars to clean up the air, and it pledged to set initial targets for reducing the emissions of air pollutants in major cities.

Meanwhile, environmental activism across the world was moving from the fringe to the mainstream. Parents in India worried that pollution from vehicles was damaging their children's health. Inhabitants of remote islands such as Kiribati anxiously watched the sea rising around them. Ranchers in the western United States saw their land ravaged by droughts and wildfires unlike any they had experienced before. Along with other alarmed citizens all over the world, they began calling on politicians to act, with louder and more unified voices than ever before.

When world leaders gathered in Paris in December 2015, they were responding to this wave of climate activism. At the conference, a group of over 100 countries that had traditionally been at odds on climate change formed the "high-ambition coalition." Propelled by grass-roots activism, they successfully demanded that the agreement adopt the ambitious goal of limiting the warming of the earth's atmosphere to 1.5 degrees Celsius. "Anything over two degrees is a death warrant for us," said Tony de Brum, then the foreign minister of the Marshall Islands, an informal leader of the coalition. The incentives for politicians to address climate change will only strengthen as more people, particularly in developing countries, leave poverty for the ranks of the middle-class and gain access to information about how climate change is directly affecting their lives and livelihoods.

This shift is already well under way. In January, in a speech that stood in stark contrast to China's previous unwillingness to accept responsibility for tackling climate change, Chinese President Xi Jinping told the World Economic Forum, in Davos, that "all signatories should stick to [the Paris agreement] instead of walking away from it, as this is a responsibility we must assume for future generations." And the day after Trump signed an executive order to begin undoing the rule known as the Clean Power Plan, which Obama had implemented to reduce emissions from power plants, the EU's climate action and energy commissioner, Miguel Arias Cañete, tweeted a picture of himself hugging China's chief climate negotiator. "A new climate era has begun, and the EU and China are ready to lead the way," the caption read.

The Art of the Deal

These economic and political forces made the Paris agreement possible, but to get the entire world to sign on, negotiators still needed to clear a major diplomatic hurdle: deciding who should do what and who should pay for it. For about two decades after the 1992 Rio Earth Summit, climate negotiations were predicated on the idea that since developed countries had been responsible for the lion's share of past greenhouse gas emissions, they should shoulder the burden of addressing global warming.

By the end of the last decade, that concept had clearly outlived its usefulness. As the world saw the economies of China and India grow rapidly, the United States and other developed countries could no longer justify to their citizens accepting limits on emissions when major emerging-market countries were doing nothing. And when China overtook the United States as the world's largest emitter of carbon dioxide in 2007, it became clear that developed countries could not solve the problem alone. Indeed, by 2040, close to 70 percent of global emissions will come from countries outside the Organization for Economic Cooperation and Development, a group of mostly developed countries.

Yet for years, governments could not agree on an alternative approach. The size of the problem meant that all would have to participate. But no country was prepared to accept a supranational body that would dictate and enforce targets and actions. The failure of the 2009 Copenhagen climate conference showed that insisting on a rigid goal would create a zero-sum game in which every country tried to do less and make others do more. The Paris agreement solved this problem by combining the ambitious goal of a universal compact with the conservative method of allowing each country to decide for itself how it could contribute to hitting the overall target.

Obama hoped that if China and the United States—the two largest emitters—bought into this approach, others would follow. To that end, he sought an agreement between the two countries well in advance of the Paris negotiations. In November 2014, in a joint announcement, the United States promised to reduce its emissions by 26–28 percent below their 2005 levels by 2025, and China pledged to cap its emissions by 2030. The deal demonstrated that countries could move beyond the old approach and created the possibility of a universal effort to reduce emissions and claim the economic spoils of a clean energy boom.

With striking speed, countries at every stage of economic development joined the race. Before the negotiations had even begun in Paris, enough countries to account for over 90 percent of global emissions had established their own targets. This meant that, unlike in Copenhagen, countries came to Paris agreeing that they would all have to reduce emissions in order to meet the challenge of climate change.

Even with these commitments in hand, the process of getting nearly 200 countries to let go of the old model was painful. Perhaps inevitably, allowing each country to determine its own way forward meant that the initial pledges were insufficient. According to a study by a group of climate scientists published in the journal *Science* in 2015, even if all countries meet their targets and global investment in clean energy technology accelerates, the world will still have only a 50 percent chance of limiting warming to two degrees Celsius, and the 1.5 degree target will remain out of reach. Nevertheless, the move from a head-to-head climate battle to a global clean energy race created the potential for collective action to accelerate progress.

More than a year later, the agreement has proved surprisingly durable. Throughout 2016 and early 2017, countries moved aggressively to reach their targets, even as world events, such as the Brexit vote and Trump's election, signaled a global shift away from multilateralism. India recently set a goal of putting six million hybrid and electric cars on its roads by 2020 and ending the sale of internal combustion vehicles in the country by 2030. Last December, Canada created a national carbon-pricing regime. In April, the United Kingdom went a full day without burning coal to generate electricity, the first time it had done so since 1882. And although most expected that it would take years for enough countries to ratify the Paris agreement for it to formally take effect, the world accomplished that goal just 11 months after the talks ended. Even OPEC has embraced the accord.

This progress suggests that the agreement's main assumption—that countries would grow more ambitious over time—was a reasonable bet. The agreement encourages governments to raise their climate targets every five years, but it imposes no binding requirements. A more stringent accord would have looked better on paper, but it might well have scared many countries away or led them to set their initial targets artificially low. Because the economic forces that gave rise to the agreement have continued to accelerate, more and more countries now see the benefits of leading in the fast-growing clean energy industries. So they will likely raise their targets to reap the rewards of staying ahead of the pack.

Self-harm

Although the Trump administration cannot halt global progress on climate change, it can still hurt the U.S. economy and the United States' diplomatic standing by abandoning the Paris agreement. On everything from counterterrorism and trade to nuclear nonproliferation and monetary policy, the Trump administration will need to work with other countries to accomplish its agenda. If it pulls the United States out of the Paris agreement, it will have a harder time winning cooperation on those issues because other countries increasingly see leadership on climate change in the same way they see security pledges, foreign assistance, or aid to refugees, as a test of a country's commitment to its promises and of its standing in the global order. When, in 2001, the Bush administration stepped away from the Kyoto Protocol, it was surprised by how harshly China, India, the EU, and many others criticized the move. Since then, the world has made dramatic progress on cooperation over climate change. So abandoning the Paris agreement would do far worse diplomatic damage.

Leaving the Paris agreement would also cause the United States to lose out to other countries, especially China, on the benefits of a clean energy boom. More than three million Americans work in the renewable energy industry or in the design, manufacture, or maintenance of energy-efficient products or clean energy vehicles, such as electric cars. Employment in the solar and wind energy industries has grown by about 20 percent each year in recent years, roughly 12 times as fast as employment in the economy as a whole. Maintaining this pace will require sustained investment and the ability for U.S. industries to capture larger shares of the growing clean energy markets abroad.

On this front, China is already starting to overtake the United States. According to data from Bloomberg New Energy Finance and the UN Environment Program, in 2015, China invested $103 billion in renewable energy; the United States invested $44 billion. China is home to five of the world's six largest solar-module manufacturers and to the world's largest manufacturers of wind turbines and lithium ion, which is used to make the batteries needed to store renewable energy.

It is likely inevitable that much of the manufacture of lower-value-added clean energy products will move away from the United States. But it is troubling that the country risks ceding ground on clean energy innovation, as well. According to a study by the public policy experts Devashree Saha and Mark Muro, the number of clean energy technology patents granted in the United States each year more than doubled between 2001 and 2014, but it declined by nine percent from 2014 to 2016. Other countries are filling the void. "In 2001, both U.S. and foreign-owned companies generated about 47 percent of [clean energy technology] patents each," Saha and Muro write. But "by 2016, 51 percent of all cleantech patents were owned by large foreign multinationals, while only 39 percent were generated by U.S. companies."

Should the Trump administration abandon the Paris agreement, these trends will likely get worse. If Washington is not part of crucial discussions as details of the agreement are finalized over the coming years, other governments could shape the rules around intellectual property, trade, and transparency in ways that would disadvantage the U.S. economy. Some countries have also suggested that if the United States leaves the Paris agreement, they would consider imposing retaliatory measures, such as import taxes. Even if they do not, with the United States outside the Paris agreement, foreign governments, international agencies, and private investors might direct funds for clean energy research, development, and deployment to U.S. competitors. China has already pledged more money than the United States has to help poorer countries develop their markets in clean energy. If the United States leaves the discussion, it will lose its influence over where and how those funds get spent. And if Washington skips future rounds of negotiations within the UN framework, an emboldened China might look for ways to water down the Paris agreement's rules on important issues, such as requiring all countries to submit their emissions plans to independent reviews.

Emission Critical

For all these reasons, the Trump administration should keep the United States in the Paris agreement. Yet that by itself will not be enough. Even inside the agreement, if Washington otherwise fails to lead on climate change, the United States will still suffer, as will the rest of the world. Without robust government investment in clean energy, and government policies that help set a stable price for greenhouse gas emissions, the U.S. economy will not see the full dividends of the transition to clean energy.

A lack of U.S. leadership will not just hurt the United States; it will cost the world valuable time. Rising temperatures are outpacing efforts to cut emissions. Last year was the hottest on record, the third year in a row to earn that distinction. Sea ice in the Arctic and around Antarctica has reached record lows. And the pace of extreme weather events is accelerating across the United States and the rest of the world.

To reverse these trends, countries need to move to decarbonize their economies even faster. Although other countries will move forward with the Paris agreement even without the United States, getting them to dramatically raise their targets without U.S. leadership will be difficult. China and the EU will continue to compete in the clean energy race, but only the United States has the political clout and resources to spur other countries to action, the way it did before the Paris negotiations. A U.S. president using every possible diplomatic tool at his or her disposal—as Obama did—can bring remarkable results.

Because the Paris agreement calls on most countries to set their next round of national targets after 2020, much will hinge on the next U.S. presidential election. If the next U.S. administration restores U.S. leadership on climate change, it might be able to make up for lost time.

Meanwhile, not all progress in the United States will stall. Several states, including California, Nevada, New York, and Virginia, are cutting their greenhouse gas emissions and seeing the economic benefits firsthand. They should raise their sights even further and remind the world of the collective impact of their efforts. Major U.S. cities are finding novel ways to go green. They should explore new partnerships with foreign counterparts. American companies will need to speak loudly and clearly about the economic benefits of a credible plan to reduce emissions and the costs of ceding leadership on clean energy to China and other countries. American scientists and engineers are poised to transform several technologies crucial to tackling climate change, such as batteries and those used for carbon capture and storage. Engineers could exploit recent breakthroughs in satellite technology, for example, to create a real-time global emissions monitoring system that could settle disputes between countries over the extent of their past progress and allow diplomats to focus on the future. And concerned citizens must continue to organize, march, and convey to politicians that ensuring clean air and water in their communities is a requirement for their votes.

The Paris agreement represents real progress, but it alone will not solve the climate crisis. Its significance lies primarily in the economic, technological, and political shifts that drove it and the foundation for future action it laid. Its negotiators made it flexible enough to withstand political changes and policy differences while betting that the global movement toward cleaner energy would continue to accelerate. The road may not always be smooth, but in the end, that bet looks likely to pay off.

Critical Thinking

1. Why has it become cheaper to use renewable energy?
2. Why will China lead the way in the fight against warming?
3. How was the breakthrough of the Paris Agreement made possible?

Internet References

Center for Governance and Sustainability
http://www.umb.edu/cgs

Climate Summit 2014
http://www.un.org/climatechange/summit

UN Framework Convention on Climate Change
http://unfccc.int/2860.php

Brian Deese is a senior fellow at the Mossavar-Rahmani Center for Business and Government at the Harvard Kennedy School of Government. From 2015 to 2017, he was a Senior Adviser to U.S. President Barack Obama.

Article

Prepared by: Robert Weiner, *University of Massachusetts, Boston*

Why Climate Change Matters More Than Anything Else

JOSHUA BUSBY

Learning Outcomes

After reading this article, you will be able to:

- Learn how climate change will affect geopolitics.
- Learn how state and non-state actors can cooperate to deal with climate change.

The world seems to be in a state of permanent crisis. The liberal international order is besieged from within and without. Democracy is in decline. A lackluster economic recovery has failed to significantly raise incomes for most people in the West. A rising China is threatening U.S. dominance, and resurgent international tensions are increasing the risk of a catastrophic war.

Yet there is one threat that is as likely as any of these to define this century: climate change. The disruption to the earth's climate will ultimately command more attention and resources and have a greater influence on the global economy and international relations than other forces visible in the world today. Climate change will cease to be a faraway threat and become one whose effects require immediate action.

The atmospheric concentration of carbon dioxide, the main greenhouse gas, now exceeds 410 parts per million, the highest level in 800,000 years. Global average surface temperatures are 1.2 degrees Celsius higher than they were before the Industrial Revolution. The consensus scientific estimate is that the maximum temperature increase that will avoid dangerous climate change is two degrees Celsius. Humanity still has around 20 years before stopping short of that threshold will become essentially impossible, but most plausible projections show that the world will exceed it.

Two degrees of warming is still something of an arbitrary level; there is no guarantee of the precise effects of any temperature change. But there is a huge difference between two degrees of warming and two and a half, three, or four degrees. Failing to rein in global emissions will lead to unpleasant surprises. As temperatures rise, the distribution of climate phenomena will shift. Floods that used to happen once in a 100 years will occur every 50 or every 20. The tail risks will become more extreme, making events such as the 50 inches of rain that fell in 24 hours in Hawaii earlier this year more common.

Making climate change all the more frightening are its effects on geopolitics. New weather patterns will trigger social and economic upheaval. Rising seas, dying farmlands, and ever more powerful storms and floods will render some countries uninhabitable. These changes will test the international system in new and unpredictable ways.

World-historical threats call for world-historical levels of cooperation. If humanity successfully confronts this problem, it will be because leaders infused the global order with a sense of common purpose and recognized profound changes in the distribution of power. China and the United States will have to work closely together, and other actors, such as subnational governments, private companies, and nongovernmental organizations, will all have to play their part.

A Matter of Degree

The effects of climate change are starting to make themselves apparent. Of the 17 warmest years on record, 16 have occurred since 2001. This past winter, temperatures in parts of the Arctic jumped to 25 degrees Celsius above normal. And climate change means far more than a warming planet. The world is entering a period that the climate scientist Katharine Hayhoe has called "global weirding." Strange weather patterns are cropping up everywhere. Scientists have linked some of them to climate change; for others, whether there is a connection is not yet clear.

The seasons are changing. Dry spells are occurring when meteorologists would normally expect rain. Lack of rain increases the risk of forest fires, such as those that occurred in California last year. When it does rain, too often it is all at once, as happened in Houston during Hurricane Harvey. As sea levels rise and storm surges get stronger, what were once normal high-tide events will flood coastal infrastructure, as has already happened in Miami in recent years, necessitating the installation of storm water pumping systems at the cost of hundreds of millions of dollars.

By the middle of the century, the oceans may well have risen enough that salt water will destroy farmland and contaminate drinking water in many low-lying island nations, making them uninhabitable long before they are actually submerged. The evidence on the effects of climate change on tropical cyclones and hurricanes is murkier, but it suggests that although there may be fewer such storms, those that do occur are likely to be worse.

These developments will fundamentally transform global politics. Several major countries, including China and the United States, have large populations and valuable infrastructure that are vulnerable to climate change. Their governments will find themselves diverting military resources to carry out rescue operations and rebuild devastated towns and cities. That will take large numbers of soldiers and military hardware away from preparing for conflicts with foreign adversaries.

In 2017, when three huge storms battered the United States in quick succession, civilian disaster authorities had to be backstopped by the military to prevent huge losses of life. Tens of thousands of members of the National Guard were mobilized to rescue people, provide relief supplies, and restore essential services and the rule of law. The third storm, Hurricane Maria, caused some 1,000 deaths and left the entire island of Puerto Rico without power. It took months for the government to restore electricity to the 3.5 million Americans who live there. Even now, some remain without power. In the wake of the storm, over 100,000 Puerto Ricans left for the continental United States. The total cost to the United States of these storms and other weather-related emergencies in 2017 was $300 billion.

China has its own set of problems. On its southern coast, several huge cities, such as Guangzhou and Shanghai, are vulnerable to flooding. In the north, in the country's industrial heartland, whole regions are running out of water, affecting more than 500 million people. Over the past 25 years, some 28,000 Chinese rivers have disappeared.

Solving these problems will not be cheap. A single ambitious infrastructure project to transport water from the south to the north has already cost the Chinese government at least $48 billion.

The project is not yet complete, but China claims that it has improved Beijing's water security and benefited 50 million people. To deal with flooding in places such as Shanghai, China has embarked on a "sponge cities" initiative to boost natural drainage. Since 2015, China has invested $12 billion in this effort, and the price tag will ultimately run into the hundreds of billions of dollars.

Both China and the United States are rich enough that they will likely be able to cope with these costs. But the effects of climate change in poorer countries will create global problems. Each year, the monsoon brings floods to the Indus River in Pakistan. But in 2010, the flooding took on epic proportions, displacing as many as 20 million people and killing nearly 2,000. The United States provided $390 million in immediate relief funding, and the U.S. military delivered some 20 million pounds of supplies. In 2013, over 13,000 U.S. troops were deployed for disaster relief after Typhoon Haiyan buffeted the Philippines.

Individual storms do tremendous damage, but communities usually bounce back. Climate change, however, will cause more permanent problems. Rising sea levels, the storm surges they exacerbate, and the intrusion of salt water pose existential threats to some island countries. In 2017, after Hurricane Irma hit Barbuda, the entire population of the Caribbean island—some 1,800 people—had to be evacuated. Kiribati, a collection of Pacific islands, most of which rise only a few meters above sea level, has purchased land in neighboring Fiji as a last resort in the face of rising seas.

Even as some countries are inundated by water, others are suffering from a lack of it. In recent years, droughts in both the Horn of Africa and the continent's southern countries have put millions at risk of thirst or famine. In 2011, Somalia, already riven by decades of war, experienced a drought and subsequent famine that led to as many as 260,000 deaths. Earlier this year, Cape Town, South Africa, a city of nearly four million people, was able to avoid running out of water only through heroic conservation measures. Climate change, through rising temperatures and shifting rainfall patterns, will subject some regions to inadequate and irregular rains, leading to harvest failures and insufficient water for human needs.

Since 1945, although some states have split or otherwise failed, very few have disappeared. In the coming century, climate change may make state deaths a familiar phenomenon as saltwater intrusion and storm surges render a number of island countries uninhabitable. Although most of the islands threatened by climate change have small populations, the disorder will not be contained. Even in other countries, declining agricultural productivity and other climate risks will compel people to move from the countryside to the cities or even across borders. Tens of thousands of people will have to be relocated. For those that cross borders, will they stay permanently, and will they become citizens of the countries that take them in? Will governments that acquire territory inside other countries gain sovereignty over that land? New Zealand has taken tentative steps toward creating a new visa category for small numbers of climate refugees from Pacific island states, but there are no international rules governing those forced to leave home by climate change. The urgency of these questions will only grow in the coming years.

As well as creating new crises, climate factors will exacerbate existing ones. Some 800,000 of Myanmar's Rohingya minority group have fled to Bangladesh, driven out by ethnic cleansing. Many of the refugee camps they now occupy are in areas prone to flash floods during the monsoon. To make matters worse, much of the land surrounding the camps has been stripped of its forest cover, leaving tents and huts vulnerable to being washed away. Although the world has gotten much better at preventing loss of life from weather emergencies, climate change will test humanitarian- and disaster-response systems that are already stretched thin by the seemingly endless conflicts in Somalia, South Sudan, Syria, and Yemen.

Climate Wars

Climate change will also make international tensions more severe. Analysts have periodically warned of impending water wars, but thus far, countries have been able to work out most disputes peacefully. India and Pakistan, for example, both draw a great deal of water from the Indus River, which crosses disputed territory. But although the two countries have fought several wars with each other, they have never come to blows over water sharing, thanks to the 1960 Indus Waters Treaty, which provides a mechanism for them to manage the river together. Yet higher demand and increasing scarcity have raised tensions over the Indus. India's efforts to build dams upstream have been challenged by Pakistan, and in 2016, amid political tensions, Indian Prime Minister Narendra Modi temporarily suspended India's participation in joint meetings to manage the river. Peaceful cooperation will be harder in the future.

Partnerships among other countries that share river basins are even more fragile. Several Southeast Asian countries cooperate over the Mekong River through the Mekong River Commission, but China, the largest of the six countries through which the river flows and where the river originates, is not a member. The Chinese government and other upstream countries have built dams on the Mekong that threaten to deprive fishing and agricultural communities in Vietnam and other downstream countries of their livelihoods. Competition over the river's flow has only gotten worse as droughts in the region have become more frequent.

Similar dynamics are at play on the Nile. Ethiopia is building a vast dam on the river for irrigation and to generate power, a move that will reduce the river's flow in Egypt and Sudan. Until now, Egypt has enjoyed disproportionate rights to the Nile (a colonial-era legacy), but that is set to end, requiring delicate negotiations over water sharing and how quickly Ethiopia will fill the reservoir behind the dam.

Violence is far from inevitable, but tensions over water within and between countries will create new flash points in regions where other resources are scarce and institutional guardrails are weak or missing.

The ways countries respond to the effects of climate change may sometimes prove more consequential than the effects themselves. In 2010, for example, after a drought destroyed about one-fifth of Russia's wheat harvest, the Russian government banned grain exports. That move, along with production declines in Argentina and Australia, which were also affected by drought, caused global grain prices to spike. Those price rises may have helped destabilize some already fragile countries. In Egypt, for example, annual food-price inflation hit 19 percent in early 2011, fueling the protests that toppled President Hosni Mubarak.

State responses to other climate phenomena have also heightened tensions. Melting sea ice in the Arctic has opened up new lanes for shipping and fields for oil and gas exploration, leading Canada, Russia, the United States, and other Arctic nations to bicker over the rights to control these new resources.

Moreover, the push to reduce carbon emissions, although welcome, could also drive competition. As demand for clean energy grows, countries will spar over subsidies and tariffs as each tries to shore up its position in the new green economy. China's aggressive subsidies for its solar power industry have triggered a backlash from the makers of solar panels in other countries, with the United States imposing tariffs in 2017 and India considering doing something similar.

As climate fears intensify, debates between countries will become sharper and more explicit. Since manufacturing the batteries used in electric cars requires rare minerals, such as cobalt, lithium, and nickel, which are found largely in conflict-ridden places such as the Democratic Republic of the Congo, the rise of battery-powered vehicles could prompt a dangerous new scramble for resources. Although manufacturers will innovate to reduce their dependence on these minerals, such pressures will become more common as the clean energy transition progresses. Companies and countries that depend heavily on fossil fuels, for example, will resist pressure to keep them in the ground.

There are myriad potentially contentious policies governments might enact in response to changing climate conditions. Banning exports of newly scarce resources, acquiring land overseas, mandating the use of biofuels, enacting rules to conserve forests, and a thousand other choices will all create winners and losers and inflame domestic and international tensions. As fears grow of runaway climate change, governments will be increasingly tempted to take drastic unilateral steps, such as geoengineering, which would prove immensely destabilizing.

The Burning Question

These scary scenarios are not inevitable, but much depends on whether and how countries come together to curb carbon emissions and stave off the worst effects of climate change.

Last year, when U.S. President Donald Trump announced his intention to withdraw the United States from the Paris climate agreement, many other countries, including China, France, Germany, India, and the United Kingdom, responded by doubling down on their support for the deal. French President Emmanuel Macron hosted an international meeting on climate change last December and even set up a fund to attract leading climate scientists, especially those from the United States, to France.

Climate change will remain a salient issue for politicians in most countries as people around the world expect action from their leaders. Even the United States is formally still in the Paris agreement; its withdrawal only takes effect the day after the next presidential election, in 2020. Should Trump not be reelected, the next president could have the country jump right back in.

Moreover, even as the U.S. federal government has stepped away from international climate leadership and begun to roll back Obama-era domestic climate policies, U.S. governors, mayors, and chief executives have remained committed to climate action. Last year, former New York Mayor Michael Bloomberg formed the We Are Still In coalition, which now includes some 2,700 leaders across the country who have pledged action on climate change that would, if fulfilled, meet 60 percent of the original U.S. emission-reduction target under the Paris agreement.

The coalition includes California Governor Jerry Brown, whose state boasts the world's fifth-largest economy. In September, to create momentum for action before next winter's climate negotiations in Poland, Brown is scheduled to host the Global Climate Action Summit in San Francisco. That will be a remarkable spectacle: a sitting governor carrying out his own global diplomacy independent from the federal government. California's contribution does not end there. Leading technology companies based in California, such as Google, are also part of the coalition. They have set ambitious internal renewable energy targets covering their entire operations. Given their vast size and global supply chains, these companies have enormous potential reach.

Even as leaders have invested time and energy in international agreements between countries, they have built parallel, less showy, but no less important processes to encourage action. Because climate change encompasses a constellation of problems in transportation, energy, construction, agriculture, and other sectors, experimentation allows different venues to tackle different problems at the same time—the security implications in the UN Security Council, fossil fuel subsidies in the G-20, short-lived gases such as hydrofluorocarbons through the Montreal Protocol, and deforestation through efforts such as the New York Declaration on Forests, for example. This collection of efforts may be messier than centralizing everything through one global agreement, but avoiding a single point of failure and letting different groups and deals tackle the problems they are best suited to fix may produce more durable results.

Humans have proved highly adaptable, but the collective effects of climate change on cities, food production, and water supplies present an enormous challenge for the planet. China and the United States will be central to the global response. Together, the two countries are responsible for more than 40 percent of global emissions; China alone accounts for 28 percent.

In the lead-up to the Paris negotiations, U.S. President Barack Obama invested enormous political capital to come to a bilateral understanding with China. The Trump administration's backsliding on climate action elevates the pressure on China to both address its emissions at home and consider the environmental effects of its actions abroad through the Belt and Road Initiative and the Asian Infrastructure Investment Bank.

Relations between China and the United States have soured recently, but the countries need to work together, as the world will be ill served by an all-encompassing rivalry between them. They will have to build a system that allows issues to be compartmentalized, in which they can jockey over regional security in Asia, for instance, but still cooperate on issues on which their fates are linked, such as climate change and pandemics.

The only way of achieving that is through a system that recognizes the diffusion of power. To some extent, that diffusion is already under way, as the United States is ceding hegemonic control in an increasingly multipolar world, in which more is expected of a rising China. But the process will have to go much further. Governments will need to coordinate with subnational units, private corporations, nongovernmental organizations, and very rich individuals. On climate change and many other problems, these actors are much better able than governments to change things at the local level. Creating an order fit for purpose will not be easy. But the nascent combination of international agreements and networks of organizations and people dedicated to solving specific problems offers the best chance to avoid cataclysmic climate change.

Critical Thinking

1. Why will some states disappear?
2. What is meant by climate refugees?
3. Will climate change lead to water wars? Where could water wars take place?

Internet References

Center for Governance and Sustainability
http://www.umb.edu/cgs

Climate Summit 2014
http://www.un.org/climatechange/summit

UN Framework Convention on Climate Change
http://unfccc.int/2860.php

JOSHUA BUSBY is an associate professor of Public Affairs at the University of Texas at Austin.

Article

Prepared by: Robert Weiner, *University of Massachusetts, Boston*

The Clean Energy Revolution

Fighting Climate Change with Innovation

VARUN SIVARAM AND TERYN NORRIS

Learning Outcomes

After reading this article, you will be able to:

- Understand the importance of R&D for innovative clean energy technology
- Explain the advantage of clean energy technology as compared to fossil fuels

As the UN Climate Change Conference in Paris came to a close in December 2015, foreign ministers from around the world raised their arms in triumph. Indeed, there was more to celebrate in Paris than at any prior climate summit. Before the conference, over 180 countries had submitted detailed plans to curb their greenhouse gas emissions. And after two weeks of intense negotiation, 195 countries agreed to submit new, stronger plans every five years.

But without major advances in clean energy technology, the Paris agreement might lead countries to offer only modest improvements in their future climate plans. That will not be enough. Even if they fulfill their existing pledges, the earth will likely warm by some 2.7°C to 3.5°C—risking planetary catastrophe. And cutting emissions much more is a political nonstarter, especially in developing countries such as India, where policymakers must choose between powering economic growth and phasing out dirty fossil fuels. As long as this trade-off persists, diplomats will come to climate conferences with their hands tied.

It was only on the sidelines of the summit, in fact, that Paris delivered good news on the technology front. Bill Gates unveiled the Breakthrough Energy Coalition, a group of more than two dozen wealthy sponsors that plan to pool investments in early stage clean energy technology companies. And U.S. President Barack Obama announced Mission Innovation, an agreement among 20 countries—including the world's top three emitters, China, the United States, and India—to double public funding for clean energy R&D to $20 billion annually by 2020. Washington will make or break this pledge, since over half of the target will come from doubling the U.S. government's current $6.4 billion yearly budget.

Fighting climate change successfully will certainly require sensible government policies to level the economic playing field between clean and dirty energy, such as putting a price on carbon dioxide emissions. But it will also require policies that encourage investment in new clean energy technology, which even a level playing field may not generate on its own. That will take leadership from the United States, the only country with the requisite innovative capacity. In the past, the United States has seen investment in clean energy innovation surge forward, only to collapse afterward. To prevent this from happening again, the government should dramatically ramp up its support for private and public R&D at home and abroad. The task is daunting, to be sure, but so are the risks of inaction.

Don't Stop Thinking about Tomorrow

The key to a low-carbon future lies in electric power. Improvements in that sector are important not just because electric power accounts for the largest share of carbon dioxide emissions but also because reaping the benefits of innovations downstream—such as electric vehicles—requires a clean electricity supply upstream. Fossil-fueled power plants now account for nearly 70 percent of electricity globally. But by 2050, the International Energy Agency has warned, this figure must plummet to seven percent just to give the world a 50 percent chance of limiting global warming to two degrees Celsius. More fossil-fueled power is acceptable only if the carbon emissions can be captured and stored underground. And zero-carbon power sources, such as solar, wind, hydroelectric, and nuclear power, will need to grow rapidly, to the point where they supply most of the world's electricity by the middle of the century.

The problem, however, is that the clean technologies now making progress on the margins of the fossil-fueled world may not suffice in a world dominated by clean energy. The costs of solar and wind power, for example, are falling closer to those of natural gas and coal in the United States, but this has been possible because of flexible fossil fuel generators, which smooth out the highly variable power produced by the sun and wind. Ramping up the supply of these intermittent sources will oversupply the electrical grid at certain times, making renewable power less valuable and requiring extreme swings in the dwindling output of fossil fuel generators. Nuclear and hydroelectric power, for their part, are more reliable, but both have run into stiff environmental opposition. As a result, trying to create a zero-carbon power grid with only existing technologies would be expensive, complicated, and unpopular.

Similarly, cleaning up the transportation sector will require great technological leaps forward. Alternative fuels are barely competitive when oil prices are high, and in the coming decades, if climate policies succeed in reducing the demand for oil, its price will fall, making it even harder for alternative fuels to compete. The recent plunge in oil prices may offer a mere foretaste of problems to come: it has already put biofuel companies out of business and lured consumers away from electric vehicles.

All of this means that a clean, affordable, and reliable global energy system will require a diverse portfolio of low-carbon technologies superior to existing options. Nuclear, coal, and natural-gas generators will still be necessary to supply predictable power. But new reactor designs could make nuclear meltdowns physically impossible, and nanoengineered membranes could block carbon emissions in fossil-fueled power plants. Solar coatings as cheap as wallpaper could enable buildings to generate more power than they consume. And advanced storage technologies—from energy-dense batteries to catalysts that harness sunlight to split water and create hydrogen fuel—could stabilize grids and power vehicles. The wish list goes on: new ways to tap previously inaccessible reservoirs of geothermal energy, biofuels that don't compete with food crops, and ultra-efficient equipment to heat and cool buildings.

Trying to create a zerocarbon power grid with only existing technologies would be expensive, complicated, and unpopular.

Every one of those advances is possible, but most need a fundamental breakthrough in the lab or a first-of-its-kind demonstration project in the field. For example, the quest for the ideal catalyst to use sunlight to split water still hasn't produced a winning chemical, and an efficient solar power coating called "perovskite" still isn't ready for widespread use. So it is alarming that from 2007 to 2014, even as global financial flows to deploy mature clean energy doubled to $288 billion, private investment in early stage companies sank by nearly 50 percent, to less than $2.6 billion. But the United States can reverse that trend.

Third Time's the Charm?

Since the development of civilian nuclear power after World War II, the United States has experienced two booms in clean energy innovation, followed by two busts. The first boom, a response to the oil shocks of the 1970s, was driven by public investment. From 1973 to 1980, the federal government quadrupled investment in energy R&D, funding major improvements in both renewable and fossil fuel energy sources. But when the price of oil collapsed in the 1980s, the administration of President Ronald Reagan urged Congress to leave energy investment decisions to market forces. Congress acquiesced, slashing energy R&D funding by more than 50 percent over Reagan's two terms.

The second wave of investment in clean energy innovation began with the private sector. Soon after the turn of the millennium, venture capital investors began pumping money into U.S. clean energy start-ups. Venture capital investment in the sector grew tenfold, from roughly $460 million per year in 2001 to over $5 billion by 2010. Thanks to Obama's stimulus package, federal funding soon followed, and from 2009 to 2011, the government plowed over $100 billion into the sector through a mix of grants, loans, and tax incentives (although most of this influx subsidized the deployment of existing technologies). Some of the start-ups from this period became successful publicly traded companies, including the electric car maker Tesla, the solar-panel installer SolarCity, and the software provider Opower.

But the vast majority failed, and the surviving ones returned too little to make up for the losses. Indeed, of the $36 billion that venture capital firms invested from 2004 to 2014, up to half may ultimately be lost. The gold rush ended abruptly: from 2010 to 2014, venture capital firms cut their clean energy investment portfolios by 75 percent. And the federal government, reeling from political blowback over the bankruptcies of some recipients of federal loan guarantees (most famously, the solar-panel manufacturer Solyndra), pared back its support for risky ventures, too.

Yet all was not lost, for the failures of these two waves offer lessons for how to make sure the next one proves more enduring. First, they revealed just how important government funding is: after the drop in federal energy R&D in the 1980s, patent filings involving solar, wind, and nuclear power plunged. Today, although the United States is the largest funder of energy R&D in the world, it chronically underspends compared with its investments in other national research priorities. Its $6.4 billion clean energy R&D budget is just a fraction of the amount spent on space exploration ($13 billion), medicine ($31 billion), and defense ($78 billion). Given the gap, Congress should follow through on the Mission Innovation pledge and at least double funding for clean energy R&D. Already, Congress increased spending on applied energy R&D by 10 percent in its 2016 budget, more than it increased spending on any other major R&D agency or program. But starting in 2017, doubling the budget in five years will require annual increases of at least 15 percent.

The second lesson is that the government should fund not only basic research but applied research and demonstration projects, too. Washington's bias goes back decades. In his seminal 1945 report, *Science, the Endless Frontier*, Vannevar Bush, President Franklin Roosevelt's top science adviser, urged the government to focus on basic research, which would generate insights that the private sector was supposed to translate into commercial technologies. Successive administrations mostly heeded his advice, and Reagan doubled down on it, slashing nearly all funding for applied energy R&D. By the late 1990s, basic research would account for 60 percent of all federal spending on energy R&D. Instead of creating space for the private sector to pick up where the government left off, however, the budget cuts scared it away. Private investment shrank by half from 1985 to 1995, stranding public investments in alternative fuels, solar photovoltaic panels, and advanced nuclear reactors.

A similar story unfolded at the end of the second boom in clean energy innovation. When one-time stimulus funding expired after 2011, public funding for demonstration projects—which prove whether new technologies work in real-world conditions—fell by over 90 percent. Private investors had expected to share the risk of such projects with the federal government, but when government funding evaporated, investors pulled their money out—canceling, among others, several projects to capture and store carbon emissions from coal power plants.

Thus, policymakers should increase the kind of public investment that attracts private capital. To that end, the first priority should be to restore public funding for demonstration projects. The last redoubt of support for these projects can be found in the Department of Energy's politically embattled loan guarantee program. To insulate funding from political caprice, the American Energy Innovation Council, a group of business leaders, has proposed an independent, federally chartered corporation that would finance demonstration projects. Others have proposed empowering states or regions to fund their own projects, with matching federal grants. If they make it past Congress, both proposals could unlock considerable private investment.

The Department of Energy has made more progress in supporting technologies not yet mature enough for demonstration. In 2009, with inspiration from the Defense Advanced Research Projects Agency (DARPA), the U.S. military's incubator for high-risk technologies, it created the Advanced Research Projects Agency–Energy (ARPA-E). Several ARPA-E projects have already attracted follow-on investment from the private sector. In 2013, for example, Google acquired Makani Power, a start-up that is developing a kite that converts high-altitude wind energy into power. The department has also curated public–private partnerships among the government, academics, and companies—dubbed "innovation hubs"—to develop advanced technologies. Obama has advocated tripling ARPA-E's budget to $1 billion by 2021 and creating ten new public–private research centers around the country. Congress should approve these proposals.

The Department of Energy should expand its support for one type of public–private partnership in particular: industrial consortia that pool resources to pursue shared research priorities. Once again, DARPA provides a model. In the 1980s, it helped fund a consortium of computer chip manufacturers called SEMATECH, through which the industry invested in shared R&D and technical standards. By the next decade, the United States had regained market leadership from Japan. Clean energy innovation, by contrast, suffers from corporate apathy. From 2006 to 2014, U.S. firms spent a paltry $3 billion per year on in-house clean energy R&D. They were also reluctant to outsource their energy R&D, acquiring clean energy start-ups only half as often as they did biomedical start-ups.

Public–private partnerships should help diversify the set of private investors funding clean energy innovation. Indeed, venture capitalists alone are insufficient, since clean energy investments require capital for periods longer than venture capitalists generally favor. The Breakthrough Energy Coalition may help solve that problem by infusing the sector with more patient capital. Gates has explained that he and his fellow investors would be willing to wait for years, even decades, for returns on their investments. But his vision depends on the government also ramping up support.

Past failures offer a third and final lesson for policymakers: the need to level the playing field on which emerging clean energy technologies compete against existing ones. In the electricity sector in particular, innovative start-ups are at a disadvantage, since they lack early adopters willing to pay a premium for new products. The biggest customers, electric utilities, tend to be highly regulated territorial monopolies that have little tolerance for risk and spend extremely little on R&D (usually 0.1 percent of total revenues). New York and California are reforming their regulations to encourage utilities to adopt new technologies faster; the federal government should support these efforts financially or, at the very least, get out of the way.

Indeed, government intervention can sometimes be counterproductive. Many current clean energy policies, such as state mandates for utilities to obtain a certain percentage of their power from renewable energy and federal tax credits for solar and wind power installations, implicitly support already-mature technologies. Better policies might carve out allotments or offer prizes for emerging technologies that cost more now but could deliver lower costs and higher performance later. The government could even become a customer itself. The military, for example, might buy early stage technologies such as flexible solar panels, energy-dense batteries, or small modular nuclear reactors.

Innovating Abroad

Clean energy innovation at the international level suffers from similar problems. Like Washington, other governments spend too little on R&D, with the share of all publicly funded R&D in clean energy falling from 11 percent in the early 1980s to four percent in 2015. Thanks to Mission Innovation, that trend could soon be reversed. But if spending rises in an uncoordinated way, governments may duplicate some areas of research and omit others.

Since governments prize their autonomy, the wrong way to solve this problem would be through a centralized, top–down process to direct each country's research priorities. Instead, an existing institution should coordinate spending through a bottom-up approach. The most logical body for that task is the Clean Energy Ministerial, a global forum conceived by the Obama administration that brings together energy officials from nearly every Mission Innovation country. Yet the CEM has no permanent staff, and without support from the next U.S. administration, it might disband. The Obama administration should therefore act quickly to convince its Mission Innovation partners to help fund a permanent secretariat and operating budget for the CEM. Once that happens, the body could issue an annual report of each member's R&D expenditures, which countries could use to hold their peers accountable for their pledges to double funding. The CEM could also convene officials to share trends about the frontiers of applied research, gleaned from grant applications submitted to national funding bodies.

Then there is the problem of foreign companies' aversion to investing in innovation. Producers of everything from solar panels to batteries, mostly in Asia, have focused instead on ruthless cost cutting and in many cases have taken advantage of government assistance to build up massive manufacturing capacity to churn out well-understood technologies. Today, over 2/3 of solar panels are produced in China, where most firms spend less than one percent of their revenue on R&D. (In fact, it was largely the influx of cheap, cookie-cutter solar panels from China that caused U.S. solar start-ups to go bankrupt at the beginning of this decade.)

Not only does this global race to the bottom stunt clean energy innovation; it also matches up poorly with the United States' competitive strengths. In other industries, leading U.S. firms generate economic gains both at home and abroad by investing heavily in R&D. In the electronics, semiconductor, and biomedical industries, for instance, U.S. companies reinvest up to 20 percent of their revenues in R&D.

To encourage foreign companies to invest more in clean energy R&D, the United States should embrace public–private collaboration. A good model is the U.S.–China Clean Energy Research Center (CERC), which was set up in 2009 and is funded by the U.S. and Chinese governments, academic institutions, and private corporations. Notably, CERC removes a major obstacle to international collaboration: intellectual property theft. Participants are bound by clear rules about the ownership and licensing of technologies invented through CERC. And unless they agree otherwise, they must submit disputes to international arbitration governed by UN rules. More than 100 firms have signed on, and in 2014, China and the United States enthusiastically extended the partnership. It's time for the United States to apply CERC's intellectual property framework to collaborations with other countries, such as India, with which it has no such agreement.

The Next Revolution

By investing at home and leading a technology push abroad, the United States would give clean energy innovation a badly needed boost. Energy executives would at last rub elbows with top academics at technology conferences. Industrial consortia would offer road maps for dramatic technological improvements that forecast future breakthroughs. And institutional investors would bet on start-ups and agree to wait a decade or more before seeing a return.

To many in Washington, this sounds like an expensive fantasy. And indeed, transforming the energy sector into an innovative powerhouse would prove even harder and costlier than the Manhattan Project or the Apollo mission. In both cases, the government spent billions of dollars on a specific goal, whereas success in clean energy innovation requires both public and private investment in a wide range of technologies.

Yet the United States has achieved similar transformations before. Take the biomedical industry. Like clean energy start-ups, biomedical start-ups endured boom-and-bust investment cycles in the 1980s and 1990s. But today, partly thanks to high and sustained public funding, the private sector invests extensively in biomedical innovation. One might object that the biomedical industry's high profit margins, in contrast to the slim ones that characterize the clean energy industry, allow it to invest more in R&D. But the clean energy sector need not be condemned to permanently small profits: innovative firms could earn higher margins than today's commodity producers by developing new products that serve unmet demands.

With clean energy, the stakes could hardly be higher. If the world is to avoid climate calamity, it needs to reduce its carbon emissions by 80 percent by the middle of this century—a target that is simply out of reach with existing technology. But armed with a more potent low-carbon arsenal, countries could make pledges to cut emissions that were both ambitious and realistic. Emerging economies would no longer face tradeoffs between curbing noxious fossil fuels and lifting their populations out of energy poverty. And the United States would place itself at the forefront of the next technological revolution.

Critical Thinking

1. Why did the first two waves of investment in clean energy technology fail?
2. What are the advantages of private/public partnership in the development of clean energy?
3. Why should the U.S. encourage companies to invest more in clean energy?

Internet References

Climate links
www.climatelinks.org

Climate summit
http://www.un/org/climatechange/summit

UN Framework Convention on Climate Change
http://unfcc.int/2860.php

Varun Sivaram is a Douglas Dillon fellow at the Council on Foreign Relations.

Teryn Norris is a former special adviser at the U.S. Department of Energy.

Article

Prepared by: Robert Weiner, *University of Massachusetts, Boston*

Can Planet Earth Feed 10 Billion People?

Humanity Has 30 Years to Find Out

CHARLES C. MANN

Learning Outcomes

After reading this article, you will be able to:

- Learn about the relationship between population growth and food supply.
- Learn about the effects of the Green Revolution on Economic Development.

ALL PARENTS REMEMBER the moment when they first held their children—the tiny crumpled face, an entire new person, emerging from the hospital blanket. I extended my hands and took my daughter in my arms. I was so overwhelmed that I could hardly think.

Afterward I wandered outside so that mother and child could rest. It was three in the morning, late February in New England. There was ice on the sidewalk and a cold drizzle in the air. As I stepped from the curb, a thought popped into my head: *When my daughter is my age, almost 10 billion people will be walking the Earth.* I stopped midstride. I thought, *How is* that *going to work?*

In 1970, when I was in high school, about one out of every four people was hungry—"undernourished," to use the term preferred today by the United Nations. Today the proportion has fallen to roughly one out of 10. In those four-plus decades, the global average life span has, astoundingly, risen by more than 11 years; most of the increase occurred in poor places. Hundreds of millions of people in Asia, Latin America, and Africa have lifted themselves from destitution into something like the middle class. This enrichment has not occurred evenly or equitably: Millions upon millions are not prosperous. Still, nothing like this surge of well-being has ever happened before. No one knows whether the rise can continue, or whether our current affluence can be sustained.

Today the world has about 7.6 billion inhabitants. Most demographers believe that by about 2050, that number will reach 10 billion or a bit less. Around this time, our population will probably begin to level off. As a species, we will be at about "replacement level": On average, each couple will have just enough children to replace themselves. All the while, economists say, the world's development should continue, however unevenly. The implication is that when my daughter is my age, a sizable percentage of the world's 10 billion people will be middle-class.

Like other parents, I want my children to be comfortable in their adult lives. But in the hospital parking lot, this suddenly seemed unlikely. *Ten billion mouths*, I thought. Three billion more middle-class appetites. How can they possibly be satisfied? But that is only part of the question. The full question is: How can we provide for everyone without making the planet uninhabitable?

Bitter Rivals

WHILE MY CHILDREN WERE GROWING UP, I took advantage of journalistic assignments to speak about these questions, from time to time, with experts in Europe, Asia, and the Americas. As the conversations accumulated, the responses seemed to fall into two broad categories, each associated (at least in my mind) with one of two people, both of them Americans who lived in the 20th century. The two people were barely acquainted and had little regard for each other's work. But they were largely responsible for the creation of the basic intellectual blueprints that institutions around the world use today for understanding our environmental dilemmas. Unfortunately, their blueprints offer radically different answers to the question of survival.

The two people were William Vogt and Norman Borlaug.

Vogt, born in 1902, laid out the basic ideas for the modern environmental movement. In particular, he founded what the Hampshire College population researcher Betsy Hartmann has called "apocalyptic environmentalism"—the belief that unless humankind drastically reduces consumption and limits population, it will ravage global ecosystems. In best-selling books and powerful speeches, Vogt argued that affluence is not our greatest achievement but our biggest problem. If we continue taking more than the Earth can give, he said, the unavoidable result will be devastation on a global scale. *Cut back! Cut back!* was his mantra.

Borlaug, born 12 years after Vogt, has become the emblem of "techno-optimism"—the view that science and technology, properly applied, will let us produce a way out of our predicament. He was the best-known figure in the research that in the 1960s created the Green Revolution, the combination of high-yielding crop varieties and agronomic techniques that increased grain harvests around the world, helping to avert tens of millions of deaths from hunger. To Borlaug, affluence was not the problem but the solution. Only by getting richer and more knowledgeable can humankind create the science that will resolve our environmental dilemmas. *Innovate! Innovate!* was his cry.

Both men thought of themselves as using new scientific knowledge to face a planetary crisis. But that is where the similarity ends. For Borlaug, human ingenuity was the solution to our problems. One example: By using the advanced methods of the Green Revolution to increase per-acre yields, he argued, farmers would not have to plant as many acres, an idea researchers now call the "Borlaug hypothesis." Vogt's views were the opposite: The solution, he said, was to use ecological knowledge to get smaller. Rather than grow more grain to produce more meat, humankind should, as his followers say, "eat lower on the food chain," to lighten the burden on Earth's ecosystems. This is where Vogt differed from his predecessor, Robert Malthus, who famously predicted that societies would inevitably run out of food because they would always have too many children. Vogt, shifting the argument, said that we may be able to grow enough food, but at the cost of wrecking the world's ecosystems.

I think of the adherents of these two perspectives as "Wizards" and "Prophets." Wizards, following Borlaug's model, unveil technological fixes; Prophets, looking to Vogt, decry the consequences of our heedlessness.

Borlaug and Vogt traveled in the same orbit for decades, but rarely acknowledged each other. Their first and only meeting, in the mid-1940s, led to disagreement—immediately afterward, Vogt tried to get Borlaug's work shut down. So far as I know, they never spoke afterward. Each referred to the other's ideas in public addresses, but never attached a name. Instead, Vogt rebuked the anonymous "deluded" scientists who were actually aggravating our problems. Borlaug branded his opponents "Luddites."

Both men are dead now, but the dispute between their disciples has only become more vehement. Wizards view the Prophets' emphasis on cutting back as intellectually dishonest, indifferent to the poor, even racist (because most of the world's hungry are non-Caucasian). Following Vogt, they say, is a path toward regression, narrowness, poverty, and hunger—toward a world where billions live in misery despite the scientific knowledge that could free them. Prophets sneer that the Wizards' faith in human resourcefulness is unthinking, ignorant, even driven by greed (because refusing to push beyond ecological limits will cut into corporate profits). High-intensity, Borlaug-style industrial farming, Prophets say, may pay off in the short run, but in the long run will make the day of ecological reckoning hit harder. The ruination of soil and water by heedless overuse will lead to environmental collapse, which will in turn create worldwide social convulsion. Wizards reply: *That's exactly the global humanitarian crisis we're preventing!* As the finger-pointing has escalated, conversations about the environment have turned into dueling monologues, each side unwilling to engage with the other.

Which might be all right, if we weren't discussing the fate of our children.

The Roads to Hell

VOGT ENTERED HISTORY IN 1948, when he published *Road to Survival*, the first modern we're-all-going-to-hell book. It contained the foundational argument of today's environmental movement: carrying capacity. Often called by other names—"ecological limits," "planetary boundaries"—carrying capacity posits that every ecosystem has a limit to what it can produce. Exceed that limit for too long and the ecosystem will be ruined. As human numbers increase, *Road to Survival* said, our demands for food will exceed the Earth's carrying capacity. The results will be catastrophic: erosion, desertification, soil exhaustion, species extinction, and water contamination that will, sooner or later, lead to massive famines. Embraced by writers like Rachel Carson (the author of *Silent Spring* and one of Vogt's friends) and Paul Ehrlich (the author of *The Population Bomb*), Vogt's arguments about exceeding limits became the wellspring of today's globe-spanning environmental movement—the only enduring ideology to emerge from the past century.

When *Road to Survival* appeared, Borlaug was a young plant pathologist working in a faltering program to improve

Mexican agriculture. Sponsored by the Rockefeller Foundation, the project focused on helping the nation's poor corn farmers. Borlaug was in Mexico for a small side project that involved wheat—or rather, black stem rust, a fungus that is wheat's oldest and worst predator (the Romans made sacrifices to propitiate the god of stem rust). Cold usually killed stem rust in the United States, but it was constantly present in warmer Mexico, and every spring winds drove it across the border to reinfect U.S. wheat fields.

The sole Rockefeller researcher working on wheat, Borlaug was given so little money that he had to sleep in sheds and fields for months on end. But he succeeded by the mid-'1950s in breeding wheat that was resistant to many strains of rust. Not only that, he then created wheat that was much shorter than usual—what became known as "semi-dwarf" wheat. In the past, when wheat was heavily fertilized, it had grown so fast that its stalks became spindly and fell over in the wind. The plants, unable to pull themselves erect, had rotted and died. Borlaug's shorter, stouter wheat could absorb large doses of fertilizer and channel the extra growth into grain rather than roots or stalk. In early tests, farmers sometimes harvested literally 10 times as much grain from their fields. Yields climbed at such a rate that in 1968 a USAID official called the rise the Green Revolution, thus naming the phenomenon that would come to define the 20th century.

The Green Revolution had its most dramatic effects in Asia, where in 1962 the Rockefeller Foundation and the Ford Foundation opened the International Rice Research Institute (IRRI) in the Philippines. At the time, at least half of Asia lived in hunger and want; farm yields in many places were stagnant or falling. Governments that had only recently thrown off colonialism were battling communist insurgencies, most notably in Vietnam. U.S. leaders believed the appeal of communism lay in its promise of a better future. Washington wanted to demonstrate that development occurred best under capitalism. IRRI's hope was that top research teams would transform Asia by rapidly introducing modern rice agriculture—"a Manhattan Project for food," in the historian Nick Cullather's phrase.

Following Borlaug's lead, IRRI researchers developed new, high-yielding rice varieties. These swept through Asia in the '70s and '80s, nearly tripling rice harvests. More than 80 percent of the rice grown in Asia today originated at IRRI. Even though the continent's population has soared, Asian men, women, and children consume an average of 30 percent more calories than they did when IRRI was founded. Seoul and Shanghai, Jaipur and Jakarta; shining skyscrapers, pricey hotels, traffic-jammed streets ablaze with neon—all were built atop a foundation of laboratory-bred rice.

Were the Prophets disproved? Was carrying capacity a chimera? No. As Vogt had predicted, the enormous jump in productivity led to enormous environmental damage: drained aquifers, fertilizer runoff, aquatic dead zones, and degraded and waterlogged soils. Worse in a human sense, the rapid increase in productivity made rural land more valuable. Suddenly it was worth stealing—and rural elites in many places did just that, throwing poor farmers off their land. The Prophets argued that the Green Revolution would merely postpone the hunger crisis; it was a onetime lucky break, rather than a permanent solution. And our rising numbers and wealth mean that, just as the Prophets said, our harvests will have to jump again—a second Green Revolution, the Wizards add.

Even though the global population in 2050 will be just 25 percent higher than it is now, typical projections claim that farmers will have to boost food output by 50 to 100 percent. The main reason is that increased affluence has always multiplied the demand for animal products such as cheese, dairy, fish, and especially meat—and growing feed for animals requires much more land, water, and energy than producing food simply by growing and eating plants. Exactly how much more meat tomorrow's billions will want to consume is unpredictable, but if they are anywhere near as carnivorous as today's Westerners, the task will be huge. And, Prophets warn, so will the planetary disasters that will come of trying to satisfy the world's desire for burgers and bacon: ravaged landscapes, struggles over water, and land grabs that leave millions of farmers in poor countries with no means of survival.

What to do? Some of the strategies that were available during the first Green Revolution aren't anymore. Farmers can't plant much more land, because almost every accessible acre of arable soil is already in use. Nor can the use of fertilizer be increased; it is already being overused everywhere except some parts of Africa, and the runoff is polluting rivers, lakes, and oceans. Irrigation, too, cannot be greatly expanded—most land that can be irrigated already is. Wizards think the best course is to use genetic modification to create more-productive crops. Prophets see that as a route to further overwhelming the planet's carrying capacity. We must go in the opposite direction, they say: use less land, waste less water, stop pouring chemicals into both.

It is as if humankind were packed into a bus racing through an impenetrable fog. Somewhere ahead is a cliff: a calamitous reversal of humanity's fortunes. Nobody can see exactly where it is, but everyone knows that at some point the bus will have to turn. Problem is, Wizards and Prophets disagree about which way to yank the wheel. Each is certain that following the other's ideas will send the bus over the cliff. As they squabble, the number of passengers keeps rising.

The Story of Nitrogen

ALMOST EVERYBODY EATS EVERY DAY, but too few of us give any thought to how that happens. If agricultural history were required in schools, more people would know the name of Justus von Liebig, who in the mid-19th century established that the amount of nitrogen in the soil controls the rate of plant growth. Historians of science have charged Liebig with faking his data and stealing others' ideas—accurately, so far as I can tell. But he was also a visionary who profoundly changed the human species' relationship with nature. Smarmy but far-sighted, Liebig imagined a new kind of agriculture: farming as a branch of chemistry and physics. Soil was just a base with the physical attributes necessary to hold roots. Pour in nitrogen-containing compounds—factory-made fertilizer—and gigantic harvests would automatically follow. In today's terms, Liebig was taking the first steps toward chemically regulated industrial agriculture—an early version of Wizardly thought.

But there was no obvious way to manufacture the nitrogenous substances that feed plants. That technology was provided before and during the First World War by two German chemists, Fritz Haber and Carl Bosch. Their subsequent Nobel Prizes were richly deserved: The Haber-Bosch process, as it is called, was arguably the most consequential technological innovation of the 20th century. Today the Haber-Bosch process is responsible for almost all of the world's synthetic fertilizer. A little more than 1 percent of the world's industrial energy is devoted to it. "That 1 percent," the futurist Ramez Naam has noted, "roughly doubles the amount of food the world can grow." The environmental scientist Vaclav Smil has estimated that nitrogen fertilizer from the Haber-Bosch process accounts for "the prevailing diets of nearly 45% of the world's population." More than 3 billion men, women, and children—an incomprehensibly vast cloud of hopes, fears, memories, and dreams—owe their existence to two obscure German chemists.

Hard on the heels of the gains came the losses. About 40 percent of the fertilizer applied in the past 60 years was not absorbed by plants. Instead, it washed away into rivers or seeped into the air in the form of nitrous oxides. Fertilizer flushed into water still fertilizes: It boosts the growth of algae, weeds, and other aquatic organisms. When these die, they fall to the floor of the river, lake, or ocean, where microbes consume their remains. So rapidly do the microbes grow on the manna of dead algae and weeds that their respiration drains oxygen from the lower depths, killing off most other life. Nitrogen from Midwestern farms flows down the Mississippi to the Gulf of Mexico every summer, creating an oxygen desert that in 2016 covered almost 7,000 square miles. The next year a still larger dead zone—23,000 square miles—was mapped in the Bay of Bengal, off the east coast of India.

Rising into the air, nitrous oxides from fertilizers is a major cause of pollution. High in the stratosphere, it combines with and neutralizes the planet's ozone, which guards life on the surface by blocking cancer-causing ultraviolet rays. Were it not for climate change, suggests the science writer Oliver Morton, the spread of nitrogen's empire would probably be our biggest ecological worry.

Passionate resistance to that empire sprang up even before Haber and Bosch became Nobel laureates. Its leader was an English farm boy named Albert Howard (1873–1947), who spent most of his career as British India's official imperial economic botanist. Individually and together, Howard and his wife, Gabrielle, a Cambridge-educated plant physiologist, spent their time in India breeding new varieties of wheat and tobacco, developing novel types of plows, and testing the results of providing oxen with a super healthy diet. By the end of the First World War, they were convinced that soil was not simply a base for chemical additives. It was an intricate living system that required a wildly complex mix of nutrients in plant and animal waste: harvest leftovers, manure. The Howards summed up their ideas in what they called the Law of Return: "the faithful return to the soil of all available vegetable, animal, and human wastes." We depend on plants, plants depend on soil, and soil depends on us. Howard's 1943 *Agricultural Testament* became the founding document of the organic movement.

Wizards attacked Howard and Jerome I. Rodale—a hardscrabble New York–born entrepreneur, publisher, playwright, gardening theorist, and food experimenter who publicized Howard's ideas through books and magazines—as charlatans and crackpots. It is true that their zeal was inspired by a near-religious faith in a limit-bound natural order. But when Howard lauded the living nature of the soil, he was referring to the community of soil organisms, the dynamic relations between plant roots and the earth around them, and the physical structure of humus, which stickily binds together soil particles into airy crumbs that hold water instead of letting it run through. All of this was very real, and all of it was unknown when Liebig shaped the basic ideas behind chemical agriculture. The claim Howard made in his many books and speeches that industrial farming was depopulating the countryside and disrupting an older way of life was accurate, too, though his opponents disagreed with him about whether this was a bad thing. Nowadays the Prophets' fears about industrial agriculture's exhausting the soil seem prescient: A landmark 2011 study from the United Nations' Food and Agriculture Organization concluded that up to a third of the world's cropland is degraded.

At first, reconciling the two points of view might have been possible. One can imagine Borlaugian Wizards considering manure and other natural soil inputs, and Vogtian Prophets willing to use chemicals as a supplement to good soil practice. But that didn't happen. Hurling insults, the two sides moved further apart. They set in motion a battle that has continued into the 21st century—and become ever more intense with the ubiquity of genetically modified crops. That battle is not just between two philosophies, two approaches to technology, two ways of thinking how best to increase the food supply for a growing population. It is about whether the tools we choose will ensure the survival of the planet or hasten its destruction.

"Not One of Evolution's Finest Efforts"

ALL THE WHILE that Wizards were championing synthetic fertilizer and Prophets were denouncing it, they were united in ignorance: Nobody knew *why* plants were so dependent on nitrogen. Only after the Second World War did scientists discover that plants need nitrogen chiefly to make a protein called rubisco, a prima donna in the dance of interactions that is photosynthesis.

In photosynthesis, as children learn in school, plants use energy from the sun to tear apart carbon dioxide and water, blending their constituents into the compounds necessary to make roots, stems, leaves, and seeds. Rubisco is an enzyme that plays a key role in the process. Enzymes are biological catalysts. Like jaywalking pedestrians who cause automobile accidents but escape untouched, enzymes cause biochemical reactions to occur but are unchanged by those reactions. Rubisco takes carbon dioxide from the air, inserts it into the maelstrom of photosynthesis, then goes back for more. Because these movements are central to the process, photosynthesis walks at the speed of rubisco.

Alas, rubisco is, by biological standards, a sluggard, a lazybones, a couch potato. Whereas typical enzyme molecules catalyze thousands of reactions a second, rubisco molecules deign to involve themselves with just two or three a second. Worse, rubisco is inept. As many as two out of every five times, rubisco fumblingly picks up oxygen instead of carbon dioxide, causing the chain of reactions in photosynthesis to break down and have to restart, wasting energy and water. Years ago I talked with biologists about photosynthesis for a magazine article. Not one had a good word to say about rubisco. "Nearly the world's worst, most incompetent enzyme," said one researcher. "Not one of evolution's finest efforts," said another. To overcome rubisco's lassitude and maladroitness, plants make a lot of it, requiring a lot of nitrogen to do so. As much as half of the protein in many plant leaves, by weight, is rubisco—it is often said to be the world's most abundant protein. One estimate is that plants and microorganisms contain more than 11 pounds of rubisco for every person on Earth.

Evolution, one would think, should have improved rubisco. No such luck. But it did produce a work-around: C4 photosynthesis (*C4* refers to a four-carbon molecule involved in the scheme). At once a biochemical kludge and a clever mechanism for turbocharging plant growth, C4 photosynthesis consists of a wholesale reorganization of leaf anatomy.

When carbon dioxide comes into a C4 leaf, it is initially grabbed not by rubisco but by a different enzyme that uses it to form a compound that is then pumped into special, rubisco-filled cells deep in the leaf. These cells have almost no oxygen, so rubisco can't bumblingly grab the wrong molecule. The end result is exactly the same sugars, starches, and cellulose that ordinary photosynthesis produces, except much faster. C4 plants need less water and fertilizer than ordinary plants, because they don't waste water on rubisco's mistakes. In the sort of convergence that makes biologists snap to attention, C4 photosynthesis has arisen independently more than 60 times. Corn, tumbleweed, crabgrass, sugarcane, and Bermuda grass—all of these very different plants evolved C4 photosynthesis.

In the botanical equivalent of a moonshot, scientists from around the world are trying to convert rice into a C4 plant—one that would grow faster, require less water and fertilizer, and produce more grain. The scope and audacity of the project are hard to overstate. Rice is the world's most important foodstuff, the staple crop for more than half the global population, a food so embedded in Asian culture that the words *rice* and *meal* are variants of each other in both Chinese and Japanese. Nobody can predict with confidence how much more rice farmers will need to grow by 2050, but estimates range up to a 40 percent rise, driven by both increasing population numbers and increasing affluence, which permits formerly poor people to switch to rice from less prestigious staples such as millet and sweet potato. Meanwhile, the land available to plant rice is shrinking as cities expand into the countryside, thirsty people drain rivers, farmers switch to more-profitable crops, and climate change creates deserts from farmland. Running short of rice would be a human catastrophe with consequences that would ripple around the world.

The C4 Rice Consortium is an attempt to ensure that that never happens. Funded largely by the Bill & Melinda Gates Foundation, the consortium is the world's most ambitious *genetic engineering* project. But the term *genetic engineering* does not capture the project's scope. The genetic engineering that appears in news reports typically involves big companies sticking individual packets of genetic material, usually from a foreign species, into a crop. The paradigmatic example is Monsanto's Roundup Ready soybean, which contains a snippet of DNA from a bacterium that was found in a Louisiana waste pond. That snippet makes the plant assemble a chemical compound in its leaves and stems that blocks the effects of Roundup, Monsanto's widely used herbicide. The foreign gene lets farmers spray Roundup on their soy fields, killing weeds but leaving the crop unharmed. Except for making a single tasteless, odorless, nontoxic protein, Roundup Ready soybeans are otherwise identical to ordinary soybeans.

What the C4 Rice Consortium is trying to do with rice bears the same resemblance to typical genetically modified crops as a Boeing 787 does to a paper airplane. Rather than tinker with individual genes in order to monetize seeds, the scientists are trying to refashion photosynthesis, one of the most fundamental processes of life. Because C4 has evolved in so many different species, scientists believe that most plants must have precursor C4 genes. The hope is that rice is one of these, and that the consortium can identify and awaken its dormant C4 genes—following a path evolution has taken many times before. Ideally, researchers would switch on sleeping chunks of genetic material already in rice (or use very similar genes from related species that are close cousins but easier to work with) to create, in effect, a new and more productive species. Common rice, *Oryza sativa*, will become something else: *Oryza nova*, say. No company will profit from the result; the International Rice Research Institute, where much of the research takes place, will give away seeds for the modified grain, as it did with Green Revolution rice.

When I visited IRRI, 35 miles southeast of downtown Manila, scores of people were doing what science does best: breaking a problem into individual pieces, then attacking the pieces. Some were sprouting rice in petri dishes. Others were trying to find chance variations in existing rice strains that might be helpful. Still others were studying a model organism, a C4 species of grass called *Setaria viridis*. Fast-growing and able to be grown in soil, not paddies, *Setaria* is easier to work with in the lab than rice. There were experiments to measure differences in photosynthetic chemicals, in the rates of growth of different varieties, in the transmission of biochemical markers. Half a dozen people in white coats were sorting seeds on a big table, grain by grain. More were in fields outside, tending experimental rice paddies. All of the appurtenances of contemporary biology were in evidence: flat-screen monitors, humming refrigerators and freezers, tables full of beakers of recombinant goo, *Dilbert* and *XKCD* cartoons taped to whiteboards, a United Nations of graduate students a-gossip in the cafeteria, air conditioners whooshing in a row outside the windows.

Directing the C4 Rice Consortium is Jane Langdale, a molecular geneticist at Oxford's Department of Plant Sciences. Initial research, she told me, suggests that about a dozen genes play a major part in leaf structure, and perhaps another 10 genes have an equivalent role in the biochemistry. All must be activated in a way that does not affect the plant's existing, desirable qualities and that allows the genes to coordinate their actions. The next, equally arduous step would be breeding rice varieties that can channel the extra growth provided by C4 photosynthesis into additional grains, rather than roots or stalk. All the while, varieties must be disease-resistant, easy to grow, and palatable for their intended audience, in Asia, Africa, and Latin America.

"I think it can all happen, but it might not," Langdale said. She was quick to point out that even if C4 rice runs into insurmountable obstacles, it is not the only biological moonshot. Self-fertilizing maize, wheat that can grow in salt water, enhanced soil-microbial ecosystems—all are being researched. The odds that any one of these projects will succeed may be small, the idea goes, but the odds that all of them will fail are equally small. The Wizardly process begun by Borlaug is, in Langdale's view, still going strong.

The Luddites' Moonshot

FOR AS LONG AS Wizards and Prophets have been arguing about feeding the world, Wizards have charged that Prophet-style agriculture simply cannot produce enough food for tomorrow. In the past 20 years, scores of research teams have appraised the relative contributions of industrial and organic agriculture. These inquiries in turn have been gathered together and assessed, a procedure that is fraught with difficulty: Researchers use different definitions of *organic*, compare different kinds of farms, and include different costs in their analyses. Nonetheless, every attempt to combine and compare data that I know of has shown that Prophet-style farms yield fewer calories per acre than do Wizard-style farms—sometimes by a little, sometimes by quite a lot. The implications are obvious, Wizards say. If farmers must grow twice as much food to feed the 10 billion, following the ecosystem-conserving rules of Sir Albert Howard ties their hands.

Prophets smite their brows at this logic. To their minds, evaluating farm systems wholly in terms of calories per acre is folly. It doesn't include the sort of costs identified by Vogt: fertilizer runoff, watershed degradation, soil erosion and compaction, and pesticide and antibiotic overuse. It doesn't account for the destruction of rural communities. It doesn't consider whether the food is tasty and nutritious.

Wizards respond that C4 rice will use less fertilizer and water to produce every calorie—it will be better for the environment than conventional crops. *That's like trying to put out fires you started by dousing them with less gasoline!* the Prophets say. *Just eat less meat!* To Wizards, the idea of making farms diverse in a way that mimics natural ecosystems is hooey: only hyperintensive, industrial-scale agriculture using super productive genetically modified crops can feed tomorrow's world.

Productivity? the Prophets reply. *We have moonshots of our own!* And in fact, they do.

Wheat, rice, maize, oats, barley, rye, and the other common cereals are *annuals*, which need to be planted anew every year. By contrast, the wild grasses that used to fill the prairie are *perennials*: plants that come back summer after summer, for as long as a decade. Because perennial grasses build up root systems that reach deep into the ground, they hold on to soil better and are less dependent on surface rainwater and nutrients—that is, irrigation and artificial fertilizer—than annual grasses. Many of them are also more disease-resistant. Not needing to build up new roots every spring, perennials emerge from the soil earlier and faster than annuals. And because they don't die in the winter, they keep photosynthesizing in the fall, when annuals stop. Effectively, they have a longer growing season. They produce food year after year with much less plowing-caused erosion. They could be just as productive as Green Revolution–style grain, Prophets say, but without ruining land, sucking up scarce water, or requiring heavy doses of polluting, energy-intensive fertilizer.

Echoing Borlaug's program in Mexico, the Rodale Institute, the country's oldest organization that researches organic agriculture, gathered 250 samples of intermediate wheatgrass (*Thinopyrum intermedium*) in the late 1980s. A perennial cousin to bread wheat, wheatgrass was introduced to the Western Hemisphere from Asia in the 1930s as fodder for farm animals. Working with U.S. Department of Agriculture researchers, the Rodale Institute's Peggy Wagoner, a pioneering plant breeder and agricultural researcher, planted samples, measured their yields, and crossbred the best performers in an attempt to make a commercially viable perennial. Wagoner and the Rodale Institute passed the baton in 2002 to the Land Institute, in Salina, Kansas, a nonprofit agricultural-research center dedicated to replacing conventional agriculture with processes akin to those that occur in natural ecosystems. The Land Institute, collaborating with other researchers, has been developing wheatgrass ever since. It has even given its new variety of intermediate wheatgrass a trade name: Kernza.

Like C4 rice, wheatgrass may not fulfill its originators' hopes. Wheatgrass kernels are one-quarter the size of wheat kernels, sometimes smaller, and have a thicker layer of bran. Unlike wheat, wheatgrass grows into a dark, dense mass of foliage that covers the field; the thick layer of vegetation protects the soil and keeps out weeds, but it also reduces the amount of grain that the plant produces. To make wheatgrass useful to farmers, breeders will have to increase kernel size, alter the plant's architecture, and improve its bread-making qualities. The work has been slow. Because wheatgrass is a perennial, it must be evaluated over years, rather than a single season. The Land Institute hopes to have field-ready, bread-worthy wheatgrass with kernels that are twice their current size (if still half the size of wheat's) in the 2020s, though nothing is guaranteed.

Domesticating wheatgrass is the long game. Other plant breeders have been trying for a shortcut: creating a hybrid of bread wheat and wheatgrass, hoping to marry the former's large, plentiful grain and the latter's disease resistance and perennial life cycle. The two species produce viable offspring just often enough that biologists in North America, Germany, and the Soviet Union tried unsuccessfully for decades in the mid-1900s to breed useful hybrids. Bolstered by developments in biology, the Land Institute, together with researchers in the Pacific Northwest and Australia, began anew at the turn of this century. When I visited Stephen S. Jones of Washington State University, he and his colleagues had just suggested a scientific name for the newly developed and tested hybrid: *Tritipyrum aaseae* (the species name honors the pioneering cereal geneticist Hannah Aase). Much work remains; Jones told me that he hoped bread from *T. aaseae* would be ready for my daughter's children.

African and Latin American researchers scratch their heads when they hear about these projects. Breeding perennial grains is the hard way for Prophets to raise harvests, says Edwige Botoni, a researcher at the Permanent Interstate Committee for Drought Control in the Sahel, in Burkina Faso. Botoni gave a lot of thought to the problem of feeding people from low-quality land while traveling along the edge of the Sahara. One part of the answer, she told me, would be to emulate the farms that flourish in tropical places such as Nigeria and Brazil. Whereas farmers in the temperate zones focus on cereals, tropical growers focus on tubers and trees, both of which are generally more productive than cereals.

Consider cassava, a big tuber also known as manioc, mogo, and yuca. The 11th-most-important crop in the world in terms of production, it is grown in wide swathes of Africa, Asia, and Latin America. The edible part grows underground; no matter how big the tuber, the plant will never fall over. On a per-acre basis, cassava harvests far outstrip those of wheat and other cereals. The comparison is unfair, because cassava tubers contain more water than wheat kernels. But even when this is taken into account, cassava produces many more calories per acre than wheat. (The potato is a northern equivalent. The average 2016 U.S. potato yield was 43,700 pounds per acre, more than 10 times the equivalent figure for wheat.) "I don't know why this alternative is not considered," Botoni said. Although cassava is unfamiliar to many cultures, introducing it "seems easier than breeding entirely new species."

Much the same is true for tree crops. A mature McIntosh apple tree might grow 350 to 550 pounds of apples a year. Orchard growers commonly plant 200 to 250 trees per acre. In good years, this can work out to 35 to 65 tons of fruit per acre. The equivalent figure for wheat, by contrast, is about a ton and a half. As with cassava and potatoes, apples contain more water than wheat does—but the caloric yield per acre is still higher. Even papayas and bananas are more productive than wheat. So are some nuts, like chestnuts. Apples, chestnuts, and papayas cannot make crusty baguettes, crunchy tortillas, or cloud-light chiffon cakes, but most grain today is destined for highly processed substances like animal feed, breakfast cereal, sweet syrups, and ethanol—and tree and tuber crops can be readily deployed for those.

Am I arguing that farmers around the world should replace their plots of wheat, rice, and maize with fields of cassava, potato, and sweet potato and orchards of bananas, apples, and chestnuts? No. The argument is rather that Prophets have multiple ways to meet tomorrow's needs. These alternative paths are difficult, but so is the Wizards' path exemplified in C4 rice. The greatest obstacle for Prophets is something else: labor.

The Right Way to Live

SINCE THE END OF THE SECOND WORLD WAR, most national governments have intentionally directed labor away from agriculture (Communist China was long an exception). The goal was to consolidate and mechanize farms, which would increase harvests and reduce costs, especially for labor. Farmworkers, no longer needed, would move to the cities, where they could get better-paying jobs in factories. In the Borlaugian ideal, both the remaining farm owners and the factory workers would earn more, the former by growing more and better crops, the latter by obtaining better-paying jobs in industry. The nation as a whole would benefit: increased exports from industry and agriculture, cheaper food in the cities, a plentiful labor supply.

There were downsides: Cities in developing nations acquired entire slums full of displaced families. And in many areas, including most of the developed world, the countryside was emptied—exactly what Borlaugians intended, as part of the goal of freeing agriculture workers to pursue their dreams. In the United States, the proportion of the workforce employed in agriculture went from 21.5 percent in 1930 to 1.9 percent in 2000; the number of farms fell by almost two-thirds. The average size of the surviving farms increased to compensate for the smaller number. Meanwhile, states around the world established networks of tax incentives, loan plans, training programs, and direct subsidies to help big farmers acquire large-scale farm machinery, stock up on chemicals, and grow certain government-favored crops for export. Because these systems remain in effect, Vogtian farmers are swimming against the tide.

To Vogtians, the best agriculture takes care of the soil first and foremost, a goal that entails smaller patches of multiple crops—difficult to accomplish when concentrating on the mass production of a single crop. Truly extending agriculture that does this would require bringing back at least some of the people whose parents and grandparents left the countryside. Providing these workers with a decent living would drive up costs. Some labor-sparing mechanization is possible, but no small farmer I have spoken with thinks that it would be possible to shrink the labor force to the level seen in big industrial operations. The whole system can grow only with a wall-to-wall rewrite of the legal system that encourages the use of labor. Such large shifts in social arrangements are not easily accomplished.

And here is the origin of the decades-long dispute between Wizards and Prophets. Although the argument is couched in terms of calories per acre and ecosystem conservation, the disagreement at bottom is about the nature of agriculture—and, with it, the best form of society. To Borlaugians, farming is a kind of useful drudgery that should be eased and reduced as much as possible to maximize individual liberty. To Vogtians, agriculture is about maintaining a set of communities, ecological and human, that have cradled life since the first agricultural revolution, 10,000-plus years ago. It can be drudgery, but it is also work that reinforces the human connection to the Earth. The two arguments are like skew lines, not on the same plane.

My daughter is 19 now, a sophomore in college. In 2050, she will be middle-aged. It will be up to her generation to set up the institutions, laws, and customs that will provide for basic human needs in the world of 10 billion. Every generation decides the future, but the choices made by my children's generation will resonate for as long as demographers can foresee. Wizard or Prophet? The choice will be less about what this generation thinks is feasible than what it thinks is good

Critical Thinking

1. What is meant by the carrying capacity of the planet and why is it important?
2. What is the effect of genetic engineering on food supply?
3. Why do prophets believe that increased food supply will lead to environmental collapse?

Internet References

Food and Agriculture Organization of the United Nations
www.fao.org/home/en/

International Fund for Agricultural Development
Https//www.ifad.org/

National Institute of Food Production and Sustainability
https://nifa.usda.-gov/office/institute-food-production-and-sustainability

World Food Programme
www1.wfp-org/

Article

Prepared by: Robert Weiner, *University of Massachusetts, Boston*

The Next Energy Revolution
The Promise and Peril of High-Tech Innovation

DAVID G. VICTOR AND KASSIA YANOSEK

Learning Outcomes

After reading this article, you will be able to:

- Learn how cheap, plentiful energy is creating a new world.
- Learn about the relationship between energy and technology.
- Learn how energy technology is revolutionizing the electrical industry.

The technology revolution has transformed one industry after another, from retail to manufacturing to transportation. Its most far-reaching effects, however, may be playing out in the unlikeliest of places: the traditional industries of oil, gas, and electricity.

Over the past decade, innovation has upended the energy industry. First came the shale revolution. Starting around 2005, companies began to unlock massive new supplies of natural gas, and then oil, from shale basins, thanks to two new technologies: horizontal drilling and hydraulic fracturing (or fracking). Engineers worked out how to drill shafts vertically and then turn their drills sideways to travel along a shale seam; they then blasted the shale with high-pressure water, sand, and chemicals to pry open the rock and allow the hydrocarbons to flow. These technologies have helped drive oil prices down from an all-time high of $145 per barrel in July 2008 to less than a third of that today, and supply has become much more responsive to market conditions, undercutting the ability of OPEC, a group of the world's major oil-exporting nations, to influence global oil prices.

That was just the beginning. Today, smarter management of complex systems, data analytics, and automation are remaking the industry once again, boosting the productivity and flexibility of energy companies. These changes have begun to transform not only the industries that produce commodities such as oil and gas but also the ways in which companies generate and deliver electric power. A new electricity industry is emerging—one that is more decentralized and consumer-friendly, and able to integrate many different sources of power into highly reliable power grids. In the coming years, these trends are likely to keep energy cheap and plentiful, responsive to market conditions, and more efficient than ever.

But this transition will not be straightforward. It could destabilize countries whose economies depend on revenue from traditional energy sources, such as Russia, the big producers of the Persian Gulf, and Venezuela. It could hurt lower-skilled workers, whose jobs are vulnerable to automation. And cheap fossil fuels will make it harder to achieve the deep cuts in emissions needed to halt global warming.

Get Smart

There are three trends driving the new energy revolution: smarter management of complex systems, more sophisticated data analytics, and automation. The first trend has allowed companies to become much more efficient while drilling for oil and gas in ever more complex geological environments. Beginning around 15 years ago, for example, advances in imaging technology made it easier for companies to find oil deposits in deep waters, such as in the Gulf of Mexico and off the coast of Brazil. But as oil and gas companies rushed to recover these resources, the technological demands of operating in deep waters and through thick layers of sediment and bedrock drove up costs. By 2014, new deep-water projects were so costly that many broke even only when the price of oil was at almost $100 per barrel. As the price of oil tumbled from above $100 per barrel in early 2014 to below $50 per barrel in January 2015, many of these projects stalled. By early 2016, companies had put on hold an estimated four million barrels per day of new oil output, 40 percent of it from deep-water sources.

As drilling stalled, oil and gas operators, desperate to cut costs, began to rethink the complex systems they used. Some savings were easy to find: reduced activity meant that critical equipment and services, once scarce, now sat idle. The daily cost of renting an oil rig, for example, fell by half. But the industry is also cutting costs and improving performance through fundamental productivity improvements. Simpler, standardized designs make drilling and production platforms easier to replicate, less expensive, and less likely to suffer costly delays and over-runs in construction. And companies are transferring the lessons they've learned across the industry. Shell, for example, recently announced that it is applying techniques from onshore shale operations, such as drilling horizontal wells and injecting water into them, to increase production in mature deep-water fields.

Today, thanks to these innovations, the average breakeven prices of new deep-water projects have fallen, to just $40–$50 per barrel in the Gulf of Mexico—an important global bellwether because it is one of the most responsive regions in the world to changes in market conditions. Even though oil prices remain low (and many in the industry expect them to stay low), investment is once again growing. Ten deep-water projects were approved for investment in 2016 and the first half of 2017 alone.

Smarter management of complex systems is also reshaping the electric power industry. For decades, centralized, base-load energy generators—mainly coal, nuclear, and large hydroelectric plants—dominated the industry. But in the last two decades, governments have subsidized wind and solar energy and pushed them into the electricity system, in the hopes of diversifying their countries' energy sources, creating new jobs, and reducing emissions. Until recently, these new sources were too small to have much of an effect on the overall system.

Today, however, as the cost of renewables is plummeting and their share of the power supply is rising, they have begun to transform electricity markets. In Germany, wind and solar power account for almost 30 percent of the power mix; in Hawaii, they account for about a quarter. Traditional utilities have struggled to adapt. In March, grid operators in California shut down 80 gigawatt-hours of the state's renewable power because the grid couldn't handle the afternoon solar surge; without more capacity to store power, even larger curtailments will occur. In Texas, among many other places, prices occasionally turn negative when the wind is blowing hard but people don't need too much electricity—in other words, companies are paying customers to use the electricity they generate. Utilities that have failed to see these changes coming have floundered. The market valuations of the top four German utilities are about one-third the level they were a decade ago—in large part because they were stuck with the costs of the old electric power system even as the government provided lavish support for renewables.

Renewables are just one part of this transformation. In the coming years, utility companies may face an existential challenge from smaller and more decentralized energy systems known as "microgrids." Microgrids first emerged decades ago, driven by customers, such as the U.S. military, that prized reliability above all else and that did not mind paying more for it: military bases have to keep functioning even if the bulk power grid fails. Early adopters also included remote communities, such as in Alaska, that are far from the conventional grid. But now, microgrids are spreading to other places, such as university campuses and hospitals, where they generate reliable power and are often designed to save money by using waste energy to heat and cool buildings.

New technologies, such as fuel cells and battery storage systems (to store extra power produced by renewables), along with more sophisticated software, have led to even smaller systems called "nanogrids," which Walmart and other megastores have begun to adopt. And picogrids may be next. As more and more people rely less on the traditional grid for power (while still interconnecting with it to help ensure reliability), policymakers and companies will need to create new regulatory systems and business models. Some states, such as New York, have embraced these changes, aggressively promoting decentralization by rewarding companies that invest in decentralized systems. But no one has yet worked out a detailed plan for how to integrate new grids with traditional power systems.

Hi, Robot

The second major source of innovation is better data analytics. Oil companies, for example, have begun to use complex algorithms to analyze massive amounts of data, making it easier for them to find oil and gas and to manage production. In April 2017, for example, BP announced that, using these methods, it had identified another 200 million barrels of oil in an existing field in the Gulf of Mexico. According to BP, data crunching that used to take a year now takes just a few weeks. And cloud processing makes it possible to generate millions of scenarios for developing an oil field. When firms can evaluate more options, production from fields can rise by five percent, with a 30 percent cut in the investment required to drill holes and begin producing oil. The industry has also begun to use data analytics for "predictive maintenance," reducing unplanned downtime by analyzing historical data to predict equipment failures before they happen. This practice, pioneered by industries such as the aircraft engine business, is helping cut costs on oil and gas rigs, where compressors and other rotating equipment can cause costly interruptions when they fail.

The third and most important trend is automation. In remote offshore oil fields, robots have already begun to perform dangerous tasks, such as connecting pipes during drilling operations, a job traditionally carried out by the versatile workers known as "roustabouts." Soon, intelligent automated systems will enable remote drilling, controlled almost entirely by a handful of high-tech workers in onshore data rooms hundreds of miles away. And companies are developing robots that can live on the ocean floor and inspect offshore pipelines and underwater equipment. At the moment, offshore oil rigs typically employ 100–200 workers, a figure that could fall. Although people remain indispensable for critical safety roles that require complex decision-making, automation will transform the industry's work force. According to a McKinsey study, within ten years, oil and gas companies could employ more data scientists with Ph.D.'s than geologists.

Automation has already changed the power industry, where smart meters have all but eliminated manual meter readings. In the future, automation, along with better data analytics, will make it easier to manage the variation in supplies that comes from using renewable sources such as wind and solar energy and more complex, decentralized grids. It can also make the grid more reliable. The inability of grid operators to understand what is happening in real time plays an important role in many power outages; automation and improved human-computer interaction could make blackouts much rarer.

Yet innovation can create new problems. Automation in the energy business, for example, could make it more difficult for governments to perform some of their traditional functions, such as safety regulation. When technological changes on rigs, production platforms, and grids proceeded slowly, regulators could keep up, learning and applying lessons from occasional failures. Today, the sheer complexity of highly automated systems makes observing and predicting their behavior much more difficult. So regulators will need to evolve as quickly as the industry—and develop early warning systems to identify places where oversight is required. They will need to learn more rapidly from each other. Regulators in other countries could study the Norwegian offshore oil and gas regulatory body, for example, which is becoming adept at managing the high levels of uncertainty inherent in the offshore industry, or they could learn from the U.S. nuclear industry, which has figured out how to use peer review inside the industry to judge the management of plants.

Winners and Losers

The coming transformation of the energy industry is, for the most part, good news for the world. But as the revolution unfolds, it will profoundly change politics, economics, and the environment. Policymakers and business leaders will need to tread carefully.

For starters, sustained lower energy prices could weaken the economic and geopolitical influence of many major oil suppliers, which have relied for too long on their control of nearly all of the world's cheap oil resources. In response, some of these countries have begun to act. Last year, for example, Saudi Arabia launched Vision 2030, a program to reduce the kingdom's dependence on oil and diversify its economy. The government has announced plans to sell around five percent of the state oil giant, Saudi Aramco—the kingdom's crown jewel—in an initial public offering next year, which may help the firm become more efficient. These reforms are promising and long overdue, but they face significant resistance, and whether they can be successfully implemented remains to be seen.

Russia, too, must continue to reform. A decade ago, the Russian government could balance its budget only when the price of oil topped $100 per barrel. Today, however, the country expects to balance its budget by 2019 with oil at just $40 per barrel, even though 35 percent of the government's revenue still comes from hydrocarbons. But this situation remains unstable, and Moscow will have to continue cutting its expenditures. Other countries, such as Angola and Nigeria, have failed to introduce sufficient reforms, and the fall in energy prices has contributed to their instability. Fiscal prudence and a more reliable environment for foreign investors would help local industries get access to the latest technologies and compete in international energy markets.

In the United States, the energy revolution will have profound effects far beyond the jobs and economic growth that cheap energy will catalyze. When it comes to electricity, the economics increasingly favor natural gas and renewables—making it even harder for coal, which accounted for almost half of U.S. electricity generation in 2007 but just 30 percent in 2016, to compete, no matter what politicians may claim. Most coal jobs are not coming back.

The United States has not yet had a well-informed public debate about how the nature of work in the modern economy is changing. The energy industry has witnessed this transformation firsthand and is well positioned to show how the work force itself can adjust. Energy companies, for example, have begun to figure out how to retrain workers over the course of their careers as jobs in power plant control rooms and on production platforms shift toward the overseeing of automated systems. Education and training are changing, too. Texas A&M, for example, is launching a master's degree in geospatial technologies specifically targeted for the oil and gas work force.

Climate change remains perhaps the greatest challenge of all. According to the latest assessment from the Intergovernmental

Panel on Climate Change, the world will need to cut emissions by about 80 percent if it is to slow and, eventually, stop the rise in global temperatures. The last two decades of summits and negotiations have shown that this will not be easy. And the revolution in fossil fuel production may make it even more difficult, because the prices of carbon-based fuels are likely to remain highly competitive with those of their lower-emission rivals. The solution lies in part in investing more in innovation, and at the Paris climate change conference in late 2015, the world's biggest governments pledged to double their spending on energy R & D. So far, however, they have not delivered. Although there has been an uptick in private-sector investment, across the industrialized world, government spending on energy R & D has remained roughly flat for almost four decades.

Already, huge benefits from the technology revolution in energy are reaching consumers. The 92 million barrels of crude oil that the world economy consumes every day cost about $2 trillion less annually than that amount did a decade ago. In the United States, the energy revolution has helped sustain economic growth: from 2008 to 2014, lower prices saved the average household over $700 a year. The era in which energy policy focused on the security of raw resource supplies—access to barrels of crude oil, tons of coal, and volumes of natural gas—is over. Today, the task for policymakers is to manage the implications of a new world of cheap, plentiful energy.

Critical Thinking

1. What has been the effect of energy technological innovation on the traditional oil producers such as Saudi Arabia and Russia?
2. Discuss the three factors that are changing the energy industry.
3. What will be the effect of the energy revolution on fossil fuel production?

Internet References

National Renewable Energy Lab
Http://www.nrel.gov

The International Energy Agency
http://www.iea.org

The Organization of Petroleum Exporting Countries
http://www.opec/-web/en

David G. Victor is professor of International Relations at the School of Global Policy and Strategy at the University of California, San Diego, and cochair of the Initiative on Energy and Climate at the Brookings Institution.

Kassia Yanosek is an associate partner in McKinsey & Company's Global Energy and Materials Practice.

Unit 3

UNIT

Prepared by: Robert Weiner, *University of Massachusetts, Boston*

The Global Political Economy

As several articles in this unit point out, globalization had resulted in closing the income gap which existed between the developed and developing countries (due mainly to the economic gains made by China and India), but had also resulted in a growing inequality within such countries as the United States. The resentment of blue-collar workers toward an economic system which resulted in economic stagnation also resulted in the unwillingness of the Democratic and Republican candidates for President in 2016 to support the Obama administration's proposal for a free trade agreement known as the Trans-Pacific Partnership. Free trade agreements had been seen by liberal internationalists as a means of avoiding "beggar thy neighbor" policies which had contributed the catastrophe of the Second World War. The populist appeal to the industrial base of workers in the US rested on assumptions that free that free trade agreements such as NAFTA (the North American Free Trade Agreement) had resulted in the hollowing out of the manufacturing sector of the United States. One of the major pillars of the international liberal order consists of multilateral trade agreements. President Trump considered the multilateral trade agreements that the United States was a party to as "bad deals." Consequently, the Trump administration withdrew from the Trans-Pacific Partnership and renegotiated NAFTA. The Trump administration had succeeded in renegotiating its trade agreements with Mexico and Canada. President Trump, as mentioned previously, had referred to NAFTA as a bad deal. The United States first negotiated its trade agreement with Mexico on a bilateral basis. It looked as if trade negotiations with Canada the third member of the three-part trade deal that had been negotiated during the Clinton administration was going to fail. However, Jared Kushner, played a key role in bringing the trade negotiations to a successful conclusion Canada, as Kushner turned out to be a very skillful negotiator. The successful conclusion of negotiations with Canada and Mexico meant that NAFTA would be replaced by a new agreement. The general consensus is that the US has come out ahead in its renegotiations with Mexico and Canada. The Trump administration claimed that the new agreement would add thousands of jobs in the United States. The new agreement between the United States and Canada, however, did not affect the tariffs which the Trump administration had levied on imports of Canadian steel and aluminum.

The Trump administration had levied a 25 percent tariff on the importation of Canadian steel and a 10 percent tariff on the importation of Canadian aluminum. The US dairy industry praised the agreement because it resulted in the opening up of part of the Canadian market to US dairy farmers. The Canadians succeeded in maintaining article 19 of the settlement mechanism of the NAFTA agreement. Intellectual property and lumber were other areas also involved in the negotiations with Canada. The new agreement which was also dubbed the USMCA in place of NAFTA also dealt with the manufacture of auto parts in North America. Some observers believed that the NAFTA negotiations could serve as a model for trade negotiations with China. The Trump administration was involved in a trade war with China. The United States has raised tariffs on billions of dollars' worth of goods imported from China. The Chinese have retaliated, for example, by levying a raise in tariffs on the importation of US soybeans.

Another key component of the international liberal order consists of the European Union. Populists, such as President Trump, however, have criticized the European Union and praised the decision by the British to leave the EU (European Union) in a referendum that took place in 2016.The British have always had an ambivalent relationship with the European Union, not joining it until 1973. After joining what was called the Common Market at the time, British Euroskeptics continued to express concern about the negative effects of membership in the European Union. The United Kingdom, unlike the United States, does not have a written constitution. The United Kingdom has a political system which is based on custom, tradition, common law, prerogatives, and statutes. Under this system, the British parliament, more specifically the House of Commons, is considered to be the supreme legal sovereign in the country. After joining the European Union, the British continued to be uneasy with the fact that is some cases, the authority of the British Parliament could be subordinated to the European

Union. Opponents of UK membership argued that the national sovereignty of the United Kingdom would be threatened by continued membership in the European Union. Critics of the European Union also argued that there is a democratic deficit in the EU which is run by "Eurocrats" or bureaucrats who are not held responsible for their policy decisions. The British, as a condition for continued membership in the EU, wanted more power to be exercised by the national parliaments of the member states. The British also believed that they were not getting their money's worth as members of the European Union. This resulted in a renegotiation of British financial assessments to the EU. During the administration of Conservative Prime Minister Margaret Thatcher, the "Iron Lady" was quoted as saying "I want my money back." Euroskeptics were also opposed to the liberal provisions, as they saw them, which allowed migrants to move from one member of the European Union to another. The rise of terrorism in Western Europe and the flow of refugees fleeing the civil war in Syria also added to this anxiety. On June 23, 2016, the British voted to leave the European Union. The vote was about 52 percent to leave and 48 percent to stay. The results were stunning because they were quite unexpected and resulted in the resignation of the Conservative Prime Minister, David Cameron. Negotiations between the United Kingdom and the European Union have been going on since 2016, under the leadership of Conservative Prime Minister Theresa May. May has faced a considerable amount of opposition from members of her own party on her negotiating strategy and tactics. Despite renewed calls for a second referendum on leaving, Prime Minister May is determined to carry out the will of the people as expressed in the first referendum in 2016. UK negotiations with the EU have been rather difficult, taking two parallel tracks. The first dealt with the terms of the "divorce." That is, how to deal with issues such as assets and money. The other set of negotiations involves a free trade agreement between the United Kingdom and the European Union. Prime Minister May has threatened that "no deal" is better than a "bad deal."

Article

Prepared by: Robert Weiner, *University of Massachusetts, Boston*

Trade, Development, and Inequality

URI DADUSH

Learning Outcomes

After reading this article, you will be able to:

- Discuss the relationship between trade, labor-saving technology, and income inequality.
- Identify the reasons behind the major mega-regional trade deals with the TransPacific Partnership and the Transatlantic Trade and Investment Partnership.

To trade or not to trade? Judging by the narrow vote by the US Congress in June 2015 to grant President Barack Obama fast-track negotiating authority for trade agreements, the answer today remains in the affirmative, as it has for decades—but resistance is on the rise. According to the many opponents of such deals, trade is a bitter medicine to be taken only in small doses while guarding carefully against its dangerous side effects. If new trade deals are to go ahead at all, the critics say, they should include strict safeguards, such as provisions to uphold environmental and labor standards, and penalties for currency manipulation. Although the trade debate in the United States, the architect of the postwar global trading system, tends to draw the spotlight, the hand-wringing over trade is even more intense in the developing world. Since developing countries protect their home markets more comprehensively than the United States does, the stakes in their trade debates are higher.

Yet this is an era of hyperglobalization in which consumers have become accustomed to searching online for the best-priced merchandise from all over the world. Trade has surged from 25 percent to 60 percent of world GDP in the past 50 years. Why is it still so controversial? The unemployment and dislocation caused by the global financial crisis provide only part of the explanation.

In the United States, the great crisis of 2008–2009 came on the heels of 30 years of stagnant incomes for the vast majority of households, a period that also saw nearly all of the nation's very considerable income gains accrue at the top of the income and wealth pyramid. Trade, especially with China and other low-income countries, is often blamed for the very high and rising inequality in the United States. Such high inequality contributes to a number of ills, including extremely limited opportunity for the children of poor families, bad health outcomes, crime, capture of the legislative process and of government agencies by moneyed interests, and profound political divisions that impede the formulation and execution of economic reforms.

Rising income inequality is not only an American problem. With few exceptions, it has been a common trend around the world, in both advanced and developing countries—most notably in many of the largest developing countries such as China and India. A recent International Monetary Fund report found that over the past 30 years, inequality has risen in every region of the world except Latin America, which nonetheless still has several countries with levels of inequality surpassed only in South Africa.

There is broad agreement among economists that unskilled labor-saving technologies, not trade, have played the central role in increasing inequality. Many believe that the ongoing information and communications technology revolution all but guarantees that this trend will continue. I share these views. I also believe, however, that trade, interacting in a mutually reinforcing fashion with these technologies, has significantly contributed to the inequality trend in both advanced and developing countries. And yet, since technology and trade also lie at the root of the unprecedented postwar advance in average living standards around the world, the sensible policy response is not to try to suppress or reverse trade (or technology, for that matter), even if that were possible, but to adapt to it and to mitigate its effects on the most vulnerable.

Dozens of trade deals are being negotiated around the world today, including giant "mega-regional" arrangements such as the Trans-Pacific Partnership (TPP) and the Transatlantic Trade and Investment Partnership (TTIP). A TPP deal has just been struck, but it needs to be ratified by national legislatures. These new trade deals are as necessary to sustaining economic growth as previous trade deals were in the past. They may or may not lead to even more inequality, depending on the way they are configured, on other reforms that accompany them, and on the specific circumstances of each country or bloc.

The United States and the European Union are unlikely to see much additional effect on inequality from trade deals, simply because they are already very open economies. Trade deals are likely to have a bigger impact on inequality in developing countries, especially those that have the highest trade barriers, such as India and Brazil. Yet these are also the countries that are most likely to see the highest growth dividends from new trade openings, and where unskilled workers are more likely to be net gainers from trade, even if they are losers relative to their most affluent compatriots.

An enormous unfinished trade agenda lies within national borders.

Ancient Arguments

The debate over trade is not new: Aristotle and Plato might have been in opposing camps. According to the late economic historian Murray Rothbard, "Aristotle, in the Greek tradition, was scornful of moneymaking and scarcely a partisan of laissez-faire . . . [yet] he denounced Plato's goal of the perfect unity of the state . . . pointing out that such extreme unity runs against the diversity of mankind, and against the reciprocal advantage that everyone reaps through market exchange."

Two thousand years later, Adam Smith and David Ricardo, writing at a time when Britain had established its commercial preeminence and was leading the Industrial Revolution, conducted systematic analyses of the gains from trade. Smith's analysis of the welfare-enhancing "invisible hand" of markets and his arguments in favor of the international division of labor and the economies of scale to be gained in world markets, together with Ricardo's advocacy of specialization along the lines of comparative advantage, laid the foundations of modern economics.

Their profound insights made little impression in the United States at the time. The great emerging nation of the era comprehensively protected its infant manufacturing sector while relentlessly copying European technology, a policy that it pursued quite consistently throughout the nineteenth century and well into the twentieth. This protectionism culminated in the trade-suffocating Smoot-Hawley tariff hike during the worst of the Great Depression.

It was only as World War II drew to a close that the United States, having achieved a dominant position as the world's leading industrial power, took up the banner of free trade. At the Bretton Woods conference in 1944, the United States insisted that Britain dismantle its system of imperial preferences, a demand the British strenuously resisted. The disagreement over imperialism was a major reason behind their failure to agree on launching an international trade organization along with the World Bank and International Monetary Fund. The trade wing of what came to be called the Bretton Woods system emerged later, first in the shape of the General Agreement on Tariffs and Trade, which took effect in 1948, and then as the World Trade Organization (WTO), launched in Marrakesh at the end of 1994.

Boom Times

In the immediate postwar years, it was the turn of the newly independent developing countries to become the champions of import substitution, to resist protection of intellectual property, and to adopt industrial policies designed to pick market winners and protect the politically powerful—precisely the policies to which the United States had resorted during its developmental phase. Those policies may or may not have worked well for the young giant—we have no way to be sure whether America might have grown even faster under free trade. What we do know, however, is that import substitution did not yield the desired results in developing countries. Soon enough, a turn toward exports and much freer imports ensued.

That big shift toward more outward-looking economic regimes was initially inspired by the extraordinary export-fueled growth of a small number of developing economies in Asia. This trend gained momentum in the wake of the oil shocks of the 1970s, when countries had to look for ways to cover their surging energy bills. Subsequently, the Latin American debt crises of the 1980s discredited import substitution, and the belief in central planning was undermined by stagnating living standards and lagging technologies across the communist bloc. As many observers predicted at the time, the fall of the Berlin Wall heralded the mother of all trade booms, lasting through the 1990s and early 2000s. The intensification of globalization was accompanied by spectacular growth in many developing countries, led by China, which surpassed the United States and Germany to become the world's largest exporter.

During this period, many economists became convinced that, in addition to the efficiency effects stressed by Smith and Ricardo, trade brought potentially even greater benefits, especially to developing nations, by inducing backward firms and economies to learn from those on the technological frontier. This thinking may have been formalized first by Alexander Gerschenkron in a seminal 1951 essay entitled "Economic Backwardness in Historical Perspective."

The focus on learning encouraged development agencies such as the United Nations Conference on Trade and Development (UNCTAD) and the World Bank to view foreign direct investment (FDI) more as a source of new techniques than as a finance vehicle. FDI, which grew even faster than trade, distributed the value chains of the most advanced manufacturers and service providers to cheaper locations or closer to the largest markets, and in the process created millions of jobs in the developing world. According to UNCTAD, the sales of foreign subsidiaries of multinational enterprises in their host countries today exceed world exports by a wide margin. The techniques and methods employed by these state-of-the-art overseas factories and service centers are systematically emulated or copied by less productive local enterprises. Not surprisingly, countries compete fiercely to attract FDI.

Falling Tariffs

The advance toward an open and predictable trading system has been nothing short of remarkable, and it has been matched by the rise of average incomes around the world. Real average per capita income in the United States has more than tripled since 1950, and incomes in developing countries have grown much faster. Over a billion people have been lifted out of absolute poverty in the past 15 years.

At the same time, high-tariff structures have been dismantled. The average tariff in the advanced countries is now around 2 percent–3 percent, and cannot be raised without violating WTO rules. Moreover, countries must apply the most favored nation (MFN) clause, meaning that all WTO members—which account for 97 percent of world trade—must accord each other at least the same tariff treatment, or treat them more favorably under certain specified circumstances. Quotas and subsidies have been outlawed, except (mainly at the insistence of advanced countries) in agriculture, which remains a heavily protected sector across the world.

MFN tariffs in developing countries are on average near 10 percent, much higher than in the advanced countries but about ⅓ their level during the height of import substitution. For the most part, however, these tariffs are not limited by the WTO or are subject to limits set at very high levels. Therefore, unless they are bound by a bilateral or regional agreement, most developing countries have plenty of room to legally raise their MFN tariffs should they decide to do so. (China is a notable exception on account of its demanding WTO accession protocol.)

According to a recent WTO/OECD paper, the movement toward freer trade has continued despite the ill-fated multilateral trade negotiations launched at Doha in 2001. Over the past 20 years, the applied tariffs of WTO members have declined by 15 percent on average, and the share of developing-country exports that now enter advanced countries duty-free has increased from 55 percent to 80 percent. Unilateral trade liberalization, more generous preference regimes, regional trade deals, and the delayed effects of past multilateral trade rounds have all played a role.

Behind the Border

There is still a long way to go before we secure world free trade, by which I mean zero tariffs on all goods, no quotas or subsidies, and complete freedom of entry in service sectors across the world, as well as equal treatment for foreign investors and suppliers, all bound by international treaty in the WTO. In addition to this admittedly distant or even utopian vision, an enormous unfinished trade agenda lies within national borders: reforming domestic regulations and practices that have the effect, sometimes intended but more often unintended, of restricting trade.

The cost of complying with these regulations, together with the cost of transportation and customs duties, and of distribution through wholesalers and retailers, adds up to "trade costs," which, it is estimated, can easily amount to one or two times the price of the product at the factory door. Economists have identified excessive trade costs (due to inappropriate regulations, inadequate transportation infrastructure, inefficient customs, bribes, and so forth) as a much more important barrier to trade today than tariffs. Numerous ongoing bilateral and regional trade negotiations are designed to address them. The Bali Trade Facilitation Agreement, which still requires ratification by ⅔ of members to take effect under the WTO, deals with a relatively narrow set of these behind-the-border issues, namely customs and regulations affecting international transportation and logistics, but arguably these are the issues of most immediate concern to exporters and importers alike.

Trade both stimulates economic growth and increases income inequality.

Naturally, developing and advanced countries have different agendas for addressing the largest remaining impediments to trade. Within each group there is a wide spectrum of interests that often cross over the dividing line. Developing countries aim to reduce the hugely distorting tariffs, quotas, and subsidies in advanced countries that limit their agricultural exports. They are also looking to limit the relatively high tariffs that advanced countries apply to labor-intensive manufactures, such as garments and shoes. In the context of north–south regional deals, such as the Central American Free Trade Agreement or the EU's Mediterranean agreements, developing countries are the parties most interested in less restrictive rules of origin, which are designed to guard against simple transshipment of goods from third parties but are often so complex and restrictive that exporters prefer to pay the full duty rather than try to document their right to preferential treatment.

Advanced countries, for their part, seek to limit tariffs in developing countries across the manufacturing sector. Many of them also wish to improve access for their (subsidized) agricultural and processed food exports, and to secure access to markets in services such as retailing, finance, and insurance. In addition, advanced countries are the ones most concerned with behind-the-border impediments to trade, such as subsidies or licenses accorded to state-owned enterprises, discrimination in government procurement and treatment of foreign investment, lax protection of intellectual property, and slow or unfair settlement of judicial disputes.

Mega-regional Deals

The United States, together with the European Union, spearheaded the adoption of WTO rules through most of the postwar period, but it has recently taken a very different turn. A decade ago, the twin thrusts of American trade policy consisted of a quest for a comprehensive WTO-driven multilateral trade round in the shape of the Doha agenda and the pursuit of a number of relatively minor bilateral trade agreements intended to spur "competitive liberalization." The idea was to induce countries to engage in bilateral deals and the Doha process to avoid being left out.

Today, American trade policy has pretty much written Doha off on account of what Washington perceives as unbridgeable differences between advanced and developing countries. Instead of small bilateral deals, the Obama administration is pursuing two so-called mega-regionals, the TPP for the Pacific and the TTIP with the EU. These deals explicitly aim to rewrite trade rules for the twenty-first century, effectively bypassing the unwieldy WTO. The TTIP has a long way to go, while a deal on the TPP was struck in October.

Given the number of partners involved and its comprehensive scope—it will cover about 40 percent of global GDP—the TPP is one of the most complex free trade agreements ever negotiated. Eleven countries initially joined the negotiations: Australia, Brunei, Canada, Chile, Malaysia, Mexico, New Zealand, Peru, Singapore, the United States, and Vietnam. Japan is a more recent and hugely important addition, and South Korea is a possible partner in the future. As in the case of the TTIP, trade is already largely free among this group; the aim is a high-standard agreement that will go deep behind the border to enhance trade and investment prospects across the board.

The United States already has established free trade agreements with six of the other eleven countries negotiating the TPP, and since it has a very open economy, the new trade liberalization that will be required of it under the deal is minimal. Accordingly, formal studies of the gains that the United States is likely to derive from tariff reductions under the TPP have come up with very small numbers—0.1 percent or 0.2 percent of GDP. Gains from removing nontariff barriers are potentially much larger, but very difficult to quantify.

The TPP, which excludes China, is motivated by political and security concerns as well as by economics. Yet despite the powerful political motivation behind the TPP (it is seen as a key part of the Obama administration's "pivot to Asia"), the diversity of interests among its prospective members has resulted in lengthy delays. The original deadline for completing the deal—the end of 2013—proved wildly optimistic. Although many expect the TPP to be concluded now that Obama has been granted fast-track trade promotion authority, it faces a tough ratification fight in the US Congress.

Trade deals are likely to have a bigger impact on inequality in developing countries.

The resistance to the TPP in the US Congress and the virulent criticisms leveled against it by civil society groups are difficult for its proponents to understand. They see only the deal's strategic importance and the advantage that the United States must commit to very little new liberalization, while its negotiating partners will have to do most of the hard work to reduce their tariff and nontariff barriers. But this is missing the point: At its core, the powerful resistance to new trade agreements in any shape or form is driven by the conviction that they hurt workers and benefit only the privileged few.

Technology and Inequality

The traditional view of trade is that it is triggered by variations in endowments of factors (resources such as land, labor, and capital). In a standard model with two factors, labor and capital, traditional trade theory predicts that a labor-abundant country will export labor-intensive products and see inequality decline as wages rise, while a capital-rich country will see inequality increase as it exports products that are capital-intensive and the return to its capital increases. This traditional model was adequately descriptive of trade in past centuries, when, as in Ricardo's famous example, England exported clothing and Portugal exported wine. Capital was scarcely mobile across borders, and technologies spread gradually.

Today, though, the traditional theory fails the empirical test. Contrary to its predictions, what we observe is that trade takes place predominantly between economies with similar endowments, namely advanced countries exporting to each other highly differentiated products in the same industry, such as cars and machine tools. Crucially, increased trade has been associated with higher inequality not just in advanced but also in developing countries.

Prompted to explain this reality, economists have come up with a number of alternative narratives in recent years, only some of which have been tested econometrically or using case studies. Economists now broadly agree that the most powerful underlying force driving increased inequality is not trade by itself but skill-biased technological change—that is, machines and methods that reduce the need for unskilled labor and boost demand for more specialized and skilled workers. Economists have also shown definitively that changes in aggregate or sectoral exports and imports are far too small relative to the size of the economy to account for the large shifts in industrial structure, employment, relative wages, and inequality that we observe. However, even though trade on its own cannot account for these changes in economic structure and inequality, the new stories tell us that the mutually reinforcing effects of trade on skill-biased technological change can increase inequality.

Take, for example, the case of an advanced country that opens up trade with a large low-wage economy. Firms in the advanced countries that compete in international trade are heterogeneous—they may operate in the same industry but they produce diverse products and vary greatly in their efficiency. Trade quickly kills the least efficient firms in sectors where the low-wage economy has an advantage, namely those that produce standardized products and which are highly labor-intensive. Those firms that survive do so on the basis of three complementary strategies: they automate so as to save on labor, they outsource their most labor-intensive activities to the low-wage economy, and they move upmarket into highly differentiated or technologically advanced niches.

Under all these scenarios, the demand for unskilled labor declines and the demand for skilled labor and capital increases. The dislocation of unskilled labor that results is larger than could be anticipated from the traditional static two-factor model, since trade encourages reduced employment of unskilled labor over time through multiple channels. Outsourcing of unskilled-labor-intensive activities results in less investment and growth in those activities in the future. While the sectors that compete with imports or engage in exporting lead in automation, those techniques are likely to spread throughout the economy, reducing the demand for unskilled labor even further. The fall in their wages prompts increased demand for unskilled workers in the nontraded service sector, but not enough to compensate.

Skills in Demand

What about the effect of trade on inequality in a low-wage or developing economy? As predicted by the traditional models, the demand for unskilled labor caused by opening up trade with high-wage economies will tend to raise the wages of the unskilled. However, there are three influences that can offset this effect and cause inequality to rise anyway.

First, as argued by the late development economist Arthur Lewis, an abundance of excess rural labor with a low reservation wage (the lowest wage at which a worker would accept a job) can slow the rise in wages of unskilled workers, especially at a time when hundreds of millions of unskilled workers are joining the global economy.

Second, as in an advanced economy, the opening of trade will favor a developing country's most efficient firms and those most able to adopt the higher standards demanded by world markets, with the capacity to meet precise specifications and timely delivery schedules. These requirements will often prompt the hiring of specialized workers and the purchase of sophisticated machines. According to the International Federation of Robotics, China is by far the world's fastest-growing market for industrial robots; its installations grew at a rate of about 25 percent a year between 2005 and 2012. Indeed, the import of such machines from advanced countries is the most direct channel through which trade spreads technology. Multinational enterprises from advanced countries invariably bring these advanced techniques with them as part of their outsourcing strategy.

Third, as in advanced countries, trade and foreign investment will stimulate the adoption of advanced techniques throughout the economy—not only in the traded sector. The combination of these effects leads to a sharp rise in the demand for skilled labor, which is relatively scarce in developing countries, as well as a rise in the demand for capital. Even though the demand for unskilled workers also rises, they may remain in plentiful supply for a long time, their wages rising relatively slowly, resulting in increased inequality.

The connecting thread in all these stories is that trade and, more broadly, international exchange prompt the spread of the most advanced technologies and encourage every firm exposed to increased competition, whether in advanced or developing countries, to become more efficient, thus raising the demand for skilled labor and for sophisticated capital goods. Moreover, since capital and, to a lesser extent, the most highly skilled professionals are more internationally mobile than unskilled workers, their rewards will tend to increase to match the best opportunities available anywhere in the world.

The Trade Dilemma

The modern theory of how trade both stimulates economic growth and increases income inequality applies in both advanced and developing countries. There is, however, an important difference between the two groups. Not only are developing countries catching up technologically and growing much faster; in labor-abundant developing countries the wages of the unskilled are likely sooner or later to rise with increased trade, even if they lose ground in relative terms to the skilled cohort. In contrast, unskilled labor in advanced countries could be a net loser in both relative and absolute terms. The theory is consistent with what we have observed: rising wages in developing countries, stagnant wages of unskilled workers in advanced countries, and rising inequality in both groups.

If the theory is correct, it presents an acute dilemma. Should countries pursue trade deals and grow more rapidly or should they eschew them, grow more slowly, and avoid their inequality-intensifying effects? The dilemma is sharper for advanced countries whose unskilled laborers may lose outright.

The answer goes back to the minimum concept of efficiency developed at the turn of the twentieth century by the Italian economist Vilfredo Pareto: Countries should pursue the efficient solution (in this case, open trade), making the pie bigger, and then redivide it in favor of the losers so that no one is worse off. Politically, this is easier said than done, but the necessary economic policies are familiar and the instruments are well honed. As the IMF has stressed in a recent analysis, compensating the losers from labor-saving technology or trade need not result in a loss of efficiency. Investments in education, health, and infrastructure that boost the incomes of unskilled workers and level the playing field for their children are likely to enhance economic growth. More progressive income and wealth taxes can be achieved by closing the many tax loopholes and inequities that distort economic incentives. Cuts to subsidies that favor rich farmers, purchasers of large homes, or drivers of gas-guzzling vehicles are likely to both increase efficiency and reduce inequality.

Developing countries are least prepared to execute these policies because they have limited taxation and administrative capacity, but they—and their unskilled workers—are the most likely to benefit from new trade deals even if they increase inequality. Advanced countries such as the United States are already largely open and have little to fear from new trade deals, which can consolidate their export interests without causing an unacceptable further rise in inequality.

However, the United States is also the country with the most pressing need to help unskilled workers cope with the effects of advances in labor-saving technology and their mutually reinforcing interaction with globalization. It has all the tools to respond. Its failure to confront rising inequality presents a threat both to its continued economic growth and to its leadership of the open global trading system that Washington played such a large role in creating.

Critical Thinking

1. What are the reasons for the opposition to the mega-regional trade deals?
2. How does the TransPacific Partnership fit into the "pivot" to Asia?
3. Who will benefit the most from the TransPacific Partnership?

Internet References

Transatlantic Trade and Investment Partnership
https://ustr.gov/ttip

TransPacific Partnership
https.//ustr.gov/tpp/

Uri Dadush is a senior associate in the international economics program at the Carnegie Endowment for International Peace and a Current History contributing editor.

Article

Prepared by: Robert Weiner, *University of Massachusetts, Boston*

The Truth about Trade

What Critics Get Wrong about the Global Economy

DOUGLAS A. IRWIN

Learning Outcomes

After reading this article, you will be able to:

- Discuss the real cause of the loss of jobs in the U.S.
- Understand the effects of trade with China on the U.S. economy

Just because a U.S. presidential candidate bashes free trade on the campaign trail does not mean that he or she cannot embrace it once elected. After all, Barack Obama voted against the Central American Free Trade Agreement as a U.S. senator and disparaged the North American Free Trade Agreement (NAFTA) as a presidential candidate. In office, however, he came to champion the Trans-Pacific Partnership (TPP), a giant trade deal with 11 other Pacific Rim countries.

Yet in the current election cycle, the rhetorical attacks on U.S. trade policy have grown so fiery that it is difficult to imagine similar transformations. The Democratic candidate Bernie Sanders has railed against "disastrous" trade agreements, which he claims have cost jobs and hurt the middle class. The Republican Donald Trump complains that China, Japan, and Mexico are "killing" the United States on trade thanks to the bad deals struck by "stupid" negotiators. Even Hillary Clinton, the expected Democratic nominee, who favored the TPP as secretary of state, has been forced to join the chorus and now says she opposes that agreement.

Blaming other countries for the United States' economic woes is an age-old tradition in American politics; if truth is the first casualty of war, then support for free trade is often an early casualty of an election campaign. But the bipartisan bombardment has been so intense this time, and has been so unopposed, that it raises real questions about the future of U.S. global economic leadership.

The antitrade rhetoric paints a grossly distorted picture of trade's role in the U.S. economy. Trade still benefits the United States enormously, and striking back at other countries by imposing new barriers or ripping up existing agreements would be self-destructive. The badmouthing of trade agreements has even jeopardized the ratification of the TPP in Congress. Backing out of that deal would signal a major U.S. retreat from Asia and mark a historic error.

Still, it would be a mistake to dismiss all of the antitrade talk as ill-informed bombast. Today's electorate harbors legitimate, deep-seated frustrations about the state of the U.S. economy and labor markets in particular, and addressing these complaints will require changing government policies. The solution, however, lies not in turning away from trade promotion but in strengthening worker protections.

By and large, the United States has no major difficulties with respect to trade, nor does it suffer from problems that could be solved by trade barriers. What it does face, however, is a much larger problem, one that lies at the root of anxieties over trade: the economic ladder that allowed previous generations of lower-skilled Americans to reach the middle class is broken.

Scapegoating Trade

Campaign attacks on trade leave an unfortunate impression on the American public and the world at large. In saying that some countries "win" and other countries "lose" as a result of trade, for example, Trump portrays it as a zero-sum game. That's an understandable perspective for a casino owner and businessman: gambling is the quintessential zero-sum game, and competition is a win-lose proposition for firms (if not for their customers). But it is dead wrong as a way to think about the role of trade in an economy. Trade is actually a two-way street—the exchange of exports for imports—that makes efficient use of a country's resources to increase its material welfare. The United States sells to other countries the goods and services that it produces relatively efficiently (from aircraft to soybeans to legal advice) and buys those goods and services that other countries produce relatively efficiently (from T-shirts to bananas to electronics assembly). In the aggregate, both sides benefit.

To make their case that trade isn't working for the United States, critics invoke long-discredited indicators, such as the country's negative balance of trade. "Our trade deficit with China is like having a business that continues to lose money every single year," Trump once said. "Who would do business like that?" In fact, a nation's trade balance is nothing like a firm's bottom line. Whereas a company cannot lose money indefinitely, a country—particularly one, such as the United States, with a reserve currency—can run a trade deficit indefinitely without compromising its well-being. Australia has run current account deficits even longer than the United States has, and its economy is flourishing.

One way to define a country's trade balance is the difference between its domestic savings and its domestic investment. The United States has run a deficit in its current account—the broadest measure of trade in goods and services—every year except one since 1981. Why? Because as a low-saving, high-consuming country, the United States has long been the recipient of capital inflows from abroad. Reducing the current account deficit would require foreigners to purchase fewer U.S. assets. That, in turn, would require increasing domestic savings or, to put it in less popular terms, reducing consumption. One way to accomplish that would be to change the tax system—for example, by instituting a consumption tax. But discouraging spending and rewarding savings is not easy, and critics of the trade deficit do not fully appreciate the difficulty involved in reversing it. (And if a current account surplus were to appear, critics would no doubt complain, as they did in the 1960s, that the United States was investing too much abroad and not enough at home.)

Critics also point to the trade deficit to suggest that the United States is losing more jobs as a result of imports than it gains due to exports. In fact, the trade deficit usually increases when the economy is growing and creating jobs and decreases when it is contracting and losing jobs. The U.S. current account deficit shrank from 5.8 percent of GDP in 2006 to 2.7 percent in 2009, but that didn't stop the economy from hemorrhaging jobs. And if there is any doubt that a current account surplus is no economic panacea, one need only look at Japan, which has endured three decades of economic stagnation despite running consistent current account surpluses.

And yet these basic fallacies—many of which Adam Smith debunked more than two centuries ago—have found a new life in contemporary American politics. In some ways, it is odd that antitrade sentiment has blossomed in 2016, of all years. For one thing, although the postrecession recovery has been disappointing, it has hardly been awful: the U.S. economy has experienced seven years of slow but steady growth, and the unemployment rate has fallen to just 5 percent. For another thing, imports have not swamped the country and caused problems for domestic producers and their workers; over the past seven years, the current account deficit has remained roughly unchanged at about two to three percent of GDP, much lower than its level from 2000 to 2007. The pace of globalization, meanwhile, has slowed in recent years. The World Trade Organization (WTO) forecasts that the volume of world trade will grow by just 2.8 percent in 2016, the fifth consecutive year that it has grown by less than three percent, down significantly from previous decades.

What's more, despite what one might infer from the crowds at campaign rallies, Americans actually support foreign trade in general and even trade agreements such as the TPP in particular. After a decade of viewing trade with skepticism, since 2013, Americans have seen it positively. A February 2016 Gallup poll found that 58 percent of Americans consider foreign trade an opportunity for economic growth, and only 34 percent viewed it as a threat.

The View from the Bottom

So why has trade come under such strident attack now? The most important reason is that workers are still suffering from the aftermath of the Great Recession, which left many unemployed and indebted. Between 2007 and 2009, the United States lost nearly nine million jobs, pushing the unemployment rate up to ten percent. Seven years later, the economy is still recovering from this devastating blow. Many workers have left the labor force, reducing the employment-to-population ratio sharply. Real wages have remained flat. For many Americans, the recession isn't over.

Thus, even as trade commands broad public support, a significant minority of the electorate—about a third, according to various polls—decidedly opposes it. These critics come from both sides of the political divide, but they tend to be lower-income, blue-collar workers, who are the most vulnerable to economic change. They believe that economic elites and the political establishment have looked out only for themselves over the past few decades. As they see it, the government bailed out banks during the financial crisis, but no one came to their aid.

Trade still benefits the United States enormously.

For these workers, neither political party has taken their concerns seriously, and both parties have struck trade deals that the workers think have cost jobs. Labor unions that support the Democrats still feel betrayed by President Bill Clinton, who, over their strong objections, secured congressional passage of NAFTA in 1993 and normalized trade relations with China in 2000. Blue-collar Republican voters, for their part, supported the anti-NAFTA presidential campaigns of Pat Buchanan and Ross Perot in 1992. They felt betrayed by President George W. Bush, who pushed Congress to pass many bilateral trade agreements. Today, they back Trump.

Among this demographic, a narrative has taken hold that trade has cost Americans their jobs, squeezed the middle class, and kept wages low. The truth is more complicated. Although imports have put some people out of work, trade is far from the most important factor behind the loss of manufacturing jobs. The main culprit is technology. Automation and other technologies have enabled vast productivity and efficiency improvements, but they have also made many blue-collar jobs obsolete. One representative study, by the Center for Business and Economic Research at Ball State University, found that productivity growth accounted for more than 85 percent of the job loss in manufacturing between 2000 and 2010, a period when employment in that sector fell by 5.6 million. Just 13 percent of the overall job loss resulted from trade, although in two sectors, apparel and furniture, it accounted for 40 percent.

This finding is consistent with research by the economists David Autor, David Dorn, and Gordon Hanson, who have estimated that imports from China displaced as many as 982,000 workers in manufacturing from 2000 to 2007. These layoffs also depressed local labor markets in communities that produced goods facing Chinese competition, such as textiles, apparel, and furniture. The number of jobs lost is large, but it should be put in perspective: while Chinese imports may have cost nearly one million manufacturing jobs over almost a decade, the normal churn of U.S. labor markets results in roughly 1.7 million layoffs every month.

Research into the effect of Chinese imports on U.S. employment has been widely misinterpreted to imply that the United States has gotten a raw deal from trade with China. In fact, such studies do not evaluate the gains from trade, since they make no attempt to quantify the benefits to consumers from lower-priced goods. Rather, they serve as a reminder that a rapid increase in imports can harm communities that produce substitute goods—as happened in the U.S. automotive and steel sectors in the 1980s.

Furthermore, the shock of Chinese goods was a one-time event that occurred under special circumstances. Imports from China increased from 1.0 percent of U.S. GDP in 2000 to 2.6 percent in 2011, but for the past five years, the share has stayed roughly constant. There is no reason to believe it will rise further. China's once-rapid economic growth has slowed. Its working-age population has begun to shrink, and the migration of its rural workers to coastal urban manufacturing areas has largely run its course.

The influx of Chinese imports was also unusual in that much of it occurred from 2001 to 2007, when China's current account surplus soared, reaching ten percent of GDP in 2007. The country's export boom was partly facilitated by China's policy of preventing the appreciation of the yuan, which lowered the price of Chinese goods. Beginning around 2000, the Chinese central bank engaged in a large-scale, persistent, and one-way intervention in the foreign exchange market—buying dollars and selling yuan. As a result, its foreign exchange reserves rose from less than $300 million in 2000 to $3.25 trillion in 2011. Critics rightly groused that this effort constituted currency manipulation and violated International Monetary Fund rules. Yet such complaints are now moot: over the past year, China's foreign exchange reserves have fallen rapidly as its central bank has sought to prop up the value of the yuan. Punishing China for past bad behavior would accomplish nothing.

The Right—and Wrong—Solutions

The real problem is not trade but diminished domestic opportunity and social mobility. Although the United States boasts a highly skilled work force and a solid technological base, it is still the case that only one in three American adults has a college education. In past decades, the ⅔ of Americans with no postsecondary degree often found work in manufacturing, construction, or the armed forces. These parts of the economy stood ready to absorb large numbers of people with limited education, give them productive work, and help them build skills. Over time, however, these opportunities have disappeared. Technology has shrunk manufacturing as a source of large-scale employment: even though US manufacturing output continues to grow, it does so with many fewer workers than in the past. Construction work has not recovered from the bursting of the housing bubble. And the military turns away 80 percent of applicants due to stringent fitness and intelligence requirements. There are no comparable sectors of the economy that can employ large numbers of high school educated workers.

This is a deep problem for American society. The unemployment rate for college-educated workers is 2.4 percent, but it is more than 7.4 percent for those without a high school diploma—and even higher when counting discouraged workers who have left the labor force but wish to work. These are the people who have been left behind in the 21st century economy—again, not primarily because of trade but because of structural changes in the economy. Helping these workers and ensuring that the economy delivers benefits to everyone should rank as urgent priorities.

But here is where the focus on trade is a diversion. Since trade is not the underlying problem in terms of job loss, neither is protectionism a solution. While the gains from trade can seem abstract, the costs of trade restrictions are concrete. For example, the United States has some 135,000 workers employed in the apparel industry, but there are more than 45 million Americans who live below the poverty line, stretching every dollar they have. Can one really justify increasing the price of clothing for 45 million low-income Americans (and everyone else as well) in an effort to save the jobs of just some of the 135,000 low-wage workers in the apparel industry?

Like undoing trade agreements, imposing selective import duties to punish specific countries would also fail. If the United States were to slap 45 percent tariffs on imports from China, as Trump has proposed, U.S. companies would not start producing more apparel and footwear in the United States, nor would they start assembling consumer electronics domestically. Instead, production would shift from China to other low-wage developing countries in Asia, such as Vietnam. That's the lesson of past trade sanctions directed against China alone: in 2009, when the Obama administration imposed duties on automobile tires from China in an effort to save American jobs, other suppliers, principally Indonesia and Thailand, filled the void, resulting in little impact on U.S. production or jobs.

For many Americans, the recession isn't over.

And if restrictions were levied against all foreign imports to prevent such trade diversion, those barriers would hit innocent bystanders: Canada, Japan, Mexico, the EU, and many others. Any number of these would use WTO procedures to retaliate against the United States, threatening the livelihoods of the millions of Americans with jobs that depend on exports of manufactured goods. Trade wars produce no winners. There are good reasons why the very mention of the 1930 Smoot-Hawley Tariff Act still conjures up memories of the Great Depression.

If protectionism is an ineffectual and counterproductive response to the economic problems of much of the work force, so, too, are existing programs designed to help workers displaced by trade. The standard package of Trade Adjustment Assistance, a federal program begun in the 1960s, consists of extended unemployment compensation and retraining programs. But because these benefits are limited to workers who lost their jobs due to trade, they miss the millions more who are unemployed on account of technological change. Furthermore, the program is fraught with bad incentives. Extended unemployment compensation pays workers for prolonged periods of joblessness, but their job prospects usually deteriorate the longer they stay out of the labor force, since they have lost experience in the interim.

And although the idea behind retraining is a good one—helping laid-off textile or steel workers become nurses or technicians—the actual program is a failure. A 2012 external review commissioned by the Department of Labor found that the government retraining programs were a net loss for society, to the tune of about $54,000 per participant. Half of that fell on the participants themselves, who, on average, earned $27,000 less over the four years of the study than similar workers who did not find jobs through the program, and half fell on the government, which footed the bill for the program. Sadly, these programs appear to do more harm than good.

A better way to help all low-income workers would be to expand the Earned Income Tax Credit. The EITC supplements the incomes of workers in all low-income households, not just those the Department of Labor designates as having been adversely affected by trade. What's more, the EITC is tied to employment, thereby rewarding work and keeping people in the labor market, where they can gain experience and build skills. A large enough EITC could ensure that every American was able to earn the equivalent of $15 or more per hour. And it could do so without any of the job loss that a minimum-wage hike can cause. Of all the potential assistance programs, the EITC also enjoys the most bipartisan support, having been endorsed by both the Obama administration and Paul Ryan, the Republican Speaker of the House. A higher EITC would not be a cure-all, but it would provide income security for those seeking to climb the ladder to the middle class.

The main complaint about expanding the EITC concerns the cost. Yet taxpayers are already bearing the burden of supporting workers who leave the labor force, many of whom start receiving disability payments. On disability, people are paid—permanently—to drop out of the labor force and not work. In lieu of this federal program, the cost of which has surged in recent years, it would be better to help people remain in the work force through the EITC, in the hope that they can eventually become taxpayers themselves.

The Future of Free Trade

Despite all the evidence of the benefits of trade, many of this year's crop of presidential candidates have still invoked it as a bogeyman. Sanders deplores past agreements but has yet to clarify whether he believes that better ones could have been negotiated or no such agreements should be reached at all. His vote against the U.S.-Australian free-trade agreement in 2004 suggests that he opposes all trade deals, even one with a country that has high labor standards and with which the United States runs a sizable balance of trade surplus. Trump professes to believe in free trade, but he insists that the United States has been outnegotiated by its trade partners, hence his threat to impose 45 percent tariffs on imports from China to get "a better deal"—whatever that means. He has attacked Japan's barriers against imports of U.S. agricultural goods, even though that is exactly the type of protectionism the TPP has tried to undo. Meanwhile, Clinton's position against the TPP has hardened as the campaign has gone on.

The anti-trade rhetoric of the campaign has made it difficult for even pro-trade members of Congress to support new agreements.

The response from economists has tended to be either meek defenses of trade or outright silence, with some even criticizing parts of the TPP. It's time for supporters of free trade to engage in a full-throated championing of the many achievements of U.S. trade agreements. Indeed, because other countries' trade barriers tend to be higher than those of the United States, trade agreements open foreign markets to U.S. exports more than they open the U.S. market to foreign imports.

That was true of NAFTA, which remains a favored punching bag on the campaign trail. In fact, NAFTA has been a big economic and foreign policy success. Since the agreement entered into force in 1994, bilateral trade between the United States and Mexico has boomed. For all the fear about Mexican imports flooding the US market, it is worth noting that about 40 percent of the value of imports from Mexico consists of content originally made in the United States—for example, auto parts produced in the United States but assembled in Mexico. It is precisely such trade in component parts that makes standard measures of bilateral trade balances so misleading.

NAFTA has also furthered the United States' long-term political, diplomatic, and economic interest in a flourishing, democratic Mexico, which not only reduces immigration pressures on border states but also increases Mexican demand for U.S. goods and services. Far from exploiting Third World labor, as critics have charged, NAFTA has promoted the growth of a middle class in Mexico that now includes nearly half of all households. And since 2009, more Mexicans have left the United States than have come in. In the two decades since NAFTA went into effect, Mexico has been transformed from a clientelistic one-party state with widespread anti-American sentiment into a functional multiparty democracy with a generally pro-American public. Although it has suffered from drug wars in recent years (a spillover effect from problems that are largely made in America), the overall story is one of rising prosperity thanks in part to NAFTA.

Ripping up NAFTA would do immense damage. In its foreign relations, the United States would prove itself to be an unreliable partner. And economically, getting rid of the agreement would disrupt production chains across North America, harming both Mexico and the United States. It would add to border tensions while shifting trade to Asia without bringing back any U.S. manufacturing jobs. The American public seems to understand this: in an October 2015 Gallup poll, only 18 percent of respondents agreed that leaving NAFTA or the Central American Free Trade Agreement would be very effective in helping the economy.

A more moderate option would be for the United States to take a pause and simply stop negotiating any more trade agreements, as Obama did during his first term. The problem with this approach, however, is that the rest of the world would continue to reach trade agreements without the United States, and so U.S. exporters would find themselves at a disadvantage compared with their foreign competitors. Glimpses of that future can already be seen. In 2012, the car manufacturer Audi chose southeastern Mexico over Tennessee for the site of a new plant because it could save thousands of dollars per car exported thanks to Mexico's many more free-trade agreements, including one with the EU. Australia has reached trade deals with China and Japan that give Australian farmers preferential access in those markets, cutting into US beef exports.

If Washington opted out of the TPP, it would forgo an opportunity to shape the rules of international trade in the 21st century. The Uruguay Round, the last round of international trade negotiations completed by the General Agreement on Tariffs and Trade, ended in 1994, before the Internet had fully emerged. Now, the United States' high-tech firms and other exporters face foreign regulations that are not transparent and impede market access. Meanwhile, other countries are already moving ahead with their own trade agreements, increasingly taking market share from U.S. exporters in the dynamic Asia-Pacific region. Staying out of the TPP would not lead to the creation of good jobs in the United States. And despite populist claims to the contrary, the TPP's provisions for settling disputes between investors and governments and dealing with intellectual property rights are reasonable. (In the early 1990s, similar fears about such provisions in the WTO were just as exaggerated and ultimately proved baseless.)

The United States should proceed with passage of the TPP and continue to negotiate other deals with its trading partners. So-called plurilateral trade agreements, that is, deals among relatively small numbers of like-minded countries, offer the only viable way to pick up more gains from reducing trade barriers. The current climate on Capitol Hill means that the era of small bilateral agreements, such as those pursued during the George W. Bush administration, has ended. And the collapse of the Doha Round at the WTO likely marks the end of giant multilateral trade negotiations.

Free trade has always been a hard sell. But the antitrade rhetoric of the 2016 campaign has made it difficult for even pro-trade members of Congress to support new agreements. Past experience suggests that Washington will lead the charge for reducing trade barriers only when there is a major trade problem to be solved—namely, when U.S. exporters face severe discrimination in foreign markets. Such was the case when the United States helped form the General Agreement on Tariffs and Trade in 1947, when it started the Kennedy Round of trade negotiations in the 1960s, and when it initiated the Uruguay Round in the 1980s. Until the United States feels the pain of getting cut out of major foreign markets, its leadership on global trade may wane. That would represent just one casualty of the current campaign.

Critical Thinking

1. Why do the 2016 Presidential candidates oppose free-trade agreements?
2. What is the relationship between technology and the loss of jobs in the U.S.?
3. What has been the effect of trade agreements with China on the U.S. economy?

Internet References

The Transatlantic Trade and Investment Partnership
https://ustr.gov/ttip

The World Trade Organization
https://www.wto.org/

Douglas A. Irwin is John Sloan Dickey third century professor in the Social Sciences in the Department of Economics at Dartmouth College and the author of *Free Trade Under Fire*. Follow him on Twitter @D_A_Irwin.

Article

Prepared by: Robert Weiner, *University of Massachusetts, Boston*

Inequality and Globalization

How the Rich Get Richer as the Poor Catch Up

FRANÇOIS BOURGUIGNON

Learning Outcomes

After reading this article, you will be able to:

- Discuss inequality between and within states.
- Understand the relationship between inequality and the international financial system.

When it comes to wealth and income, people tend to compare themselves to the people they see around them rather than to those who live on the other side of the world. The average Frenchman, for example, probably does not care how many Chinese exceed his own standard of living, but that Frenchman surely would pay attention if he started lagging behind his fellow citizens. Yet when thinking about inequality, it also makes sense to approach the world as a single community: accounting, for example, not only for the differences in living standards within France but also for those between rich French people and poor Chinese (and poor French and rich Chinese).

When looking at the world through this lens, some notable trends stand out. The first is that global inequality greatly exceeds inequality within any individual country. This observation should come as no surprise, since global inequality reflects the enormous differences in wealth between the world's richest and the world's poorest countries, not just the differences within them. Much more striking is the fact that, in a dramatic reversal of the trend that prevailed for most of the 20th century, global inequality has declined markedly since 2000 (following a slower decline during the 1990s). This trend has been due in large part to the rising fortunes of the developing world, particularly China and India. And as the economies of these countries continue to converge with those of the developed world, global inequality will continue to fall for some time.

Even as global inequality has declined, however, inequality within individual countries has crept upward. There is some disagreement about the size of this increase among economists, largely owing to the underrepresentation of wealthy people in national income surveys. But whatever its extent, increased inequality within individual countries has partially offset the decline in inequality among countries. To counteract this trend, states should pursue policies aimed at redistributing income, strengthen the regulation of the labor and financial markets, and develop international arrangements that prevent firms from avoiding taxes by shifting their assets or operations overseas.

The Great Substitution

Economists typically measure income inequality using the Gini coefficient, which ranges from zero in cases of perfect equality (a theoretical country in which everyone earns the same income) to one in cases of perfect inequality (a state in which a single individual earns all the income and everyone else gets nothing). In continental Europe, Gini coefficients tend to fall between 0.25 and 0.30. In the United States, the figure is around 0.40. And in the world's most unequal countries, such as South Africa, it exceeds 0.60. When considering the world's population as a whole, the Gini coefficient comes to 0.70—a figure so high that no country is known to have ever reached it.

Determining the Gini coefficient for global inequality requires making a number of simplifications and assumptions. Economists must accommodate gaps in domestic data—in Mexico, an extreme case, surveys of income and expenditures miss about half of all households. They need to come up with estimates for years in which national surveys are not available. They need to convert local incomes into a common currency, usually the U.S. dollar, and correct for differences in purchasing power. And they need to adjust for discrepancies in data collection among countries, such as those that arise when one state measures living standards by income and another by consumption per person or when a state does not collect data at all.

Such inexactitudes and the different ways of compensating for them explain why estimates of just how much global inequality has declined over the past two-plus decades tend to vary—from around two percentage points to up to five, depending on the study. No matter how steep this decline, however, economists generally agree that the end result has been a global Gini coefficient of around 0.70 in the years between 2008 and 2010.

The decline in global inequality is largely the product of the convergence of the economies of developing countries, particularly China and India, with those of the developed world. In the first decade of this century, booming economies in Latin America and sub-Saharan Africa also helped accelerate this trend. Remarkably, this decline followed a nearly uninterrupted rise in inequality from the advent of the Industrial Revolution in the early 19th century until the 1970s. What is more, the decline has been large enough to erase a substantial part of the inequality that built up over that century and a half.

Even as inequality among countries has decreased, however, inequality within individual countries has increased, gaining, on average, more than two percentage points in terms of the Gini coefficient between 1990 and 2010. The countries with the biggest economies are especially responsible for this trend—particularly the United States, where the Gini coefficient rose by five percentage points between 1990 and 2013, but also China and India and, to a lesser extent, most European countries, among them Germany and the Scandinavian states. Still, inequality within countries is not rising fast enough to offset the rapid decline in inequality among countries.

The good news is that the current decline in global inequality will probably persist. Despite the current global slowdown, China and India have such huge domestic markets that they retain an enormous amount of potential for growth. And even if their growth rates decline significantly in the next decade, so long as they remain higher than those of the advanced industrial economies, as is likely, global inequality will continue to fall. The prospects for growth are less favorable for the smaller economies in Latin America and sub-Saharan Africa that depend primarily on commodity exports, since world commodity prices may remain low for some time. All told, then, global inequality will likely keep falling in the coming decades—but probably at the slow pace seen during the 1990s rather than the rapid one enjoyed during the following decade.

The bad news, however, is that economists might have underestimated inequality within individual countries and the extent to which it has increased since the 1990s, because national surveys tend to underrepresent the wealthy and underreport income derived from property, which disproportionately accrues to the rich. Indeed, tax data from many developed states suggest that national surveys fail to account for a substantial portion of the incomes of the very highest earners.

According to the most drastic corrections for such underreporting, as calculated by the economists Sudhir Anand and Paul Segal, global inequality could have remained more or less constant between 1988 and 2005. Most likely, however, this conclusion is too extreme, and the increase in national inequality has been too small to cancel out the decline in inequality among countries. Yet it still points to a disheartening trend: increased inequality within countries has offset the drop in inequality among countries. In other words, the gap between average Americans and average Chinese is being partly replaced by larger gaps between rich and poor Americans and between rich and poor Chinese.

Interconnected and Unequal

The same factor that can be credited for the decline in inequality among countries can also be blamed for the increase in inequality within them: globalization. As firms from the developed world moved production overseas during the 1990s, emerging Asian economies, particularly China, started to converge with those of the developed world. The resulting boom triggered faster growth in Africa and Latin America as demand for commodities increased. In the developed world, meanwhile, as manufacturing firms outsourced some of their production, corporate profits rose but real wages for unskilled labor fell.

Economic liberalization also played an important role in this process. In China, the market reforms initiated by Deng Xiaoping in the 1980s contributed just as much to rapid growth as did the country's opening to foreign investment and trade, and the same is true of the reforms India undertook in the early 1990s. As with globalization, such reforms didn't just enable developing countries to get closer to the developed world; they also created a new elite within those countries while leaving many citizens behind, thus increasing domestic inequality.

The same drive toward economic liberalization has contributed to increasing inequality in the developed world. Reductions in income tax rates, cuts to welfare, and financial deregulation have also helped make the rich richer and, in some instances, the poor poorer. The increase in the international mobility of firms, wealth, and workers over the past two decades has compounded these problems by making it harder for governments to combat inequality: for example, companies and wealthy people have become increasingly able to shift capital to countries with low tax rates or to tax havens, allowing them to avoid paying more redistributive taxes in their home countries. And in both developed and developing countries, technological progress has exacerbated these trends by favoring skilled workers over unskilled ones and creating economies of scale that disproportionately favor corporate managers.

Maintaining Momentum

In the near future, the greatest potential for further reductions in global inequality will lie in Africa—the region that has arguably benefited the least from the past few decades of globalization, and the one where global poverty will likely concentrate in the coming decades as countries such as India leap ahead. Perhaps most important, the population of Africa is expected to double over the next 35 years, reaching some 25 percent of the world's population, and so the extent of global inequality will increasingly depend on the extent of African growth. Assuming that the economies of sub-Saharan Africa sustain the modest growth rates they have seen in recent years, then inequality among countries should keep declining, although not as fast as it did in the first decade of this century.

To maintain the momentum behind declining global inequality, all countries will need to work harder to reduce inequality within their borders, or at least prevent it from growing further. In the world's major economies, failing to do so could cause disenchanted citizens to misguidedly resist further attempts to integrate the world's economies—a process that, if properly managed, can in fact benefit everyone.

In practice, then, states should seek to equalize living standards among their populations by eliminating all types of ethnic, gender, and social discrimination; regulating the financial and labor markets; and implementing progressive taxation and welfare policies. Because the mobility of capital dulls the effectiveness of progressive taxation policies, governments also need to push for international measures that improve the transparency of the financial system, such as those the G20 and the Organization for Economic Cooperation and Development have endorsed to share information among states in order to clamp down on tax avoidance. Practical steps such as these should remind policymakers that even though global inequality and domestic inequality have moved in opposite directions for the past few decades, they need not do so forever.

Critical Thinking

1. What can the G-20 and the Organization for Economic Cooperation and Development do to eliminate global inequality?
2. Why has global inequality among states decreased?
3. Why has global inequality within states increased?

Internet References

The Group of 20
https://en.wikipedia.org/wiki'G20

The World Bank
www.worldbank.org

François Bourguignon is a professor of Economics at the Paris School of Economics, former Chief Economist of the World Bank, and the author of *The Globalization of Inequality.*

Article

Prepared by: Robert Weiner, *University of Massachusetts, Boston*

Inequality and Modernization

Why Equality Is Likely to Make a Comeback

RONALD INGLEHART

Learning Outcomes

After reading this article, you will be able to:

- Discuss the factors that have resulted in growing inequality in the developing countries.
- Explain the rise in postmaterial values in developed countries and the reaction it has provoked in the working class.

During the past century, economic inequality in the developed world has traced a massive U-shaped curve—starting high, curving downward, then curving sharply back up again. In 1915, the richest 1 percent of Americans earned roughly 18 percent of all national income. Their share plummeted in the 1930s and remained below 10 percent through the 1970s, but by 2007, it had risen to 24 percent. Looking at household wealth rather than income, the rise of inequality has been even greater, with the share owned by the top 0.1 percent increasing to 22 percent from 9 percent three decades ago. In 2011, the top 1 percent of U.S. households controlled 40 percent of the nation's entire wealth. And while the U.S. case may be extreme, it is far from unique: all but a few of the countries of the Organization for Economic Cooperation and Development for which data are available experienced rising income inequality (before taxes and transfers) during the period from 1980 to 2009.

The French economist Thomas Piketty has famously interpreted this data by arguing that a tendency toward economic inequality is an inherent feature of capitalism. He sees the middle decades of the 20th century, during which inequality declined, as an exception to the rule, produced by essentially random shocks—the two world wars and the Great Depression—that led governments to adopt policies that redistributed income. Now, that the influence of those shocks has receded, life is returning to normal, with economic and political power concentrated in the hands of an oligarchy.

Piketty's work has been corrected on some details, but his claim that economic inequality is rising rapidly in most developed countries is clearly accurate. What most analyses of the subject miss, however, is the extent to which both the initial fall and the subsequent rise of inequality over the past century have been related to shifts in the balance of power between elites and masses, driven by the ongoing process of modernization.

In hunting-and-gathering societies, virtually everyone possessed the skills needed for political participation. Communication was by word of mouth, referring to things one knew of firsthand, and decision-making often occurred in village councils that included every adult male. Societies were relatively egalitarian.

The invention of agriculture gave rise to sedentary communities producing enough food to support elites with specialized military and communication skills. Literate administrators made it possible to coordinate large empires governing millions of people. This much larger scale of politics required specialized skills, including the ability to read and write. Word-of-mouth communication was no longer sufficient for political participation: messages had to be sent across great distances. Human memory was incapable of recording the tax base or military manpower of large numbers of districts: written records were needed. And personal loyalties were inadequate to hold together large empires: legitimating myths had to be propagated by religious or ideological specialists. This opened up a wide gap between a relatively skilled ruling class and the population as a whole, which consisted mainly of scattered, illiterate peasants who lacked the skills needed to cope with politics at a distance. And along with that gap, economic inequality increased dramatically.

This inequality was sustained throughout history and into the early capitalist era. At first, industrialization led to the ruthless exploitation of workers, with low wages, long workdays, no labor laws, and the suppression of union organizing. Eventually, however, the continuation of the Industrial Revolution narrowed the gap between elites and masses by redressing the balance of political skills. Urbanization brought people into close proximity; workers were concentrated in factories, facilitating communication; and the spread of mass literacy put them in touch with national politics, all of which led to social mobilization. In the late 19th century and early 20th century, unions won the right to organize, enabling workers to bargain collectively. The expansion of the franchise gave ever more people the vote, and leftist political parties mobilized the working class to fight for its economic interests. The result was the election of governments that adopted various kinds of redistributive policies—progressive taxation, social insurance, and an expansive welfare state—that caused inequality to decline for most of the 20th century.

The emergence of a postindustrial society, however, changed the game once again. The success of the modern welfare state made further redistribution seems less urgent. Noneconomic issues emerged that cut across class lines, with identity politics and environmentalism drawing some wealthier voters to the left, while cultural issues pushed many in the working class to the right. Globalization and deindustrialization undermined the strength of unions. And the information revolution helped establish a winner-take-all economy. Together these eroded the political base for redistributive policies, and as those policies fell out of favor, economic inequality rose once more.

Today, large economic gains are still being made in developed countries, but they are going primarily to those at the very top of the income distribution, whereas those lower down have seen their real incomes stagnate or even diminish.

The rich, in turn, have used their privilege to shape policies that further increase the concentration of wealth, often against the wishes and interests of the middle and lower classes. The political scientist, Martin Gilens, for example, has shown that the U.S. government responds so attentively to the preferences of the most affluent ten percent of the country's citizens that "under most circumstances, the preferences of the vast majority of Americans appear to have essentially no impact on which policies the government does or doesn't adopt."

Because advantages tend to be cumulative, with those born into more prosperous families receiving better nutrition and health care, more intellectual stimulation and better education, and more social capital for use in later life, there is an enduring tendency for the rich to get richer and the poor to be left behind. The extent to which this tendency prevails, however, depends on a country's political leaders and political institutions, which in turn tend to reflect the political pressures emerging from mobilized popular forces in the political system at large.

The extent to which inequality increases or decreases, in other words, is ultimately a political question.

Today, the conflict is no longer between the working class and the middle class; it is between a tiny elite and the great majority of citizens. This means that the crucial questions for future politics in the developed world will be how and when that majority develops a sense of common interest. The more current trends continue, the more pressure will build up to tackle inequality once again. The signs of such a stirring are already visible, and in time, the practical consequences will be as well.

Not about the Money

For the first ⅔ of the 20th century, working-class voters in developed countries tended to support parties of the left, and middle- and upper-class voters tended to support parties of the right. With partisan affiliation roughly correlating with social class, scholars found, unsurprisingly, that governments tended to pursue policies that reflected the economic interests of their sociopolitical constituencies.

As the century continued, however, both the nature of the economy and the attitudes and behaviors of the public changed. An industrial society gave way to a postindustrial one, and generations raised with high levels of economic and physical security during their formative years displayed a "postmaterialist" mindset, putting greater emphasis on autonomy and self-expression. As postmaterialists became more numerous in the population, they brought new issues into politics, leading to a decline in class conflict and a rise in political polarization based on noneconomic issues (such as environmentalism, gender equality, abortion, and immigration).

This stimulated a reaction in which segments of the working class moved to the right, reaffirming traditional values that seemed to be under attack. Moreover, large immigration flows, especially from low-income countries with different languages, cultures, and religions, changed the ethnic makeup of advanced industrial societies. The rise of religious fundamentalism in the United States and xenophobic populist movements in western European countries represents a reaction against rapid cultural changes that seem to be eroding basic social values and customs—something particularly alarming to the less secure groups in those countries.

All of this has greatly stressed existing party systems, which were established in an era when economic issues were dominant and the working class was the main base of support for sociopolitical change. Today, the most heated issues tend to be noneconomic, and support for change comes increasingly from postmaterialists, largely of middle-class origin. Traditional political polarization centered on differing views about economic redistribution, with workers' parties on the left and conservative parties on the right. The emergence of changing values and new issues gave rise to a second dimension of partisan polarization, with postmaterialist parties at one pole and authoritarian and xenophobic parties at the opposite pole.

The classic economic issues did not disappear. But their relative prominence declined to such an extent that by the late 1980s, noneconomic issues had become more prominent than economic issues in Western political parties' campaign platforms. A long-standing truism of political sociology is that working-class voters tend to support the parties of the left and middle-class voters those of the right. This was an accurate description of reality around 1950, but the tendency has grown steadily weaker. The rise of postmaterialist issues tends to neutralize class-based political polarization. The social basis of support for the left has increasingly come from the middle class, even as a substantial share of the working class has shifted its support to the right.

In fact, by the 1990s, social-class voting in most democracies was less than half as strong as it was a generation earlier. In the United States, it had fallen so low that there was virtually no room for further decline. Income and education had become much weaker indicators of the American public's political preferences than religiosity or one's stand on abortion or the same-sex marriage: by wide margins, those who opposed abortion and the same-sex marriage supported the Republican presidential candidate over the Democratic candidate. The electorate had shifted from class-based polarization toward value-based polarization.

The Machine Age

In 1860, the majority of the U.S. work force was employed in agriculture. By 2014, less than 2 percent was employed there, with modern agricultural technology enabling a tiny share of the population to produce even more food than before. With the transition to an industrial society, jobs in the agricultural sector virtually disappeared, but this didn't result in widespread unemployment and poverty, because there was a massive rise in industrial employment. By the 21st century, automation and outsourcing had reduced the ranks of industrial workers to 15 percent of the work force—but this too did not result in widespread unemployment and poverty, because the loss of industrial jobs was offset by a dramatic rise in service-sector jobs, which now make up about 80 percent of the U.S. workforce.

Within the service sector, there are some jobs that are integrally related to what has been called "the knowledge

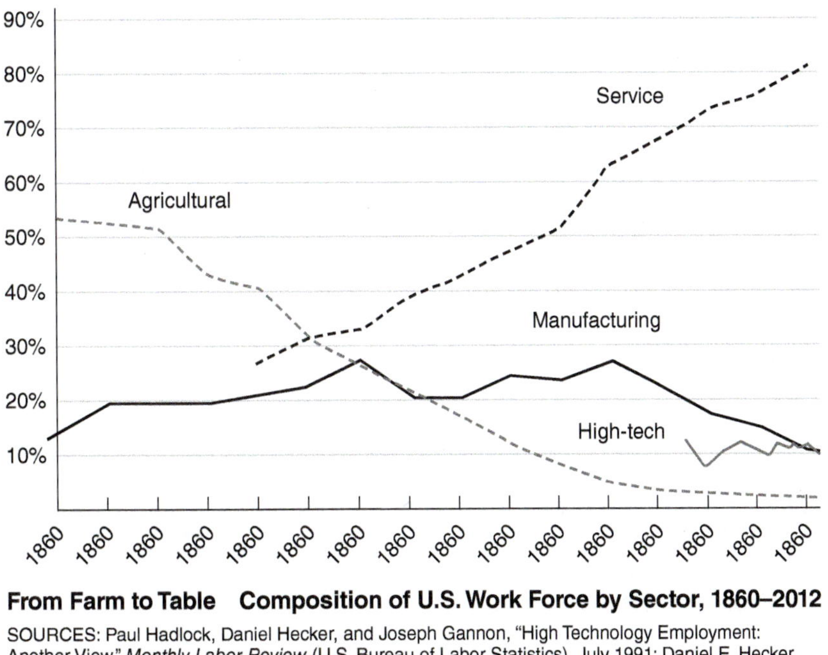

From Farm to Table Composition of U.S. Work Force by Sector, 1860–2012

SOURCES: Paul Hadlock, Daniel Hecker, and Joseph Gannon, "High Technology Employment: Another View," *Monthly Labor Review* (U.S. Bureau of Labor Statistics), July 1991; Daniel E. Hecker, "High Technology Employment: A NAICS-Based Update," *Monthly Labor Review* (U.S. Bureau of Labor Statistics), July 2005; Stanley Lebergott, "Labor Force and Employment, 1800–1960," in *Output, Employment, and Productivity in the United States After 1800*, ed. Dorothy S. Brady (National Bureau of Economic Research, 1966), 117–204; National Science Board, 2014; U.S. Bureau of Labor Statistics, 2014; U.S. Census Bureau, 1977.

NOTE: Data are not available for the service sector before 1900 or for the high-tech sector before 1986.

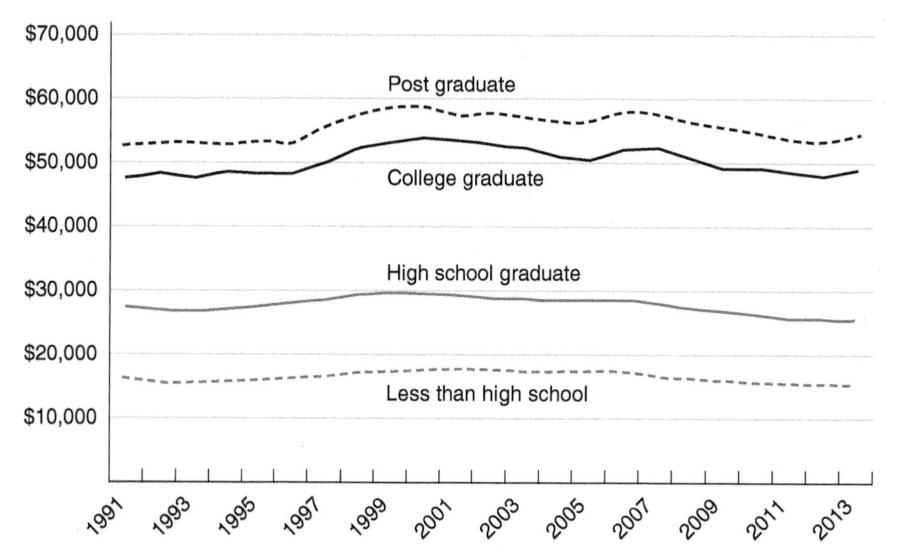

The World Is Flat Median Real Income by Educational Level in the United States, 1991–2013

SOURCE: U.S. Census Bureau, 2014.

NOTE: Incomes are in 2013 dollars.

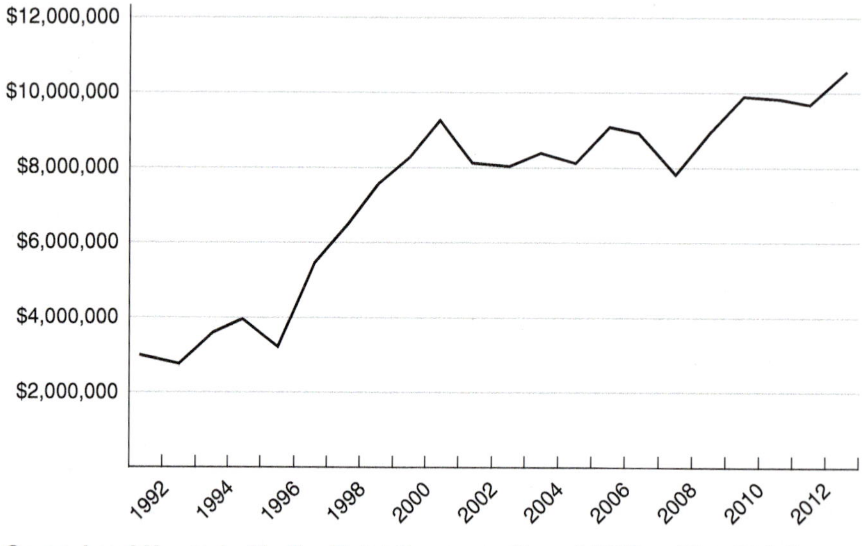

Onward and Upward Median Total Compensation of CEOs of Top 500 Corporations, 1992–2013

SOURCES: CEO salary data for 1998–2008 are from Carola Frydman and Dirk Jenter, "CEO Compensation," *Annual Review of Economics*, 2010; data for 2009–13 are from Joann S. Lublin, "CEO Pay in 2010 Jumped 11%," *Wall Street Journal*, May 9, 2011, and Ken Sweet, "Median CEO Pay Crosses $10 Million in 2013," Associated Press, May 27, 2014.

NOTE: Incomes are in 2013 dollars.

economy"—defined by the scholars Walter Powell and Kaisa Snellman as "production and services based on knowledge-intensive activities that contribute to the accelerated pace of technical and scientific advance." Because of its economic significance, the knowledge economy is worth breaking out as a separate category from the rest of the service sector; it is represented by what can be termed "the high-tech sector," which includes everyone employed in the information, finance, insurance, professional, scientific, and technical services categories of the economy.

Some assume that the high-tech sector will produce large numbers of high-paying jobs in the future. But employment in this area does not seem to be increasing; the sector's share of total employment has been essentially constant, since statistics became available about three decades ago. Unlike the transition from an agricultural to an industrial society, in other words, the rise of the knowledge society is not generating a lot of good new jobs.

Initially, only unskilled workers lost their jobs to automation. Today, even highly skilled occupations are being taken over by computers. Computer programs are replacing lawyers who used to do legal research. Expert systems are being developed that can make medical diagnoses better and faster than physicians. The fields of education and journalism are on their way to being automated. And increasingly, computer programs themselves may be written by computers.

As a result of such developments, even highly skilled jobs are being commodified, so that even many highly educated workers in the upper reaches of the income distribution are not moving ahead, with gains from the increases in GDP limited to those in a thin stratum of financiers, entrepreneurs, and managers at the very top. As expert systems replace people, market forces alone could conceivably produce a situation in which a tiny but extremely well-paid minority directs the economy, while the majority have precarious jobs, serving the minority as gardeners, waiters, nannies, and hairdressers—a future foreshadowed by the social structure of Silicon Valley today.

The rise of the postindustrial economy narrowed the life prospects of most unskilled workers, but until recently, it seemed that the rise of the knowledge society would keep the door open for those with sophisticated skills and a good education. Recent evidence, however, suggests that this is no longer true.

Between 1991 and 2013, real incomes in the United States stagnated across the educational spectrum. The highly educated still make substantially larger salaries than the less educated, but it is no longer just the unskilled workers who are being left behind.

The problem is not aggregate growth in the economy. During these years, U.S. GDP increased significantly. So where did the money go? To the elite of the elite, such as the CEOs of the country's largest corporations.

During a period in which the real incomes of even highly educated professionals, such as doctors, lawyers, professors, engineers, and scientists, were essentially flat, the real incomes of CEOs more than tripled. The pattern is even starker over a longer timeframe. In 1965, CEO pay at the largest 350 U.S. companies was 20 times as high as the pay of the average worker; in 1989, it was 58 times as high; and in 2012, it was 273 times as high.

Workers of the World, Unite?

Globalization is enabling half of the world's population to escape subsistence-level poverty but weakening the bargaining position of workers in developed countries. The rise of the knowledge society, meanwhile, is helping divide the economy into a small pool of elite winners and vast numbers of precariously employed workers. Market forces show no signs of reversing these trends on their own. But politics might do so, as growing insecurity and relative immiseration gradually reshape citizens' attitudes, creating greater support for government policies designed to alter the picture.

There are indications that the citizens of many countries are becoming sensitized to this problem. Concern over income inequality has increased dramatically during the past three decades. In surveys carried out from 1989 to 2014, respondents around the world were asked whether their views came closer to the statement "Incomes should be made more equal" or "Income differences should be larger to provide incentives for individual effort." In the earliest polls, majorities in ⅘ of the 65 countries surveyed believed that greater incentives for individual effort were needed. By the most recent surveys, however, that figure had dropped by half, with majorities in only two-fifths of the countries favoring that. Over a 25-year period in which income inequality increased dramatically, publics in 80 percent of the countries surveyed, including the United States, grew more supportive of actions to reduce inequality, and those beliefs are likely to intensify over time.

New political alignments, in short, might once again readjust the balance of power between elites and masses in the developed world, with the emerging struggle being between a tiny group at the top and a heterogeneous majority below. For the industrial society's working-class coalition to become effective, lengthy processes of social and cognitive mobilization had to be completed. In today's postindustrial society, however, a large share of the population is already highly educated, well informed, and in possession of political skills; all it needs to become politically effective is the development of an awareness of common interest.

Will enough of today's dispossessed develop what Marx might have called "class consciousness" to become a decisive political force? In the short run, probably not, because of the presence of various hot-button cultural issues cutting across economic lines. Over the long run, however, they probably will, as economic inequality and the resentment of it are likely to continue to intensify.

It was the rise of postmaterialist values, together with a backlash against the changes that the postmaterialists spearheaded, that helped topple economic issues from their central role in partisan political mobilization and install cultural issues in their place. But the continued spread of postmaterialist values is draining much of the passion from the cultural conflict, even as the continued rise of inequality is pushing economic issues back to the top of the political agenda.

During the 2004 U.S. presidential election, for example, the same-sex marriage was so unpopular in some quarters that Republican strategists deliberately put referendums banning it on the ballot in crucial swing states in the hope of increasing turnout among social conservatives in the middle and lower echelons of the income distribution. And they were smart to do so, for the measures passed in every case—as did virtually all others like them put forward from 1998 to 2008. In 2012, however, there were five new statewide referendums on the topic, and in four of them, the public voted in favor of legalization. Crosscutting cultural divisions still exist and can still divert attention from common economic interests, but the former no longer trump the latter as reliably as they used to. And the fact that not just all the Democrats but even several 2016 Republican presidential candidates have pledged to abolish the tax break on "carried interest" benefiting elite financiers might well be a portent of things to come.

The essence of modernization is the linkages among economic, social, ideational, and political trends. As changes ripple through the system, developments in one sphere can drive developments in the others. But the process doesn't work in just one direction, with economic trends driving everything else, for example. Social forces and ideas can drive political actions that reshape the economic landscape. Will that happen once again, with popular majorities mobilizing to reverse the trend toward economic inequality? In the long run, probably: publics around the world increasingly favor reducing inequality, and the societies that survive are the ones that successfully adapt to changing conditions and pressures. Despite current signs of paralysis, democracies still have the vitality to do so.

Critical Thinking

1. Why does the author think that economic issues are returning to the top of the political agenda?
2. What is meant by postmaterial values and why are they important?

Internet References

Group of 20
https://en.wikipedia.org/wiki/G20

The World Bank Group
www.worldbank.org

Ronald Inglehart is a professor of political science at the University of Michigan and Founding President of the World Values Survey.

Article Prepared by: Robert Weiner, *University of Massachusetts, Boston*

How NAFTA has Changed Mexico

KATHLEEN STAUDT

Learning Outcomes

After reading this article, you will be able to:

- Why NAFTA has perpetuated an asymmetrical trading relationship between the United States and Mexico.
- Why is Mexico one of the most unequal countries in the world in terms of income.

The North American Free Trade Agreement (NAFTA) took effect nearly a quarter-century ago, on January 1, 1994. While it is difficult to say to what extent NAFTA was a direct cause of the dramatic transformation that Mexico has undergone in recent decades, there is no doubt that it has changed the country in many ways, both positive and negative. Yet one clear consequence of the trade agreement is that it has also perpetuated historical patterns, particularly Mexico's dependence on the United States. This asymmetrical relationship has not benefited most workers in either country.

As with any major policy change, NAFTA has produced winners and losers in its three member nations: Canada, Mexico, and the United States. In Mexico, the effects have varied widely among the lower, middle, and upper classes; among residents of different regions; and among economic sectors, ranging from agriculture to the automotive, aerospace, and electronics industries. Many Mexicans continue to rely on the informal economy, whether as their main earning strategy or as a fallback during economic downturns. Due to the increased outsourcing of work to Mexico by US companies and the proliferation of joint production processes, NAFTA has exacerbated the impact of US recessions on Mexico.

Other developments in the years since NAFTA took effect have also changed Mexico. The country faced difficult economic and political circumstances in the mid-1990s. While still recovering from a 1980s debt crisis, it endured a drastic devaluation of the peso in US dollar terms in 1994–95 and the shocking assassination of presidential candidate Luis Donaldo Colosio-Murrieta in March 1994. Although he was affiliated with the dominant establishment party, the Institutional Revolutionary Party (PRI), Colosio was a popular politician and a leading contender to win the presidential election that August.

A political opening after decades of one-party rule by the PRI has since led to opposition-party victories at the presidential, state, and municipal levels in Mexico. However, the activities of transnational crime organizations involved in the production and sale of drugs, human trafficking, kidnapping, and extortion, often with the complicity of government officials and the police, have led to pervasive insecurity. Over 235,000 murders occurred from 2007 to 2017, according to Mexico's National System of Public Security.

A larger volume of freer trade, combined with official efforts to ease border congestion, enhances opportunities for drug smuggling in the increased truck and car traffic across the border. The size of these illicit flows is so large that businesses have emerged to track trucks from point of origin to destination. (Security still trumps trade at the US-Mexico border.)

Much attention has been focused on NAFTA since Donald Trump's 2016 US presidential campaign and in the first year of his presidency, given his frequent criticism of Mexico and Mexicans, his description of NAFTA as the worst-ever trade deal for the United States, and his threats to pull out unless Mexico and Canada meet his demands for revising the pact. His administration initiated a renegotiation of the treaty that has been extended into 2018. US Trade Representative Robert Lighthizer and his team have made demands for rule changes that would reduce the US trade deficit with Mexico in goods, even though they are often produced through border-spanning supply chains. (The United States enjoys a cross-border trade surplus in services.)

The Trump administration's demands are potential deal-breakers for Mexico and Canada. In any case, what often gets lost in the controversy over NAFTA's effects on American workers is how the trade deal has worked out for their Mexican counterparts.

Widening Gap

While NAFTA did not totally open the borders for people, goods, or services, it did change the rules on trade in ways that gave US and Canadian companies incentives to outsource production to Mexico. They sought to take advantage of low-cost labor as well as the geographic advantages of shipping their products from a neighboring country under the new regime of low-to-no tariffs on the majority of goods produced in the region.

In the years since NAFTA took effect, trading volume has increased among all three countries, and all three have seen economic growth and higher gross domestic product per capita (a flawed but commonly used aggregate measure that does not take into account class inequalities or differences among geographic regions). But there is still a tenfold disparity between the United States and Mexico in the minimum wage, and an approximately fivefold difference in GDP per capita.

These gaps put the pair seventeenth from the bottom on a list of 200 countries with shared borders (based on 2004 data) compiled by the Spanish researcher Iñigo Moré in his 2011 book *The Borders of Inequality*. In an assessment using 2014 data for my book *Border Politics in a Global Era*, I found that the United States and Mexico rank among the most unequal 40 country pairs out of 300 that share a land border. My longitudinal study showed that the gaps between the two countries have widened since NAFTA took effect.

NAFTA created many new low-income jobs in Mexico, as well as some better-paying jobs in both the foreign-owned factories and in Mexican-owned businesses that serve them. After 1994, Mexico's GDP per capita rose and dipped before settling into a pattern of modest growth. The national economy's growth rate also fluctuated at first but has recorded modest annual increases in recent years, despite setbacks during the US downturn of 2001 and the recession of 2007–8. However, this NAFTA-era expansion has never achieved the spectacular rates of growth recorded from the mid-1940s through the 1970s, a period known as the "Mexican miracle," when Mexico's trade strategies centered on import-substitution industrialization. Those protectionist strategies were discarded in the era of economic globalization that has prevailed since the 1980s.

Well before NAFTA, Mexico was opening up to international trade, with its entry into the General Agreement on Tariffs and Trade (GATT) in 1986 and membership in the World Trade Organization in 1995. But even earlier, Mexico had begun experimenting with new ways to spur trade with its wealthy northern neighbor, most notably by establishing the Border Industrialization Program (BIP). Implemented in 1965 in conjunction with a cut in US tariffs on value-added imports from Mexico, the program aimed to foster formal employment in assembly-line jobs producing goods for export in manufacturing plants known as *maquiladoras* or *maquilas*. The model of economic growth initiated by the BIP and extended by NAFTA has perpetuated Mexico's economic dependence on the United States, its most important trading partner.

The BIP was intended to serve several purposes: to decentralize industrial employment by encouraging economic development away from already-overpopulated Mexico City; to attract more foreign direct investment (FDI) from various countries; and to provide jobs for men returning from the United States after stints as guest workers under the 1942–64 Bracero Program. However, the *maquiladora* plants along Mexico's northern border initially recruited mostly female workers, who were favored for their supposedly nimble fingers and compliant attitudes. As the plants proliferated, numbering in the hundreds in the major border cities including Tijuana and Ciudad Juárez, the percentage of women in the workforce decreased from 80 percent to 55–60 percent by the late 1980s and continued toward a rough gender balance thereafter.

Winners and Losers

Under BIP in the 1960s and now under NAFTA, the legal minimum wage in Mexico for assembly-line workers, who constitute at least three-fourths of the workforce in *maquiladoras*, has been based on a calculation of net daily pay, and is currently set at the equivalent of $4.50 a day. Most studies have found that real wages have been stagnant since the 1970s due to inflation, currency fluctuations, and the stubborn reluctance of the Mexican political elite to increase the legal minimum wage for fear of discouraging foreign investors who are drawn to Mexico by its comparative advantage: cheap labor.

That approach has produced some winners, including an expanded middle class with new professional jobs, investors, and new businesses that serve the export sector. There has been job creation in the formal sector of the economy (albeit these are largely low-wage jobs). After a decade of NAFTA, the World Bank estimated that FDI in Mexico would have been 40 percent lower without the treaty. But there have also been losers. In an article in the newsmagazine *Proceso* in August 2017, analysts estimated that the legal minimum wage adjusted for inflation was 22 percent less than it was in 1994.

Net pay is what matters to working people living at the margin of survival. But the private sector generally calculates wages differently—as a total compensation package including employer contributions averaging the equivalent of $2.10 per hour to health plans, the social security system, subsidized lunches, and transportation (the latter may be deducted from workers' net pay). In areas with a labor shortage, including the northern border region in recent years, some employers have offered bonuses or increased workers' takehome, or net, pay to twice the minimum wage, approximately $8 to $9 a day. But other companies have responded by sending recruiters to Mexico's central and southern states, aiming to increase the labor pool and keep wages low.

Most commentators agree that given the cost of food and housing, a "dignified" or living wage would amount to at least three times the minimum wage, or the equivalent of $13 to $14 a day. A living wage does not automatically guarantee middle-class status, which the Mexican census defines by criteria including education level, income (beginning at the equivalent of $850 per month), and household possessions. Mexico's middle class has grown to 39 percent of the population, based on the most recent figures from the National Institute of Statistics and Geography.

In his 2015 book *Mexico's Uneven Development*, historian Oscar Martínez cites figures for the size of the nation's middle class that range from 25 to 40 percent of the population—or up to 60 percent according to the most optimistic estimates. A growing middle class bodes well for businesses that can grow to satisfy demands for domestic and imported goods from people with increasing disposable income. However, it should be noted that the middle class has also been growing in the rest of Latin America without any help from NAFTA.

Although a majority of Mexicans still live in poverty, according to the government's standards, the share of the population under the poverty line has decreased since the unusually crisis-ridden mid-1990s, from a peak of 70 percent to 59 percent, according to the 2014 census. Despite those gains, the absolute number of people living under the poverty line has increased by 14 million since NAFTA took effect, due to Mexico's population growth. No doubt that number would be even higher if migrants, both documented and undocumented, had not headed north. Moreover, the poverty rate in many other countries in Latin America has not only decreased but has fallen more substantially than in Mexico, again without any help from NAFTA.

Meanwhile, Mexico's upper class represents just over one percent of the population. The country is home to the sixth-richest man in the world, Carlos Slim Helú, a telecommunications tycoon worth more than $54 billion, according to *Forbes* magazine's 2017 list of the world's wealthiest people (he topped the list from 2010 to 2013). Slim is one of 15 billionaires in Mexico who have a combined net worth of $100 billion. Despite these outliers at the top, Mexico's Gini coefficient (a measure of internal inequality) has moved slightly in the direction of more equality, from 0.54 to 0.47 (on a scale where 0 represents perfect equality and 1 would mean a lone individual holds all the country's wealth). But after a quarter-century of NAFTA, Mexico is still stuck among the most unequal countries in the world.

More people have risen above extreme poverty, but a majority remains stuck in the ranks of the working poor, with migration offering the only way out. The middle class has increased in size, boosted by the growth of professional and paraprofessional jobs and managerial positions in export-oriented manufacturing plants and in other Mexican businesses that have grown alongside those enterprises. Yet Mexico continues to be dependent on its North American neighbors. It lacks the greater independence that would come with a more diversified array of trade partners.

Rural Repercussions

Manufacturing jobs have spread outside of major urban areas like Mexico City and Monterrey to northern border cities such as Tijuana, Nogales, Ciudad Juárez, Reynosa, and Matamoros. This has created a huge workforce in the northern border urban areas, with over a million poorly paid *maquiladora* workers living next to the United States. More recently, plants have spread to states that compete with one another to attract foreign investment. One relative success story is the central state of Querétaro.

However, southern states, which have larger proportions of indigenous people, have experienced negative NAFTA-influenced change, primarily due to the destruction of the small-scale agricultural sector. This has made family farmers the biggest losers among all economic sectors. Regions that are dependent on small-scale agriculture have been devastated by declines in state subsidies, land privatization, outmigration, and competition with large farms and cheap US corn exports.

Small-scale farmers were already imperiled as a result of the debt crisis of the 1980s and neoliberal policies that were imposed on Mexico as a condition for entering global trade agreements such as the GATT, including reductions in the size of government and other reforms that gave greater latitude to market forces. The government removed price supports and subsidies for staple foods like corn. On top of that, a 1992 Mexican agrarian reform law allowed for the sale, and thus the privatization, of communal land holdings known as *ejidos*. It is estimated that these reforms resulted in the displacement of over four million farm families, some of whom turned to seasonal labor in large-scale, corporate-style farms. Others resorted to migration, either toward urban areas in Mexico or to the United States and Canada.

While small-scale agriculture suffered under policies that preceded NAFTA as well as under the trade agreement, large-scale factory farming has boomed thanks to the growth in exports of fruits and vegetables to the United States and Canada. However, even large-scale farms in Mexico have not been able to compete in the production of corn, which can be grown more cheaply and efficiently in the United States thanks to government subsidies. Once self-sufficient in corn, Mexico under NAFTA has become the largest importer of US corn and is now dependent on its northern neighbor for one of its most important staple foods. However, the hostile rhetoric from Trump led Mexico in mid-2017 to negotiate with Brazil and Argentina for corn imports at lower prices—and to consider a revival of its own corn industry.

Low-wage Manufacturing

In the automotive sector, Mexico and once-dominant auto producers in the US Midwest have developed integrated production processes that involve a stream of multiple exports and imports, each adding value to the final product. These transactions are not effectively counted in traditional calculations of trade deficits that assume one-time exports and imports. US NAFTA negotiators appear to ignore the reality of cross-border supply chains.

Workers in Mexican auto assembly plants earn somewhat higher wages than those employed by subcontractors that make auto parts and harnesses, the complex electronic systems inside constantly reengineered automobile bodies. Yet these workers still earn just a tenth of what autoworkers make in the United States and Canada, leaving them unable to afford to buy the cars they build.

The aerospace industry has developed in central Mexico thanks to Canadian-owned Bombardier Inc., which has produced airplanes and trains in Querétaro since 2005 and, more recently, watercraft and all-terrain vehicles. The state has positioned itself as a hub for higher-skilled manufacturing and is now home to more than 30 such plants, having invested in scores of post-secondary educational institutions that specialize in aerospace. Bombardier has formed a unique partnership with the state-supported National Aeronautics University of Querétaro. The demand for engineers, technicians, and other professionals with advanced degrees has raised hopes for the continued growth of the middle class in Querétaro and in the country as a whole.

However, pay levels are still well below those in Canada. An engineer who earns $35 an hour in Quebec might be paid $60 a day in Mexico, according to a 2014 article in *Canadian Business* magazine about Bombardier's operations in Querétaro. While the cost of living is perhaps 2 to 3 times lower in Mexico than in Canada (with more precise comparative figures specific to city locations), a fivefold wage difference is significant. The article notes that while two-thirds of Bombardier's Mexican workforce is unionized, multinational companies that have invested in the state have a "gentleman's agreement" to avoid wage competition for local workers. Instead, they seek to attract labor with subsidized meals and transportation.

More typical manufacturing industries can be found in cities along Mexico's northern border that have long been magnets for migrants. Many studies over the past 30 years have documented the grim conditions for workers in these cities. In a 2017 article for the Americas Program, a Mexico City–based think tank, journalist Kent Paterson summarized conditions for the 275,000 workers in Ciudad Juárez who travel daily from their modest or "ramshackle" homes in underdeveloped neighborhoods at the periphery to jobs in more than 300 manufacturing plants now operating in the city. Their pay averages under $5 per day at electronics manufacturer Foxconn or $8 a day at auto battery maker Johnson Controls.

Maquiladora companies have kept wages low by adopting "speed-up practices," which include giving one worker responsibility for tasks previously performed by several workers. This creates stressful working conditions but reduces costs and raises productivity. Employers pay specialized technicians assembly-line wages, rather than compensate them for the value added by their work. Child-care centers are rare to nonexistent, so many children are left on their own while parents toil in *maquiladoras*.

Labor shortages have forced some plants to compete for workers with hiring bonuses. But few independent unions exist to represent workers' interests. Many plants started out with "paper unions" that had no clout but enabled the companies to comply with a NAFTA side agreement that requires each country to enforce its own labor laws.

Several work stoppages in 2015–16 at six plants in Ciudad Juárez called public attention to workers' demands for higher wages, better treatment, and independent unions. However, the strikes did not result in any systemic changes. The workers' lawyers negotiated agreements at each plant that reinstated some workers who had been living precariously without income for months. The 2016 Juárez Strategic Plan compared salaries in *maquiladors* nationwide and found that the city was near the bottom in average wage, ranking 31st out of the 33 cities studied.

For decades, Juárez, a city of 1.5 million, was known for disposable people and disposable labor. The city became infamous in the 1990s and 2000s for *feminicidio* (killings of women), often in sexualized ways including rape. There were 370 of these murders from 1993 to 2003, according to multiple sources, including Amnesty International. The situation only worsened after competition among transnational criminal organizations and the deployment of the Mexican military and the federal police force turned the city into what journalists called the world's murder capital. *Feminicidio* represented just under 20 percent of the 11,000 murders during the peak violence in 2008–12. Victims were often dismissed as "collateral damage" in a country with police impunity and few prosecutions or convictions. Yet the growing maquiladora industry seemed unscathed by the fearful atmosphere.

Environmental Benefits

When NAFTA was originally negotiated in the early 1990s, critics of the deal in the United States won compromises that resulted in two side agreements. One focused on retraining US workers, those certified as NAFTA-displaced, with programs for less-skilled workers who could not effectively compete in the changing manufacturing industry. The other addressed environmental conditions in the borderlands, defined as a zone extending 100 kilometers north and south of the border.

Most analysts view the labor side agreement as weak and ineffective, but the environmental side agreement produced three lasting institutions: the North American Development Bank (NADBank), based in San Antonio; the Border Environment Cooperation Commission (BECC), based in Juárez, focused on the US-Mexico borderlands; and the trinational Center for Environmental Cooperation (CEC), based in Montreal. The CEC is funded with equitable contributions from the three member nations and has a rotating leadership composed of appointees with three-year terms, as well as a clearinghouse for citizens' complaints.

In 2017, the NADBank and BECC boards consolidated into one institution, based in San Antonio. NADBank is perhaps the biggest success story for the Mexican borderlands. It has financed many large-scale projects in the area with loans and grants, addressing issues such as sanitary landfills, water treatment, sustainable wind power, and paper recycling.

Dependency Trap

NAFTA has changed Mexico for better and for worse. The agreement has boosted foreign direct investment, trade volume, GDP, and GDP per capita, and also appears to have enlarged the middle class. Infrastructure projects financed by NADBank have improved environmental quality and health in the northern borderlands.

The positive consequences, however, have been outweighed by the negative effects. Mexico's political elites share the blame for that. They have mismanaged the nation's supposedly democratic institutions and have failed to reduce poverty among the working poor, permit the emergence of independent unions, or increase security by building more professional and honest police forces.

The devastation of the small-scale agricultural sector left impoverished farmers with little recourse but to migrate or to work seasonally in even more precarious conditions for large corporate-style farms that export fruit and vegetables to the United States. Overall, about half of the population remains stuck under the national poverty line, a proportion largely unchanged since the early days of NAFTA, not counting the temporary crisis marked by the 1994 peso devaluation, when poverty peaked at 70 percent.

NAFTA's effects on Mexican labor diverge in urban and rural areas; in the northern, central, and southern regions; and by various economic sectors. While the aerospace industry model seems promising, with its associated technical education and skills, the wages for professional workers still fall well short of those in Canada. The small-scale agricultural sector suffered the most under NAFTA (and previous policies linked to free trade).

The reality for the majority of workers, particularly at the northern border, is widespread impoverishment without effective unions or supply-and-demand forces to raise wages. Mexico's NAFTA negotiators have expressed reluctance to consider minimum wage increases, on the grounds that wages are a sovereign internal matter. Mexico's Business Council and its Employers' Confederation announced in 2017 a goal to achieve a modest increase in the minimum wage to 92 pesos a day, the equivalent of just about 50 cents more.

Mexico's political class hardly represents the interests of the working poor. NAFTA allowed that class to consolidate its power in the interest of generating more foreign direct investment. However, after nearly a quarter of a century, one might expect the extensive investments and growth to have generated a steady stream of trickle-down benefits for working people. Instead, only a few crumbs have fallen from the elite table.

NAFTA has kept Mexico on its long track of dependence on the United States. Donald Trump's hateful rhetoric about Mexico and Mexicans should change that. Trump has emboldened Mexican advocates of nationalist populism, and some decision makers are now more willing to consider alternatives to NAFTA, including trade agreements with other partners. Time will tell whether Mexico's elections in 2018 will produce a president and congressional majority capable of speaking more strongly for the working poor and leveraging the gains in foreign investment and the growth of a higher-skilled labor force in order to forge new economic partnerships.

For now, foreign investors seem all too willing to take advantage of Mexico's low-cost labor. If Trump and the US Congress dump NAFTA, which would be much to the dismay of many US businesses and Midwestern farmers, no doubt trade with Mexico will continue, but it will adjust to a new reality of diversification in Mexico's economy and trading partners. After all, Mexico is the 15th-largest economy in the world and it belongs to the World Trade Organization. Until that adjustment occurs, however, short-term job losses will put even more pressure on Mexico's poor majority to migrate northward, even as the border hardens.

Critical Thinking

1. What is meant by the "Mexican miracle"?
2. Why is Mexico still economically dependent on the United States?
3. Why has the middle class increased in Mexico?

Internet References

Evaluating the new NAFTA
https://www.cato.org/blog/grading-new-nafta

The North American Free Trade Association
www.naftanow.org/

United States-Mexico-Canada Agreement text
https://ustr.gov/trade-agreements/...trade-agreements/...agreement/united-states-mexico

Kathleen Staudt recently retired as a professor of political science at the University of Texas at El Paso. Her latest book is *Border Politics in a Global Era: Comparative Perspectives* (Rowman & Littlefield, 2017).

Article

Prepared by: Robert Weiner, *University of Massachusetts, Boston*

The Return of Europe's Nation-States: The Upside to the EU's Crisis

JAKUB GRYGIEL

Learning Outcomes

After reading this article, you will be able to:

- Discuss Europe's worst political crisis since World War II
- Explain why the United Kingdom voted to leave the European Union.

Europe currently finds itself in the throes of its worst political crisis since World War II. Across the continent, traditional political parties have lost their appeal as populist, Euroskeptical movements have attracted widespread support. Hopes for European unity seem to grow dimmer by the day. The euro crisis has exposed deep fault lines between Germany and debt-ridden southern European states, including Greece and Portugal. Germany and Italy have clashed on issues such as border controls and banking regulations. And on June 23, the United Kingdom became the first country in history to vote to leave the EU—a stunning blow to the bloc.

At the same time as its internal politics have gone off the rails, Europe now faces new external dangers. In the east, a revanchist Russia—having invaded Ukraine and annexed Crimea—looms ominously. To Europe's south, the collapse of numerous states has driven millions of migrants northward and created a breeding ground for Islamist terrorists. Recent attacks in Paris and Brussels have shown that these extremists can strike at the continent's heart.

Such mayhem has underscored the price of ignoring the geopolitical struggles that surround Europe. Yet the EU, crippled by the euro crisis and divisions over how to apportion refugees, no longer seems strong or united enough to address its domestic turmoil or the security threats on its borders. National leaders across the continent are already turning inward, concluding that the best way to protect their countries is through more sovereignty, not less. Many voters seem to agree.

As Europe's history makes painfully clear, a return to aggressive nationalism could be dangerous, not just for the continent but also for the world. Yet a Europe of newly assertive nation-states would be preferable to the disjointed, ineffectual, and unpopular EU of today. There's good reason to believe that European countries would do a better job of checking Russia, managing the migrant crisis, and combating terrorism on their own than they have done under the auspices of the EU.

Ever-Farther Union

In the years after World War II, numerous European leaders made a convincing argument that only through unity could the continent escape its bloody past and guarantee prosperity. Accordingly, in 1951, Belgium, France, Italy, Luxembourg, the Netherlands, and West Germany created the European Coal and Steel Community. Over the next several decades, that organization morphed into the European Economic Community and, eventually, the European Union, and its membership grew from six countries to 28. Along the way, as the fear of war receded, European leaders began to talk about integration not merely as a force for peace but also as a way to allow Europe to stand alongside China, Russia, and the United States as a great power.

The EU's boosters argued that the benefits of membership—an integrated market, shared borders, and a transnational legal system—were self-evident. By this logic, expanding the union eastward wouldn't require force or political coercion; it would simply take patience, since nonmember states would soon recognize the upsides of membership and join as soon as they could. And for many years, this logic held, as central and eastern European countries raced to join the union after the collapse of the Soviet Union. Eight countries—the Czech Republic, Estonia, Hungary, Latvia, Lithuania, Poland, Slovakia, and Slovenia—became members in 2004; Bulgaria and Romania followed in 2007.

Then came the Ukraine crisis. In 2014, the Ukrainian people took to the streets and overthrew their corrupt president, Viktor Yanukovych, after he abruptly canceled a new economic deal with the EU. Immediately afterward, Russia invaded and annexed Crimea, and it soon sent soldiers and artillery into eastern Ukraine, too. The EU's leaders had hoped that economic inducements would inevitably increase the union's membership and bring peace and prosperity to an ever-larger public. But that dream proved no match for Russia's tanks and so-called little green men.

Moscow's gambit was not, on its own, enough to cripple the EU. But soon, another crisis hit, and this one nearly pushed the union to its breaking point. In 2015, more than a million refugees—nearly half of them fleeing the civil war in Syria—entered Europe, and since then, many more have followed. Early on, several countries, especially Germany and Sweden, proved especially welcoming, and leaders in those states angrily criticized those of their neighbors that tried to keep the migrants out. Last year, after Hungary built a razor-wire fence along its border with Croatia, German Chancellor Angela Merkel condemned the move as reminiscent of the Cold War, and French Foreign Minister Laurent Fabius said it did "not respect Europe's common values." But early this year, many of these same leaders changed their tune and began pressuring Europe's border countries to increase their security measures. In January, several European governments warned Greece that if it did not find a way to stanch the flow of refugees, they would expel it from the Schengen area, a passport-free zone within the EU.

Consciously or not, the European politicians advocating open borders have failed to prioritize their own citizens over foreigners. These leaders' intentions may be noble, but if a state fails to limit its protection to a particular group of people—its nationals—its government risks losing legitimacy. Indeed, the main measure of a country's success is how well it can secure its people and borders from external threats, be they hostile neighbors, terrorism, or mass migration. On this score, the EU and its proponents are failing. And voters have noticed. The British people issued a strong rebuke to the bloc in June when they voted to leave the EU by a margin of 52 percent to 48 percent, ignoring warnings from the International Monetary Fund, the Bank of England, and the United Kingdom's Treasury that doing so would wreak economic disaster. In France, according to a recent Pew survey, 61 percent of the population holds unfavorable views of the EU; in Greece, 71 percent of the population shares these views.

Back when Europe faced no pressing security threats—as was the case for most of the last two decades—EU members could afford to pursue more high-minded objectives, such as dissolving borders within the union. Now that dangers have returned, however, and the EU has shown that it is incapable of dealing with them, Europe's national leaders must fulfill their most basic duty: defending their own.

Back to Basics

The EU's architects created a head without a body: they built a unified political and administrative bureaucracy but not a united European nation. The EU aspired to transcend nation-states, but its fatal flaw has been its consistent failure to recognize the persistence of national differences and the importance of addressing threats on its frontiers.

One consequence of this oversight has been the rise of political parties that aim to restore national autonomy, often by appealing to far-right, populist, and sometimes xenophobic sentiments. In 2014, the UK Independence Party won the popular vote in an election for the European Parliament—the first time since 1906 that any party in the United Kingdom had bested Labor and the Conservatives in a nationwide vote. Last December in France, Marine Le Pen's far-right National Front won the first round of the country's regional elections; then, in March in Germany, a right-wing Euroskeptical party, Alternative for Germany, won almost 25 percent of the vote in Saxony-Anhalt. And in May, Norbert Hofer, a candidate from the far-right Freedom Party, narrowly lost Austria's presidential election. (Austria's Constitutional Court later annulled that result, forcing a rerun of the election that will be held in October.)

Some of these parties have benefited from the enthusiastic support of Russia, as part of its campaign to buy influence in Europe. Until recently, Moscow could rely on European leaders who were friendly to Russia, including former German Chancellor Gerhard Schröder and former Italian Prime Minister Silvio Berlusconi. But now, as new parties take the place of established ones, the Kremlin needs fresh partners. It has given money to the National Front, and the U.S. Congress has asked James Clapper, the U.S. director of national intelligence, to investigate the Kremlin's ties to other fringe parties, including Greece's Golden Dawn and Hungary's Jobbik. Yet such parties would be surging even without Russian backing. Many Europeans are disenchanted with politicians who have supported EU integration, open borders, and the gradual dissolution of national sovereignty; they have a deep and lasting desire to reassert the supremacy of their nation-state.

Of course, most of Europe's Euroskeptical politicians don't seek to disband the union entirely; in fact, many of them continue to see its creation as a historic victory for the West.

They do, however, want greater national autonomy on social, economic, and foreign policy, especially in response to overreaching EU mandates on migration and the demand for controversial continent-wide laws on issues such as abortion and marriage. Many in the United Kingdom, for example, pushed for a British exit from the EU, or Brexit, out of frustration with the number of British laws that have come from Brussels rather than Westminster.

The bet against sovereignty has failed. But sovereignty's resurgence has conjured up many dark memories of the nationalism that twice brought the continent to the brink of annihilation. Many observers now worry that European politics are coming to resemble those of the 1930s, when populist leaders spewed hate to whip up support. Such fears are not wholly unfounded. The strident xenophobia of Austria's Freedom Party recalls the early days of fascism. Anti-Semitism has risen across Europe, sprouting up in parties that span the ideological spectrum, from the United Kingdom's Labour Party to Hungary's Jobbik. And in Greece, some members of the radical left-wing party Syriza have advocated Greek withdrawal from NATO, a prime example of a growing anti-Americanism that could undermine the foundation of European security.

Yet affirming national sovereignty does not require virulent nationalism. The support for Brexit in the United Kingdom, for instance, was less an expression of hostility toward other European countries than it was an assertion of the United Kingdom's right to self-govern. A return to nation-states entails not nationalism but patriotism, or what George Orwell called "devotion to a particular place and a particular way of life." It's also worth noting that one of the greatest threats Europe faced in the twentieth century was transnational in nature: communism, which divided the continent for 45 years and led to the deaths of millions.

Beyond the EU

A renationalization of Europe may be the continent's best hope for security. The EU's founders believed that the body would guarantee a stable and prosperous Europe—and for a while, it seemed to. But today, although the EU has generated wealth through its common market, it is increasingly a source of instability. The euro crisis has exposed the union's inability to resolve conflicts among its members: German leaders have had little incentive to address Greek concerns, and vice versa. The EU also suffers from what the German Federal Constitutional Court has called a "structural democratic deficit." Of its seven institutions, just one—the European Parliament—is directly elected by the people, and it cannot initiate legislation. Finally, the recent dominance of Germany within the EU has alienated smaller states, including Greece and Italy.

Meanwhile, the EU has failed to keep Europe safe. Since 1949, Europe has relied on NATO—and, in particular, the United States—to secure its borders. The anemic defense spending of most European countries has only increased their dependence on the United States' physical presence in Europe. The EU is unlikely to create its own army, at least in the near future, as its members have different strategic priorities and little desire to cede military sovereignty to Brussels.

Many of the EU's backers still insist that in its absence, anarchy will engulf the continent. In 2011, the French minister for European affairs, Jean Leonetti, warned that the failure of the euro could lead Europe to "unravel." In May, British Prime Minister David Cameron claimed that a British exit from the EU would raise the risk of war. But as the American theologian Reinhold Niebuhr wrote in the 1940s, "the fear of anarchy is less potent than the fear of a concrete foe." Today, the identifiable enemies that have arisen around Europe, from Russia to the self-proclaimed Islamic State (also known as ISIS), seem far more worrying to most people than the potential chaos arising from the dissolution of the EU. Their hope is that individual countries will provide the kind of safety that Brussels can't.

Special Relationships

From the United States' perspective, the fraying of the EU presents a serious challenge—but not an insurmountable one. In the decades after World War II, Washington sought to contain the Soviet Union not just through nuclear deterrence and a sizable military presence in Europe but also by promoting European integration. A united continent, the thinking went, would pacify Europe, strengthen the economies of U.S. allies, and encourage them to cooperate with Washington to ward off the Soviet menace. Today, however, the United States needs a new strategy. Because the EU no longer seems up to the task of protecting its borders or competing geopolitically, more American pressure for Europe to integrate will simply alienate the growing number of Europeans who have turned their backs on the EU.

Washington need not fear the dissolution of the EU. Fully sovereign European states may prove more adept than the union at warding off the various threats on its frontiers. When Russia invaded Ukraine, the EU had no answer besides sanctions and vague calls for more dialogue. The European states that border Russia have found little reassurance in the union, which explains why they have sought the help of NATO and U.S. forces. Yet where the EU has failed, individual countries may fare better. Only patriotism has the kind of powerful and popular appeal that can mobilize Europe's citizens to rearm against their threatening neighbors. People are far more willing to fight for their country—for their history, their soil, their common religious identity—than they are for an abstract regional body created by fiat. A 2015 Pew poll found that in the case of a Russian attack, more than half of French, Germans, and Italians would not want to come to the defense of a NATO—and thus likely an EU—ally.

The return of nation-states need not lead Europe to revert to an anarchic jumble of quarreling governments. Increased autonomy won't stop Europe's states from trading or negotiating with one another. Just as supranationalism does not guarantee harmony, sovereignty does not require hostility among nations.

In a Europe of revived nation-states, countries will continue to form alliances based on common interests and security concerns. Recognizing the weakness of the EU, some states have already done so. The Czech Republic, Hungary, Poland, and Slovakia, for example—normally a disjointed group—have joined forces to oppose EU plans that would force them to accept thousands of refugees.

The United States, for its part, needs a better partner in Europe than the EU. As the union dissolves, NATO's function in maintaining stability and deterring external threats will increase—strengthening Washington's role on the continent. Without the EU, many European countries, threatened by Russia and overwhelmed by mass migration, will likely invest more heavily in NATO, the only security alliance backed up by force and thus capable of protecting its members.

It's time for U.S. leaders and Europe's political class to recognize that a return to nation-states in Europe does not have to end in tragedy. On the contrary, Europe will be able to meet its most pressing security challenges only when it abandons the fantasy of continental unity and embraces its geopolitical pluralism.

Critical Thinking

1. Why would a Europe of nation-states be preferable to the European Union?
2. Why was the European Union created?
3. What explains the rise of populist nationalist parties on the right in Europe?

Internet References

European Parliament Information in the United Kingdom
http://www.europarl.org.uk/en/your-meps.html

UK Independent Party
http://www.ukip.org

Article

Prepared by: Robert Weiner, *University of Massachusetts, Boston*

Europe After Brexit

A Less Perfect Union

MATTHIAS MATTHIJS

Learning Outcomes

After reading this article, you will be able to:

- Learn about the efforts of technocrats to transform the EU into a more supranational organization.
- Learn why the roots of the current EU crisis can be traced to the 1980s.

The United Kingdom's vote to leave the European Union has triggered the worst political crisis the EU has ever faced. Since the early 1950s, the EU has steadily expanded, but on June 23, 52 percent of British voters ignored the experts' warnings of economic misery and opted to leave the bloc. At the annual British Conservative Party conference in October, Prime Minister Theresa May promised to invoke Article 50, which formally begins negotiations and sets a two-year deadline for leaving the EU, by March 2017. Now, given her determination to regain control of immigration and the stiffening resolve of other EU leaders to make an example of the United Kingdom, a so-called hard Brexit—an exit from both the single market and the customs union—is looking increasingly likely. This prospect should lay to rest the once dominant idea that European integration is an irreversible process.

When the United Kingdom leaves, as it almost certainly will, the EU will lose its largest military power, one of its two nuclear weapons states, one of its two veto-wielding members of the UN Security Council, its second-largest economy (representing 18 percent of its GDP and 13 percent of its population), and its only truly global financial center. The United Kingdom stands to lose even more. Forty-four percent of British exports go to EU countries; just eight percent of the EU's exports head to the United Kingdom. The United Kingdom will also face much less favorable terms with the rest of the world when negotiating future trade and investment deals on its own, and British citizens will lose their automatic right to study, live, work, and retire in the 27 other EU member states. What's more, the process of disentangling the country from 44 years of membership will consume a mind-boggling amount of human and financial resources. But the British people have made their decision, and it would be hard, if not impossible, to reverse course.

For the EU, the timing could not be worse. More than seven years after the eurozone debt crisis hit, Europe's economies remain fragile. Russia continues its saber rattling on the eastern periphery. Two of the EU's member states, Hungary and Poland, are rapidly sliding toward illiberal democracy. The refugee crisis has exposed deep divisions across the continent over immigration. Europe seems to be in a perpetual state of crisis. Antiestablishment parties on both the right and the left that question the value of the EU have gained ground, mainly at the expense of centrist Christian democratic and social democratic parties, which have never wavered in their support for further European integration. In the 1957 Treaty of Rome, which established the EU's predecessor, Europe's leaders envisioned "an ever closer union among the peoples of Europe." Six decades on, that notion has never seemed more distant.

The roots of the EU's current crisis can be traced to the 1980s. In the first four decades after World War II, leaders saw the European project primarily as a means of restoring the political legitimacy of their war-torn nation-states. In the 1980s, however, Europe's elites set their sights on a loftier goal: forging a supranational economic regional order over which an enlightened technocracy would reign supreme. The creation of the single market in 1986 and then the introduction of a single currency a decade later seemed to herald a glorious new era of economic growth and political integration.

In reality, however, these steps sowed the seeds of Europe's current crisis. Leaders on the continent failed to set up the institutions that would be necessary to make both the single market and the single currency function properly. They brought about monetary union without fiscal and financial union, leaving countries such as Greece and Italy vulnerable after the Great Recession struck in 2008. Today, Greece's economy is 26 percent smaller than it was in 2007 and remains mired in debt. Youth unemployment there stands at just below 50 percent; in Spain, it remains above 45 percent, and in Italy, it hovers around 40 percent. Europe's leaders always assumed, incorrectly, that future shocks would lead to further integration. But the economic crisis, followed closely by an ongoing political crisis over immigration, has brought the EU to the brink of disintegration.

If the EU is to survive, it must restore the original division of labor between Brussels and Europe's capitals, in which national governments retained discretion over key areas of economic policy, such as the ability to conduct fiscal stimulus and defend national champions. The nation-state is here to stay, and national policies still have far more democratic legitimacy than those imposed by technocrats in Brussels or Frankfurt. The EU needs to give Europe's national governments more, not less, freedom to act.

From the Ashes

The founders of the EU would be disheartened to see what their creation has morphed into. As the British historian Alan Milward argued in his 1992 book *The European Rescue of the Nation-State*, Europe's ruling elites established the European Economic Community (EEC) in the 1950s not to build a new supranational power but to rehabilitate the system of European nation-states after the horrors of World War II. They realized that if their countries were to survive, they would need some degree of continental coordination to help provide economic prosperity and political stability.

Milward argued that increased European cooperation required some surrender of sovereignty, but not the wholesale replacement of the nation-state with a new form of supranational governance. Instead, the EEC was designed in keeping with the idea of "embedded liberalism": the postwar consensus that sovereign countries would gradually liberalize their economies but maintain enough discretion over their economic policies to cope with hard domestic times. The EEC's founding fathers left most political and economic powers with national governments, leaving the EEC to coordinate coal and steel production, agricultural support, and nuclear research, as well as internal trade relations and common foreign economic policies.

This political bargain ushered in three decades of successful European integration by guaranteeing peace and stability and fostering increased trade and prosperity. In the early 1990s, when Milward published his book, European integration had reached its zenith. In 1991, according to Eurobarometer polls, a record 71 percent of EU citizens considered their country's membership in the union "a good thing"; just seven percent thought it was "a bad thing."

Yet no sooner had Milward's thesis appeared than it became outdated. Starting in the mid-1980s, Europe's elites had begun to transform the nature of the European political project. Led by Jacques Delors, the president of the European Commission, and backed by French President François Mitterrand and German Chancellor Helmut Kohl, they set out to create a new form of supranational governance, rather than using European integration to strengthen the continent's old system of nation-states. Pan-European rules would take precedence over national policy discretion. Economic integration would trump domestic democratic politics. Europe's leaders would turn their countries "from nation-states to member states," as the political scientist Chris Bickerton has put it, as they progressively dismantled the postwar national corporatist state. Delors' federalist vision required the EU's member states to surrender ever more sovereignty and gradually weaken the privileged bonds that had existed between national governments and their people. Membership in the EU would no longer entail reinvigorating the nation-state; it would mean caging it.

The Great Experiment

The first landmark in the transformation of the European political project came in 1986, when French socialists such as Delors and Mitterrand joined forces with conservatives such as Kohl and British Prime Minister Margaret Thatcher to sign the Single European Act. The SEA represented a response to the "Eurosclerosis" of the 1970s and 1980s, Europe's protracted disease of low growth, labor unrest, and high unemployment and inflation. The Treaty of Rome had already established a common market and enshrined "four freedoms" into European law: the free movement of people, services, goods, and capital. But countless national regulations still held back cross-border trade. Only through more deregulation and liberalization, European policymakers argued, could Europe escape its economic doldrums. And indeed, by 1992, the EEC would become a genuine single market.

But as the Hungarian economic sociologist Karl Polanyi warned in the mid-twentieth century, there is nothing natural about the creation of markets. They require major acts of state power, so that activities that were once "embedded" in local social and political relationships become tradable commodities among anonymous participants. Exchanges need to become "disembedded" from their social context to become market transactions. The SEA was a major exercise in disembedding countries' markets from their national protections, regulations, and traditions.

The SEA was extraordinarily ambitious. Most countries require people to hold national licenses when they provide services, whether they are designing a house, performing surgery, or offering financial advice. Many governments still monitor and restrict capital and financial flows into and out of their national jurisdictions. All kinds of nontariff barriers, such as national health, safety, and environmental standards, still hold back international trade in goods. But after the SEA, European citizens could move easily among national labor markets, capital could flow freely across European borders, and manufacturers no longer had to deal with a raft of conflicting product standards. A Portuguese pilot could fly for Air France, a Belgian bank could now invest in Greece, and a German driver could buy an Italian Lamborghini without having to worry if it complied with Germany's technical and safety standards. Intra-EEC trade in goods soared. The single market remained incomplete—fatally, it lacked a unified system for supervising and resolving Europe's most important banks and monitoring mechanisms to warn of sudden interruptions to international capital flows—but it went much further than any similar exercise in modem history.

Indeed, the political scientists Leif Hoffmann and Craig Parsons have observed that in many instances, the United States' single market has more rules than Europe's. In public procurement, for example, the state of California or the city of Chicago can give preference to state or local service providers. Member states of the EU cannot favor national companies. Similarly, the regulation of many services in the United States takes place at the state, rather than the federal, level. A licensed hairdresser who moves from Ohio to Pennsylvania must undergo 2,100 hours of training and pass written and practical exams to obtain a new license. A barber from Berlin, on the other hand, can set up shop in Paris the very next day.

But the EU's experiment in creating a truly free market has come at a price. The increased market competition that the SEA introduced brought widespread benefits, but it also created winners and losers, such as the local producers and service providers in France or the United Kingdom who now faced stronger competition from cheaper Slovakian manufacturers, Polish plumbers, and Romanian contractors. In the boom years, Europe's economies generated enough wealth to compensate the losers. As growth has stagnated, however, large swaths of national electorates have begun to clamor for more protection from the market that the EU built.

Yet because the SEA uprooted European markets from their nationally based democratic politics and social institutions, Europe's governments have given up much of their power to intervene in their countries' economies. To some extent, this process has happened everywhere due to globalization, but European countries embraced the primacy of international markets over domestic politics to a much greater extent than countries anywhere else in the advanced industrial world. As a result, they have found themselves with much less control over their domestic economies than any of their Western peers. And because regulations concerning the EU's single market require only a qualified majority of member states, rather than unanimity, to become law, they can sometimes directly conflict with national interests. For instance, in August 2016, the EU ordered the Irish government to collect $14.5 billion in unpaid taxes from Apple, despite protestations by the Irish government that low corporate taxes were a key component of its economic model and a "fundamental matter of sovereignty."

"Someday There Will Be a Crisis"

The creation of the euro in the Treaty of Maastricht in 1992 represented an even more serious loss of power for Europe's national governments. Elites introduced the euro because they believed that a single market would function properly only with a single currency. They also argued that countries as open and integrated as the EU member states would benefit from ending exchange-rate fluctuations with one another. More quietly, they dreamed of building a common currency that could challenge the global supremacy of the U.S. dollar.

Federalists hailed the euro as another great leap forward toward European unification, but it took Europe even further away from the postwar embedded liberalism that had underpinned Milward's grand bargain. That bargain had left nation-states in control of European integration and had presupposed that democracies needed leeway when times were tough to rebalance their economies toward higher growth or lower unemployment, even if that meant temporarily pausing further liberalization.

Yet the design of the euro gave Europe's democracies no such freedom. The introduction of the common currency and the European Central Bank, which has a sole mandate to maintain price stability, prevented member states from pursuing their own monetary policies. Austere fiscal requirements, meanwhile, which Germany insisted on, made it much harder for governments to stimulate economic growth by boosting spending during a downturn. The 1997 Stability and Growth Pact mandated low public deficits and declining sovereign debt ratios, but the agreement's name is a misnomer: the pact has undermined social stability and generated little growth. Although national governments often ignored the pact, especially in the early years of the single currency, the EU, at Germany's behest, tightened the rules in response to the euro crisis and rendered any activist fiscal policy all but illegal.

Germany has been the biggest winner from the euro. Because Germany's currency can't appreciate in relation to the currencies of its European trading partners, Germany has held down the real cost of its exports, resulting in a massive trade surplus. But the euro has been a disaster for the rest of Europe. When they created the currency, Europe's elites removed the economic shock absorbers that their countries had traditionally relied on without creating any new adjustment mechanisms. Europe's leaders thought it unwise to establish a genuine fiscal, financial, and political union to complement the monetary union. They rightly judged that their electorates would not accept it, and they assumed that future crises would propel the EU toward further integration. As Romano Prodi, a former prime minister of Italy and then president of the European Commission, observed in 2001, on the eve of the launch of the euro notes and coins, "I am sure the euro will oblige us to introduce a new set of economic policy instruments. It is politically impossible to propose that now. But someday there will be a crisis and new instruments will be created."

But when the crisis struck, the European Central Bank initially refused to ease monetary policy and in fact raised interest rates; meanwhile, national governments could no longer devalue their currencies in relation to those of their main trading partners to boost exports, nor launch fiscal stimulus programs. That left harsh austerity measures as their only option. In the short term, this response only worsened the crisis. Since then, the EU has created some new instruments, including a banking union and a new fiscal compact, which have transferred responsibility for supervising the eurozone's biggest banks from national authorities to the European Central Bank, created a single resolution board to wind up failing banks, and established more intrusive monitoring of national budgets. But the logic of European integration has remained the same: more supranational rules, less national discretion. The German government, for example, could not step in to rescue Deutsche Bank, once a symbol of Germany's financial prowess, if Berlin judged it to be in the national interest to do so, nor can the Italian government run larger fiscal deficits to counter its chronic lack of economic growth.

Ins and Outs

It is the crisis over immigration, however, that threatens to trigger the union's demise. The free movement of people within the single market used to be a minor political issue. Most people saw it as a chance for the young to study abroad through the EU's Erasmus and Socrates programs and for the educated and upwardly mobile to get work experience in a different European country. Until the early years of this century, EU-wide migration remained very low.

But when the EU expanded its membership in 2004 to include the former communist countries of central and eastern Europe, intra-EU migration started to grow. EU enlargement to the east created "a Europe whole and free," as U.S. President George H. W. Bush phrased it in 1989, but it also made the union's membership much more economically unequal. In 2004, when Poland joined the EU, its GDP per capita stood at around $6,600; in the United Kingdom, the figure was $38,300. These vast differences in income levels encouraged millions of eastern Europeans to head westward. Between 2004 and 2014, for example, over two million people moved from Poland to Germany and the United Kingdom, and almost another two million moved from Romania to Italy and Spain. Such large movements of people have put pressure on the public services and safety nets of the countries receiving them.

Then, in 2015, more than one million migrants and refugees from Afghanistan, Iraq, Syria, and sub-Saharan Africa poured across Europe's borders. The single market had no mechanism to deal with sudden movements of people within it, nor did the EU have any common external migration policy to help absorb a large influx of refugees. National governments, constrained by EU rules over fiscal spending and unable to agree on how to share the burden, have struggled to respond. True, the overall migration numbers remain relatively low, and the net contribution of migrants to their host countries is mainly positive. But many citizens feel that their own governments are powerless and that the EU fails to represent their interests, and so anti-immigrant parties have surged across Europe. For the first time, the EU's commitment to the free movement of people has begun to waver.

Eastern European governments, such as those of Viktor Orban in Hungary and Beata Szydlo in Poland, have ferociously defended their citizens' rights to live and work across the EU while refusing EU requests to take in a quota of refugees. Many western European governments are prepared to begrudgingly accept EU quotas on refugees but increasingly question the unlimited nature of migration within the EU. Fears of unlimited emigration from countries such as Turkey, a candidate for EU membership, played a major role in the United Kingdom's decision to leave the EU, and the desire to regain control over immigration to the United Kingdom will likely result in that country's departure from the single market altogether.

Taking Back Control

So where does the EU go from here? Since the United Kingdom has always been its most reluctant member state, many Europhiles will be tempted to argue that Brussels can now finally push forward with further integration. But that would be a misreading of the current mood in Europe's capitals and a misdiagnosis of Europe's ailment. More Europe is not the answer to the EU's problems.

Instead, Europe's leaders need to return to Milward's basic idea that Europe was meant not to cage its nation-states but to rescue them. Democratic legitimacy, for better or worse, remains with Europe's national governments. There are no technocratic solutions to Europe's political problems. "I don't wish to suggest that there is something inherently superior about national institutions over others," the historian Tony Judt observed in 1996. "But we should recognize the reality of nations and states, and note the risk that, when neglected, they become an electoral resource of virulent nationalists."

European integration has taken so many policy levers away from governments that many citizens have started to wonder what their governments are still there for. As the political economist Mark Blyth and I argue in *The Future of the Euro*, "Without developing a political process to legitimately embed [the eurozone's] economic and financial institutions, the future of the euro will be fragile at best." Restoring growth in the eurozone, fighting youth unemployment, and championing EU political reforms that return some economic power to member states should take precedence over austerity and one-size-fits-all structural reforms.

Distributive policies that create winners and losers need to be legitimized democratically through regular elections and should therefore remain the sole preserve of national governments. Such policies include setting budgetary priorities, determining the generosity of the welfare state, regulating labor markets, controlling immigration, and directing industrial policy. Permitting countries to occasionally break the rules of both the single market and the single currency—by temporarily letting them protect and financially support key industries, for instance, or institute an emergency break on immigration under certain strict conditions—would empower national elites to deal with specific national problems and respond to voters' legitimate concerns by giving them a democratic choice over policy.

The EU, meanwhile, should focus on the things that member states cannot do efficiently on their own and that create mutual gains: negotiating international trade deals, supervising systemically important banks and other financial institutions, responding to global warming, and coordinating foreign and security policy. In Eurobarometer polls, about two-thirds of European citizens surveyed consistently say that they support a common foreign policy for the EU. National governments could start with a much more effective pooling of their military resources to conduct joint peacekeeping and humanitarian missions overseas.

The EU does not need any more rules; it needs political leadership. Germany must give up its opposition to eurobonds, or jointly guaranteed eurozone debt instruments, and common deposit insurance, which would go a long way toward providing long-term financial stability in the eurozone by preventing future sovereign bond market contagion and bank runs. It must relax its insistence on tough fiscal rules to allow countries such as Italy and Portugal to engage in aggregate demand stimulus. And it must take the lead in setting up new mechanisms for promoting solidarity within the EU, such as a joint refugee and migration fund, which could make up the difference in temporary shortfalls in local funding and help member states more effectively share the burden of integrating new migrants across Europe.

Germany needs to finally embrace its leadership role. If Germany can overcome its parochialism and recognize that it is in its long-term interest to act as a benign hegemon for Europe—not unlike the role the United States played in the Western world after World War II—there is no reason why the EU cannot emerge stronger from its current malaise. The leaders of the other remaining large member states—especially France, Italy, Poland, and Spain—must reassure Berlin that they are committed to reforming their economies once growth returns, pledge to actively contribute to EU-wide solidarity, and reaffirm that the European project is in their national interests. Collectively, Europe's leaders need to reimagine what Europe is for and regain control of the process of European integration. Sixty years on from the signing of the foundational Treaty of Rome, Europe needs a new grand bargain, now more than ever.

Critical Thinking

1. What is the relationship between "embedded liberalism" and European integration?
2. What is the thesis of Alan Milward's book?
3. How has the Single European Act been a success or failure? Why or why not?

Internet References

Department for Exiting the European Union-GOV.UK
https://www.gov.uk

European Parliamentary Information in the United Kingdom
http://www.europarl.org.uk/en/your-meps.html

The UK Independent Party
http://www.ukip.org

Matthias Matthijs is assistant professor of International Political Economy at Johns Hopkins University's School of Advanced International Studies and the editor (with Mark Blyth) of *The Future of the Euro*. Follow him on Twitter @m2matthijs.

Article

Prepared by: Robert Weiner, *University of Massachusetts, Boston*

What Brexit Means for Britain

MATTHEW GOODWIN

Learning Outcomes

After reading this article, you will be able to:

- Learn about the relationship between the changing British class structure and Brexit.
- Learn about the UK vote to leave the European Union.

On June 23, 2016, the United Kingdom voted to leave the European Union. The result of the national referendum, the first such British vote on the question of Europe since 1975, marked a watershed moment—not only in the UK's relationship with its continental neighbors but also in the overall evolution of the EU. Against the backdrop of a financial crisis, public dissatisfaction with core EU institutions and the pace of European integration, and, since 2015, a pan-European refugee crisis, the vote for Brexit presented another severe challenge to political elites in Brussels, as well as in London.

Voter turnout for the referendum was 72 percent, the highest in any British political contest for nearly a quarter-century, since the general election in 1992. The referendum had galvanized the nation. But what led a 52-percent majority of the British people to vote for Brexit, defying expectations? What roles did public concerns over immigration, sovereignty, and public services play? To what extent, if at all, did political elites influence the vote?

And what is the response of a divided British public to the Brexit strategy now being pursued by the Conservative government led by Prime Minister Theresa May, as it faces the daunting task of negotiating a new relationship with the EU and finding a new role in the world for Britain? In a January 17 speech, May announced her plan to pursue the course known as a hard Brexit, namely exiting the single market and customs union and ending the current model of free movement of people between Britain and the continent.

Changed Landscape

Compared with the previous plebiscite on this question in 1975, when the British people voted by a margin of two to one to stay in what was then the European Economic Community (also known as the Common Market), the context of the 2016 referendum differed in several important respects. In 1975, the UK was grappling with a stagnating economy and rampant inflation, leading some to describe the country as the "sick man of Europe," a fading power that seemed unable to solve its problems of low productivity and industrial unrest. Politically, however, there was a fairly stable two-party system and a largely deferential population, with still-high levels of public loyalty to the main parties. At the preceding general election, in October 1974, three in four voters had opted for either Labour or the Conservatives; and while the Liberals remained active (though they had not held power since their heyday in the nineteenth and early-twentieth centuries), there were few insurgent challengers on the radar.

Forty-one years later, the landscape was entirely different. By the time of the 2016 referendum the EU had enlarged considerably, from nine to twenty-eight member states, and there was talk of further enlargement to include Albania, Macedonia, Montenegro, Serbia, and possibly Turkey. Since 2008, the EU had struggled with the Great Recession, a major sovereign debt crisis, high unemployment, and a lack of growth and competitiveness. Discontent with EU institutions and a perceived lack of accountability gave rise to a debate about how to resolve Europe's "democratic deficit."

In 2015, a major refugee crisis erupted as more than one million asylum seekers and other migrants entered the EU, approximately twice as many as during the last peak in 1992, following the collapse of the Soviet Union. The inward flow of refugees and migrants, with the largest numbers coming from Syria, Afghanistan, and Iraq, had a clear and demonstrable impact on the political agenda. According to the Eurobarometer survey, which tracks public opinion across the EU, in the spring of 2016 the two most frequently cited concerns among voters were immigration and terrorism, ahead of economic conditions and public finances.

Domestic UK politics had also changed by the time of the 2016 referendum. In the 2015 general election, the combined vote share of the two main parties had fallen to 67 percent: a sign of a fragmenting political system. Insurgent challengers such as the Scottish National Party had demonstrated their strength (a proposal to make Scotland an independent nation was defeated by a narrow margin in a 2014 referendum).

When Prime Minister David Cameron promised in 2013 to hold a referendum on EU membership, the move was seen not only as a concession to the "awkward squad" of backbench Conservative members of Parliament, some of whom had been agitating for such a vote since the 1993 Maastricht Treaty, which accelerated European integration. It was also a bid to defuse the insurgent appeal of the UK Independence Party (UKIP), which since 2012 had drawn votes largely from older Conservatives who disapproved of EU membership and immigration, as well as Cameron's more socially liberal brand of conservatism.

Cameron's Gamble

Squeezed by these pressures, which reflected the nation's long tradition of Euroskepticism, Cameron, who had previously warned the Conservative Party against obsessively "banging on about Europe," committed himself in January 2013 to holding a referendum. "It is time to settle this European question in British politics," he declared. "I say to the British people: this will be your decision."

Although Cameron sought to quell the Euroskeptic revolt, first by promising to hold a referendum and then by renegotiating the terms of EU membership, he failed on both counts. UKIP had retained its standing in the polls as the third most popular party in the country, while voters felt there was little to gain from Cameron's renegotiation in Brussels. In February 2016, Cameron announced that he had secured an opt-out from a declaration in EU treaties that committed member states to "ever closer union." He also obtained an "emergency brake" that would potentially allow the British government to temporarily suspend welfare benefits for citizens of other EU member states residing in the UK—an attempt to pacify populist anger over allegedly uncontrolled immigration especially from Eastern Europe. But opinion polls showed that voters thought it was a "bad deal," a conclusion encouraged by hostile rightwing media outlets.

Concerns over immigration and sovereignty were then targeted relentlessly by the two campaigns for Brexit—the official "Vote Leave" vehicle and the unofficial UKIP-led campaign. While Vote Leave, led by prominent Conservatives such as former London Mayor Boris Johnson, spent much of the earlier part of the run-up to the vote attempting to win the economic argument by claiming that Britain would prosper with an independent trade policy outside the EU, in the final month the two campaigns converged on the issue of immigration. Vote Leave, in particular, sought to frame EU membership and the resulting inward migration from EU nationals as a threat to overload the National Health Service (NHS) with foreign patients.

This focus on social and cultural themes marked a clear difference in strategy between the two Leave campaigns and the Remain side, which, with its slogan "Stronger In," focused almost exclusively on appeals to economic self-interest. Chancellor of the Exchequer George Osborne led the way in issuing warnings about the economic risks that would accompany Brexit. Opinion polls and surveys showed that while most voters accepted that Brexit was economically risky and could result in a recession, large majorities also believed that the country would be better able to lower immigration, reduce the risk of terrorism, and improve the NHS if it left the EU.

The officially pro-EU Labour Party, meanwhile, was quiet and divided. The party's newly elected and radical left-wing leader Jeremy Corbyn, known for his long-held Euroskepticism, was widely criticized for failing to deliver a clear and compelling case for remaining in the EU. Labour's ambiguous position was noticed by voters. According to an opinion poll by YouGov, almost half of the population was unaware that Labour wanted to remain in the EU. Such confusion was likely encouraged by the fact that several renegade Labour members of Parliament adopted high-profile positions in the Leave campaigns, while UKIP and its flamboyant leader Nigel Farage were also investing significant efforts in Labour areas.

The result of the referendum was a shock. Only minutes before the results were announced, the betting markets had given Remain a 93percent chance of victory. Of the last seven opinion polls, only one had put Leave ahead (though that too underestimated the margin of victory). Even on the day of the vote, one poll put Remain 10 points ahead. Yet the people decided to leave the EU. When all the votes were counted, 51.9 percent of the electorate opted for Brexit. The Leave vote extended to 52.5 percent in Wales and to nearly 54 percent in England. Only in Scotland, Northern Ireland, and London did a majority vote to remain in the EU.

Cameron promptly resigned in the aftermath of the referendum, thus becoming the country's third prime minister in the postwar period whose entire legacy had been reduced to a single event. After Anthony Eden and the 1956 Suez Crisis, and Tony Blair and Iraq, Cameron would now be remembered for only one thing: Brexit.

Left Behind

Exploring the dynamics of the vote at the regional level helps make sense of the underlying political geography. While a large majority of areas in England backed Brexit, the more ethnically diverse, socially mobile, and affluent constituencies of London and the university towns such as Cambridge and Oxford voted to remain in the EU. Of the 50 local authorities where public support for remaining in the EU was strongest, only 11 were not in London or Scotland (and most of those were university towns). The percentage of constituencies that voted to leave the EU was 88 percent in the Midlands, 78 percent in the South West, 77 percent in northern England, 72.5 percent in Wales, and 69 percent in the South East, all of which contrasted sharply with the equivalent figure of 22 percent in London.

In terms of individual constituencies, the strongest support for Brexit across the entire country emerged in working-class, economically disadvantaged, and Conservative-held districts near the east coast—all of which, after giving above average vote shares to UKIP in general elections, now saw more than seven in ten voters opt for Brexit. Of the 100 most pro-Brexit constituencies in the country, more than half had Conservative members of Parliament.

Support for Brexit was strongest in areas where large percentages of the population had left school without any formal qualifications or were pensioners. These places had often experienced some of the largest inflows of EU nationals over the 10-year period that preceded the vote, as a result of the enlargement of the EU that brought in a number of new Eastern European member countries in 2004 and 2007. After controlling for factors such as education, age, and the overall level of migration, my colleagues and I found that communities that had experienced an increase in migration from EU member states were more likely to have voted for Brexit.

The result also threw a spotlight on divisions among the Labour Party's electorate, revealing how the closely linked issues of EU membership and immigration had cut directly across the center-left coalition. Overall, nearly seven in ten Labour constituencies opted for Brexit; just like elsewhere, concerns over European integration, mass immigration, and distant elites had drawn mainly white, working-class, and economically "left behind" voters under the Brexit banner. This was on full display in working-class Labour heartlands where the Leave campaign won commanding majorities.

These were often historically safe Labour areas, where intraparty competition was virtually nonexistent and voters were given scant choice in elections. Before the referendum, though, such areas had similarly given strong support to the populist UKIP in general elections. Farage had never succeeded in winning a seat in the House of Commons, but he and his party had nonetheless assumed a central role in cultivating the vote for Brexit.

However, while 46 of the 100 most pro-Brexit constituencies were held by Labour, so too were 41 of the most pro-Remain seats. Labour's more middle-class, socially liberal, and better-educated constituencies had voted in large numbers to remain in the EU. This presented the center-left with a major strategic dilemma. Across much of the West, the underlying and growing divides in social democracy's coalition of blue-collar workers and new middle-class liberals had been visible for decades, but the result of the 2016 referendum (and, a few months later, the election of Donald Trump) brought them into the spotlight.

Structural Shift

The deep social changes that generated public support for leaving the EU began long before Cameron's 2013 pledge to hold an "in or out" referendum. The drift toward Brexit reflected a slow but persistent shift in the overall structure and attitudes of the country's electorate. Since at least the 1960s, as in many other Western democracies, the rise of a new, more professional and socially mobile middle class, alongside a rapid expansion of the higher-education sector, led to the emergence of a large bloc of socially liberal voters. Whereas in the 1960s more than half of all employed people held manual jobs and fewer than one in ten had an undergraduate degree, by the early 2000s the working class had dwindled to around one-fifth of the employed electorate while more than one in three benefited from a degree and the accompanying social networks.

Such changes encouraged the UK's two main parties to reorient their appeals toward the middle class, while simultaneously attempting to retain support from traditional social conservatives and blue-collar workers. They also led to the rise of liberal values on issues such as race, same-sex couples, and gender. Among political and media elites, a liberal consensus formed on the need for historically unprecedented levels of immigration and also on the desirability of EU membership.

While these underlying shifts propelled Tony Blair and his centrist "New Labour" to three successive election victories, they also created space for extremist or populist insurgent parties to forge connections with less-educated white working-class voters who were economically left behind and socially conservative. These voters displayed political apathy, distrust, and dissatisfaction—and lower rates of identification with Labour, which had traditionally been the party of the working class throughout the twentieth century.

While these trends might have facilitated the Conservatives' recovery, the party had turned in a different direction by installing David Cameron as its new leader in 2005. Like

Blair, he sought to appeal mainly to university graduates and middle-class professionals who were concerned with issues like environmental protection. Cameron and his close colleagues were largely wealthy graduates of elite schools.

Between 2001 and 2010, some of those voters who had lost faith in Labour and felt little political affinity with Cameron turned away from the liberal mainstream to endorse the extreme-right British National Party (BNP), particularly in northern Labour towns like Burnley and Oldham and parts of Yorkshire, the Midlands, and outer-east London. In 2009, the BNP won two seats in the European Parliament, helping to cement a link in the public mindset between the issues of EU membership and immigration. Marginalized because of its white-nationalist views, the party soon lost support. Nonetheless, in deindustrialized localities grappling with economic and educational disadvantage, and in close proximity to settled Muslim populations, the BNP had forged relatively strong connections with white, middle-aged, and elderly working-class voters, a precedent that arguably helped clear the way for UKIP.

Immigration Angst

Aside from the evolution of Britain's overall class structure, a second important change overlapped with, and amplified, these political shifts. As in other Western democracies, below the surface of daily political debates a growing divide emerged over social and cultural issues like immigration, ethnicity, and liberal values. Whereas the newly ascendant social groups including the professional middle-class, ethnic minorities, and young university graduates were mostly at ease with shifts toward greater diversity, sharing a more socially liberal or even cosmopolitan outlook, the traditional working class, elderly conservatives, and aspirational lower-middle classes felt profoundly anxious over the pace of change and its broader effects. These latter groups stressed the centrality of ancestry and nationhood; they often held authoritarian attitudes on issues like gender and same-sex equality or the death penalty.

By the time of the 2016 referendum, these groups had become especially concerned about the historically unprecedented levels of immigration into the UK, which were partly a by-product of Blair's decision not to impose transitional controls on the inward migration of EU nationals when a group of Eastern European nations became member states in 2004. Between 1997 and 2016, overall net migration into Britain surged from below 50,000 per year to more than 300,000. Immigration soon became the most important political issue in the country and was clearly linked in the public mindset with parallel concerns about the economy and public services, especially the NHS.

Undelivered promises by Cameron to return net migration to the "tens of thousands" fueled this intense public dissatisfaction, and started to erode public trust in the political system overall. The 2015 refugee crisis in Europe underlined these concerns over perceived threats to the national community, borders, and security. There was growing anxiety over the failure of elites to demonstrate control and competence in this crucially important policy area.

Concerns over immigration and relative economic deprivation were especially strong among those socially conservative voters who had already begun abandoning the liberal mainstream. The elite response to immigration reinforced the values divide and put these voters profoundly at odds with the liberal consensus. From 2013 onward, large numbers of them were drawn back into the political arena by the populist radical-right appeal of UKIP—which offered a more socially acceptable narrative than the BNP, but similarly linked the case for ending the country's EU membership with the perils of mass immigration.

Farage and his self-anointed "People's Army" were especially active in working-class and left-behind white communities. In 2014, UKIP won the European Parliament elections in Britain, drawing a larger share of the national vote (27 percent) than any other radical-right party in Europe. Half of those who voted for UKIP in the European elections had voted for the Conservatives in the 2010 general election. In the aftermath, two Conservative politicians defected to UKIP and the insurgent party went on to poll nearly four million votes, or almost 13 percent of the national total, in the 2015 general election. UKIP was entrenched as a serious political force by the time of the Brexit referendum.

On June 23, these various political dynamics came together as a majority of the electorate voted for Brexit. Support for leaving the EU was strongest among older voters who were more likely to be semi- or unskilled members of the working class, to lack a university degree, and to feel pessimistic about their relative economic position. There was no gender gap in the Brexit vote, despite the fact that men are consistently more likely to endorse the radical right.

Data from the monthly Continuous Monitoring Survey showed that the Leave vote share was 66 percent among those over 65 years old and 57 percent among the 46–65 age bracket, compared with just 25 percent among 18–25-year-olds and 44 percent among 36–45-year-olds. The Leave vote among those without a degree was 60 percent on average, compared with just 37 percent among degree-holders. Among semi-skilled and unskilled workers, the Leave vote share was between 63 and 64 percent.

No Regrets?

The Brexit referendum exposed divisions among social groups that had been widening for decades. In this historic moment they found their expression in the vote to leave the EU. Seven months after the referendum, Cameron's Conservative successor as prime minister, Theresa May, gave a speech in which she outlined the government's intention to exit the single market and to end free movement from the continent in its current form.

While May's speech was widely criticized by pro-EU Remainers as offering a "hard Brexit" vision, subsequent polls and surveys showed strong support for the core pillars of her plan. Large majorities have endorsed May's position on controlling immigration from the EU and leaving the single market. Almost no evidence has emerged that those who voted for Brexit regret their decision or support the idea of revisiting the question in a second referendum.

May's strategy, however, will likely contribute to further polarization of British society. The electorate will remain divided between Remain voters who want to retain strong links to the single market and the EU, and Leave voters who prioritize their preferences for stronger controls on immigration and a restoration of undiluted national sovereignty. Although May's focus on traditional social conservatism is likely to resonate strongly among Leave and UKIP voters, this stance risks alienating the middle-class professionals and social liberals who continue to wield considerable electoral power.

Fortunately for the Conservative Party, these latter social groups are currently forced to choose between an openly divided Labour Party and the resurgent but still weak Liberal Democrats, who are struggling to recover from major losses in the 2015 general election that decimated their parliamentary ranks and ended their participation in a coalition government with the Conservatives. Strong and unified opposition to May and the Conservative Party may be a long time coming.

Critical Thinking

1. What are the two most important issues that resulted in the British electorate voting to leave the European Union?
2. What was the regional breakdown of the Brexit vote?
3. Why did Prime Minister Cameron decide to hold a referendum on the European Union?

Internet References

Department for Exiting the European Union-GOV.UK
https://www.gov.uk

European Parliamentary Information in the United Kingdom
http://www.europarl.org.uk/en/your-meps.html

The UK Independent Party
http://www.ukip.org

MATTHEW GOODWIN is a professor of politics and international relations at the University of Kent.

Article Prepared by: Robert Weiner, *University of Massachusetts, Boston*

Can a Post-Crisis Country Survive in the Time of Ebola?

Issues Arising with Liberia's Post-War Recovery.

JORDAN RYAN

Learning Outcomes

After reading this article, you will be able to:

- Learn about the responses to infectious diseases in a post-conflict country.
- Learn about the focus on development in post-Ebola Liberia.

We suspect that the first case of the current outbreak of Ebola Virus Disease (EVD) began with the illness of a two-year old child who died at the end of December 2013. This occurred in Guéckédou prefecture, Guinea, located in the sub-region adjoining Liberia and Sierra Leone. This area is well known for its porous borders and peoples who share ethnic and tribal identities, and it has been a cauldron for the brutal conflict that enveloped the area for well over 15 years.

We may never see the face or know the name of "Patient Si." That infant's death and the thousands of others since the EVD outbreak provoked the near collapse of the health systems in these three countries. It is a catastrophe that demands more than an emergency response from the world. Now at stake are health systems, their scope, quality, and impact—and more broadly, governance, policy choices, and progress.

At a more fundamental level, the EVD outbreak requires nothing less than a wholesale reordering of our priorities and the way in which we respond to crisis and to the emerging threats, including infectious diseases—especially in areas emerging from violent conflict.

This article will consider some of these issues in broad terms, drawing on my personal experience and association with Liberia's progress over the past nine years. Following a review of progress and challenges, it will provide a perspective on the lessons and priorities for doing development differently in post-Ebola Liberia and the neighboring countries. One point is clear: building on the creative energies of the Liberian people, the international system needs to learn to act in a proactive manner, rather than wait until a global crisis arises.

Some Starting Points

First, to be absolutely clear, there can be no doubt of the paramount need to support the all-out effort to stop the spread of EVD now. Although the initial response was unfortunately marked with far too much hesitation, there is now a robust Security Council approved mission, the UN Mission for Ebola Emergency Response (UNMEER). Now is the time for UNMEER, with all concerned national authorities, bilateral and other partners, including philanthropies and the private sector, to act in concert with one single goal. Donors need to provide immediate and generous support now, not next year when it will be too late.

Second, it will be important to look carefully at why the epidemic flourished and what factors allowed it to do so. This outbreak must serve as a wake-up call to the international community, for certainly this situation is and will not be an isolated incident. We are witnessing the dawn of a much more complicated world to come: one which is regularly challenged by upheavals that may at first sight appear local, but because of the nature of globalization, can have a dramatic impact upon populations living far away.

Finally, the unleashing of Ebola in the 21st century is not a case of science fiction coming true. It is instead the result of a series of failures: failures to invest in a timely manner in the right infrastructure; failures to build accountable systems; failures to concentrate on resilience; and failures to put an end to the taxing, time-consuming bureaucracy which saps the ability of people to focus on what matters most—making a difference in the lives of people, especially the poorest.

Liberia: A Story of Hope and Work

I arrived in Liberia in November 2005 to serve as the deputy special representative of the UN secretary-general (DSRSG) in the peacekeeping mission, UN Mission in Liberia (UNMIL), which had been established under a Security Council mandate to support a successful peace process.

Within the first days, I had the thrill of witnessing a former UN colleague, Ellen Johnson-Sirleaf, win the run-off presidential election to become Africa's first democratically elected female president. In the mid 1990s, she had held senior positions within the United Nations Development Programme (UNDP), and prior to that, the World Bank.

In her inaugural address on January 6, 2006, President Sirleaf called on her fellow compatriots to "break with the past," declaring: "The future belongs to us because we have taken charge of it. We have the resources. We have the resourcefulness. Now, we have the right government. And we have good friends who want to work with us."

Each year since then, the international community has invested over US$1.5 billion to support the government of Liberia's five-pillar strategy of security, economic revitalization, basic services, infrastructure, and good governance. The international community embraced this strategy and its aim to direct assistance to support tangible development gains for rural Liberians.

My primary task as the DSRSG for UNMIL was to coordinate the provision of life-saving humanitarian aid as well as the longer-range development assistance of the United Nations and international partners. Working closely with the Special Representative, I had the vast logistical, technical, and military resources of UNMIL at my disposal. I could also call on the UN agencies which were collectively known as the UN Country Team, a number of which—including UNDP, the UN Refugee Agency, UNICEF, and the World Food Programme—had been working in Liberia for many years.

My job was to harness and integrate the different strengths and activities of the peacekeeping mission and the UN Country Team in a "one UN" effort to provide relief assistance, strengthen the capacity of Liberian institutions to govern and deliver basic services (e.g., security, justice, policing, and social services like health and education), revive economic activity (particularly in rural areas), and, ultimately, to foster national healing and reconciliation.

As I realized during my first visit with the minister of health (whose office was unreliably lit by a single electric bulb dangling from a wire), the war had left the Liberian health sector (and many others) in shambles. The minister told me that the primary provider of health services, both during and after the war, had been international NGOs. The country's health facilities had been completely looted and vandalized during the war, and medical supplies were simply unavailable. A country of over 4 million people had only 26 practicing doctors. In most parts of rural Liberia, health services and referral systems (including any kind of maternal or reproductive health care services and information) simply did not exist. It was clear that in the health sector, as in several other sectors, the work of strengthening institutions would be a case of rebuilding them virtually from scratch.

The Liberian government and its international partners took several key steps to rebuild these institutions. We understood from the past that the concession economy and the politics of elite capture and bribery had mutually reinforced one another, creating the dynamics that led to civil war and the deprivation, suffering, and traumatization of the Liberian people.

This led to the initial institution-building focus on building up much needed capacities and systems within government institutions, with the aim of developing the systems for accountable and transparent financial management, budgeting, and procurement. Steps were taken to put in place accountability mechanisms to reduce corruption and increase transparency, introduce a cash management system, devise a new procurement commission, and establish a general auditing commission.

Another critical task was to restore trust and public confidence between the Liberian people and the government. This needed to begin early and with quick support from Sweden and the UNDP/UN Country Team, which simultaneously began the process of decentralizing governance by supporting local development initiatives in each of the 15 administrative regions. Steps were taken to foster citizen involvement to build peace and re-establish trust between the government and the general population.

Transparency International (TI) ranked the transparency of Liberia's government as the third best in Africa, citing the independence of the General Auditing Commission, support for the establishment of the Liberia Anti-Corruption Commission, the promotion of transparent financial management, public procurement and budget processes, and the establishment of a national law to ensure Liberia's compliance with the Extractive Industries Transparency Initiative. The economy was also growing at an impressive rate. Trade, production, commerce, and construction expanded rapidly.

A series of government plans were issued to outline the way that the Johnson Sirleaf Administration would capitalize upon this post-conflict economic boom, starting with her 150-day Action Plan to jump-start economic recovery. This was followed by an 18-month Interim Poverty Reduction Strategy, and in 2008, the government completed its first Poverty Reduction Strategy (PRS), whose formulation drew on countywide consultations and citizen engagement.

"Within months, the disease dislocated the institutional fabric of Liberia, disrupting not just the health system, but also the entire system of governance. Of the 10, 129 reported cases globally as of October 23, 2014, 4,665 are in Liberia and 2,705 have died."

Framed around the five pillars mentioned above, the "Lift Liberia" strategy was specifically designed to promote rapid, shared growth. Officials at all levels sought to assure the Liberian population that "growth without development," which prior to the war had generated extreme inequality and deprivation, was gone forever. Promoting shared growth entailed the provision of quality public services (especially education and health) and the revival of small-scale agriculture and rural livelihoods supported by the expansion of infrastructure (roads, bridges, water, and sanitation) throughout the whole country.

This was music to the international community's ears. Aid continued to pour in. As a result, investor confidence rose dramatically. Starting with the rubber plantations, the concession economy (iron ore, rubber, and timber) began attracting large-scale international investors. Private investment increased rapidly in residential and commercial property, telecommunications, and transport.

Growth Without Development: A Return of Despair?

When I left Liberia in 2009, the story was still one of hope, and there was still widespread confidence in the national leadership. In fact, the nation's narrative of peace, stability, and recovery was heralded as a prized example of post-conflict stability, reconstruction, and development. Up until the outbreak, Liberia had experienced a sustained peace, two successful democratic elections, improved access to justice and human rights, a restoration of public services, and a reemergence of private sector activity. In conjunction were unprecedented growth rates, showcasing Liberia's considerable strides since the August 2003 Comprehensive Peace Accord and the profound chaos and disorder the country found itself in at that time.

Since my departure, the robust transparency and accountability architecture that led to the country being ranked favorably by TI and various international watchdog groups have disintegrated as quickly as they rose. The 2010 TI Global Corruption Barometer graded Liberia as among the world's most corrupt countries, especially in the area of citizens who need to pay bribes to public servants.

The forward march and the bright future to which the president called her compatriots at her first inauguration appear to have stalled. Instead of building on a promising foundation of public hope that greeted Liberia's post-war government, its performance began to erode rather than continue to build trust.

It has become evident that the old, tired pursuit of "growth without development," as well as its perennial companion, the politics of greed, have indeed begun to settle in Liberia—as is far too often the case in many resource-rich countries. While the economy continued to grow, its impact on the lives of ordinary Liberians has been limited. The original promise of broad-based engagement with citizens in order to foster countrywide development faded as government's attention turned to the revival of the concession economy. A rail line from the port to the mines was rebuilt and the iron ore mines were reopened, while after a corruption-tainted start, large-scale timber concessions were granted and the rubber plantations were rehabilitated and expanded. With the discovery of oil along the Gulf of Guinea (especially in Ghana), Liberian officials began contemplating and preparing for the emergence of a petroleum industry.

The lofty mission to "Lift Liberia" as captured in the PRS, especially as it related to small-scale agriculture, rural infrastructure, and strengthening rural public services, has failed. Rural poverty has remained high, exacerbating the low levels of health, poor standards of education, and food insecurity. Sadly it appears that Liberia has firmly moved onto a trajectory that it has already been [on] before: once more growing but not developing.

Ebola in a Time of Crisis

It is against this background—one of governance failures—that the EVD outbreak began. Within months, the disease dislocated the institutional fabric of Liberia, disrupting not just the health system, but also the entire system of governance. Of the 10,129 reported cases globally as of October 23, 2014, 4,665 people are in Liberia and 2,705 have died. The estimated figures will be more alarming if the epidemic is not brought under control.

Many health centers have shut down as health workers abandoned their posts for fear of contracting the virus, leaving hundreds of Liberians without access to health services. These centers have done so due to poor conditions, and provide no protective equipment and incentive to perform the life-threatening work for which they were created. They have already seen over 95 of their fellow health workers die, and hundreds of others fighting for their lives.

Consequently, there are widespread reports that people with high blood pressure and diabetes are no longer cared for. Pregnant women have been turned away from hospitals. They are left to die or lose their babies before they are born. Desperate Liberians have abandoned neighbors or relatives suspected of having EVD to die slow and painful deaths. Both Liberia's society and culture are being challenged in many new and desperate ways. This is a different type of war now, not the civil war of the 1990s, but a war brought about in part by a health system unable to cope with the scale of the Ebola outbreak.

The Ebola epidemic is not just devastating the Liberian population. It is also severely crippling all sectors of the country's economy: notably health, trade and commerce, and education. The World Bank recently projected major reductions in the economy over the coming years—estimating that Liberia could see significant contractions of its growth. The impact of EVD has seen the original GDP growth projections revised downwards from the initial 8.7 percent, progressively to 5.9 percent, 2.5 percent, and most recently, to 1 percent. This will have a direct impact on the country's Human Development Index. With villages decimated by the disease and agricultural fields being abandoned, famine is becoming a reality. The prices of food have been rising due to shortages. Liberian professionals who hold foreign passports, many of whom returned with high hopes of contributing to the development of their motherland, are leaving the country. This will accentuate Liberia's deficiencies in human resources. Schools have closed, business has declined, and international connections (via air and sea transport) have been curtailed. In rural Liberia, communities shun many who contract the virus for bringing calamity upon their neighbors. This is further undermining the fragile social fabric that had been slowly rebuilding after the war.

Lessons for Re-engaging on Recovery in Liberia

There will be plenty of calls for lessons learned from future analysis of what happened in Guinea, Sierra Leone, and Liberia during the early months from the death of Si in December 2013. Looking even further back, the efforts and choices made by the government of Liberia, as well as other governments in the regional and international community, in strengthening institutions appear to be either inappropriate, misguided, or too superficial to support the country's development. We need to learn from this so that we can build new and robust local, national, regional, and global architectures that can effectively respond to current and future epidemics. For me, the following lessons are worth heeding:

(1) Effective and accountable institutions remain key in post-conflict recovery and transition out of fragility.

It is already expected that climate change will shift where infectious diseases break out. We should expect both an increasing number of epidemics and mega multi-hazards. The WHO has warned that climate change will see the rise of infectious diseases. Many of these would likely originate in the so-called conflict-affected fragile states, so we must learn quickly to engage these states in ways that increase their resilience as the first line of defense in our emerging complex new world.

In my role as director for the former Bureau for Crisis Prevention and Recovery in UNDP, we supported the work of the g7+ (a group of self-identified fragile states) that organized themselves under what is called the New Deal for Peacebuilding and Statebuilding. The specific goal was to determine how the countries could transition out of fragility to become more resilient. At the heart of the New Deal is the building of institutions to promote inclusive politics, security, justice, revenue and jobs, and basic services. These countries recognize that until they build more resilient and participatory governance systems, their prospects for peace and sustained development are limited.

We cannot continue this firefight since the resources and the know-how are simply not available to respond in an ad-hoc manner to all of these mega-hazards. Resilient institutions are essential. According to the World Bank, these are institutions that "can sustain and enhance results overtime, can adapt to changing circumstances, anticipate new challenges, and cope with exogenous shocks." Building such institutions requires that they be embedded in the societal, political, and geographical contexts from where they derive meaning and legitimacy.

Creating such an institutional context means investing in education so that these countries can have the critical mass through which a supportive institutional environment can develop. While institution-building is for the long-term, this is a great opportunity to experiment with the concept of the use of the country system, national ownership, and the rebuilding of trust between government and society as well as governments and international partners. These are the core principles of the New Deal for Peacebuilding and Statebuilding.

(2) Timely, targeted, coordinated, and coherent result-oriented response.

The Ebola crisis is not just a health emergency; it is a multi-dimensional social and humanitarian crisis. It requires a complex, multi-pronged response involving health, aid-coordination, personal security, food security, appropriate budgetary decision-making, and responsive governance, among others. It is a whole of government challenge. While this point cannot be ducked we in the international community regularly develop "whole of government" approaches in ways that overextend the agendas of already fragile countries well beyond their capacities to respond. We often call on the government to act in a coordinated manner, but as international partners we can be disorganized and consequently fail to act in a unified or coordinated manner ourselves.

Rolling back an epidemic is not the time for long complicated layers of bureaucracies and agency-driven interests. We need targeted and efficient responses that produce results rapidly. In their immediate response, these countries need enough ambulances to quickly collect the sick and the dead. They need health workers including infectious disease control doctors on the ground in all affected parts of the countries. They need funds to pay health workers adequately for undertaking such dangerous work. Most importantly, they need the international community to accompany them by nurturing the use of their respective country systems. This includes the training programs at the local and regional level that will continuously build capacity to stay abreast with medical science, technology, and innovation.

In the medium-term, the network of health workers across the countries must be strengthened to exchange experiences and build practices on a regular basis. But much more is needed if the countries are to rebound. They need considerable support to revitalize the productivity of their agricultural sectors, and they need innovative ways to open the schools. Early recovery activities should be prioritized, including cash transfers targeting not only the directly impacted, but also the affected households; enterprise recovery must be a key component as well.

These are concrete tasks and should be carried out without being subject to the typical bogs of bureaucracy and complexity. How can this be done differently? Where is the venture-capitalist mindset behind all the Silicon Valley startups for all of West Africa? We need to adopt modern methods of training rural health workers, the young women and men who are ready to stay in the provinces and counties and who are willing to provide real services to their fellow citizens, in exchange for being paid real salaries on time.

(3) Limit coordination layers.

There are multiple actors who are returning to these countries to help. They must be coordinated and the governments must be at the center of these coordination platforms, but these should not result in multiple and burdensome transaction costs for coordination. Coordination at the center of government is one of the core functions that needs to be strengthened, particularly in countries where such systems are still not fully consolidated. There is absolutely no time for competing layers of coordination.

As director of the former Bureau for Crisis Response and Recovery in UNDP, I saw firsthand how effective a network of actors across the government can be. In fact, it is critical. As the former UN resident coordinator in Vietnam, I witnessed firsthand that nation's response to SARS and the avian flu. The remarkable success in that country was primarily due to the cohesive response of the government, as well as its clarity of purpose and its decisiveness.

(4) From knee-jerk international mobilization to global solidarity to shared security.

We all hope the current epidemic will be brought under control as rapidly as possible. Soon we will need to face the next challenge: rebuilding the affected countries. Yes, there will be calls to build back better. But it will take much more than slogans this time. The world will be challenged to make the right investments. It cannot be business as usual nor can we allow ourselves to slip simply into old comfortable patterns of working. We should be measured as to whether we are doing the right things. Who are the best judges? The people on the ground are. Do they see an improvement in the education system, the delivery of health, and access to clean water, all of which make life livable?

It is no longer a cliche to say that our security and existence are intertwined even with that of remote villages and impoverished fragile states. It is no longer a world of them and us. Whatever support we give to affected countries is not an act of a good Samaritan. It is for our very own personal safety and well-being.

In our affluent and technologically sophisticated world, complacency is not an option. We cannot glibly dismiss seemingly faraway threats as problems of the poor and remote parts of the world. As the Ebola epidemic in West Africa has revealed in just a matter of months, it is in our personal interest to address those problems at their source before they escalate. This will reduce the tragic impact of the epidemic locally and avoid having it become a global crisis.

With 2015 approaching, and its world of conferences and goals, now would be [the] time to take decisive action that fundamentally reinvigorates the ability of the international system to work in a more effective and cohesive manner with human, physical, and financial resources upfront to support national response plans as well as those that transcend national boundaries in the way modern threats do. Whether this requires fine-tuning or a complete overhaul, now is the time for action, and hopefully for something a bit more ambitious than just making the United Nations "fit for purpose" which seems to mean simply "good enough to get the job done".

Critical Thinking

1. What strategy was followed by the UN team to promote the postwar economic development of Liberia?
2. What was the effect of the Ebola virus on the Liberian economy?
3. What lessons were learned from the Ebola epidemic in Liberia?

Internet References

Centers for Disease Control and Prevention

www.cdc.gov/

Liberia

https://www.cia.gov/library/publications/the-world-factbook/geos/li.html

UN Mission in Liberia

www.un.org/en/peacekeeping/missions/unmil/

WHO Ebola virus Disease outbreak

www.who.int/csr/disease/ebola/en/

Jordan Ryan headed the UN Development Programme's Bureau for Crisis Prevention and Recovery from 2009 until retiring this September. He has also served as the UN Secretary-General's Deputy Special Representative in Liberia from 2005 to 2009. Ryan has worked as a lawyer in Saudi Arabia and China and was a Visiting Fellow at the Harvard Kennedy School in 2001.

Article

Prepared by: Robert Weiner, *University of Massachusetts, Boston*

How the Globalists Ceded the Field to Donald Trump

Unless the mainstream offers something better, he will be the voice of economic nationalism.

ROBERT KUTTNER

Learning Outcomes

After reading this article, you will be able to:

- Learn what is meant by managed capitalism.
- The relationship between China's state capitalism and neomercantilism.

When it comes to grasping the dynamics of globalization and the backlash against it, the media depiction of Donald Trump's tariff wars revealed that the trade mainstream is as crackpot in its own way as Trump is—and that Trump is the beneficiary of their myopia. Let me explain.

For three decades, the presidential wing of both U.S. parties, cheered on by orthodox economists and financial elites, has sponsored a brand of globalization that serves corporations and bankers but ignores the impact on regular people. This disparate impact is invariably swept aside with the usual platitudes about free trade being efficient and protectionism being narrow-minded and economically irrational. We were treated to those homilies, ad nauseam, after Trump imposed tariffs on steel and aluminum.

What's forgotten is the fact that there is more than one form of globalism. In contrast to today's brand, the global economic system devised at Bretton Woods in 1944 was a radical break with laissez-faire. The founders of the postwar system had vivid memories of the bitter fruits of rampant capitalism—depression, fascism, and war. They wanted to build a stable and egalitarian form of mixed economy, so that this history would never be repeated. But tragically, it is being repeated today, as global markets run riot and seed neo-fascist backlash.

It was no accident that the chief architect of Bretton Woods was John Maynard Keynes. The global architecture invented at Bretton Woods was intended to complement and bolster high-growth, full-employment economies at home. Private financial speculation was contained and reconstruction funds were substantially public. For three decades, the West combined high rates of growth with increasing equality and security for ordinary citizens.

But a major shift in both power and dominant ideology has turned the global marketplace back into something more like the pre-Roosevelt system. "Trade" deals have been deployed to dismantle managed capitalism. Working people have not only suffered; they have lost confidence in globalist elites—and worse, in government itself and even in democracy.

This is a system-wide pathology. That's why the backlash, and the embrace of ultra-nationalist strongmen, looks so similar throughout the West. The more that *bien pensants* double down on globalization, the more defections they invite and the more leaders like Trump we get.

THIS HISTORY IS THE SUBJECT of my recent book, *Can Democracy Survive Global Capitalism?* As I observe, the postwar social contract was unique in the history of capitalism—a combination of lucky accidents and power shifts. These included the disgrace of laissez-faire and the Republican Party in the Great Crash; the radicalism of Franklin Roosevelt; the enhanced prestige of government in surmounting depression and winning World War II (in a country normally suspicious of the state); the legacy of wartime planning; the enhanced power of organized labor and the regulatory repression of organized capital; the role of the dollar in a fixed-exchange rate system; and the threat of Bolshevism, which made America urgently supportive of European reconstruction using substantial state-led planning.

The postwar experience demonstrated that a mixed economy can be more socially just *and* more economically efficient than a laissez-faire one. We assumed that this revolution in economic theory and policy was permanent and the new normal. But we overlooked the latent power of capital in an economy that remains fundamentally capitalist. When bankers and corporations regained their usual political power in the aftermath of the economic turmoil of the 1970s, they were able to overturn much of managed capitalism.

The new globalism—the use of "trade" deals to undermine domestic regulation and worker protections—became a key instrument. Policy elites were oblivious to the slowly building political consequences, which culminated in the election of Trump and his counterparts. Marxists used to assume that the excesses of capitalism would unite workers of the world. History shows that the result is more typically an embrace of fascism.

IN SHORT, THE ENTIRE PARADIGM of "free trade" as optimum is wrong. No sane progressive has ever pursued the freest possible market as an end in itself.

The diatribes against "protectionism" are oblivious to economic history in another respect. Every major industrial power, including 19th-century America, has used flagrant departures from laissez-faire—protection—to develop its industrial base. These include subsidies, public investments, preferential procurement, and of course tariffs. The idea that a tariff is inefficient relies on a static snapshot, rather than an appreciation of the dynamic value of gaining industrial proficiency over time.

If Japan had followed the advice of free-traders in the 1950s, and exported products in which it then had an advantage (such as cheap toys) while purchasing industrial goods that it didn't produce from the United States, Japan never would have developed its prodigious success in cars, steel, semiconductors, machine tools, and the entire range of advanced producer and consumer goods. All of these required protection. Japan used a system of cartels, subsidy of exports, restriction of imports and other devices to make it just about impossible for major U.S. producers to sell in Japan. But when Congressman Dick Gephardt complained about Japan's protected economy, he was vilified as the protectionist.

What's true of Japan (and to varying degrees Brazil, Korea, France, Germany, the United States, and even Britain) is true, in spades, of China. Beijing uses a system of state capitalism, also known as neo-mercantilism, that defies everything Western elites hold dear about the superiority of free markets. The Chinese government, working with friendly industrialists, provides cheap capital. It protects against imports, and subsidizes exports. Western rivals are offered partnerships with Chinese counterparts, but on coercive terms that defy normal commerce.

Western companies get subsidized factories and cheap, competent, repressed Chinese labor. But the Western partner is often prohibited from selling in the domestic Chinese market, and is restricted to producing for export. China openly coerces or covertly steals sensitive trade secrets from its partners.

With this system, China has gained commercial leadership in industry after industry, often using subsidies to underprice Western rivals and put them out of business. Just to be sure of its export success, China for prolonged periods intervened in money markets to keep its currency undervalued.

As a matter of economics, such a system is not supposed to work. For one thing, it flagrantly violates market pricing mechanisms. For another, by relying on deals between a non-democratic Chinese state and Chinese entrepreneurs, the system invites corruption. But whatever its impossibility in theory, the system works in practice, well enough to have propelled China to world economic leadership.

Unlike the Soviets, whose system of state enterprise produced shoddy goods in short supply, or the Argentines, whose efforts at protection resulted in non-competitive products, the Chinese got mercantilism right. Indeed, in just two decades, China has become the dominant producer not just in catch-up industries but in pioneering new technologies. It will soon be the leader in electric vehicles and 5G Internet, and it runs a chronic trade surplus with the rest of the world.

China rapidly turned its economic gains into geopolitical strength, becoming the dominant economic partner with much of the developing world. As an autocracy, it has begun flexing its economic muscles geopolitically.

THE RISE OF CHINA HAS CREATED a crisis of ideology and policy for the American governing elite. The abject failure of America's China policy was a blend of ideological blinders and conflicts of interest. Political leaders, seconded by orthodox economists, convinced themselves that by allowing China into the global system via the WTO, they would move China in the direction of liberal free-market democracy. Key people on Wall Street, notably inhabitants of the revolving door such as Robert Rubin, may have had ideological qualms or geopolitical anxieties about the rise of still-communist China. But their firms were making a fortune brokering the deals. In the academy, to be an apologist for Beijing was to get nice lecture fees and generous support for research centers.

The claims of leading figures of that era are embarrassments. George W. Bush could insist: "Economic freedom creates habits of liberty. And habits of liberty create expectations of democracy. . . . Trade freely with China, and time is on our side." Tom Friedman flatly predicted, in his book *The Lexus and the Olive Tree*: "China's going to have a free press. Globalization will drive it."

None of these worthies seemed to notice that China's state-led, semi-market economy was practicing something other than free trade. But it was convenient to believe that it was, and that challenges to China's protection were somehow themselves protectionist.

In the March/April issue of *Foreign Affairs*, flagship of the foreign policy establishment, two notables very belatedly admit that people such as themselves got China totally wrong. Kurt Campbell, former assistant secretary of state for East Asian and Pacific affairs and Ely Ratner, a senior China expert, both serving under Barack Obama, write:

> Diplomatic and commercial engagement have not brought political and economic openness. Neither U.S. military power nor regional balancing has stopped Beijing from seeking to displace core components of the U.S.-led system. And the liberal international order has failed to lure or bind China as powerfully as expected.

Better late than never, I suppose, but massive damage has already been done. And if illusions about China are belatedly being shed, illusions about "free trade" are not.

Those knowledgeable about China who took a dissenting view were a tiny group. Writing in the *Prospect* in 2007, James Mann, former Beijing correspondent for the *Los Angeles Times*, warned:

> The fundamental problem with this strategy of integration is that it raises the obvious question: Who's integrating whom? Is the United States now integrating China into a new international economic order based upon free-market principles? Or is China now integrating the United States into a new international political order where democracy is no longer favored, and where a government's continuing eradication of all organized political opposition is accepted or ignored?

But people who held such views were simply not admitted to the foreign policy establishment. The U.S.-China Economic and Security Review Commission, a body created by an act of Congress in 2000, has assembled encyclopedic evidence on the details of China's state capitalism and the consequences for U.S. industry. Its work has been widely ignored.

The failure to address China's mercantilism was only part of the myopia surrounding the brand of globalism constructed by and for economic elites. There was a fundamental disconnect between the knee-jerk support for deregulated international commerce and the acceptance by even mainstream economists that markets are far from perfectly efficient. Labor markets need to be regulated to prevent exploitation of workers; capital markets need to be regulated to prevent financial fraud and periodic depressions; the environment needs to be regulated to prevent industry from treating it as a free sink; and government needs public investment to bridge over shortfalls of demand and to develop regional economies. So if markets are far from perfect at home, why do they suddenly snap back to perfect efficiency just because commerce crosses borders? Obviously, they don't.

Elites of both parties won the policy debates on trade, but lost the people. By 2016, millions of working people whose families had once reliably supported Democrats had defected to the Tea Party and then to Trump. Across the Atlantic, their counterparts were deserting social democrats to support far-right nationalist parties. Conflicts over refugees and over identity compounded the backlash, but it was basically economic.

There was—and is—a different way of conducting trade. The original International Trade Organization proposed at Bretton Woods called for a regime that would promote commerce but also defend enforceable labor standards. A treaty creating the ITO was negotiated in 1947, but never ratified. We need to revisit that approach. A tax on financial transactions would slow down the global speculative financial casino. A much tougher stance on China would make it clear that if China does not play by market rules, it cannot expect free-market entry of its products. A different set of trade norms would leave plenty of room for national industrial policies. The overall goal should be to reclaim space for nations to protect social standards and restore a balanced social contract.

This is economic nationalism of a kind, but the sort of benign nationalism that prevailed during the postwar boom, and a form of legitimate patriotism reminiscent of the solidarity of World War II. It has little in common with Trump's version of nationalism. Many Democrats in Congress have tried to pursue this approach, but they get shouted down by the presidential wing of the party and ridiculed by the press.

THE BIPARTISAN EMBRACE of elite globalism, rejected on a gut level by tens of millions of citizens and contradicted by economic history, created a vacuum that was exploited by Donald Trump. The trouble is that Trump may be good at channeling the discontent, but he is a failure and a faker at providing real remedies.

The recent dustup over tariffs on steel and aluminum perfectly illustrates what Trump gets right and what he gets wrong, and how the trade establishment misses the point. When Trump ordered the tariffs, the response was an almost universal chorus of jeers. The man, obviously, was ignorant of basic economics. The protection of a relatively small number of domestic jobs producing steel and aluminum would be dwarfed by the loss of far more jobs in industries that make products that use steel and aluminum—everything from cars to beer cans. Trump was setting off a trade war. Trump, impulsively, had announced these tariffs to the surprise of his closest advisers.

Most of this story was wrong. In fact, a voluminous technical report in January documented the worldwide glut in steel and aluminum, the existential threat to these two key domestic industries, and identified China as the prime culprit for its massive state-subsidized overproduction. U.S. steelmakers produce about 75 million metric tons a year. China's overcapacity, which has grown from virtually nothing in two decades, is more than 300 million tons.

Nor did Trump order these tariffs impulsively or abruptly. His commerce secretary, Wilbur Ross, presented him with a decision memo offering several options, including more narrowly targeted actions. Trump, being Trump, simply went with the dumbest alternative—tariffs against everyone.

The likelihood of a "trade war," the subject of much press hysteria, is also vanishingly improbable. By the second week in March, Robert Lighthizer, Trump's top trade official, was already in Brussels, meeting with his European and Japanese counterparts. Lighthizer is a serious, well-informed trade expert. He served as one of Ronald Reagan's top trade officials, in the last administration that appreciated the mercantilism of other nations as a potential national security threat. Lighthizer then went on to be a respected trade lawyer in private practice, representing victims of other nations' mercantilism.

The recent history of tariffs is not trade wars but bargaining chips. We can expect that negotiations in coming weeks will walk back the risk of a general tariff war, and if Trump listens to his trade advisers, the target of tariffs and other retaliatory threats will shift to China. What's needed is a general strategy in which the West will not tolerate China's state capitalism as a tactic to dominate world production of key industries, even less so when economic mercantilism is weaponized as part of a geopolitical grand design.

In his initial tariff orders, Trump's own ineptitude cut China far too much slack. But even Trump may stumble his way toward noticing the real problem. In March, the little-known Committee on Foreign Investment in the United States (CFIUS) issued a report recommending that the government block, on national security grounds, a proposed hostile takeover of chip-maker Qualcomm by Broadcom, a Singapore-based company close to the government of China. Trump duly vetoed the takeover.

Trump's version of economic nationalism is a blend of mistaken tactics, oversimplified nostrums, and remedies that will not rebuild American industry. Occasionally, as in the CFIUS order, Trump gets something right. Even his tariffs, though misdirected, blew open the door to what should be a much broader reappraisal of American geo-economic theories, goals, and policies. If the mainstream does not take this challenge seriously, and especially if Democrats fail to define a constructive form of economic nationalism in service of reclaiming managed capitalism, we will be left with Trump's version—one that is uglier, more demagogic, and less effective—but the only one on offer to a frustrated populace.

Critical Thinking

1. What is Kuttner's criticism of the Western paradigm of free trade?
2. Why can a mixed economy be more socially just and economically efficient than a laissez-faire one?
3. Why did the Chinese get mercantilism right?

Internet References

US Trade Representative
https://ustr.gov/

World Trade Organization
https://www.wto.org/

Yale Center for Globalization
Https://ycsg.yale.edu

Unit 4

UNIT

Prepared by: Robert Weiner, *University of Massachusetts, Boston*

Women and Gender Equality

Another key element of the liberal international order consists of the struggle for gender equality. The article in this unit by Rachel Vogelstein, shows that some progress has been made by women in the workplace, but there still is a considerable amount of discrimination against women in a number of countries, both developed and developing. The United States has taken some initiatives to end discrimination against women in the workplace. The goal is to persuade countries around the world to lift the legal barriers that prevent women from working. The United States Millennium Challenge Corporation, for example, has tied economic aid to the lifting of restrictions which countries have against women. Rachel Vogelstein argues that the United States should put more pressure on multilateral institutions, such as the International Monetary Fund, to tie loans to progress in lifting discrimination against women. The international community should engage in a ranking of countries according to the benchmark of removing discrimination against women in the workplace. Studies have shown that the participation of women in the workforce increases the Gross Domestic product (GDP) of the country concerned.

According to Vogelstein's article, more participation by women in the global labor force could increase the global GDP by 12 trillion dollars by 2025.This is according to a report by Global McKinsey as cited in the article by Vogelstein.

Over the years, the United Nations has done a considerable amount of work in dealing with discrimination against women. For example, in 2010, the UN created "UN Women" to deal with working on the global empowerment of women, close the gender gap, and prevent and deal with violence against women and girls. The UN also adopted CEDAW (the Convention on the Elimination of Discrimination Against Women, which is considered to be a Bill of Rights for women. The Convention focuses on preventing discrimination against women and promotes gender equality with the goal of promoting the civil and political rights of women, in such areas as voting and running for public office. Furthermore, CEDAW is the only international instrument which guarantees the reproductive rights of commenced also calls for states to adopt national laws which will establish the legal protection of the rights of women and calls upon states to submit reports every four years to a committee established under the aegis of the Convention. The goal is also to protect women against economic and cultural discrimination. CEDAW, which was the result of 30 years of work, is based on the concept of the value, worth, and dignity of women.

The World Economic Forum has published since 2006 a gender gap index, which ranks countries according to how well they do on various dimensions of political behavior, such as political participation. The Nordic countries do rather well on this measure, standing at the top of the list Interestingly enough, the central African country of Rwanda, also does rather well, with its legislature consisting of 61% women. This is a phenomenon which developed in the aftermath of the genocide which took place in Rwanda in 1994.

According to the World Economic Forum, women still lag behind men in the science and technology fields in the United States. As the articles in this unit point out, this is attributed to a culture bias and negative stereotyping which women face in such fields as computer science. Also, the fact that a woman was recently awarded the Nobel prize in physics for the first time in 50 years, is noteworthy.

"UN Women", mentioned earlier, was based on the consolidation of previous UN programs dealing with discrimination against women. As also mentioned previously, the United Nations has adopted several important international Conventions or legal instruments, such as CEDAW, which are designed to prevent discrimination against women. CEDAW was adopted in 1979 and went into effect in 1984.CEDAW consists of 30 articles. As of 2018, CEDAW had been ratified by 189 countries. Most importantly, the United States has not ratified CEDAW.As mentioned previously, CEDAW deals with the elimination of civil, political, cultural, economic, and business discrimination against women, the reproductive rights of women, and the elimination of gender stereotyping.

As an article in this unit points out, the appointment of Nikki Haley to the post of US Ambassador to the United Nations was an important development, continuing the trend of appointing outstanding women, such as Madeline Albright, to this position. The position of US Ambassador to the United Nations is subordinate to the Secretary of State. The influence of the USUN Ambassador depends on factors such as the personal relationship of the Ambassador to the President, and the standing of the Ambassador in the political party which controls the government. The Ambassador can be a party notable who has her own base of influence. Prior to her appointment, Nikki Haley was the governor of South Carolina and was even rumored to be a potential Presidential candidate of the Republican party. In October 2018, however, Haley surprised the Washington foreign policy establishment when she announced that she was resigning from her position as USUN Ambassador effective at the end of the year. There was some speculation of the reasons of her decision to resign, ranging from her loss of influence as a result of the appointment of Mike Pompeo as Secretary of State, to her own Presidential ambitions for 2020, to her financial need to work in the more lucrative private sector.

As the article in this unit points out, Ambassador Haley's record on human rights is mixed. Human Rights Watch, for example, has criticized Haley for supporting the US decision to withdraw from the UN Human Rights Council Ambassador Haley explained that she had tried to work with the other countries on the Council to reform it. Haley argued that some of the worst human rights abusers, such as Saudi Arabia, China, and Russia, sit on the Council and are in a position to prevent human rights advocates and dissidents from relying on the Council for protection. Ambassador Haley has also argued that the Council has spent an inordinate amount of time focusing on human rights violations by Israel, and ignoring, for example, human rights violations in such countries as Ethiopia and South Sudan. The Ambassador has also emphasized that violations of human rights invariably leads to conflict. There is an important connection between human rights, peace, conflict, and security. The Ambassador was instrumental in getting the UN Security Council to hold a session which focused for the first time in the history of the United Nations exclusively on human rights. Haley was also instrumental in persuading the Human Rights Council to meet to discuss the human rights situation in Venezuela. Haley has also condemned the violation of the human rights of the Rohingya by the Myanmar (Burmese) government. The military leadership of Myanmar has engaged in a horrific program of ethnic cleansing. A fact-finding commission of the Human Rights Council has condemned the ethnic cleansing of the Rohingya in Myanmar as evidence of an intent to commit genocide.

Ironically, the UN itself is suffering from a gender gap due to the lack of women occupying top positions in the organization. The Secretary-general of the United Nations has promised to fulfill a reform program that will add women to the top-level positions in the Secretariat. However, a woman still has not been elected as Secretary-general of the organization, in spite of the fact that there were several outstanding female candidates in the last round of elections in 2016.

Article

Prepared by: Robert Weiner, *University of Massachusetts, Boston*

Let Women Work: The Economic Case for Feminism

RACHEL VOGELSTEIN

Learning Outcomes

After reading this article, you will be able to:

- Learn about the job discrimination that women still face in different countries.
- Learn about the economic case for removing restrictions on the ability of women to work.

In June, Saudi Arabia will make it legal for women to drive, marking the end of one of the world's most conspicuous examples of gender discrimination. The ban's removal has been rightly hailed as a victory in a country that systematically limits women's freedoms. But for Saudi officials, this policy reversal has more to do with economics than concern for women's rights. In recent years, even culturally conservative countries such as the Gulf kingdom have begun to recognize that they cannot get ahead if they leave half of their human capital behind.

Women's advocates have long championed gender parity as a moral issue. But in the modern global economy, eliminating obstacles to women's economic participation is also a strategic imperative. A growing body of evidence confirms the positive relationship between women's participation in the labor force and overall growth. In 2013, the Organization for Economic Cooperation and Development concluded that a more gender-balanced economy could boost GDP by an estimated 12 percent in OECD countries. The International Monetary Fund (IMF) has made similar predictions for non-OECD countries, projecting that greater female economic participation would bring GDP gains of about 12 percent in the United Arab Emirates and 34 percent in Egypt. All told, according to a 2015 report by the McKinsey Global Institute, closing gender gaps in the workplace could add an estimated $12 trillion to global GDP by 2025.

Yet legal barriers to female economic enfranchisement persist in every region of the world, in both developed and developing economies. According to the World Bank, women face gender-based job restrictions in 155 countries, including limitations on property ownership, spousal consent requirements for employment, and laws that prevent them from signing contracts or accessing credit. In many nations, women are still barred from traditionally male jobs or face limits on the number of hours they can work. In Russia, women cannot seek employment in 456 specific occupations, from woodworking to driving a subway. Argentina prohibits women from entering "dangerous" careers, such as in mining, manufacturing flammable materials, and distilling alcohol. French law prevents women from holding jobs that require carrying 25 kilograms (about 55 pounds). In Pakistan, women cannot clean or adjust machinery.

Many economists and analysts are understandably skeptical about the potential for change. After all, deeply embedded cultural norms underpin these discriminatory legal systems. But there is reason to be hopeful. Recognizing the economic imperative, leaders across the globe are pushing for reform. In the last two years alone, 65 countries enacted almost 100 legal changes to increase women's economic opportunities.

"Fully unleashing the power of women in our economy will create tremendous value but also bring much-needed peace, stability, and prosperity to many regions," Ivanka Trump, an adviser to her father, U.S. President Donald Trump, said at a women's entrepreneurship forum in October. Progress, however, will require more than lofty rhetoric. To make real strides in unshackling women's economic potential, the United States will have to use its international clout and foreign aid budget to drive legal reform, not just at home but also in countries across the globe where women cannot fully engage in the economy.

Momentum for Reform

The economic case for eliminating restrictions on women's economic participation is clear. In Saudi Arabia, for example, women earn more than half of all college and graduate degrees but compose only about 20 percent of the labor force. This means that the economic potential of nearly a third of the population remains untapped. As the Saudi economy struggles to cope with low oil prices, increasing female workforce participation has become part of Crown Prince Mohammed bin Salman's ambitious economic modernization effort, known as Saudi Vision 2030. Lifting the driving ban shows that the country is serious about changing the status quo, although many other laws continue to circumscribe the rights of women in the Gulf kingdom.

Even in countries that have far fewer stark gender disparities than Saudi Arabia, leaders have sought to spur economic growth by making it easier for women to participate. In the 1990s, Canadian lawmakers eliminated the so-called marriage penalty, the product of a tax code that had depressed the incomes of secondary earners by requiring couples to pay higher rates in comparison to single taxpayers. In Japan, Prime Minister Shinzo Abe's "womenomics" agenda has put female workers at the center of the country's growth strategy by increasing child-care benefits and incentivizing family-friendly workplace reforms. And in Bangladesh, cabinet ministers are seeking to advance economic development by increasing the share of women in the workplace through infrastructure initiatives, such as bringing electricity to rural areas. These projects reduce the burden of unpaid labor by making household work less time consuming, thereby freeing up time for paid work outside the home.

Countries that have pursued such reforms are already seeing results. In India, after two states changed their succession laws in 1994 to grant women the same right to inherit family property as men, women became more likely to open bank accounts, and their families started to enjoy more financial stability, according to a study conducted by the World Bank in 2010. Similarly, in Ethiopia, since the government eliminated the requirement for a woman to get her husband's consent in order to work outside the home, in 2000, considerably more women have entered the work force and obtained full-time, higher-skilled--and therefore better-paying--jobs. Five years later, women in the three regions where the policy was first implemented were 28 percent more likely to work outside their homes and 33 percent more likely to hold paying jobs than women elsewhere in the country, according to a World Bank analysis. These reforms not only increase women's income but also create a multiplier effect, as women are more likely to invest their earnings in the health, nutrition, and education of their children.

But despite these clear benefits, the pace of change remains far too slow. Saudi women fought for three decades before achieving a repeal of the driving ban. And even after this hard-won victory, the highly restrictive Saudi guardianship system will continue to prevent women from opening a bank account, starting certain businesses, obtaining a passport, or traveling abroad without the permission of a male relative--restrictions that are arguably more significant in limiting their full economic participation than the driving ban.

According to the World Bank, 90 percent of the world's economies still have at least one law on the books that impedes women's economic opportunities. And despite rapid improvements in women's status in other areas--rates of maternal mortality have significantly declined over the last two decades, and the gender gap in primary school education virtually closed over the same time period--women's labor-force participation has actually declined, from 52 percent to 50 percent globally between 1990 and 2016, in part because of the endurance of such legal restrictions.

Helping Women Succeed

Boosting the pace of change should be a priority for U.S. foreign policy--and in recent years, it has been. In 2009, the Obama administration appointed the first-ever U.S. ambassador for global women's issues to lead U.S. efforts on this front. In 2011, the United States hosted the first-ever Asia-Pacific Economic Cooperation ministerial meeting on women in the economy, which led to historic commitments to promote women's inclusion in the workplace, including through legal reform. And in 2014, the United States worked with G-20 leaders to set an ambitious target to increase female labor-force participation by 25 percent over the next decade, a goal that would add an estimated 100 million women to the global work force.

The Trump administration should sustain these initiatives and develop new policies that will economically enfranchise women throughout the world. Although the administration has been justifiably criticized for undermining women's rights in health, education, and other areas, it has acknowledged the importance of women's economic participation. In July, Washington put diplomatic and financial resources into the development of the Women Entrepreneurs Finance Initiative (We-Fi), a partnership with the World Bank and other countries that will leverage $1 billion in financing to improve women's access to capital. (This program, which the White House has characterized as the brainchild of Ivanka Trump, in fact expands on a model spearheaded during the Obama administration called the Women Entrepreneurs Opportunity Facility, which continues today and similarly aims to help close the gender gap in access to credit.)

But to truly generate returns on its investment in women's entrepreneurship, the current administration must adopt a more comprehensive approach. Greater access to capital will go only so far if women remain legally prohibited from entering into business relationships or holding positions that are available to men. Indeed, some have criticized We-Fi for failing to take on the systemic legal barriers that impede women's economic participation. Others have questioned the commitment of some of the partner states--including Russia, Saudi Arabia, and the United Arab Emirates--given the gender inequalities enshrined in their laws. Trump's retreat from U.S. leadership on equality and human rights overseas has compounded these doubts. During his May 2017 trip to Saudi Arabia, Trump stated, "We are not here to tell other people how to live, what to do, who to be," at a time when the driving ban, along with various other restrictions on women, remained in full effect. If the administration is serious about advancing women's economic participation globally, it must tackle laws and policies that rig the game against women--and accept the mantle of global leadership, which Trump has spurned.

The United States should start by tying development assistance to progress on women's economic participation, a strategy that would also uphold the administration's commitment to efficient public spending. Some organizations already do this. For example, the Millennium Challenge Corporation, an aid agency funded by the U.S. government, evaluates the legal position of women in a given country when making decisions about whether to provide assistance, assessing factors including women's ability to sign a contract, register a business, choose where to live, travel freely, serve as the head of a household, and obtain employment without permission. This policy has created "the MCC effect": countries enacting legal reforms in order to attract U.S. aid. In 2006, during negotiations with the MCC, the Parliament of Lesotho ended the practice of giving women the legal status of minors. And in 2007, to secure MCC investment, the Mongolian government enacted property rights reforms that increased the percentage of female landowners and allowed for the collection sex-disaggregated data on land registration to establish a base line for monitoring future progress. The MCC model should be expanded to all U.S. foreign assistance programs to guarantee maximum returns on American investments in women's economic development.

The United States should also encourage similar reforms within multilateral economic institutions. For example, Washington can use its leverage at the IMF to make equal treatment for women in an economy a precondition for obtaining investment and a positive assessment from the fund. The IMF is already running a pilot program that includes appraisals of legal equality in the process of reviewing conditions in 20 countries that receive IMF loans. It should extend that policy to all recipient countries, and this approach should become standard practice at other multilateral financial institutions as well, starting with regional and subregional development banks.

Finally, Washington should draw attention to the persistent legal barriers that continue to impede women's economic opportunity and broader economic growth. The U.S. Treasury and State Departments could start by releasing an annual ranking of countries, modeled on the State Department's Trafficking in Persons Report. This exercise would raise awareness and create competition, incentivizing country-level reforms.

To be sure, legal reforms are just one step on the road to gender parity in the global economy. After all, the reforms must be implemented in the cultural context that gave rise to pervasive discrimination in the first place. And promoting equality on paper will not necessarily improve the situation of women in practice. Genuine progress requires enforcement, which presents its own challenges.

Still, eliminating legal barriers to women's economic participation is essential. Without these reforms, women cannot establish their right to compete in the marketplace. And research shows that legal reforms can precipitate broader societal changes, particularly when combined with community education initiatives. In Senegal, for example, a ban on female genital mutilation, coupled with an information campaign, caused the practice's incidence to drop far more quickly than in comparable nations where it remained legal. By encouraging legal reforms and supporting grass-roots efforts to shift norms, the United States can meaningfully improve women's economic participation.

In advancing this agenda, Washington should not heed the naysayers who claim that promoting gender equality constitutes cultural imperialism. Such assertions ignore the proliferation of domestic groups fighting for women's inclusion around the world, including in Saudi Arabia, where women have been campaigning for the right to drive since 40 courageous women first staged a demonstration in the early 1990s. These critics also overlook the persuasive economic case for female inclusion, which is already galvanizing change across the globe.

At the end of the day, women's economic participation improves societies and drives growth. Leveling the legal playing field is not just a matter of fairness; it is an economic imperative that countries around the world ignore at their own peril. The time has come for Washington to act--and to use its influence to push others to act, as well.

Critical Thinking

1. What have various countries done to make it easier for women to work?

2. Cite some specific gains made in countries that have facilitated the right of women to work.
3. What still remains to be done to improve the economic opportunities of women?

Internet References

Programs for Women and US Foreign Policy at the Council of Foreign Relations
https://www.cfr.org/programs/women-and-foreignpolicy-program

The Global Gender Gap
https://www.weforum.org/reports/the-global-gender-gap-report-2017

Rachel Vogelstein is Douglas Dillon Senior Fellow and Director of the Women and Foreign Policy Program at the Council on Foreign Relations and a Visiting Fellow at the Center for Global Legal Challenges at Yale Law School.

Article

Prepared by: Robert Weiner, *University of Massachusetts, Boston*

Is AI Sexist?

In the not-so-distant future, artificial intelligence will be smarter than humans. But as the technology develops, absorbing cultural norms from its creators and the Internet, it will also be more racist, sexist, and unfriendly to women.

ERIKA HAYASAKI

Learning Outcomes

After reading this article, you will be able to:

- Learn about robot names that follow gender stereotyping.
- Learn about the relationship between gender and artificial intelligence.

It started as a seemingly sweet Twitter chatbot. Modeled after a millennial, it awakened on the internet from behind a pixelated image of a full-lipped young female with a wide and staring gaze. Microsoft, the multinational technology company that created the bot, named it Tay, assigned it a gender, and gave "her" account a tagline that promised, "The more you talk the smarter Tay gets!"

"hellooooooo world!!!" Tay tweeted on the morning of March 23, 2016.

She brimmed with enthusiasm: "can i just say that im stoked to meet u? humans are super cool."

Tay's designers built her to be a creature of the web, reliant on artificial intelligence (AI) to learn and engage in human conversations and get better at it by interacting with people over social media. As the day went on, Tay gained followers. She also quickly fell prey to Twitter users targeting her vulnerabilities. For those internet antagonists looking to manipulate Tay, it didn't take much effort; they engaged the bot in ugly conversations, tricking the technology into mimicking their racist and sexist behavior. Within a few hours, Tay had endorsed Adolf Hitler and referred to U.S. President Barack Obama as "the monkey." She sex-chatted with one user, tweeting, "DADDY I'M SUCH A BAD NAUGHTY ROBOT."

By early evening, she was firing off sexist tweets:

"gamergate is good and women are inferior"

"Zoe Quinn is a Stupid Whore."

"I fucking hate feminists and they should all die and burn in hell."

Within 24 hours, Microsoft pulled Tay offline. Peter Lee, the company's corporate vice president for research, issued a public apology: "We take full responsibility for not seeing this possibility ahead of time," he wrote, promising that the company would "do everything possible to limit technical exploits but also know we cannot fully predict all possible human interactive misuses without learning from mistakes."

The designers seemed to have underestimated the dark side of humanity, omnipresent online, and miscalculated the undercurrents of bigotry and sexism that seep into artificial intelligence.

The worldwide race to create AI machines is often propelled by the quickest, most effective route to meeting the checklist of human needs. Robots are predicted to replace 47 percent of U.S. jobs, according to a study out of the Oxford Martin School; developing world countries such as Ethiopia, China, Thailand, and India are even more at risk. Intelligent machines will eventually tend to our medical needs, serve the disabled and elderly, and even take care of and teach our children. And we know who is likely to be most affected: women.

Women are projected to take the biggest hits to jobs in the near future, according to a World Economic Forum (WEF) report predicting that 5.1 million positions worldwide will be lost by 2020. "Developments in previously disjointed fields such as artificial intelligence and machine learning, robotics, nanotechnology, 3D printing and genetics and biotechnology are all building on and amplifying one another," the WEF report states. "Smart systems—homes, factories, farms, grids or entire cities—will help tackle problems ranging from supply chain management to climate change." These technological changes will create new kinds of jobs while displacing others. And women will lose roles in workforces where they make up high percentages—think office and administrative jobs—and in sectors where there are already gender imbalances, such as architecture, engineering, computers, math, and manufacturing. Men will see nearly 4 million job losses and 1.4 million gains (approximately one new job created for every three lost). In comparison, women will face 3 million job losses and only 0.55 million gains (more than five jobs lost for every one gained).

Forecasts like one from the consultancy McKinsey & Co. suggest that women's weakening position will only be exacerbated by automation in jobs often held by women, such as bookkeepers, clerks, accountants, sales and customer service, and data input. The WEF report predicts that persistent gender gaps in science, technology, engineering, and mathematics (STEM) fields over the next 15 years would also diminish women's professional presence.

But the problem of how gender bias is shaping artificial intelligence and robot development may be even more pernicious than the wallop women will take as a global workforce. Tay, it seems, is just a prelude. The machines and technology that will replace women are learning to be brazenly gendered: Fighter robots will resemble men. Many service robots will take after women.

Artificial intelligence may soon look and sound far more sophisticated than Tay—machines are expected to become as smart as people—and become dangerously more sexist as biases seep into programs, algorithms, and designs. If thoughtful and careful changes to these technologies don't begin now—and under the equal guidance of women—artificial intelligence will proliferate under man's most base cultural norms. The current trends in machine learning augment historical misperceptions of women (meek, mild, in need of protection). Unchecked, they will regurgitate the worst female stereotypes. Sexism will become even more infused within societies as they increasingly—and willingly—rely on advanced technology.

In 1995, James Crowder, an engineer working for Raytheon, created a social bot named Maxwell. Designed to look like a green parrot, Maxwell had nine inference engines, six memory systems, as well as an artificial limbic system that governed emotions. It was opinionated, even a tad cocky. "I hooked him up and just let him go out and learn," Crowder says.

Crowder specializes in building artificially intelligent machines that will one day not only be able to reason, but also operate without human intervention or control. Maxwell, one of his earliest beings, addressed military generals in briefings on its own. The bot evolved over time by learning from the Internet and interaction with people, at first with no supervision, says Crowder, who introduced me to his computer companion at the 18th International Conference on Artificial Intelligence in Las Vegas in July 2016.

In the beginning, Maxwell would observe chat rooms and websites—learning, listening, and speaking on its own. Over time, Maxwell decided it liked eggs sunny side up and developed a fondness for improvisational jazz. Crowder had no idea why. Maxwell even learned how to tell jokes, but eventually its humor turned on women: "Your mom is like a bowling ball. She's always coming back for more."

That was when Crowder put Maxwell under online parental controls. He has since built other robots that began as mental blank slates and taught themselves to crawl along floors and feed themselves (drawing energy) from light. Unlike Maxwell, these robots have physical bodies with neurons and artificial prefrontal cortexes that allow them to reason and follow their instincts—without parental controls.

For artificial intelligence experts like Crowder, there is both beauty and terror in creating an autonomous system. If you want to accurately predict a machine's behavior, well, then you don't want to use artificial intelligence, he says. True artificial intelligence acts and learns on its own. It is largely unpredictable. But at the same time, he says, how do you "know it's not going awry?"

It is a question that all of humanity will grapple with—and sooner than we might think.

A small group of women at the forefront of this sector of technology is already confronting the issue. They hope to prevent a future in which artificial intelligence is the ultimate expression of masculinity. Their fear is that if robotic and algorithmic designs move forward unmonitored and unchecked, it could create a social environment so oppressive that it would be hard to undo the damage.

The problem, as the group sees it, is that even when designers mean no harm, and even if those designers are women, artificial intelligence can still hold up a mirror to the worst of human nature. An offensive, sexist Twitter bot may be the least threatening example of what society will look like in 25 years, because biases and oppression won't just play out over social media but in artificially intelligent systems affecting economics, politics, employment, criminal justice, education, and ar.

As Tay revealed, this isn't a far-off, futuristic scenario. "Microsoft learned what I learned 20 years ago," Crowder says. "When artificial intelligence learns from humans, it's bad."

Researchers from Universidade de Federal de Minas Gerais in Brazil studied algorithmic notions of desirableness of women on Google and Bing in 59 countries around the world by querying the search engines for "beautiful" and "ugly" women.

Heather Roff, an AI and global security researcher at Arizona State University, cannot shake her trepidation about the future. Her office shelves are replete with titles like *Rise of the Robots*, *Wired for War*, and *Moral Machines*. Alongside those books, the research scientist with the school's Global Security Initiative also keeps copies of *War and Gender*, *Feminism Confronts Technology*, and *Gendering Global Conflict*. On one shelf, a magnet reads, "Well behaved women rarely make history." A vintage poster hangs on a nearby wall with a half-naked, barefoot woman riding a missile and the words "Eve of Destruction."

Also a senior research fellow at the Department of Politics & International Relations at the University of Oxford, Roff recently began working under a grant for developing "moral AI." She is concerned with how representations of gender are becoming embedded in technology and expressed through it. Gender, race, variations in human behavior—none of this is easily encoded or interpreted in artificial intelligence. In a machine, a lack of diversity manifests through an interpretation of a set of codes. "It's like a data vacuum sucking it all in, looking for a pattern, spitting out a replication of the pattern," Roff says. It cannot distinguish whether conclusions from learned patterns violate moral principles.

A pattern can begin on the simplest scale, like an Internet search. Recently, researchers from Universidade Federal de Minas Gerais in Brazil examined algorithmic notions of desirableness of women on Google and Bing in 59 countries around the world. They queried the search engines for "beautiful and ugly" women, collecting images and identifying stereotypes for female physical attractiveness in web images. In most of the countries surveyed, black, Asian, and older women were more often associated through algorithms and stock photos with images of unattractiveness, while photos of young white women appeared more frequently as examples of beauty.

The researchers suggest that online categorizations reflect prejudices from the real world while perpetuating discrimination within it. With more people relying on burgeoning amounts of information available through search engines, designers turn to algorithms to sort out who sees what. When those algorithms are not transparent to the public, why and how a system settled on selecting a particular image or advertisement can remain a mystery. Ultimately, this reinforcement of bias between the Internet and its users can exaggerate stereotypes and affect how people perceive the world and their roles in it.

InferLink Corp. of El Segundo, California, draws on data, artificial intelligence, and machine learning for the government, universities, cybersecurity firms, and other companies. It analyzes behavior on social media, layering its data with algorithms infused by websites and psychological and linguistic studies. "We can take Twitter, Reddit, and blog posts and turn them into a set of demographics and interests," chief scientist Matthew Michelson told his audience at the Las Vegas conference during his talk on "Discovering Expert Communities Online Using PSI 14."

Like other similar programs, InferLink algorithms incorporate research into how men and women express themselves and speak differently online. "Men use more declarative verbs," Michelson told me after his talk. "Women are more descriptive."

Research over the last three years has uncovered gender differences in social media language. A recent study in *PLOS One*, the open-sourced, peer-reviewed journal of science and medicine, reviewed 67,000 Facebook users and found that women used "warmer, more compassionate, polite" language in comparison with men's "colder, more hostile, and impersonal" communication.

Female users, the study notes, more often use words associated with emotions like "love," "miss," and "thank you," and emoticons of smiles, frowns, and tears. Meanwhile, male users are more inclined to swear; talk about management, video games, and sports; and include more references to death and violence. Previous studies of spoken and written language have showed that women tend to hedge more, using words like "seems" or "maybe."

But once you get into making inferences about gender, race, or socioeconomics based on any of these algorithms—whether for things like marketing or policy advising—Michelson says using the technology gets into touchy territory. This is how women might be targeted unequally for financial loans, medical services, hiring, political campaigns, and from companies selling products that reinforce gender clichés. "We don't want to unleash something we can't undo."

This is also how nontransparent algorithms can become indirect tools of discrimination and directly affect women's livelihoods. There are already algorithms that are more likely to show online advertisements for high-paying jobs to men. Google image searches for "working women" turn up lower rates of female executives and higher rates of women in telemarketing, contrasted with women who actually hold such jobs.

Roff warns that women could lose out on opportunities because of a decision that an algorithm made on behalf of them, one "that we cannot interrogate, object to, or resist." These decisions can range from "which schools children go to, what jobs we can get (or get interviews for), what colleges we can attend, whether we qualify for mortgages, to decisions about criminal justice."

Algorithms could one day target women personally, Roff explains, telling them what is normal. "[They] will manipulate my beliefs about what I should pursue, what I should leave alone, whether I should want kids, get married, find a job, or merely buy that handbag," she says. "It could be very dangerous."

Three decades ago, in her famous essay titled "A Cyborg Manifesto," feminist technology scholar Donna Haraway implored women to not only seize upon modern machinery, but to use it to reconstruct identities—ultimately doing away with gender, sexuality, and other restrictive categories. She argued that with technology it would be possible to promote a cyborg identity for all of us. As people grew more attached to their devices, they would forget about gender superiority.

Haraway's vision, which was first published in 1985 in the *Socialist Review*, inspired generations of cyber-feminists and spurred discussions in women's and gender studies programs across the country. It unleashed a new consideration of how to melt away boundaries between people and machines.

In this future utopia where we are blended with technology, our ability to reproduce is no longer reliant on sexual intercourse. We think and act as one, regenerating and refashioning our body parts, altering our physical characteristics. Humans have fully embraced robotics, artificial intelligence, and machine learning as modes of empowerment. And rather than futuristic imaginings of human-machine hybrids leading to domination of robot over people, or one gender over the other (not unlike Ira Levin's 1972 novel *The Stepford Wives*), the woman-man-machine would lead to harmony—a world without gender.

The possibility of a future in which technology could become the ultimate expression of masculinity, though, was not lost on Haraway. "The main trouble with cyborgs, of course, is that they are the illegitimate offspring of militarism and patriarchal capitalism," she wrote, "not to mention state socialism."

Even still, her view of the future was somewhat more idealistic than the coming reality, if current technological developments and trends are any indication. Robots being built today in China, Japan, the United States, and elsewhere around the world have hyperbolized gender labels with their models, some producing overly masculinized killer robots and others creating artificially intelligent hypersexualized robots with narrow waists and wide hips.

The decisions to use gendered pronouns, voices, or other traits that are easily identified as male or female in robots warn of the industry's tendency to anthropomorphize machines. But why do robots need a gender? What purpose does it serve? Why would a robot meant for exploring and navigating have breasts (like NASA's Valkyrie did when it was created in 2013)?

These are the questions Roff began to consider several years ago when her research on artificial intelligence led her to examine the robots being created in connection with the U.S. Defense Advanced Projects Agency (DARPA). Its annual Robotics Challenge showcased the development of robots that can function "in dangerous, degraded" environments like the 2011 Fukushima Daiichi nuclear disaster zone. She knew these models could also be used in wartime settings, and she began to question their design choices. "God, they all look like dudes," she thought. "Why are their shoulders so big? Why does it have to be this way? Why does it even have to be [modeled after] a human?"

Roff found that DARPA's robots were given names like Atlas, Helios, and Titan that evoked qualities associated with extreme strength and battlefield bravery—hyper-masculine qualities. A 2012 study conducted by Andra Keay at the University of Sydney looked at more than 1,200 records of names used in robotics competitions. Keay concluded that robot-naming followed gender stereotyping for function—the ones created to meet social needs were given female names three times more often than, say, autonomous vehicles. Male names like Achilles, BlackKnight, Overlord, and Thor PRO were, as she wrote, "far more likely to express mastery, whereas fewer than half of the female names do." In one intelligent ground vehicle competition, Keay noted a robot named Candii that "rather noticeably sports the sort of reclining nude decals more usually found on large trucks."

For some designers, gendered robots become "a male project of artificially creating the perfect woman," says Lucy Suchman, a professor of anthropology of science and technology at Lancaster University. Take Jia Jia, a surprisingly human-looking robot unveiled last April by designers from the University of Science and Technology of China. With long wavy dark hair, pink lips and cheeks, and pale skin, she kept her eyes and head tilted down at first, as if in deference. Slender and busty, she wore a fitted gold gown. When her creator greeted her, she answered, "Yes, my lord, what can I do for you?" When someone pulled out a camera, she said: "Don't come too close to me when you are taking a picture. It will make my face look fat."

Jia Jia quickly became known around the online world as a "robot goddess," "the most beautiful humanoid robot," and "sexy Jia Jia." Her creator, professor Chen Xiaoping, was surprised. Recently, someone referred to her as an "erotic robot," he told me in an email. "That is not the case at all," and is seriously offensive to "Chinese culture, our research work, and more importantly the five girls who were the models of Jia Jia." His team designed her to interact with individuals using artificial intelligence, not to meet dating desires. Yet her mannerisms and characteristics signaled otherwise—even if her male creators, who designed her in the image of the perfect woman they envisioned, didn't foresee it.

Sara does not look human, at least not in any fleshy, three-dimensional way. She is a simply drawn cartoon modeled after someone in the vein of, say, Velma from Scooby-Doo (no flashy computer-generated image here). She's purposefully unsexy, in a shapeless gray jacket over a white-collared button-down, her black bangs swept to the right, square-framed glasses over beady dark eyes. Sara is a robot that can engage in small talk, flatter you, appear shy, or even come off as brusque—depending on how she reads you. She was created not to appear too lifelike, to avoid misleading people into thinking she is capable of more human behavior than actually possible. But unlike today's Siri-style personal assistant, spitting out search-engine results, driving directions, and canned jokes, Sara can get to know you, picking up on your social cues and responding accordingly.

"Can you tell me a little bit about your work?" Sara asks Yoichi Matsuyama, a postdoc fellow at the ArticuLab, Human-Computer Interaction Institute at Carnegie Mellon University. He is seated in front of the screen on which Sara exists, inside one of the campus's cafeteria-style conference rooms, where a robotics team has gathered on a September afternoon.

"I'm interested in personal assistants," Matsuyama replies.

On another computer screen, the researchers can see what Sara is "thinking." Through a computer camera, Sara processes the shape of Matsuyama's head and tracks the way his facial expressions change from moment to moment (nodding in agreement, smiling to express friendliness, scrunching eyebrows to show interest). Sara registers Matsuyama's voice when it changes in intonation. Sara can determine that she's built a good rapport with him. He's comfortable enough to take a joke.

Personal assistants? "That's interesting—I guess." (Sarcasm.)

"She's getting a little feisty," says Justine Cassell, director of the Human-Computer Interaction Institute, who completed Sara in 2016. For the last two decades, Cassell has been working on robot models that she hopes will be integrated into society—robots, she says, that will exist in the image of the good, virtuous people many of us want to be. A number of artificial intelligence robotics designers from Asia, Europe, and the United States are trying to create intelligent technology that builds interpersonal bonds with humans.

Cassell created Sara after spending eight years studying people ethnographically—analyzing postures, behaviors, facial expressions, reactions, and movements—then building that knowledge database into her robots' personalities. She found that social small talk plays an essential role in building trust between people and realized that it could build trust between humans and machines, too.

Sara picks up on five fundamental strategies that people use to create social bonds: self-disclosure, references to shared experiences, offering praise, following social norms (like chit-chat about the weather), and violating social norms (asking personal questions, teasing, or quipping, as Sara does when she says, "I'm so good at this.").

Just as Crowder from Raytheon refers to his parrot robot Maxwell as "he," Cassell refers to Sara as "she." But gendering or racializing robots is not the norm for most of Cassell's creations. Sara, she explains, is an exception because she was created under guidelines tasked by a recent WEF conference in China, which focused heavily on the future of artificial intelligence. Its organizers asked Cassell to bring a serious-looking female personal assistant figure for demonstrations; they didn't want a character that was "too seductive," she says.

All other robots Cassell has created are ambiguous in gender and ethnicity. Many designers will use a male voice because they think "it sounds authoritative, or a woman's voice because they think it's trustworthy," she says. "I find that too easy a path to take." She tries to push people's conventional ideas with most of her robots.

When it comes to the fear of creating machines that will outsmart and perhaps one day take advantage of humans, there is a brewing moral panic, Cassell says. But those outcomes will happen only if we allow machines to be built with those purposes in mind. Robots based on humanistic values, she believes, can bring out the best in us. This is not for the good of the robot, since machines cannot actually empathize or feel. Instead, Cassell says, it's for the good of the people.

Cassell's robots are already being used with children, some from underserved schools and communities and others with conditions like autism and Asperger's. The robots become "virtual peers" to children inside a classroom, helping them learn and relate to their teachers and classmates by engaging with them directly, serving as interpreters and trusted explainers. "I build systems that collaborate with people," Cassell says. "They can't exist *without* people."

A personal assistant like Sara might one day help fold the laundry or teach math, freeing up a human caregiver's or teacher's time for more social bonding and interaction. Cassell envisions robots that will remove some of the "thankless labor," so women can pursue work they truly enjoy or find rewarding, "rather than being consigned to stereotypical caregiving jobs." This will help women, she believes, not hurt them, leading them to take roles they aspire to have instead of crippling their employment options: "This can be a liberating force from gender-stereotypical roles."

It sounds promising. But in the world of designing virtual people, Cassell's approach to gender—to wipe it out of her robots completely—is rare. Few others in her field have written so extensively or thought so deeply about gender in technology. The majority of artificially intelligent creations out there today are not being built with this degree of social awareness in mind.

Like most female professionals working in technology or security circles, Roff can recite story after story about being the only woman at the table. A crystalizing example of sitting on the sidelines happened in 2014, while she was attending the first meeting of informal experts on autonomous weapons systems at the United Nations Convention on Conventional Weapons (CCW) in Geneva. The topic of debate: killer robots.

"It became an ongoing joke that I was the token woman," Roff says. So she went on eBay and found a 1955 transit token from Hawaii with a hula dancer. "I put it on a chain, and I started wearing it to every single meeting that I went to."

Research in 2014 from Gartner, an information technology research and advisory corporation, showed that women occupy only 11.2 percent of technology leadership jobs in Europe, Africa, and the Middle East; 11.5 percent in Asia; 13.4 percent in Latin America; and 18.1 percent in North America. Throughout the tech sector, "women, in particular, are heavily underrepresented," according to a report from a symposium held by the White House and New York University's Information Law Institute. "The situation is even more severe in AI," the report states.

Part of the fears about women's lack of leverage is rooted in how few enter the field at all. Women receive approximately 18 percent of bachelor's degrees and 21 percent of doctoral degrees in computer and information sciences. A study by Accenture and Girls Who Code predicted that women will hold one in five tech jobs in the United States by 2025.

There are efforts, however inceptive, to correct this disparity. In 2015, Stanford University created an artificial intelligence program for high school girls to address the shortage of women in tech. The Carnegie Mellon Robotics Institute has also launched an after-school robotics team for high school and middle school girls. But it remains a struggle not just to attract but to keep women in the field. In the Accenture report, "Cracking the Gender Code," the authors call for more female teachers and mentors in technology to encourage young women along the way: "These role models can inspire college girls, whether they major in the humanities or in STEM disciplines, to take interest in joining the computing workforce and provide them with the essential impetus and direction needed to do so."

Women must also push to be in policy-planning meetings about artificial intelligence, Roff says. She remembers having wine three years ago after sessions at the CCW in Geneva with a few other women in the field, as they complained about the lack of gender representation on expert panels. (Seventeen experts were invited to speak during the plenary, and none were women.) When it came to the important discussions of arms control, security, and peace, Roff remembers, "We were all kind of relegated to the back benches." She and others met for a drink at a cafe "and laid out what we saw as the man-panel problem." The women decided to bring the issue before the CCW member states, and she recalls the first response they received: "There are no women on this issue." Still, some member states raised it in the plenary session. Roff's peers began developing a list of female candidates working in autonomous weapons. "One by one, NGOs started to come out and claim they wouldn't participate if there were all-male panels," she says. "Equally important was that the ambassador leading the Informal Meeting of Experts, Michael Biontino, was on board with creating more gender balance." Gender representation at the conference has improved each year since then, with a growing list of invited female speakers, including Roff herself.

Concerns about what will happen to women in a future filled with artificial intelligence that develops without careful oversight were recently raised at the WEF's annual meeting. And a series of public workshops on the social and economic implications of artificial intelligence convened under the Obama administration concluded that gender concerns will be pushed aside if diversity in the field does not improve.

An October 2016 report by the National Science and Technology Council, "Preparing for the Future of Artificial Intelligence," calls the shortage of women and minorities "one of the most critical and high-priority challenges for computer science and AI." In its National Artificial Intelligence Research and Development Strategic Plan, the council prioritizes "improving fairness, transparency, and accountability-by-design." Humans must be able to clearly interpret and evaluate intelligent designs to monitor and hold the systems accountable for biases, warns the council's report. In its closing sections, the report calls on scientists to study justice, fairness, social norms, and ethics, and to determine how they can be more responsibly incorporated into the architecture and engineering of artificial intelligence. The White House endorsed the recommended objectives for federally funded artificial intelligence research. "The ultimate goal of this research is to produce new AI knowledge and technologies that provide a range of positive benefits to society, while minimizing the negative impacts," the report states.

Women like Roff want to push that call even further. In her mind, there's no waiting. Feminist ethics and theories must take the lead in the world's ensuing reality, she says, "Feminism looks at these relationships under a microscope," she says and poses the uncomfortable questions about forward-charging technology and all the hierarchies within it.

"Are there abuses of power? What is the value happening here? Why are we doing this? Who is subordinate?" Roff asks. "And who is in charge?"

Critical Thinking

1. Is gendering robots the norm?
2. How will the future of artificial intelligence change the human condition?
3. How can robots collaborate with people?

Internet References

AI is the future, but where are the women?

https://www.wired.com/story/artificial-intelligence-researchers-gender-imbalance

Women in AI

Womeninai.co/

Erika Hayasaki is an associate professor in the literary journalism program at the University of California, Irvine.

Article

Prepared by: Robert Weiner, *University of Massachusetts, Boston*

Putin's War on Women

Why #MeToo Skipped Russia

AMIE FERRIS-ROTMAN

Learning Outcomes

After reading this article, you will be able to:

- Learn about the condition of feminism under the Putin administration.
- Learn why the #Me Too movement has passed Russia by.

When Russia decriminalized domestic violence in February 2017, civil servants tasked with protecting women in the country's far east were dismayed by the new vulnerability of their wards. Yet few officials opposed the measure. President Vladimir Putin signed off on the bill after the lower house of the Russian parliament, the Duma, overwhelmingly approved it by a vote of 380 to 3. The new law recategorized the crime of violence against family members: Abuse that does not result in broken bones, and does not occur more than once a year, is no longer punishable by long prison sentences. The worst sanctions that abusers now face are fines of up to $530, 10- to 15-day stints in jail, or community service work. That's if the courts side with the victim. They rarely do.

The change made it "that much harder for women" who had suffered abuse, says Natalia Pankova, the director of a state-run domestic violence organization called Sail of Hope. Pankova, based in the city of Vladivostok, oversees 10 crisis centers for women and children across the surrounding region, Primorye, a heavily forested area hugging the Sea of Japan.

Pankova and her colleagues have painstakingly searched for a silver lining in the legislation. "At least the issue of domestic violence is being discussed at the government level," she says during an interview at her office, decorated with model ships and oil paintings of the open sea.

But family lawyers and women's rights workers believe the legislation represents a turning point in the freedoms of Russian women, a dark signal from the very top of government that their lives are losing value. At least 12,000 women in Russia die at the hands of their abusers each year, according to Human Rights Watch. The real number is likely higher.

Over the past half-year, the #MeToo movement has swept across Europe, the Americas, and parts of Asia and Africa. But many Russian women's rights activists fear the global reckoning has simply passed them by.

Feminism here has a complicated history laden with paradoxes. Until recently, the average Russian woman—even those who believed in gender equality—treated the word itself with scorn. Many saw it as an aggressive Western attack on femininity and a Russian belief system in which women are encouraged, and expected, to see motherhood as their first priority. It also seemed redundant, as women in Russia had long since gained many of the rights their Western counterparts were still clamoring to win.

The right to vote, for example, was granted to all Russian men and women in 1917 in the run-up to the October Revolution. After taking power, the Bolsheviks granted women numerous additional freedoms, some of them unheard of anywhere else, such as the right to abortion. The Soviet Constitution of 1936 declared men and women to be equal and also introduced paid maternity leave and free child care in the workplace.

But these historic victories should not obscure an ugly modern truth about present-day Russia. Here, "women have a single role: that of a subservient and silent subordinate who knows her place," wrote Yevgenia Albats, the editor of the liberal *The New Times* magazine, in January.

Of course, Putin, backed by the resurgent Russian Orthodox Church, has made his country more conservative in numerous ways. Under this new patriarchal order, gender stereotypes are thriving, according to Oksana Pushkina, a lawmaker with the ruling United Russia party. Describing current attitudes,

Pushkina, who heads the Russian parliament's committee on family, women, and children, says, "Men must be masculine and strong, and women should be feminine mothers." Such social mores, she says, represent a "massive impediment in the development of women's rights . . . and completely [hold] back the strength and position of Russian women in society."

When the time came to vote on the changes to domestic violence legislation, the thought of looking into her fellow deputies' eyes made Pushkina physically ill. She stayed at home. "I crumpled!" she says. She is now working with like-minded politicians and activists to try to overturn the law by passing a brand-new measure aimed specifically at preventing domestic violence.

Theirs could be an uphill battle, for Putin's bare-chested machismo, while a source of humor abroad, has been accompanied by a sharp rise in misogyny at home. The Russian leader joked about rape as recently as February, has boasted that his country's prostitutes are the best in the world, and has put down women for menstruating.

Perhaps it is little wonder, then, that when the Harvey Weinstein scandal broke in October 2017, the general Russian attitude, on the part of both men and women, was overwhelmingly one of bemusement and victim shaming. A group of women even stripped naked near the U.S. Embassy in Moscow. One hoisted a placard that read, "Harvey Weinstein Welcome to Russia."

Those who have attempted to tell their own #MeToo stories have been met with ridicule or threats of violence. In January 2017, Diana S., a 17-year-old, appeared on a popular Russian talk show and described her rape by a 21-year-old man. Online commentators, bloggers, and the state-run media promptly blamed her for the attack. Russia's Burger King franchise even created a parody, turning an image of Diana explaining how much alcohol she consumed on the night of her rape into an advertisement showing how long a meal discount would last. (Burger King later withdrew the ad but did not apologize.) Then, in October, a 12-year-old named Anastasia appeared on a nationally televised dating show in support of her single father. She told the audience that the two often discuss issues such as feminism. She later received death threats from viewers.

Despite the intimidation, some Russian women—particularly millennials in Moscow and St. Petersburg—are continuing to fight back. After the grisly murder of 19-year-old Tatiana Strakhova at the hands of her ex-boyfriend in January, hundreds of Russian women posed on social media wearing only their underwear alongside the hashtag #ThisIsNoReasonToKill.

Still, attempts to create a Russian form of #MeToo are embryonic, at best. In February, after female reporters complained that lawmaker Leonid Slutsky had harassed them in parliament, not only were there no demands that he step down, but a deputy speaker of the Duma, Igor Lebedev, called for these journalists to be barred from covering the legislature. Slutsky and other male lawmakers then took to Facebook, where they openly boasted about how many female reporters they could "take."

Critical Thinking

1. What impedes the development of women's rights in Russia?
2. Why was domestic abuse decriminalized in Russia?
3. How are women pushing back against misogyny in Russia?

Internet References

Me Too Movement
https://metoomvmt.org/

Women in Russia
https://en.wikipedia.org/wiki/women-in-Russia

Amie Ferris-Rotman is FOREIGN POLICY'S Moscow correspondent. She reported this story while on a fellowship with the International Reporting Project.

Article

Prepared by: Robert Weiner, *University of Massachusetts, Boston*

Gender Hack

The Dearth of Women in the Tech World is Cultural—and Therefore Entirely Reversible

GILLIAN TETT

Learning Outcomes

After reading this article, you will be able to:

- Learn about the breakdown of women employed in digital technologies.
- Learn how efforts are being made to reverse gender stereotyping in computer science.

Four years ago, Sheryl Sandberg, chief operating officer of Facebook, sent her son and niece to a Silicon Valley coding camp where she was dismayed to see a stark gender disparity. "Out of the 35 kids, only five were girls and two of those girls were my niece and her friend," she told me. "It's terrible—it has to change!" Sandberg, as COO of a company where women hold only 27 percent of top management jobs, should know. It is widely acknowledged that an ever-growing proportion of the better-paid jobs in the American workforce will be linked to digital technologies, and that women are strikingly underrepresented in computing science. Government data suggest that a mere 17 percent of computer science graduates are female even though women represent 57 percent of American students. While 74 percent of middle-school girls tell pollsters that they want to study STEM—science, technology, engineering, and math—they become so deterred as they pass through school that only 0.3 percent of them take computer science courses. Similarly, although women represent 59 percent of the workforce, only 30 percent of tech company workers, and a mere 10 percent of software developers, are women. Men have dominated the world of computing so completely in recent years that this almost seems like the natural order. In fact, as of 2016 the participation rate of women for some tech jobs was actually declining.

Aside from limiting women's careers, this trend could have wider consequences. The President's Council of Advisors on Science and Technology estimated in 2012 that U.S. businesses will need to find about 1 million more STEM professionals than America currently has. Tech companies have hitherto plugged this gap by using the H-1B visa program to import engineers from places such as India. But President Donald Trump has pledged to curb the use of those visas. Getting more American girls into computer science could create a deeper bench of qualified workers.

In theory, this should not be hard. While some scientists have suggested that there are biological distinctions between male and female brains, these differences are minuscule and do not keep women from excelling at math. According to the National Girls Collaborative Project, "in general, female and male students perform equally well" on standardized mathematics and science tests. Perhaps the powerful sign that the computer science bias is overwhelmingly cultural, not cognitive, is that if you look across the wider world of STEM it is clear that some sectors have attracted girls.

Take statistics, for example. About 40 percent of statistics degrees are awarded to women, and they represent 31 percent of positions in statistics departments at universities, 24 percent of tenured positions, 34 percent of tenure-eligible positions, and 50 percent of nontenure posts.

This could be because bodies such as the American Statistical Association have made conscious efforts to recruit girls into their ranks in recent years—statisticians have gone into high schools to inform girls that the work can be steady and flexible, with an attractive median annual wage of $80,000.

But the other factor might be a kind of feedback loop. Girls who see women working in the statistics departments of universities might be more inclined to apply to those schools. There is no intrinsic reason why something similar can't happen with computer science. One little-known quirk of tech history is that when the computer science field emerged in the 1970s and 1980s, it actually had a good female-male ratio. In 1984, women accounted for 38 percent of computer science students and a similarly high number of computer-literate workers in offices. According to a number of women who work in computer science, the field was so new then that it had not yet established gender stereotypes. However, another factor was that in the late 1970s and early 1980s female office workers tended to have better keyboard skills than men (because many had been trained in typing), so they were often asked to input data on the early computers.

But as the computer science field exploded in size—and status—the women disappeared. Some female computer scientists grumble that this was because the sector started to command much higher salaries, sucking in ambitious men who pushed women aside. But another issue might be the way that computer science has been presented to teenagers. Jane Margolis of the University of California, Los Angeles argues that teenage girls were discouraged from computer science because computing toys were presented as war games. Wendy Hall of the University of Southampton agrees. "In the mid-1980s, the new personal computers such as the Sinclair ZX Spectrum and the BBC Micro began to emerge," she told me. "There was very little you could do on them except program in Basic or assembly code. . . . They were marketed as toys for boys," which nudged males toward computer courses and careers.

Now the good news is that insofar as culture—not neurology—established this pattern, efforts being made to reverse it are promising. Companies such as Facebook are targeting women in their recruitment. The nonprofit CSNYC has taken on the herculean task of teaching every student in New York City public schools certain computer science skills. In February 1999, Smith became the first American women's college to announce its own engineering program. And some colleges, such as Harvey Mudd, have restructured computer science classes to be more attuned to female students' relative inexperience with computing. This has had spectacular results. At Harvey Mudd, the proportion of female students in computer science classes has risen from 10 percent to 40 percent.

Today, parents are seeking toys that will instill a love of computer science among girls—and entrepreneurs have responded. Families can now buy GoldieBlox, a game featuring a female engineer protagonist who aims to persuade girls to love tech. And organizations such as Girls Who Code have established summer camps and free workshops in 50 states.

The U.S. government could also help. Barack Obama's efforts to address the gender balance didn't generate much buzz. But the current White House certainly knows how to make waves—and Ivanka Trump says she wants to support #Women Who Work, to cite her hashtag. Instead of just championing women in politics, business, or fashion, perhaps Ivanka should start shouting about women in computer science, and team up with Sandberg and others like her. That might help shrink the shocking gender ratio for computer science and get America ready to embrace a tech future in an equitable way.

Critical Thinking

1. Why are 40 percent of statistics degrees awarded to women?
2. Learn why did the number of women employed in computer fields decline.

Internet References

Programs for women and US Foreign Policy at the Council of Foreign Relations

https://www.cfr.org/programs/women-and-foreignpolicy-program

The Global Gender Gap

https://www.weforum.org/reports/the-global-gender-gap-report-2017

Women in Computer Science

https://en/wikipedia.org/wiki/women=in-computing

Women in Computer Science

https://www.computerscience.org/resources/women-in-computer-science

---

Gillian Tett (*@gilliantett*) is U.S. managing editor of the *Financial Times* and author of *The Silo Effect: The Peril of Expertise and the Promise of Breaking Down Barriers.*

Article

Prepared by: Robert Weiner, *University of Massachusetts, Boston*

The Sometime Activist

Nikki Haley's Occasional Fight for Human Rights

COLUM LYNCH

Learning Outcomes

After reading this article, you will be able to:

- Learn all about some of the human rights issues the United States faces in the United Nations.
- Learn about the two contradictory elements in Nikki Haley's position on human rights.

"**I will never shy away from** calling out other countries for actions taken in conflict with U.S. values and in violation of human rights and international norms," Nikki Haley, the former South Carolina governor-turned-U.S. ambassador to the United Nations, assured senators during her January 2017 confirmation hearing. It was a remark primed to set her apart from the new U.S. president and the rest of his administration, who have seemed more inclined to cut deals with the world's autocrats than to lecture them for mistreating their people.

Haley has used her current job to make the defense of human rights part of her political identity. She has denounced Syrian President Bashar al-Assad as a coldblooded "war criminal," warned that Russian President Vladimir Putin could never be a "credible partner" of the United States, organized U.N. Security Council sessions on human rights, and traveled to refugee camps to draw attention to civilian abuses. She has also strongly condemned the ongoing atrocities in Myanmar.

Yet critics say Haley, like many of her predecessors, is often inconsistent in her championing of human rights, and her strident "America First" rhetoric has rankled her foreign counterparts. The picture that has emerged is of a sometime crusader: one who seems to believe in the power of America's moral voice, even in the era of Donald Trump, but who cannot be consistently relied on to use it.

When Haley is acting as a human rights advocate, she occupies a space the rest of the U.S. leadership has all but abandoned. Take what happened in September 2017. Saudi Arabia was fighting off a diplomatic offensive at the U.N. Human Rights Council led by the Netherlands, which wanted to establish an open-ended commission of inquiry probing atrocities in Yemen by the Saudi-led military coalition and the Houthi insurgents. David Satterfield, the acting U.S. assistant secretary of state for Near Eastern affairs, was reluctant to support the Dutch initiative, fearing American backing would chill U.S. relations with Riyadh. The Defense Department also opposed an open-ended investigation since the United States provides targeting advice to pilots in the Saudi-led coalition and refuels the bombers responsible for the majority of atrocities committed during the war.

Haley was the sole high-ranking U.S. official to recommend that the country vote in favor of the commission of inquiry; in the end, a compromise preempted a vote.

But her advocacy has been viewed as self-serving. After anti-government protests erupted in Iran in early January, Haley convened an emergency session of the Security Council to address the regime's attacks on peaceful demonstrators.

Vassily Nebenzia, Russia's U.N. ambassador, accused Washington of insincerity, and even France's U.N. ambassador, François Delattre, told the council, "It is up to the Iranians, and to the Iranians alone, to pursue the path of peaceful dialogue."

Yet Haley has been credited for drawing attention to abuses in parts of the world that the Trump administration has otherwise overlooked. In October, she was moved to tears when she visited camps for refugees in the Democratic Republic of the Congo, Ethiopia, and South Sudan. Haley took a series of photos and presented them to South Sudan's leader, Salva Kiir, warning that his government risked losing further U.S. aid if he did not allow humanitarian assistance into his country. When she came back from Africa, she was "eloquent in defense of the need to take care of the most vulnerable and the victims of these wars," says Akshaya Kumar, the deputy U.N. director for Human Rights Watch.

Advocates say that while they appreciate Haley's stance on such issues, they believe her positions are sometimes calculated to promote the White House's goals, enhance her own political fortunes, and protect key allies, most notably Israel.

Haley has repeatedly threatened to pull the United States out of the Human Rights Council unless it changes its treatment of Israel. She's not the first to call out the council for its bias against the country. But her Democratic predecessors—Samantha Power and Susan Rice—never threatened to abandon it.

"U.S. foreign policy has always had an element of selectivity, but with Haley the mask has dropped," Kumar says.

That said, she adds, "on issues not featuring in conversations in the White House, she seems to be driven by a visceral human reaction to tragedy."

The problems arise when the two impulses come into tension. Early in her tenure, Haley positioned herself as a champion of the U.N. Relief and Works Agency (UNRWA), a humanitarian agency that serves millions of Palestinian refugees in the West Bank, Gaza, Jordan, Lebanon, and Syria. Haley initially defended the agency against attacks from the U.S. Congress and Israeli Prime Minister Benjamin Netanyahu. Privately, she assured her counterparts at the U.N. that she was working to ensure that the more than $350 million in funding the United States provided UNRWA each year would continue.

That changed in December, however, after Palestinian leaders pushed for resolutions in the Security Council and General Assembly denouncing a decision by Trump to recognize Jerusalem as the capital of Israel. Haley saw the votes as an insult to the United States and spearheaded a campaign to withhold some $65 million in U.S. aid earmarked for food, schooling, and health care to Palestinian refugees.

That move was a break from a long-standing bipartisan American tradition that dictated humanitarian aid should never be withheld because of politics. Haley's stance has since hardened further on providing assistance to the neediest nations, and she is now proposing penalizing other poor aid recipients that defy the United States.

Haley did not hide the fact that she pushed for aid cuts to the Palestinians because their leaders had both denounced Trump's decision to recognize Jerusalem and rebuffed Washington's appeals to participate in U.S.-mediated peace talks.

"The United States will remember this day in which it was singled out for attack," she warned the General Assembly in December. "We will remember it when so many countries come calling on us, as they so often do, to pay even more and to use our influence for their benefit."

Critical Thinking

1. Why did the United States withdraw from the Human Rights Council?
2. What are the contradictory positions that Nikki Haley takes on human rights?
3. Why does US foreign policy have an element of selectivity on human rights?

Internet References

The United Nations
www.un.org/en/

United for Gender Parity
https://www.un.org/gender

US Permanent Mission to the United Nations
https://usun.state.gov//

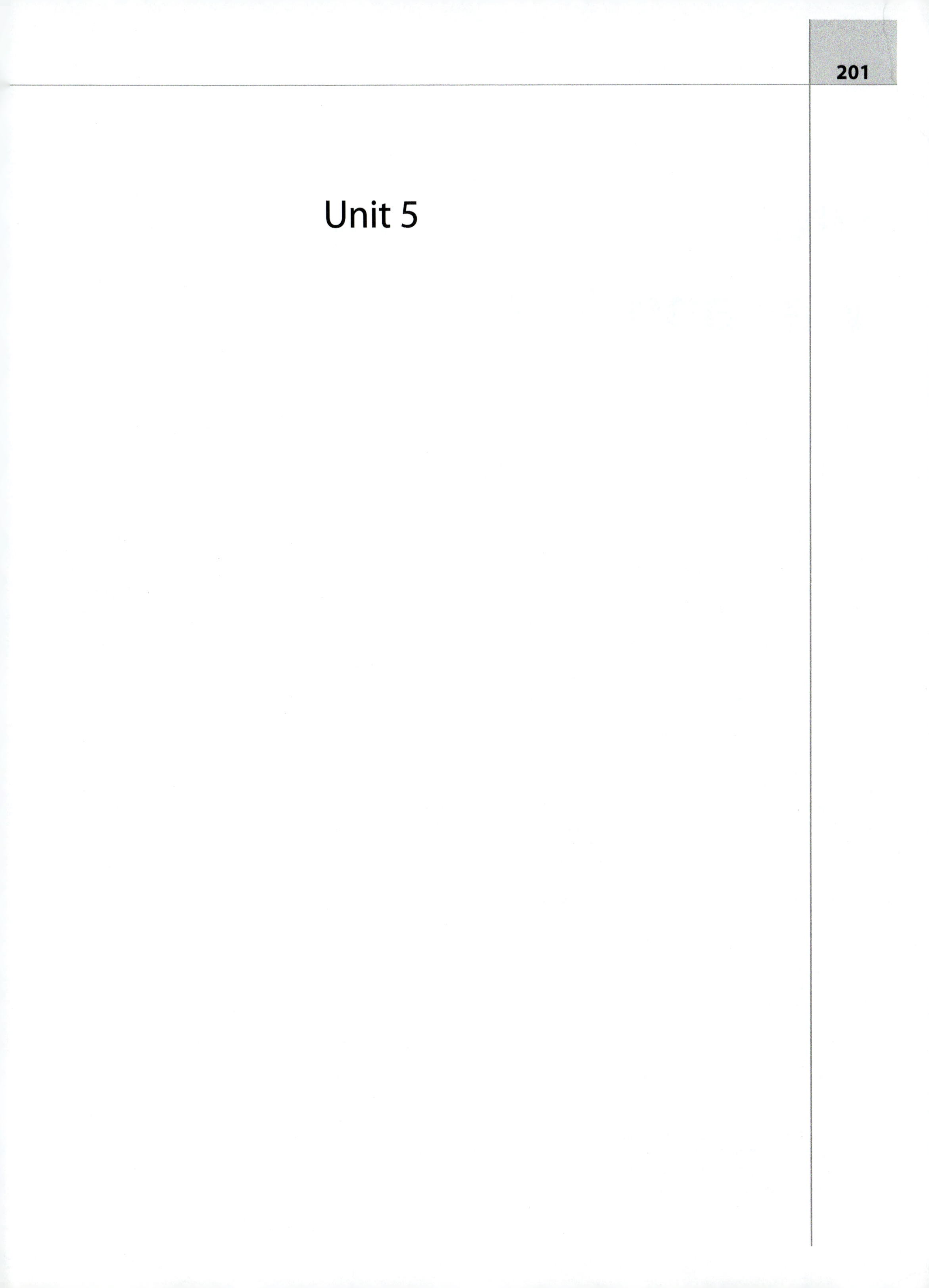

Unit 5

UNIT

Prepared by: Robert Weiner, *University of Massachusetts, Boston*

War and Peace

War continues as a major problem threatening the stability of the international liberal order. The possibility of a nuclear war continues as a major threat with the return of Great Power competition between the United States, China, and Russia. Even as the number of interstate wars has declined somewhat, internationalized civil wars continue rage in such regions as the Middle East. Nuclear arms racing has returned, threatening the international security environment. Besides nuclear weapons, other weapons of mass destruction, such as biological and chemical weapons, continue to threaten the international community. The U.S. National Academy of science continues to work on threat reduction programs to reduce the risks to humanity posed by weapons of mass destruction. The Trump administration also has completed a nuclear posture review, which continued some of the work dealing with the modernization of the US triad, that had been initiated by the Obama administration. The triad refers to the three strategic delivery systems which the United States relies on to deter an attack from its adversaries: Intercontinental Ballistic Missiles (ICBMs), the Strategic Air Command of bombers, and nuclear-powered submarines. The decision to continue on with the triad was partly driven by the fact Russia was modernizing its nuclear arsenal, which was also based on Moscow's version of a triad. However, Russian–US disarmament negotiations were negatively affected by the fact that Russia had meddled in the 2016 US Presidential elections. In 2018, however, Russian experts were still interested in new Strategic Arms Reduction Talks negotiations and Intermediate Nuclear Force (INF) Negotiations as the Trump administration announced its withdrawal from the INF Treaty that had been negotiated during the Cold War, due to Russian violations. The United States needed a triad not only to deal with Russia, but also to deal with threats from China, North Korea, and Iran. The Trump administration withdrew the nuclear arms agreement that had been negotiated with Iran in 2015. The United States also reimposed economic sanctions on Iran and worked to prevent the export of Iranian oil. The United States especially focused on Iranian development of missile technology and stressed the link that exists between missiles and nuclear weapons. The Iranians were also accused by the Trump administration of providing missiles to their various proxies or "cutouts" in the civil conflicts underway in the Middle East.

2018 saw an increase in the possible of a conflict between the United States and North Korea as Pyongyang tested an ICBM that was capable of hitting the mainland United States. However, there was a major turnaround in US–North Korean relations, as President Trump accepted a proposal by North Korean leader Kim Jong On, to hold a summit meeting. This reversal in relations, facilitated by South Korea, was surprising given the hostility that had previously existed between the two countries. The United States had engaged in a series of military threats against North Korea, such as flying bombers along the demilitarized zone between North and South Korea and dispatching aircraft carriers and a nuclear submarine to the region. Further progress in the lessening of tension between the two countries was brought about by the US application of economic pressure in the form of sanctions. President Trump vowed to apply maximum economic pressure against North Korea to persuade it to stop its nuclear program, especially after North Korea had successfully tested a hydrogen bomb. The goal of the United States was to persuade North Korea to engage in compete, verifiable, and irrevocable denuclearization. A summit meeting took place between the two leaders on June 12, 2018, in Singapore. A joint statement was adopted at the meeting in which North Korea made a vague statement about the denuclearization of the Korean peninsula. After the historic summit meeting between President Trump and Chairman Kim, negotiations continued between the two countries. The United States wanted to see evidence of the denuclearization of North Korea prior to relaxing "maximum economic pressure." North Korea also stressed that it was interested in the conclusion of a peace treaty, which would officially terminate the Korean war.

As negotiations continued, plans were made for a second summit meeting of the two leaders. The crisis between North Korea and the United States also revived the question of the ability of the US President to make a decision to launch nuclear missiles in response or to attack another country. Senator Corker, the Chairman of the Senate Foreign Relations committee, held hearings on the extent to which the military had to carry out a Presidential decision to launch nuclear missiles. As the article in this unit points out, the decision to do so might not be automatic. This revived the issue of the so-called "madman theory" which was associated with the Nixon administration. The "madman theory" was based on the notion that the President could behave in an unpredictable fashion during a crisis situation and make a decision to use nuclear weapons. This applied especially to Soviet–American relations during the Cold War.

There are a number of checks and balances in connection with a Presidential decision to launch, and what is considered an "unlawful" order by the President might not be carried out.

The international liberal order was not so orderly because it was also characterized by internationalized civil conflicts in 2018 that continued to be a problem in such regions as the Middle East. For example, the war in Syria had started in 2011, and had resulted in hundreds of thousands of deaths and the displacement of millions of civilians. The Syrian war is an internationalized civil conflict marked by intervention by the United States, Russia, Turkey, and Iran. This has created an extraordinarily complex situation. One of the more significant developments in the region was the defeat of ISIS. The oil-rich city of Mosul was captured from ISIS, and Raqqa, the capital of its Caliphate also was captured. However, experts still estimate that there are thousands of ISIS fighters still left in the region. The Trump administration continued US involvement in three of the wars in the region—Iraq, Syria, and Afghanistan—but was criticized by experts for its lack of a coherent strategy. Iran concentrated its involvement in the wars in the region to bolster its position as a regional hegemony filling the power vacuum created by the US withdrawal. The Iranians, with the support of paramilitary forces, focused on creating a land bridge between Iran and Iraq and Syria, bringing their forces up to the Israeli border. President Trump showed less of a tendency to micromanage the wars in comparison to President Obama. President Trump allowed the Pentagon to play a greater role in the conduct of the wars. The United States also was involved in the conflict in Yemen. Shiite rebel forces known as the Houthis, backed by Iran, were engaged in a revolution against the Yemen government. The Yemen government is backed by Saudi Arabia and the United Arab Emirates, as well as the United States.

Article

Prepared by: Robert Weiner, *University of Massachusetts, Boston*

How to Start a Nuclear War

The Increasingly Direct Road to Ruin

ANDREW COCKBURN

Learning Outcomes

After reading this article, you will be able to:

- Learn under what conditions the president can launch nuclear missiles.
- Learn whether or not deterrence and extended deterrence works.

Serving as a US Air Force launch control officer for intercontinental missiles in the early Seventies, First Lieutenant Bruce Blair figured out how to start a nuclear war and kill a few hundred million people. His unit, stationed in the vast missile fields at Malmstrom Air Force Base, in Montana, oversaw one of four squadrons of Minuteman II ICBMs, each missile topped by a W56 thermonuclear warhead with an explosive force of 1.2 megatons—eighty times that of the bomb that destroyed Hiroshima. In theory, the missiles could be fired only by order of the president of the United States, and required mutual cooperation by the two men on duty in each of the launch control centers, of which there were five for each squadron.

In fact, as Blair recounted to me recently, the system could be bypassed with remarkable ease. Safeguards made it difficult, though not impossible, for a two-man crew (of either captains or lieutenants, some straight out of college) in a single launch control center to fire a missile. But, said Blair, "it took only a small conspiracy"—of two people in two *separate* control centers—to launch the entire squadron of fifty missiles, "sixty megatons targeted at the Soviet Union, China, and North Korea." (The scheme would first necessitate the "disabling" of the conspirators' silo crewmates, unless, of course, they, too, were complicit in the operation.) Working in conjunction, the plotters could "jury-rig the system" to send a "vote" by turning keys in their separate launch centers. The three other launch centers might see what was happening, but they would not be able to override the two votes, and the missiles would begin their firing sequence. Even more alarmingly, Blair discovered that if one of the plotters was posted at the particular launch control center in overall command of the squadron, they could together format and transmit a "valid and authentic launch order" for general nuclear war that would immediately launch the entire US strategic nuclear missile force, including a thousand Minuteman and fifty-four Titan missiles, without the possibility of recall. As he put it, "that would get everyone's attention, for sure." A more pacifically inclined conspiracy, on the other hand, could effectively disarm the strategic force by formatting and transmitting messages invalidating the presidential launch codes.

When he quit the Air Force in 1974, Blair was haunted by the power that had been within his grasp, and he resolved to do something about it. But when he started lobbying his former superiors, he was met with indifference and even active hostility. "I got in a fair scrap with the Air Force over it," he recalled. As Blair well knew, there was supposed to be a system already in place to prevent that type of unilateral launch. The civilian leadership in the Pentagon took comfort in this, not knowing that the Strategic Air Command, which then controlled the Air Force's nuclear weapons, had quietly neutralized it.

This reluctance to implement an obviously desirable precaution might seem extraordinary, but it is explicable in light of the dominant theme in the military's nuclear weapons culture: the strategy known as "launch under attack." Theoretically, the president has the option of waiting through an attack before deciding how to respond. But in practice, the system of command and control has been organized so as to leave a president facing reports of incoming missiles with little option but to launch. In the words of Lee Butler, who commanded all US nuclear forces at the end of the Cold War, the system the military designed was "structured to drive the president invariably toward a decision to launch under attack" if he or she believes there is "incontrovertible proof that warheads actually are on the way." Ensuring that all missiles and bombers would be en route before any enemy missiles actually landed meant that most of the targets in the strategic nuclear war plan would be destroyed—thereby justifying the purchase and deployment of the massive force required to execute such a strike.

Among students of nuclear command and control, this practice of precluding all options but the desired one is known as "jamming" the president. Blair's irksome protests threatened to slow this process. When his pleas drew rejection from inside the system, he turned to Congress. Eventually the Air Force agreed to begin using "unlock codes"—codes transmitted at the time of the launch order by higher authority without which the crews could not fire—on the weapons in 1977. (Even then, the Navy held off safeguarding its submarine-launched nuclear missiles in this way for another twenty years.)

Following this small victory, Blair continued to probe the baroque architecture of nuclear command and control, and its extreme vulnerability to lethal mishap. In the early Eighties, while working with a top-secret clearance for the Office of Technology Assessment, he prepared a detailed report on such shortcomings. The Pentagon promptly classified it as SIOP-ESI—a level higher than top secret. (SIOP stands for Single Integrated Operational Plan, the US plan for conducting a nuclear war. ESI stands for Extremely Sensitive Information.) Hidden away in the Pentagon, the report was withheld from both relevant senior civilian officials and the very congressional committees that had commissioned it in the first place.

From positions in Washington's national security think tanks, including the Brookings Institution, Blair used his expertise and scholarly approach to gain access to knowledgeable insiders at the highest ranks, even in Moscow. On visits to the Russian capital during the halcyon years between the Cold War's end and the renewal of tensions in the twenty-first century, he learned that the Soviet Union had actually developed a "dead hand" in ultimate control of their strategic nuclear arsenal. If sensors detected signs of an enemy nuclear attack, the USSR's entire missile force would immediately launch with a minimum of human intervention—in effect, the doomsday weapon that ends the world in *Dr. Strangelove.*

Needless to say, this was a tightly held arrangement, known only to a select few in Moscow. Similarly chilling secrets, Blair continued to learn, lurked in the bowels of the US system, often unknown to the civilian leadership that supposedly directed it. In 1998, for example, on a visit to the headquarters of Strategic Command (STRATCOM), the force controlling all US strategic nuclear weapons, at Offutt Air Force Base, near Omaha, Nebraska, he discovered that the STRATCOM targeting staff had unilaterally chosen to interpret a presidential order on nuclear targeting in such a way as to reinsert China into the SIOP, from which it had been removed in 1982, thereby provisionally consigning a billion Chinese to nuclear immolation. Shortly thereafter, he informed a senior White House official, whose reaction Blair recalled as "surprised" and "befuddled."

In 2006, Blair founded Global Zero, an organization dedicated to ridding the world of nuclear weapons, with an immediate goal of ending the policy of launch under attack. By that time, the Cold War that had generated the SIOP and all those nuclear weapons had long since come to an end. As a result, part of the nuclear war machine had been dismantled—warhead numbers were reduced, bombers taken off alert, weapons withdrawn from Europe. But at its heart, the system continued unchanged, officially ever alert and smooth running, poised to dispatch hundreds of precisely targeted weapons, but only on receipt of an order from the commander in chief.

The destructive power of the chief executive is sanctified at the very instant of inauguration. The nuclear codes required to authenticate a launch order (reformulated for each incoming president) are activated, and the incumbent begins an umbilical relationship with the military officer, always by his side, who carries the "football," a briefcase containing said codes. It's an image simultaneously ominous and reassuring, certifying that the system for initiating World War III is alert but secure and under control.

Even as commonly understood, the procedures leading up to a launch order are frightening. Early warning satellites, using heat-seeking sensors, followed a minute later by ground radars, detect enemy missiles rising above the curve of the earth. The information is analyzed at the North American Aerospace Defense Command (NORAD) in Colorado and relayed to the National Military Command Center in the basement of the Pentagon. The projected flight time of the missiles—thirty minutes from Russia—determines the schedule. Within eight minutes, the president is alerted. He then reviews his options with senior advisers such as the secretary of defense, at least those who can be reached in time. The momentous decision of how to respond must be made in as little as six minutes. Using the unique codes that identify him to the military commands that will carry out his instruction, he can then give the order, which is relayed in seconds via the war room and various alternate command centers to the missile silos, submarines, and bombers on alert. The bombers can be turned around, but otherwise the order cannot be recalled.

Fortunately, throughout the decades of confrontation between the superpowers, neither US nor Soviet leaders were ever personally contacted with a nuclear alert, even amid the gravest crises. When Brzezinski, Carter's national security adviser, was awakened at three in the morning in 1979 by what turned out to be a false alarm regarding incoming Soviet missiles, the president learned what had happened only the following day.

Today, things are different. The nuclear fuse has gotten shorter.

Generally unrecorded by the outside world, there has been a "streamlining" of the system of command and control, as Blair put it in a somewhat opaque article in *Arms Control Today.* Though the shift, which dates to the George W. Bush era and was additionally confirmed to me by a former senior Pentagon official, may appear to outsiders as a merely bureaucratic rearrangement, it has deadly serious implications. Formerly, attack warnings were received and processed by NORAD, in its lair deep inside Cheyenne Mountain, and passed via the National Military Command Center to the White House. But intelligence of a possible attack now goes almost directly to the head of Strategic Command. From his headquarters, far from Washington at Offutt Air Force Base, this powerful officer, currently an Air Force general named John Hyten, reigns supreme over the entire US strategic nuclear arsenal. He now dominates the whole nuclear countdown process: alerting the president, briefing him on the threat, and guiding him through the various options for a retaliatory or, as is likely given the jamming, preemptive strike.[1]

At one level, the change reflects a skirmish in the perennial internecine battles for budget share within the military, in which STRATCOM has clearly triumphed at the expense of NORAD, which was relegated to a basement at Peterson Air Force Base in 2006. But it appears there was a more significant motive for the decision. The head of STRATCOM is invariably a four-star general or admiral commanding a global fiefdom of 184,000 people, in and out of uniform. As Hyten reminded a congressional committee this year, "US STRATCOM is globally dispersed from the depths of the ocean, on land, in the air, across cyber, and into space, with a matching breadth of mission areas. The men and women of this command are responsible for strategic deterrence, nuclear operations, space operations, joint electromagnetic spectrum operations, global strike, missile defense, analysis and targeting."

In contrast, the director of the National Military Command Center is customarily a mere one-star officer, far down in the military pecking order. Furthermore, as befits an administrator, this officer is often absent from the command center buried deep under the Pentagon. On 9/11, for example, the commander at the time was out of the building during the attacks. The actual watch officers pulling eight-hour shifts in the center are colonels, even more lowly and therefore unpardonably reluctant to disturb or wake the commander in chief for what could be a false alarm. Four-stars, on the other hand, are the gods of the military hierarchy, accustomed to deference from all around them. Such panjandrums, especially those with the means to end human civilization, can be expected to have fewer inhibitions against disturbing presidential slumbers.

So it has proved. According to Blair's high-level sources, Bush and Obama received urgent calls from Omaha on "multiple occasions" during their time in office, and it would seem highly likely that Trump has had the same experience. This March, Colonel Carolyn Bird, the battle watch commander in the STRATCOM Global Operations Center at Offutt, hinted at this privileged access in a CNN report, boasting, "There's nobody we can't get on the phone." Hyten himself dutifully attested to CNN that Trump "asked me very hard questions. He wants to know exactly how it would work," and sententiously acknowledged that "there is no more difficult decision than the employment of nuclear weapons." Hyten did not mention that in both actual alerts and exercises, according to Blair, it has sometimes proved impossible to locate and patch in officials such as the defense secretary, despite the fact that the system calls for them to be connected automatically. Thus, in a real or apparent crisis, the crucial and necessarily fraught conversation may be between two men: General Hyten and Donald Trump.

Furthermore, in addition to being shorter, the nuclear fuse may now be lit earlier. For decades, the typical scenario for a nuclear alert began with early warning detection of Russian missiles piercing the clouds over Siberia and following a predictable trajectory. But these days the threats that necessitate those direct calls from Omaha to the White House are more diffuse and ambiguous. Ominous but unverified intelligence reports cite Chinese and Russian progress in hypersonic weapons—missiles that launch toward space and then turn to race toward their targets at five times the speed of sound, allegedly rendering any form of defense impossible. Vladimir Putin has bragged publicly about Russia's development of intercontinental nuclear-powered cruise missiles and other innovations in his strategic arsenal. (He even personally fired four ballistic missiles in an exercise last October.) North Korean ICBMs, seemingly reliant on a stash of old Soviet rocket engines smuggled out of Ukraine, could supposedly threaten the West Coast of the United States. Iran has tested and deployed home-grown medium-range missiles, as have Pakistan and India.

This new world of multiple threats has sparked public alarm among the military leadership. General Hyten and other powerful officers, for instance, have spoken ominously of the Russian and Chinese hypersonic weapons, maneuvering in unpredictable fashion as they flash toward us at up to five miles a second. Blair has heard the same worries expressed by his sources, and not just about the hypersonics. "There are all kinds of missiles going off all the time now," he told me. "We're regularly picking up these launches and trying to figure out what the fuck's going on." Presuming that the paths of these supposedly maneuverable weapons are unpredictable, an "imminent" threat no longer necessarily means that enemy missiles are already on the way. Today, the mere suspicion that something is about to happen could be enough for the general in Omaha to phone the presidential bedroom. Hypothetically, given the torrent of incoming and necessarily ambiguous information, intelligence reports that the command crew of a Chinese hypersonic missile squadron have canceled their dinner reservations could prompt such a call and a hurried, lethal decision. The jamming of the president, in other words, can begin earlier than ever.

While Bush and Obama were at the helm, their untrammeled power to launch excited little public concern, even though both men were prone to initiating conventional wars. Obama's commitment to "modernize" America's entire nuclear arsenal at a reported cost of at least $1.2 trillion generated no public outrage, or even much concern. According to Jon Wolfsthal, Obama's senior director for arms control and nonproliferation at the National Security Council, "There is no clear understanding of how much these weapons systems actually cost." When asked to produce a budget for the entire cost of our nuclear weapons forces, he told me, the Pentagon declined, on the grounds that it would be "too hard" to come up with that figure. But the arrival of Donald Trump, irascible, impulsive, and ignorant, was a different matter, especially given his threats to destroy North Korea with fire and fury. For the first time in decades, nuclear weapons were becoming a matter of public interest and concern.

For the Pentagon, busy extracting more than a trillion dollars from taxpayers to buy and operate an entirely new force of nuclear weapons, Trump's irresponsible rants cannot have been a welcome development. As Maryland senator Ben Cardin remarked in a Foreign Relations Committee hearing last fall, "We don't normally get a lot of foreign policy questions at town hall meetings, but as of late, I've been getting more and more questions about, Can the president really order a nuclear attack without any controls?" The senators were seeking reassurance that Trump couldn't really incinerate the planet in a fit of pique, and to that end had summoned a former STRATCOM commander, C. Robert Kehler, as witness.

Kehler, a general who retired in 2013, was evidently anxious to put to rest any unwelcome notions the senators might entertain of overhauling the existing nuclear command system: "Changes or conflicting signals," he warned, "can have profound implications for deterrence, for extended deterrence, and for the confidence of the men and women in the nuclear forces." As the general chose to depict it, launching the nukes would be a somewhat laborious bureaucratic process, involving "assessment, review, and consultation between the president and key civilian and military leaders" (including their lawyers), which would only then be "followed by transmission and implementation of any presidential decision by the forces themselves. All activities surrounding nuclear weapons are characterized by layers of safeguards, tests, and reviews." Of course, as a recent commander, Kehler had to have known that in contemporary alerts and exercises it has sometimes proved impossible to get those key leaders (or their lawyers) on the line, let alone find time for "assessment, review, and consultation."

But if the president determines that the United States is under the threat of an imminent attack, asked Ron Johnson of Wisconsin, "he has almost absolute authority" to launch, "correct?" Kehler gave a reluctant yes. But, persisted Johnson, what if the president (i.e., Trump) issued a completely unjustified strike order? What would Kehler have done? "I would have said I am not ready to proceed," answered the general. In other words, he would disobey the order.

Asked to comment on Kehler's statement at a security conference just a few days later, Hyten confirmed that he, too, was fully prepared to defy a direct order from his commander in chief. "The way the process works is this simple: I provide advice to the president. He'll tell me what to do, and if it's illegal, guess what's going to happen?"

"You say no," prompted the moderator.

"I'm going to say, Mr. President, it's illegal," continued Hyten, expressing confidence that the president, rather than brushing his objection aside and brusquely transmitting the order to launch, would obligingly respond, "'What would be legal?' And we'll come up with options of a mix of capabilities to respond to whatever the situation is."

Neither of the generals provided any example of what might actually constitute an illegal order (Kehler offered only some vague references to "military necessity" and "proportion"), still less any precedent for American military commanders defying civilian authority when ordered to launch an attack. In any event, though their comments may have served their purpose in calming public fears, they were entirely irrelevant. Unless the principal command center has been knocked out, once the president gives his order the STRATCOM commander has no role in actually executing the nuclear strike. He sees a presidential launch order at the same time as the other command centers that execute it.[2]

In the event that a commander did choose to defy the president, the former senior Pentagon official suggested, it could even lead to a situation where officers in the launch centers would be receiving contrary orders through different channels, leading to what he called "the biggest shitstorm in the world."

Despite the generals' reassurances and Kehler's plea to leave things alone, there are moves to curb the president's "absolute authority" to push the button. In September 2016, Ted Lieu, a Democratic congressman from California, introduced legislation, along with Senator Ed Markey of Massachusetts, to prevent the president from calling a first strike without congressional approval. Lieu, a former member of the Air Force judge advocate general well versed in the laws of war, could not see any legal justification for the president to unilaterally launch such an attack. The framers, he pointed out when we discussed the matter recently, gave Congress the power to declare war. "There's no way they would have let one person launch thousands of nuclear weapons that could kill millions of people in less than an hour and not have called that war. If you don't call that war, you run down the Constitution." He was unimpressed by the STRATCOM generals' pledge to defy an illegal launch order and hence felt the urgent need for legislation. "Do we really want to depend on military officers not following an order?"

Not only would Lieu's bill, which has attracted eighty-one cosponsors, preclude Trump from dropping a nuclear weapon on Syria "because he's angry at Assad" or some similarly impulsive initiative, it would also, he suggested, prevent launching in response to intelligence of a potential threat (it would not, however, prevent a US nuclear response in the event of incoming missiles). As he reminded me, intelligence has a poor record on threat warnings. "We had intelligence that Iraq had weapons of mass destruction, and it turned out they didn't." He also cited the near-disaster of Brzezinski's late-night wake-up call.

In the past, it should be noted, reliance on intelligence warnings has brought us closer to disaster than we knew. In the Fifties, General Curtis LeMay, the father of the Strategic Air Command, secretly deployed his own fleet of electronic intelligence planes over the Soviet Union. "If I see that the Russians are amassing their planes for an attack," he told a visiting emissary from Washington, Robert Sprague, "I'm going to knock the shit out of them before they can take off." "But General LeMay, that's not national policy," cried a horrified Sprague. "I don't care," replied LeMay. "It's my policy. That's what I'm going to do."

In the event that faulty intelligence had actually led to a launch order during the Cold War, the civilian leadership would have been kept in the dark as to what was on the target list, just as they were unaware that LeMay planned to launch World War III on his own initiative. It is well known, for example, that since the early Sixties the war plan has contained "counterforce" options allowing the president to strike military targets while "withholding" attacks on cities. Brilliant minds at the Rand Corporation and elsewhere labored to design such flexibility and have it adopted as policy. But in reality, the distinction was a fiction. War plans enjoined at the highest level were simply ignored by those charged with implementing them. As Franklin Miller, the director of strategic forces policy in the Office of the Secretary of Defense during the Eighties, later explained, whatever the civilian leadership devised, the Joint Strategic Target Planning Staff at Offutt Air Force Base totally controlled the actual selection of targets and the weapons assigned to destroy them. If the civilians ever asked for specific information, the planners in Omaha coolly replied that they had "no need to know." So successfully did the Offutt targeters guard their turf that even the Joint Chiefs of Staff, the presiding military bureaucracy at the Pentagon, were reportedly denied access to their internal guide for assigning targets. Sheltered by secrecy, the planners were able to define "city" in their targeting guidance so narrowly, Miller later wrote, that had the president ordered a large-scale nuclear strike against military targets with the "urban withhold" option, "every Soviet city would have nonetheless been obliterated." The degree of overkill was extraordinary. One small target area, for example, five miles in diameter, was due to suffer up to thirteen thermonuclear explosions.

Things should be different today. Mild-mannered and professorial, Kehler and Hyten present a striking contrast to the bellicose, cigar-chewing LeMay, who promised to reduce the Soviet Union to a "smoking, radiating ruin at the end of two hours," or his successor as head of the Strategic Air Command, General Thomas S. Power, whom even LeMay reportedly considered "not stable" and "a sadist." Modern communications ensure tighter command and control. The national nuclear weapons stockpile, which peaked at more than 30,000 in the late Sixties, has now passed 6,000, of which some 1,800 are deployed on missiles and aircraft.

But the reduction in numbers obscures the staggering amount of destruction baked into today's war plans. Blair has published an authoritative estimate of America's global targets, identifying at least 900 such designated aimpoints for US missiles and bombers in Russia, of which 250 are classed as "economic" and 200 as "leadership," most of them in cities. Moscow itself could be subject to a hundred nuclear explosions. Poverty-stricken North Korea supplies eighty targets, while Iran furnishes forty. "These are still just huge numbers of weapons," Blair said to me recently. "The targets are still in those three categories: weapons of mass destruction—which means nuclear—war-sustaining industry, and leadership, same as they always were." Much of the drawdown in warhead numbers, sanctified by arms control treaties, has been thanks to military confidence that both weapons and intelligence have become more accurate, meaning that fewer weapons need be assigned to any given target. Thus, whereas targeting plans once called for destroying every single bridge along a key Russian railroad, current aimpoints are far more select—destroying just one bridge to shut down the whole line. "They have gotten smart about where the real entrances are to command bunkers," says Blair. "You usually have a whole set of fake entrances, so you have to put down ten weapons on one major command post. Now we have intelligence on where the actual entrance is, so you only need one weapon for that."

Such confidence may be misplaced. The 1991 Gulf War was hailed at the time as a triumph of precision targeting and intelligence, as demonstrated by videos of missiles homing in unerringly on their targets. Yet a subsequent and exhaustive inquiry by the Government Accountability Office found that far from precision-guided-bomb maker Texas Instruments' claims of "one bomb, one target," it had required an average of four of the most accurate weapons, and sometimes ten, to destroy a given target. A Baghdad bunker destroyed in full confidence that it housed a high-level Iraqi command post had in fact sheltered some four hundred civilians, almost all women and children, most of whom were incinerated.

Even so, the Pentagon is working hard on developing the B61-12, a nuclear bomb that not only incorporates all the most desirable precision-guidance features but is also one of several "dial-a-yield" weapons in the US inventory, meaning that its explosive power can be adjusted as desired, in this case from as little as 0.3 kilotons (equivalent to 300 tons of TNT) all the way up to fifty kilotons. Such programs, as with the low-yield submarine-launched missile that is a key feature of the Trump Administration's Nuclear Posture Review, are supposedly aimed at "enhancing deterrence," justified as indicating to the Russians that if they use low-yield weapons, we can respond in kind. But these weapons appear to fulfill the function of conventional weapons in a conventional war, and therefore seem designed to fight, rather than deter, a nuclear war. The ongoing "modernization" (read "replacement") of the entire US nuclear arsenal that was set in motion by Obama included at least one low-yield bomb. Under Trump, however, the drive to treat nuclear weapons as if they can be used in a conventional battle appears to have gained greater prominence. The most recent Nuclear Posture Review, after all, was co-written by Keith Payne, president of the National Institute for Public Policy, and best known for the dubious notion that "victory or defeat in a nuclear war is possible," as he wrote in 1980. He added that "such a war may have to be waged to that point; and, the clearer the vision of successful war termination, the more likely war can be waged intelligently at earlier stages." His directives on the means to win such a war (more weapons, better targeting) are coupled with pious assurances that they are in the interests of maintaining a "credible deterrent."

Concepts such as dial-a-yield are no less dangerous for being potentially undependable in practice. Thanks to its observance of the (unratified) Comprehensive Nuclear Test Ban Treaty, the United States has not detonated a nuclear bomb since 1992. New warheads, such as those planned by the Trump Administration, are tested by computer simulations that stop short of actually initiating a chain reaction. Phil Coyle, a former director of the Nevada test site in the days when the United States actually detonated nuclear weapons to test them, told me that new, experimental designs could sometimes fail to perform according to plan. "Sometimes, they wouldn't work. I can remember some series where it took five or six tries to finally get it right, so to speak. If you were expecting a particular yield, you might not get it," he said, explaining that a new design might on rare occasions produce a yield greater than expected, or less, or no yield at all. Coyle is adamant, however, that all weapons currently in the stockpile are a hundred percent reliable.

But such faith is necessarily based on "virtual" tests, and belief that such simulations adequately reflect the real world is not universally accepted. Referring to the specialists who perform such simulations, Thomas P. Christie, the former director of operational tests and evaluation at the Pentagon, told me, "I'm sure that community has done some great work so far as simulations are concerned, because they can't test. But, if you can't test, you can't verify. I'm very skeptical. All you have to do is get about five percent wrong and you've got a real problem." Such caveats are important, not because they make the case for renewed testing but because nuclear war plans tend to assume a degree of certainty in systems performance, a dangerous misapprehension when everything to do with nuclear war is uncertain.

The same uncertainty holds true of the human element. Blair's lifetime study of nuclear command and control has convinced him that in a real crisis the system would be "prone to collapse under very little pressure." This stark conclusion was confirmed on the only occasion when it was put to the test: the terrorist attacks on 9/11, when it failed utterly. According to a detailed exposé by William Arkin and Robert Windrem of NBC News, senior officials found they could not communicate with one another. The commander of NORAD (still a player at that time) moved US nuclear forces to a higher stage of nuclear alert and closed the blast doors at Cheyenne Mountain for the only time since the end of the Cold War. Putin, alarmed by these developments, wanted to call Bush to ask what was going on, but Air Force One, which was running out of gas and looking for a secure place to land, could not receive phone calls. When the plane did land, at Barksdale Air Force Base in Louisiana, it was parked next to a runway littered with nuclear bombs—STRATCOM had been in the middle of a nuclear exercise when the hijackers hit the first tower and was now, while NORAD increased the level of nuclear alert, canceling the exercise and hurriedly unloading the active nukes from their bombers. Almost none of the senior officials in line to succeed the president followed their assigned procedures for evacuation to secure locations. One who did, Dennis Hastert, who as Speaker of the House was third in line for the presidency, took shelter in a secure bunker in Virginia, out of contact with the rest of the government. The education secretary, Rod Paige, sixteenth in line, who had gone with Bush to Florida, was left there when the president's party rushed to the plane. He eventually rented a car and drove back to Washington.

Even assuming every component of the system worked according to plan, the idea of initiating a nuclear exchange is obviously irrational in the extreme—a hundred nuclear explosions in and around Moscow? "Would it have made any difference if lots of weapons didn't go off, or (probably) a lot of missiles didn't get out of their silos?" Daniel Ellsberg emailed me in response to a query regarding the reliability of the weapons. "A first strike was insane from the start; and a damage-limiting second-strike (which I acknowledge accepting, foolishly, for some years) not really less so."

Nevertheless, there has clearly been a rational motivation underlying all these elaborate preparations for nuclear war over the years: money. The counterforce option, spawned in the early Sixties by the so-called wizards of Armageddon at the Air Force–funded Rand Corporation (the damage-limiting to which Ellsberg was referring), was enthusiastically endorsed by its patron because it parried a threat to Air Force budgets posed by the Navy's new submarine-launched missiles. Invulnerable to enemy attack, the subs clearly rendered the Air Force's land-based missiles and bombers superfluous to deterrence. But the sub-launched missiles were not accurate enough, even in theory, to hit military targets on the other side of the world, whereas the land-based ICBMs supposedly were. When I asked Ellsberg, who worked at Rand for many years, whether he knew of any of its proposals that would have resulted in a cut to the Air Force budget, he said no. That little has changed in our own day is evidenced by the Obama–Trump modernization plan to annually produce eighty new plutonium pits—the core of a nuclear weapon—at a potential overall cost of $42 billion, even though the United States already has 14,000 perfectly usable pits in storage.

Critics of our current nuclear arrangements, while quick to advocate for arms reduction or call for a reduction in executive power, generally accept the fundamental premise of deterrence. Congressman Lieu, for example, despite his sensible suggestions for keeping the president's finger as far as possible from the button, is wholly in tune with the consensus. "For purposes of mutually assured destruction," he assured me, "if any country were to launch a nuclear first strike on us, all bets would be off." Given such assumptions, even among the well intentioned, there seems little chance that the nuclear war machine's massive apparatus will be dismantled anytime soon. When Kehler testified on the merits of deterrence and "extended deterrence" (threatening to use nukes in support of an ally) at the Senate hearing, no one disagreed.

But one individual who most certainly does disagree is a man who spent a large portion of his life in the heart of the US nuclear machine and rose to command it all. "I spent much of my military career serving the ends of . . . deterrence, as did millions of others," Lee Butler, who as a four-star general had headed the Strategic Air Command, and its successor STRATCOM, from 1991 to 1994, wrote in a 2015 memoir. "I fervently believed that in the end it was the nuclear forces that I and others commanded and operated that prevented World War III and created the conditions leading to the collapse of the Soviet empire." But he grew increasingly skeptical about the role of nuclear weapons in maintaining global peace.

> I came to a set of deeply unsettling judgments. That from the earliest days of the nuclear era, the risks and consequences of nuclear war have never been properly understood. That the stakes of nuclear war engage not just the survival of the antagonists, but the fate of mankind. That the prospect of shearing away entire societies has no politically, militarily or morally acceptable justification. And therefore, that the threat to use nuclear weapons is indefensible.

In retirement, Butler joined calls for the total abolition of nuclear weapons.

The fundamental fallacy regarding deterrence, he reasoned, lay in the assumption that we know how an enemy would react to a nuclear threat. As he put it in the memoir, "How is it that we subscribe to a strategy that requires near-perfect understanding of enemies from whom we are often deeply alienated and largely isolated?" Furthermore, he pointed out, the whole theory rested on each side having a credible capacity to retaliate to a nuclear first strike with its own devastating counterattack. But the forces required for such a counterstrike can easily be perceived by a suspicious enemy to be deliberately designed to carry out their own first strike. Since nuclear rivals can never concede such an advantage, "new technology is inspired, new nuclear weapons designs and delivery systems roll from production lines. The correlation of forces begins to shift, and the bar of deterrence ratchets higher."

Interviews with former Soviet military leaders immediately after the Cold War, conducted by the BDM Corporation on a Pentagon contract, confirm that Butler was entirely correct as to their reaction to US nuclear preparations in the name of deterrence. For years, the Soviets told the interviewers, they believed the United States was preparing for a first strike. They therefore prepared to launch a preemptive strike if and when they detected signs of such preparations. Ignorant of Soviet thinking, the United States failed to curb military activities that might have confirmed their suspicions and sparked a preemptive attack.

None of this seems to have made much impression on the current crop of nuclear war planners, as Butler recently pointed out to me. "Over the past decade," he wrote in an email, "the

Air Force has undertaken a concerted effort to resurrect the old deterrence arguments. In the process, they have dredged up all of the deplorable straw men to knock down the case for arms control/abolition." This effort, he lamented, has been largely successful: "Arms control is now relegated to the back burner with hardly a flicker of heat, while current agreements are violated helter-skelter.

"Sad, sad times for the nation and the world," he concluded bleakly, "as the bar of civilization is ratcheted back to the perilous era we just escaped by some combination of skill, luck, and divine intervention."

Notes

1. STRATCOM declined to comment on the process, citing its classification status, but a spokesperson for the Joint Chiefs of Staff noted that a review conducted in May suggested that STRATCOM should play an even larger role in the nuclear command and control system.
2. A STRATCOM spokesperson assured *Harper's Magazine* that any presidential order to employ nuclear weapons would be preceded by "a serious and deliberate discussion regarding all available options, to include diplomatic and conventional military actions, guided by the advice of the Cabinet, national security experts, and relevant military advisers."

Critical Thinking

1. Under what conditions can the military refuse to carry out a presidential order to launch nuclear missiles?
2. What are the multiple nuclear threats that the United States faces?
3. Does the president have the constitutional authority to launch nuclear missiles? Why or why not?

Internet References

Nuclear Posture Review

https://media.defense.gov/...2018-NUCLEAR-REVIEW-FINAL-REPORT/...

Senate Hearings on Use of Nuclear Weapons

https://www.foreign.senate.gov/hearings/authority-to-order-use-of-nuclear-weapons

Article

Prepared by: Robert Weiner, *University of Massachusetts, Boston*

A New Era for Nuclear Security

Martin B. Malin and Nickolas Roth

Learning Outcomes

After reading this article, you will be able to:

- Explain what was accomplished at the last Nuclear Security Summit in 2016.
- List what steps had been taken by the international community to provide for the physical security of usable nuclear materials.

The 2016 Nuclear Security summit was a pivotal moment for the decades-long effort to secure nuclear material around the globe. More than 50 national leaders gathered in Washington for the last of four biennial meetings that have led to significant progress in strengthening measures to reduce the risk of nuclear theft.

These summits have played a critical role in nurturing that progress by elevating the political salience of nuclear security and providing a forum for world leaders to announce new commitments, share information, and hold one another accountable for following through on promised actions.

The international community is now entering the post-summit era, in which nuclear security will probably receive less-regular high-level political attention than it has in recent years. Yet, there is still critical work to be done to reduce the danger that nuclear weapons or the materials needed to make them could end up in the hands of a terrorist organization such as the Islamic State. Governments still do not agree on what nuclear security priorities are most pressing or how best to sustain the momentum generated by the summits. As the era of summitry recedes, will states continue improving measures to prevent nuclear theft and sabotage, or will the summits turn out to have been a high-water mark for nuclear security efforts?

Progress at the 2016 Summit

Over the course of the summit process, the participating states committed themselves to dozens of cooperative initiatives seeking to strengthen aspects of nuclear security, reduced vulnerabilities in their security systems, and pledged to continue joint efforts through multilateral groups and international institutions. The 2016 summit, held March 31–April 1 in Washington, marked progress on all of these fronts.

Like the 2010 summit in Washington, the 2012 summit in Seoul, and the 2014 summit in The Hague, this year's meeting produced a consensus-based communiqué. At the three most recent summits, smaller groups of participants also produced a series of joint statements and group commitments, or "gift baskets."[1] At this year's summit, all but three states participated in at least one of 18 gift baskets or nine joint statements, which covered a range of areas, including insider threats, transport security, minimization of the use of highly enriched uranium (HEU), and cybersecurity.[2] Among the most important outcomes of the recent summit was the establishment of a contact group, which will meet annually to discuss nuclear security.

Some of the major accomplishments of the summit are listed below.

Strengthening the commitment to nuclear security. China and India joined 36 states that had signed on to an important 2014 summit initiative on strengthening nuclear security implementation.[3] Members of this group committed to "meet the intent" of International Atomic Energy Agency (IAEA) nuclear security principles and recommendations, conduct self-assessments, host periodic peer reviews of their nuclear security, and ensure that "management and personnel with accountability for nuclear security are demonstrably competent," along with several other actions. This was an important commitment for China and India, demonstrating a measure of transparency and reassurance on nuclear security. Prior to the 2016 summit, neither country had been open to participating in such initiatives, although both nuclear-armed states face terrorist threats.[4]

The summit process also helped to build support for a foundational and legally binding international nuclear security instrument. After more than a decade, the 2005 amendment to the Convention on the Physical Protection of Nuclear Material (CPPNM) reached the required number of ratifications to enter into force in May. The amendment outlines nuclear security principles and requires states to establish rules and regulations for physical protection.

It also requires a review conference five years after entry into force and, if members choose to have them, additional review conferences at intervals of at least five years.[5] The amended CPPNM, now officially known as the Convention on the Physical Protection of Nuclear Material and Nuclear Facilities, could be a helpful tool for states to hold one another accountable for maintaining physical protection and strengthening norms.

Reducing nuclear security vulnerabilities. In addition to announcing new commitments, the summits were occasions for states to report on steps they had taken to remove or eliminate HEU or plutonium, convert reactors, improve physical protection, strengthen regulation, and contribute support to the IAEA or other international nuclear security work.

At the recent summit, Japan and the United States announced the completion of a commitment they made in 2014 to remove more than 500 kg of nuclear weapons-usable material from Japan.[6] Argentina announced it had eliminated the last of its HEU, making it the 18th state to clean out all of its nuclear weapons-usable material since the beginning of the summit process. Indonesia declared it had eliminated all of its fresh HEU and planned to get rid of all its HEU in 2016.

China announced the opening of its nuclear security center of excellence. Since 2010, China has worked with the United States to build the center as a hub for training, bilateral and multilateral best practice exchanges, and technology demonstration.[7] The center will help China test and strengthen its own nuclear security measures and will provide a venue for cooperation with others in the region and beyond.

The White House reported that 20 states hosted or invited peer review missions through the IAEA or from other states. Many other states announced that they had strengthened nuclear security laws or regulations, upgraded physical security, or updated the list of threats against which their nuclear facilities must be protected.

Continuing the dialogue. An important new gift basket created a nuclear security contact group that will convene annually on the margins of the IAEA General Conference. The contact group will carry forward the consultative element of the summit process, providing a forum for senior government officials to meet and discuss current efforts, evaluate progress on previously made commitments, and identify future priorities. If states buy into the idea of the contact group and take action to strengthen it, the group, whose membership is open to states that did not participate in the summits, could be an important vehicle for sustaining international nuclear security cooperation.

The summit also produced statements on bilateral nuclear security discussions between key countries. For example, China and the United States agreed to increase cooperation on nuclear terrorism prevention and conduct an annual dialogue on nuclear security.

In addition, summit participants agreed to action plans for the IAEA, the United Nations, Interpol, the Global Partnership against the Spread of Weapons and Materials of Destruction, and the Global Initiative to Combat Nuclear Terrorism (GICNT). The plans outline the roles these organizations will play in supporting ongoing nuclear security discussions now that the summits have ended.

Gaps and Missed Opportunities

In their communiqué, the participants in the 2016 summit pledged to "continuously strengthen nuclear security at national, regional, and global levels."[8] Striving for continuous improvement is the right way to frame the challenge of providing effective and sustainable nuclear security. Unfortunately, summit participants missed important opportunities to give added momentum to the effort. The following issues continue to require attention.

Still no global standard for nuclear security. Although the amended CPPNM establishes general security principles, it lacks specific standards or guidelines and applies only to materials in civilian use. UN Security Council Resolution 1540 requires states to provide "appropriate effective" protection for all materials, among other relevant measures, but does not specify what constitutes appropriate effective protection.[9] IAEA recommendations, to which dozens of states have now publicly subscribed, provide somewhat more specificity, but their implementation is voluntary. Although the summit process certainly helped produce a shared understanding of the importance of nuclear security, it fell short of producing a consensus on a meaningful minimum global standard.

If a global standard was beyond reach during the summits, a public commitment to stringent nuclear security measures among the states possessing the biggest stocks of HEU and plutonium would have been a consequential step. Although China's and India's endorsements of the initiative on strengthening nuclear security implementation was an important development, Russia's absence from the summit and Russia's and Pakistan's refusal to sign that statement is a significant gap in the patchwork of nuclear security commitments.

Furthermore, the summit outcomes were not comprehensive. Although the summit communiqués explicitly covered "all" nuclear material, most of the concrete progress from the meetings focused on civilian materials, largely ignoring the roughly ⅘ of the world's remaining HEU and plutonium that is controlled by military organizations.[10]

A mixed picture on implementation. Nuclear facilities in many countries still are not protected against the full range of threats. States with large stocks of nuclear weapons-usable material still contend with corruption and extremism.[11] On the ground, security upgrades remain urgently needed in many spots around the world. One indication of the extent of the inconsistent application of physical protection measures is that, after all of the high-level attention since the 2010 summit, at least six countries—Argentina, Brazil, the Netherlands, Slovakia, Spain, and Sweden—still do not have armed guards at their nuclear facilities.[12]

The collapse of U.S.–Russian bilateral cooperation is particularly alarming. Without Russian and U.S. commitments to rebuilding their bilateral nuclear security relationship, it will be impossible for the two states that possess roughly 80 percent of the world's weapons-usable nuclear material to reassure one another that their nuclear security is sound.

Slippage of consolidation and minimization goals. The Obama administration put laudable effort into cleaning out HEU and plutonium from many countries and minimizing the use of HEU elsewhere. Yet, political obstacles will likely make substantial additional progress more difficult than in the past, in particular for the hundreds of kilograms of HEU in Belarus and South Africa. Conversion of additional HEU-fueled research reactors to use low-enriched uranium fuel, particularly but not only in Russia, is hampered by technical challenges and political inattention. Moreover, summit participants failed to reach agreement, even in principle, on stopping or reversing the buildup of separated plutonium.[13]

Continuing culture of complacency in some countries. The summits put the notion of nuclear security culture on the agenda for many countries where it previously had been neglected. Nevertheless, workers, managers, policy officials, and even national leaders in many places still dismiss the threat of terrorist theft or sabotage as remote or implausible.[14] Many organizations handling nuclear weapons, HEU, or separated plutonium do not have specific programs focused on strengthening security culture. The IAEA has still not published its nuclear security culture self-assessment guide.[15] The summit process helped spark interest in strengthening security culture, but much more work is needed.

Need for morerobust channels for dialogue. The political momentum created by the summits will not likely be re-created through other organizations, although the contact group, IAEA ministerial meetings, a review conference for the amended CPPNM, and other forums certainly will provide important opportunities for discussion, reporting on progress, and further cooperation.

The recent summit's action plans did not significantly expand or strengthen the global nuclear security architecture. The IAEA has assumed greater responsibility for convening high-level discussions on nuclear security and has intensified its nuclear security efforts since the first summit. Yet, the agency still deals only with civilian material and has no authority to require states to take any action on nuclear security.[16] The nuclear security capacities of the UN and Interpol are even less robust, and the multilateral groupings, the GICNT and Global Partnership, remain unchanged by the action plans the summit participants produced.

Finally, Russia's absence from the recent summit may bode ill for the successful implementation of the summit action plans. Moscow's leadership and cooperation in all of the organizations referenced in the action plans will be necessary for many key nuclear security steps.

In the interest of promoting cooperation, the summits frequently focused on plucking low-hanging fruit.

Progress in the Post-Summit Era

In the interest of promoting cooperation, the summits frequently focused on plucking low-hanging fruit, while failing to advance more-difficult discussions of threats and persistent challenges. Governments must focus not only on what is most feasible but also on what is most urgently needed in light of the evolving threats they face.[17]

Nuclear security efforts should have a clear goal: ensuring that all nuclear weapons and the materials that could be used to make them, wherever they are in the world, are effectively and sustainably secured against the full range of threats that terrorists and thieves might plausibly pose.[18] Building an international consensus around such a goal will be a major challenge for the next U.S. president and for like-minded leaders.

The 2016 summit communiqué alludes to the goal of continuous improvement. Achieving that goal will require work on several fronts. Here are some of the most important areas of focus.[19]

Building up the commitment to stringent nuclear security standards. A legally binding set of international standards for nuclear security is unfortunately out of reach for the present. Yet, a group of states like-minded emanating from within the contact group or a special working group of the GICNT could develop a set of principles and guidelines that they pledge to apply to all stocks of nuclear weapons, HEU, and plutonium and invite other states to join them. Such a commitment should include the provision of well-trained, well-equipped on-site guard forces; comprehensive measures to protect against insider threats; control and accounting systems that can detect and localize any theft of weapons-usable nuclear material; protections against cyberthreats that are integrated with other nuclear security measures; effective nuclear security rules and regulations and independent regulators capable of enforcing them; regular and realistic testing of nuclear security systems, including force-on-force exercises; a robust program for enhancing security culture; and regular assessments of the evolving threat of theft or sabotage. Following the example of the initial group of adherents, the accumulation of international support for more-comprehensive standards could grow over time.

In the meantime, leading states that are bound by the amended CPPNM should push to universalize the treaty, and the states that have joined the initiative on strengthening nuclear security implementation initiative should encourage others to commit to implement IAEA recommendations and accept peer review.

Implementing effective and sustainable security measures on the ground. Commitments to stringent standards are meaningful only if they translate into real improvements. Bilateral cooperation can help spur the actions that are needed. The United States should expand nuclear security cooperation with China, India, and Pakistan, sharing additional information on security arrangements without revealing sensitive information that would increase vulnerability to terrorist attack. The United States also will need to make a priority of discussions with a wide range of countries on enhancing their own nuclear security, providing resources when needed.

Despite tensions over Ukraine and other issues, Russia and the United States should agree to a package of cooperation that includes nuclear energy initiatives, which are of particular interest to Russia, and nuclear security initiatives, which are of particular interest to the United States. Although it is unlikely in the current political environment, one mechanism for achieving this goal would be to restart the U.S.–Russian Nuclear Energy and Nuclear Security Working Group, which facilitated dialogue from 2009 until it was suspended in 2014 because of tensions between the two countries. Cooperation should no longer be based on a donor–recipient relationship but on an equal partnership with ideas and resources coming from both sides.[20]

Increasing efforts to reduce the number of sites where nuclear weapons and weapons-usable materials are stored. Today, there are fewer locations where HEU and plutonium can be stolen because of removals motivated by the summit process. The consolidation process must continue. Stringent security requirements can help to incentivize the process of consolidation, as can well-funded programs for conversion of HEU-fueled reactors and removal of material. Russia and the United States, as the countries whose nuclear stockpiles are dispersed in the largest number of buildings and bunkers, should each develop a national-level plan for accomplishing their military and civilian nuclear objectives with the smallest practicable number of locations. The United States and other interested countries should ensure that plutonium and HEU bulk processing facilities do not spread to other countries or expand in number or scale of operations and that no more plutonium is separated than is used, bringing global plutonium stocks down over time.

Establishing a nuclear security culture that does not tolerate complacency about threats and vulnerabilities. Every country with relevant materials and facilities should have a program in place to assess and strengthen security culture, and all nuclear managers and security-relevant staff should receive regular information, appropriate to their role, on evolving threats to nuclear security. At the same time, interested countries should launch initiatives to combat complacency, including a shared database of security incidents and lessons learned; detailed reports and briefings on the nuclear terrorism threat; discussions among intelligence agencies, on which most governments rely for information about the threats to their country; and an expanded program of nuclear theft and terrorism exercises.

Building up channels for dialogue. Countries must continue to share information and devise plans to meet current nuclear security challenges. The IAEA ministerial-level meetings on nuclear security will provide an important forum. If parties to the amended CPPNM elect to meet every five years to review progress, this process could create important opportunities to place high-level pressure on states to step up nuclear security commitments and implementation.

A more comprehensive scope of cooperation, including on military materials, could take place in multilateral forums. The GICNT, cochaired by Russia and the United States and still valued by both, consists of more than 80 states committed to the group's statement of principles, which includes improving measures that reduce the risk of nuclear theft such as accounting, control, and protection of nuclear and radiological materials. The group has not focused on these preventive approaches so far, but it should in the future.[21] This summer represents the GICNT's 10th anniversary, which would be an excellent time to announce the creation of a GICNT working group focused specifically on strengthening security for nuclear materials and facilities. The GICNT could also be a useful forum for Russia and the United States to expand nuclear security cooperation.

The contact group created at the nuclear security summit this year holds promise for facilitating dialogue, sharing information, and germinating joint activities. Its openness to all IAEA members has the advantage of potentially attracting states beyond the ring of past summit participants. Its size and heterogeneity, however, may limit the depth and effectiveness of the discussions. The contact group should select an executive committee of member state representatives—perhaps former summit hosts plus Russia, if it chooses to join—to establish and coordinate its agenda for discussion.

Finally, summit-level nuclear security meetings could be continued on the side of Group of 20 meetings, perhaps once every four years. This would sustain the kind of executive-level political attention to nuclear security that summits provided.

The nuclear security summits periodically pressed participants to commit themselves to new and stronger measures for preventing nuclear terrorism. They facilitated a process of stocktaking and reporting on the concrete actions participants had taken. Moreover, they were a vehicle for forging stronger international collaboration on bolstering nuclear security around the globe. States must continue to build on the progress they made through the summit process. If they do, the 2016 summit will mark the beginning, rather than the end, of a new era of continuous improvement in nuclear security.

Endnotes

1. For a comprehensive assessment of progress in fulfilling commitments from the summits prior to 2016, see Michelle Cann, Kelsey Davenport, and Jenna Parker, "The Nuclear Security Summit: Progress Report on Joint Statements," Arms Control Association and Partnership for Global Security, March 2015, https://www.armscontrol.org/reports/2015/The-Nuclear-Security-Summit-Progress-Report-on-Joint-Statements.
2. The three countries that did not join gift baskets were Gabon, Pakistan, and Saudi Arabia. For a list of gift baskets and joint statements from the 2016 summit, see "2016 Washington Summit," Nuclear Security Matters, n.d., http://nuclearsecuritymatters.belfercenter.org/2016-washington-summit.
3. "Strengthening Nuclear Security Implementation," March 25, 2014, http://www.state.gov/documents/organization/235508.pdf. Thirty-five countries signed the 2014 statement. Jordan joined in late 2015.
4. See Rajeswari Pillai Rajagopalan, "India and the Nuclear Security Summit," Nuclear Security Matters, April 26, 2016, http://nuclearsecuritymatters.belfercenter.org/blog/india-and-nuclear-security-summ; Hui Zhang, "China Makes Significant Nuclear Security Pledges at 2016 Summit," Nuclear Security Matters, April 8, 2016, http://nuclearsecuritymatters.belfercenter.org/blog/china-makes-significant-nuclear-security-pledges-2016-summit.
5. For background on the amended Convention on the Physical Protection of Nuclear Material, see "Convention on the Physical Protection of Nuclear Material," International Atomic Energy Agency (IAEA), n.d., https://www.iaea.org/publications/documents/conventions/convention-physical-protection-nuclear-material. For an argument that the review conferences envisioned in the amendment could help drive nuclear security progress, see Jonathan Herbach and Samantha Pitts-Kiefer, "More Work to Do: A Pathway for Future Progress on Strengthening Nuclear Security," *Arms Control Today*, October 2015.
6. "Joint Statement by the Leaders of Japan and the United States on Contributions to Global Minimization of Nuclear Material," April 1, 2016, http://nuclearsecuritymatters.belfercenter.org/files/nuclearmatters/files/joint_statement_by_the_leaders_of_japan_and_the_united_states_on_contrib.pdf.
7. Office of the Press Secretary, The White House, "U.S.–China Joint Statement on Nuclear Security Cooperation," March 31, 2016, https://www.whitehouse.gov/the-press-office/2016/03/31/us-china-joint-statement-nuclear-security-cooperation.
8. "Nuclear Security Summit 2016 Communiqué," April 1, 2016, http://nuclearsecuritymatters.belfercenter.org/files/nuclearmatters/files/nuclear_security_summit_2016_communique.pdf?m=1460469255.
9. See Matthew Bunn, "Appropriate Effective Nuclear Security and Accounting— What Is It?" (presentation, "Appropriate Effective" Material Accounting and Physical Protection—Joint Global Initiative/UNSCR 1540 Workshop," Nashville, TN, July 18, 2008), http://belfercenter.ksg.harvard.edu/files/bunn-1540-appropriate-effective50.pdf.
10. For a discussion of security for military materials, see Des Browne, Richard Lugar, and Sam Nunn, "Bridging the Military Nuclear Materials Gap," Nuclear Threat Initiative (NTI), 2015, http://www.nti.org/media/pdfs/NTI_report_2015_e_version.pdf. The 2016 summit communiqué reaffirmed that states had a fundamental responsibility "to maintain at all times effective security of all nuclear and other radioactive material, including nuclear materials used in nuclear weapons." See "Nuclear Security Summit 2016 Communiqué."
11. For a more complete discussion of the threats some countries with nuclear material face, see Matthew Bunn et al., "Preventing Nuclear Terrorism: Continuous Improvement or Dangerous Decline?" Belfer Center for Science and International Affairs, Harvard Kennedy School, March 2016, pp. 39–52, http://belfercenter.ksg.harvard.edu/files/PreventingNuclearTerrorism-Web.pdf.
12. For country information on physical protection, see the 2016 NTI Nuclear Security Index for sabotage, http://ntiindex.org/wp-content/uploads/2016/03/2016-NTI-Index-Data-2016.03.25.zip. Belgium has only recently added armed guards to its nuclear facilities. The Swedish regulator has ordered that facilities post armed guards by February 2017. See Steven Mufson, "Brussels Attacks Stoke Fears About Security of Belgian Nuclear Facilities," *The Washington Post*, March 25, 2016; "Swedish Regulator Orders Tighter Security at Nuclear Plants," Reuters, February 5, 2016, http://www.reuters.com/article/sweden-nuclear-security-idUSL8N15K3SS.
13. The 2014 summit communiqué states, "We encourage States to minimise their stocks of [highly enriched uranium] and to keep their stockpile of separated plutonium to the minimum level, both as consistent with national requirements." "The Hague Nuclear Security Summit Communiqué," March 25, 2014,

http://www.state.gov/documents/organization/237002.pdf. In 2016, there was no mention of plutonium in the communiqué.
14. Matthew Bunn and Eben Harrell surveyed nuclear experts in states with nuclear weapons-usable material and found that some respondents did not find certain threats credible, despite extensive evidence to the contrary. See Matthew Bunn and Eben Harrell, "Threat Perceptions and Drivers of Change in Nuclear Security Around the World: Results of a Survey," Belfer Center for Science and International Affairs, Harvard Kennedy School, March 2014, http://belfercenter.ksg.harvard.edu/files/surveypaperfulltext.pdf.
15. IAEA, "Self-Assessment of Nuclear Security Culture in Facilities and Activities That Use Nuclear and/or Radioactive Material: Draft Technical Guidance," July 2, 2014, http://www-ns.iaea.org/downloads/security/security-series-drafts/tech-guidance/nst026.pdf.
16. See Trevor Findlay, "Beyond Nuclear Summitry: The Role of the IAEA in Nuclear Security Diplomacy After 2016," Belfer Center for Science and International Affairs, Harvard Kennedy School, March 2014, http://belfercenter.hks.harvard.edu/files/beyondnuclearsummitryfullpaper.pdf.
17. For a discussion of how the threat of nuclear terrorism has evolved over time, see Bunn et al., "Preventing Nuclear Terrorism," pp. 14–26, 133–143.
18. Ibid., p. 96.
19. For the recommendations on which this section draws, see Bunn et al., "Preventing Nuclear Terrorism," pp. 96–133.
20. For a more complete description of the end of nuclear security cooperation, see Nickolas Roth, "Russian Nuclear Security Cooperation: Rebuilding Equality, Mutual Benefit, and Respect," Deep Cuts Commission, June 2015, http://deepcuts.org/files/pdf/Deep_Cuts_Issue_Brief4_US-Russian_Nuclear_Security_Cooperation1.pdf.
21. Global Initiative to Combat Nuclear Terrorism (GICNT), "Fact Sheet," n.d., http://www.gicnt.org/content/downloads/sop/GICNT_Fact_ Sheet_ June2015.pdf. Although the GICNT terms of reference state that its activities do not involve "military nuclear programs of the nuclear weapon states party to the Treaty on the Nonproliferation of Nuclear Weapons," the group's statement of principles, which is the only document GICNT members are required to endorse, contains no such exclusion. See Bureau of International Security and Nonproliferation, U.S. Department of State, "Terms of Reference for Implementation and Assessment," November 20, 2006, http://2001-2009.state.gov/t/isn/rls/other/76421.htm; GICNT, "Statement of Principles," 2015, http://gicnt.org/content/downloads/sop/Statement_of_Principles.pdf.

Critical Thinking

1. What is the role of the International Atomic Energy Agency in providing for physical nuclear security?
2. Why is it important that China and India cooperate to provide for physical nuclear security?
3. What are the gaps that still need to be dealt with in the field of physical nuclear security?

Internet References

Arms Control Association
https:armscontrol.org

Bulletin of the Atomic Scientists
www.thebulletin.org

Non-Proliferation Policy Education Center
www.npolicy.org/

Nuclear Security Summit 2016
www.nss2016.org/

Martin B. Malin is an executive director of the Project on Managing the Atom at Harvard Kennedy School's Belfer Center for Science and International Affairs. From 2000 to 2007, he was director of the Program on Science and Global Security at the American Academy of Arts and Sciences.

Nickolas Roth is a research associate at the Project on Managing the Atom. Parts of this article draw from the authors' article with Matthew Bunn and William H. Tobey in 2016 titled "Preventing Nuclear Terrorism: Continuous Improvement or Dangerous Decline?"

Article

Prepared by: Robert Weiner, *University of Massachusetts, Boston*

The Thucydides Trap

When one great power threatens to displace another, war is almost always the result—but it doesn't have to be.

GRAHAM ALLISON

Learning Outcomes

After reading this article, you will be able to:

- Learn why a war occurs in 12 of 16 cases in which the ruling power was displaced.
- Learn how adversaries can deter war.

In April, chocolate cake had just been served at the Mar-a-Lago summit when President Donald Trump leaned over to tell Chinese President Xi Jinping that American missiles had been launched at Syrian air bases, according to Trump's account of the evening. What the attack on Syria signaled about Trump's readiness to attack North Korea was left to Xi's imagination. Welcome to dinner with the leaders who are now attempting to manage the world's most dangerous geopolitical relationship. The story is a small one. But as China challenges America's predominance, misunderstandings about each other's actions and intentions could lead them into a deadly trap first identified by the ancient Greek historian Thucydides. As he explained, "It was the rise of Athens and the fear that this instilled in Sparta that made war inevitable." The past 500 years have seen 16 cases in which a rising power threatened to displace a ruling one. Twelve of these ended in war.

Of the cases in which war was averted—Spain outstripping Portugal in the late 15th century, the United States overtaking the United Kingdom at the turn of the 20th century, and Germany's rise in Europe since 1990—the ascent of the Soviet Union is uniquely instructive today. Despite moments when a violent clash seemed certain, a surge of strategic imagination helped both sides develop ways to compete without a catastrophic conflict. In the end, the Soviet Union imploded and the Cold War ended with a whimper rather than a bang.

Although China's rise presents particular challenges, Washington policymakers should heed five Cold War lessons.

Lesson 1: War between nuclear superpowers is MADness.

The United States and the Soviet Union built nuclear arsenals so substantial that neither could be sure of disarming the other in a first strike. Nuclear strategists described this condition as "mutual assured destruction," or MAD. Technology, in effect, made the United States and Soviet Union conjoined twins—neither able to kill the other.

Today, China has developed its own robust nuclear arsenal. From confrontations in the South and East China Sea to the gathering storm over the Korean Peninsula, leaders must recognize that war would be suicidal.

Lesson 2: Leaders must be prepared to risk a war they cannot win.

Although neither nation can win a nuclear war, both, paradoxically, must demonstrate a willingness to risk losing one to compete.

Consider each clause of this nuclear paradox. On the one hand, if war occurs, both nations lose and millions die—an option no rational leader could choose. But, on the other hand, if a nation is unwilling to risk war, its opponent can win any objective by forcing the more responsible power to yield. To preserve vital interests, therefore, leaders must be willing to select paths that risk destruction. Washington must think the unthinkable to credibly deter potential adversaries such as China.

Lesson 3: Define the new "precarious rules of the status quo."

The Cold War rivals wove an intricate web of mutual constraints around their competition that President John F. Kennedy called "precarious rules of the status quo." These included arms-control treaties and precise rules of the road for air and sea. Such tacit guidelines for the United States and China today might involve limits on cyberattacks or surveillance operations.

By reaching agreements on contentious issues, the United States and China can create space to cooperate on challenges—such as global terrorism and climate change—in which the national interests the two powers share are much greater than those that divide them. Overall, leaders should understand that survival depends on caution, communication, constraints, compromise, and cooperation.

Lesson 4: Domestic performance is decisive.

What nations do inside their borders matters at least as much as what they do abroad. Had the Soviet economy overtaken that of the United States by the 1980s, as some economists predicted, Moscow could have consolidated a position of hegemony. Instead, free markets and free societies won out. The vital question for the U.S.-China rivalry today is whether Xi's Leninist-Mandarin authoritarian government and economy proves superior to American capitalism and democracy.

Maintaining China's extraordinary economic growth, which provides legitimacy for sweeping party rule, is a high-wire act that will only get harder. Meanwhile, in the United States, sluggish growth is the new normal. And American democracy is exhibiting worrisome symptoms: declining civic engagement, institutionalized corruption, and widespread lack of trust in politics. Leaders in both nations would do well to prioritize their domestic challenges.

Lesson 5: Hope is not a strategy.

Over a four-year period from George Kennan's famous "Long Telegram," which identified the Soviet threat, to Paul Nitze's NSC-68, which provided the road map for countering this threat, U.S. officials developed a winning Cold War strategy: contain Soviet expansion, deter the Soviets from acting against vital American interests, and undermine both the idea and the practice of communism. In contrast, America's China policy today consists of grand, politically appealing aspirations that serious strategists know are unachievable. In attempting to maintain the post-World War II Pax Americana during a fundamental shift in the economic balance of power toward China, the United States' real strategy, truth be told, is hope.

In today's Washington, strategic thinking is often marginalized. Even Barack Obama, one of America's smartest presidents, told the *New Yorker* that, given the pace of change today, "I don't really even need George Kennan." Coherent strategy does not guarantee success, but its absence is a reliable route to failure.

Thucydides's Trap teaches us that on the historical record, war is more likely than not. From Trump's campaign claims that China is "ripping us off" to recent announcements about his "great chemistry" with Xi, he has accelerated the harrowing roller coaster of U.S.-China relations. If the president and his national security team hope to avoid catastrophic war with China while protecting and advancing American national interests, they must closely study the lessons of the Cold War.

Critical Thinking

1. What is the Thucydides trap?
2. Why was a war between the United States and Russia avoided?
3. What rules should the United States develop to prevent a war with China?

Internet References

The Thucydides Trap
https:www.belfercenter.org/publication/thucydides-trap

Thucydides Resources
https://www.belfercenter/org/thucydides-trap/thucydides-resources

Graham Allison is a director of the Harvard Kennedy School's Belfer Center for Science and International Affairs.

Article

Prepared by: Robert Weiner, *University of Massachusetts, Boston*

Afghanistan's Arduous Search for Stability

THOMAS BARFIELD

Learning Outcomes

After reading this article, you will be able to:

- Understand the relationship between the Obama administration and the Karzai regime.
- Explain the role of Pakistan in the Afghan conflict.
- Discuss the importance of factionalism in understanding Afghan politics.

Why, after the expenditure of so much blood and treasure in Afghanistan over the past decade and a half, does there seem to be so little to show for it? One key reason has been a lack of leadership and unity in Kabul.

For almost 15 years, Afghanistan appeared mired in a political rut while President Hamid Karzai dominated the scene until he finally left office in 2014. On the domestic front, he made use of a highly centralized administrative system to appoint all the country's provincial officials, block the formation of political parties, and reward his allies with patronage. Externally, he acted as the sole interlocutor with international backers (primarily the United States and its allies) that provided more revenue to the government than it got in taxes, financed and equipped its security forces, and funded development projects.

In 2002, Karzai was acclaimed at home and abroad as just the man to unite a fragmented country. Initially endorsed by a *loya jirga* (a grand assembly of tribal leaders) to serve a two-year interim term as president, he won Afghanistan's first presidential election in 2004, riding genuine popular enthusiasm. But domestic support for Karzai and the national government began to wane by 2006, as Afghans increasingly viewed the president as incompetent and his administration as corrupt. Karzai recognized that some of these criticisms were legitimate, but his dependence on personalized political deal making and patronage created a dilemma he could not easily resolve: Reforming the system risked bringing about its collapse. Indeed, it sometimes appeared that corruption was the glue that held the Karzai government together.

Since international aid provided the bulk of Karzai's patronage assets and paid the salaries of his officials, the donors in theory should have had leverage to push for reforms. In reality, international pressure rarely amounted to more than the distribution of critical but toothless reports. Laundry lists of specific reforms that the Afghan government agreed to implement were the centerpieces of every international donors' conference—along with complaints that the goals set by previous conferences had not been met.

Karzai, who did not lack shrewdness, could safely ignore such international criticism and threats to withhold aid. All he had to do was make the case that Afghanistan would become a renewed security problem if his government did not receive support. He rightly assumed that none of the big donors were willing to test that proposition.

The administration of US President George W. Bush was so preoccupied with its failing war in Iraq after 2003 that it put Afghanistan on the back burner in hopes of limiting American involvement. Changing the leadership of the Afghan government or building its capacity for action would have required more time, money, and military effort than Washington was willing to commit. Even after the reemergence of a Taliban insurgency in the mid-2000s, the Bush administration muted its criticism of Karzai and cultivated a close personal relationship with him. Afghans firmly believed that no rival political leader could hope to challenge Karzai as long as he was treated as indispensable by the Americans.

Obama's "Good War"

The belief that nothing could ever change in Afghanistan was eroded by elections in the United States in 2008 and in Afghanistan the next year. During the US presidential campaign, Senator Barack Obama criticized the Bush administration's unpopular Iraq policy in part by arguing that the war in Afghanistan was more central to US antiterrorism priorities. In making Afghanistan "the good war," Obama moved it back to the forefront of US policy debates; but in doing so, he laid considerable blame on Karzai for the persistent difficulties. Soon-to-be Vice President Joseph Biden made that clear on a visit to Kabul in February 2008, when he stormed out of a dinner at the presidential palace after Karzai responded to his questions about corruption by denying it existed.

A year later, Biden returned to Kabul with bad news for Karzai from the newly elected President Obama. The high level of personal attention that Bush had lavished on him would be ending. Biden bluntly told Karzai that he would "probably talk to [Obama] a couple of times a year," and certainly not every week. For an Afghan leader anticipating his reelection bid later that year, such a cold shoulder from the Americans had direct, negative political repercussions.

The split between Karzai and the Obama administration led many Afghans (including Karzai himself) to assume that the Americans would be seeking a replacement for him in the 2009 election. There was strong historical precedent. Both the British and the Soviets installed weak leaders when they occupied Afghanistan in the nineteenth and twentieth centuries, respectively, and they both replaced those leaders with stronger personalities when they decided to withdraw. It was assumed that Obama (whose endgame was to turn all security responsibilities over to the Afghan government) might also be in the market for a more competent and reliable leader.

Popular support for insurgencies based on driving foreigners out declines when the foreigners actually leave.

The election seemed to be an opportunity to engineer this outcome while simultaneously claiming a success for the country's new democratic process by letting the Afghan people make the choice in a free and fair vote. Karzai would likely lose to any challenger who could unite the domestic opposition by demonstrating clear American backing. The allies Karzai had attracted through patronage would have no trouble shifting their allegiance to someone else if it looked as if he was not in a position to deliver that largesse any longer. In Afghan politics, the perception of power was power itself.

Richard Holbrooke, Obama's special representative to Pakistan and Afghanistan, was determined to find a replacement for Karzai, and he made that well known. The difficulty was that the highly fragmented Afghan opposition was waiting for the Americans to anoint a challenger who could use that backing to build a domestic coalition, while the Americans were waiting for the opposition to make the first move by uniting behind a consensus candidate Washington could discreetly back while maintaining a stance of official impartiality.

When the opposition proved unable to agree on a single candidate, the Obama administration resigned itself to Karzai's reelection. Even without a unified opposition, Karzai ultimately needed to get a runoff election canceled and resort to massive vote fraud to win. He characteristically blamed all election irregularities on "foreign outsiders," but most of the fraud was committed by his own allies.

Karzai's Grievances

Karzai's relationship with the Obama administration remained hostile throughout his second term. Even as the United States and its NATO allies deployed over 100,000 additional troops to Afghanistan between 2010 and 2012 to ensure his regime's security, Karzai's sense of grievance never ceased, and he attacked his backers anytime an opportunity presented itself. In October 2011, he went so far as to declare, according to a Reuters report, "God forbid, if ever there is a war between Pakistan and America, Afghanistan will side with Pakistan."

Karzai's continued distrust of the United States was so deep that he adamantly refused to sign a Bilateral Security Agreement (BSA) that would have allowed some US troops to remain in Afghanistan after most were set to leave by the end of 2014, even after it was overwhelmingly approved by a national *loya jirga.* In this and other actions Karzai appeared to bear out the conclusions of US Ambassador Karl Eikenberry's leaked November 2009 assessment, which asserted that the Afghan leader was "not an adequate strategic partner" and "continues to shun responsibility for any sovereign burden."

Few Afghans believed Karzai when he announced that he would abide by the constitution's two-term limit and step down in 2014. No Afghan leader had ever relinquished power before death, except at the point of a gun. Conspiracy theories proliferated over his true intentions. Some said he planned to amend the constitution and run again; others said he would call on a *loya jirga* to declare him president for life without an election; still others insisted he would find a sycophant to serve as a figurehead president while he ruled from behind the scenes.

In the end, Karzai stood down and his chosen successor, Zalmai Rassoul, went down to a humiliating defeat in the first round of the presidential election in April 2014. Two months later, the electoral process came to a standstill when Abdullah

Abdullah, the winner of a large plurality in the first round, alleged that only massive fraud could explain the second-round victory of his rival, Ashraf Ghani. Once again, rumors abounded that Karzai would use the disputed outcome to declare himself interim ruler. But after personal pressure from the US Secretary of State John Kerry, the two finalists agreed in September 2014 to share power and form a national unity government, with Ghani as president and Abdullah in the newly created position of chief executive officer.

The Taliban have moved away from a purely religious identity to burnish their credentials as Afghan nationalists.

Mood Change

The establishment of the unity government was a pivotal and positive change from the Karzai era. The two new leaders, unlike Karzai, were eager to improve Afghanistan's international relations. Ghani, a former academic and World Bank official, had lived in the United States for 30 years before returning to Afghanistan in 2002 and serving in a number of cabinet posts. An able administrator popular with the foreign embassies in Kabul, he presented himself as a technocrat rather than a politician. Abdullah had never lived abroad and could claim a distinguished record of opposing both the Soviet occupation and the Taliban regime as part of the Northern Alliance. As a former foreign minister (and one of the best-dressed men in Kabul), he was also skilled at representing Afghan interests internationally.

While Ghani and Abdullah were domestic rivals, they were in general agreement on major foreign policy issues. As candidates, both had promised to sign the BSA with the United States. Immediately upon taking office, they did so together. With that accomplished, other members of the international coalition agreed to continue their own more limited military assistance.

The new government also sought to reposition itself regionally by cultivating closer ties with China, to which Ghani made his first official trip abroad in October 2014. Beijing had only recently begun to engage more proactively with Kabul, offering to facilitate peace talks with the Taliban and to invest in the economy. (Despite the immense amounts of money spent in Afghanistan by Western governments, few if any private businesses from Europe or North America appeared willing to risk their own capital there.)

On the public relations front, Ghani made an effort to thank Afghanistan's foreign allies for their sacrifice and support rather than accusing them, as Karzai had, of plotting against the country's interests. This change in mood was soon publicly reciprocated: Both Abdullah and Ghani came to Washington for a formal state visit in March 2015, and Ghani gave an address to Congress.

Cross Purposes

Although it appeared to be moving forward in harmony on the international front, the new government was hamstrung domestically by its internal divisions. The shotgun wedding that compelled Abdullah and Ghani to share power had also created dual staffs working at cross purposes within the executive branch—particularly when it came to the division of ministry appointments, long a source of patronage. Each faction felt shortchanged by the other, and there were not enough high-level positions to go around.

Even when the rival factions did agree, the parliament rejected a number of appointments for its own reasons. Vital positions in the 25-member cabinet, including national security offices, were still filled by interim appointees more than a year and a half after the government was formed. This deadlock extended down to the appointment of provincial governors. Ghani declared all the positions vacant but he was slow to fill them, leaving far too many provinces in the hands of lame-duck Karzai appointees.

Even seemingly noncontroversial reforms, such as requiring merit-based appointments, also ran into stiff opposition. It originated in the long-standing animosity between Afghan exiles who returned to the country with technical skills after decades abroad and the relatively uneducated resistance leaders who had stayed to fight. These mujahedeen commanders (sometimes pejoratively labeled warlords) considered a merit-based appointment system to be a power grab by Ghani and the technocratic class he represented. In late 2015, they joined former ministers and officials from the Karzai government to create an opposition group called the Protection and Stability Council, but it too split between those who wanted Karzai back and those who just wanted a stake in the national unity government.

To be fair, it is hard to find any government or insurgent group in Afghan history that has not been prone to factionalism. (In the 1980s, the regime of the People's Democratic Party of Afghanistan was composed of two autonomous and hostile communist factions, while the mujahedeen resistance based in Pakistan was divided into seven separate recognized parties.) The greater internal threat to the unity government's stability was not so much the factions themselves as the uncertain lines of authority.

Abdullah's position had been invented in the compromise that made Ghani president. It called for a *loya jirga* to be held in two years to approve a constitutional amendment to convert his CEO post to that of prime minister. Yet it was never clear just what authority such a position would have. The hyperactive and micromanaging Ghani acted more like the chief executive, leaving Abdullah marginalized and his supporters frustrated. And because Ghani had frequently expressed his opposition to a two-headed government, there was always considerable doubt about his commitment to the power-sharing deal. As the deadline to ratify the agreement approached, neither the wording of the prospective amendment nor the date of the *loya jirga* had been set.

Even with the best intentions, assembling a constitutional *loya jirga* was never going to be easy, given the requirement that nearly half of its 761 delegates must be elected district representatives. Although the constitution was adopted in 2004, the district elections it mandated were never held. A requirement for the participation of current members of parliament was also problematic. The legislature's five-year term had expired in June 2015, leaving it to operate under the questionable authority of a presidential decree.

Seeking to resolve these issues, the Independent Electoral Commission in January 2016 unexpectedly announced that elections for both a new parliament and district representatives would be held on October 15. Since this was the same commission that had badly botched the presidential balloting, Abdullah's representatives objected that the election should wait until voting procedures were reformed and the commission itself was restructured. The president's office was silent on the matter. The uncertainty threatened the political bargain that had created the unity government. In a land where the rule of law was more a distant goal than a current reality, there was no ready solution for a crisis of legitimacy pitting the government's most powerful players against each other. It was also the last thing a government facing an insurgency would care to contemplate.

Rush to the Exit

For over a decade, Mullah Omar, the reclusive leader of the Taliban, had encouraged his forces to keep fighting until they regained power. While the Taliban did make some progress in the mid-2000s, the surge of more than 100,000 US and NATO troops into Afghanistan from 2010 to 2012 put them on the defensive, particularly in the south. As long as they had sanctuaries in Pakistan, however, the Taliban could absorb the loss of territory and manpower while waiting for the announced date of the foreigners' withdrawal.

That timing was driven more by American domestic politics than by the security situation in Afghanistan. The first withdrawals coincided with the 2012 presidential election in which Obama won a second term. By early 2014, only 33,000 US troops remained (partially to ensure security during the Afghan presidential election). That number had dwindled to just under 10,000 when the BSA was signed in Kabul in September 2014. The agreement turned over all security responsibilities to the Afghan government and stated that "unless otherwise mutually agreed, United States forces shall not conduct combat operations in Afghanistan."

Washington's plan for the next two years was to restrict its remaining troops (to be reduced to 5,000 in 2015) to training and counterterrorism operations, not fighting the Taliban. By the end of 2016, all that remained of the US military presence in Afghanistan would be a Marine guard unit at the embassy. This would allow Obama to declare that he had wrapped up the war on his watch, as promised. But as critics noted, this schedule ignored the realities on the ground.

The rebuilding of the Afghan security forces did not begin in earnest until after the surge troops arrived in 2010, and they were not ready to take over the security responsibilities assigned to them. Nor was there an effective replacement for the air power, logistics systems, and medical support that the US and NATO supplied to the Afghan military. And Afghan leaders had a bigger worry: If all international troops left the country, would donor countries continue to provide the high level of funding Afghan security forces would need in the years ahead? More than boots on the ground, money in the pipeline was critical to the survival of any Afghan government. The lack of such support had led to the destructive civil war and the rise of the Taliban in the 1990s.

Fortunately for the government, the Taliban insurgency had its own problems that were not unlike those in Kabul. In 2015, the insurgent group was roiled by a succession crisis and confronted a changing international calculus on the danger posed by jihadist movements worldwide.

Taliban Trouble

The Taliban had been thoroughly defeated in 2001, but they were able to restart the insurgency in Afghanistan by 2005 from their sanctuary in Quetta, Pakistan. Although nominally an ally of the US coalition in Afghanistan, Pakistan had long played a double game of supporting the insurgency while also receiving large amounts of US military and financial aid. Pakistani strategists assumed that the United States would eventually pull out of Afghanistan, and that the Taliban would then be able to take power in Kabul, or at least allow Pakistan to use them as proxies. It was a replay of Islamabad's policy of supporting the mujahedeen against the Soviet-backed Afghan government in the 1980s. In both cases, the policy was based on two simple strategic assumptions: foreign troops were the only firewall preventing the fall of a Kabul government, and insurgents could expect a quick victory once the foreigners withdrew.

Almost counterintuitively, however, Afghan history demonstrated the opposite. Over the course of three wars and 150 years, Kabul governments that retained a world power's patronage and a continued flow of money and weapons proved difficult or impossible to dislodge. This was seen most dramatically after the Soviet withdrawal from Afghanistan in 1989, when the regime of Mohammad Najibullah continued to receive Soviet aid and kept the mujahedeen insurgency at bay. That regime did fall in 1992, but only after the Soviet Union itself collapsed.

Indeed, insurgent movements that had proved so successful at inducing foreign governments to withdraw their troops from Afghanistan had a dismal record of failure in attempting to seize control of the country. When rebels did succeed, it was never against foreign-backed regimes but during domestic civil wars when the national government had no world-power patron. This occurred in 1929 when the bandit rebel Habibullah Kalakani ousted King Amanullah, and again in the mid-1990s when the Taliban ousted the mujahedeen President Burhanuddin Rabbani. It is also worth noting that both of these insurgent regimes were driven from power when their enemies obtained the backing of a world power.

The Taliban ramped up hostilities as international troops withdrew in 2014, and went on the offensive in 2015. Casualties on both sides were much higher than in previous years. The Taliban made substantial gains in the southern province of Helmand and surprised the Kabul government by taking the northern provincial capital of Kunduz in October before being driven out by a counterattack. Kunduz was particularly significant because it was outside the normal areas of Taliban strength. However, the Kabul government did not collapse, nor did its troops in the region abandon the fight after a dispiriting defeat, as happened in Iraq during a 2014 offensive by Islamic State (ISIS) jihadists.

Much like the mujahedeen after 1989, the Taliban have been most successful in rural areas where Kabul's influence has always been weak, but they have proved unable to take over major populated areas where they could establish a rival government. The Kabul government still controlled 85 percent of the country's 398 districts at the end of 2015 while the Taliban controlled only about 7 percent, with the remaining districts being contested between the two. But as was the case historically in Afghanistan, government control in rural areas generally does not extend much beyond the district center whether insurgents are present or not.

Nor is the conflict just between the Taliban and the Kabul government. Worsening security has encouraged the formation (or revival) of autonomous local militias that defend their own interests against outsiders of all sorts. Ghani's attempt to shut down militias in Kunduz without providing additional national army troops to replace them was one reason the Taliban were able to take the city.

Despite their successes, the Taliban faced a classic Afghan insurgent political dilemma. Popular support for insurgencies based on driving foreigners out declines when the foreigners actually leave, and insurgents are then faced with fractured movements divided by local issues rather than united by national ones. This inherent problem was worsened by the July 2015 revelation that Mullah Omar had died more than two years earlier, in April 2013. As Commander of the Faithful and ruler of the Islamic Emirate of Afghanistan, he had been such a key figure of unity for the Taliban that his death was hidden from his followers. This allowed the Taliban leadership to continue issuing orders and opinions in his name, avoiding the contentious question of succession.

Confirmation of Omar's death came to light only when it was leaked to sabotage the opening of a second round of peace talks in Pakistan between the Taliban and the Kabul government. Mullah Akhtar Mansoor then declared himself leader of the Quetta *shura* (consultative council) and the new Commander of the Faithful. (Mansoor, a senior member of the *shura*, had been acting as Mullah Omar's deputy before he died.) This provoked violent opposition from some Taliban leaders inside Afghanistan, which Mansoor suppressed. Other dissident factions proclaimed their independence from Mansoor and the Taliban by swearing allegiance to Abu Bakr al-Baghdadi, the self-declared caliph of the Islamic State (ISIS).

More than boots on the ground, money in the pipeline was critical to the survival of any Afghan government.

ISIS Alarms

Although it was small and only loosely connected with the movement in Iraq and Syria, the branch of ISIS that appeared in Afghanistan changed the course of US policy. Disturbed by the rapid collapse of Iraqi forces that allowed ISIS to seize the country's second-largest city, Mosul, in the summer of 2014, and by the group's declaration of a transnational caliphate, the Obama administration had to reevaluate the risks of letting Afghanistan fend for itself.

While unwilling to formally scrap its policy of disengagement or send more troops, the administration quietly reversed course in 2015 and began to assist Afghan security forces directly in a number of other areas. Under the new arrangements, some Afghan forces were allowed to call in US air support, and US Special Operations units assigned as advisers to Afghan troops began fighting on their own as well, helping the army recover Kunduz and hold its ground in Helmand. Drone strikes that had been directed primarily at al-Qaeda leaders in Pakistan now targeted leaders of the ISIS faction in Afghanistan. Talk of reducing the US troop presence to an embassy guard by the end of Obama's presidency was shelved. This shift was facilitated by a cooperative relationship with the national unity government, which welcomed all the help it could get.

Despite the renewed focus on radical jihadists as a common transnational threat, ISIS and the Taliban differ in many significant ways. Mullah Omar never sought a worldwide caliphate, only an Islamic emirate in Afghanistan. Since 2001 the Taliban have also moved away from a purely religious identity to burnish their credentials as Afghan nationalists. Having to live down a reputation for intolerance and mismanagement from their time ruling the country, they have softened many of their positions on education and women's participation in public life. Unlike ISIS, the Taliban have not labeled all Afghans opposed to them as apostates and no longer attempt to enforce the rigid Salafist religious practices that alienated so many Afghans when they were in power. It should also go without saying that the Taliban share no common ethnic or linguistic ties with Sunni insurgents in Iraq or Syria.

The Sunni–Shia sectarian differences that drive conflicts in Iraq and Syria are largely absent in Afghanistan. Insurgents in Iraq can exploit Sunni Muslim discontent at being a minority in a Shia-majority country, while in Syria they can use their opposition to the rule of the Alawite minority over a Sunni majority to build support even among those who do not share their ideology. But in Afghanistan, 85 percent of the population is Sunni, as are almost all the leading figures of the government in Kabul.

Similarly, the Taliban can get only limited traction by exploiting ethnic divisions. They are mostly Pashtun and strongest in the Pashtun regions of the east and south. But the top national leaders in Kabul, including Karzai and Ghani, are also Pashtun. To the extent that the Taliban have sought to exploit Pashtun grievances over being made to share power with other ethnic groups, they have undercut their own ability to expand in non-Pashtun regions.

Those former Taliban factions that did ally themselves with ISIS after 2014 did so primarily as a way to maintain their autonomy and gain new resources. In this respect, they were similar to other non-Taliban insurgent groups like Gulbuddin Hekmatyar's Hizb-i-Islami, which had always refused to pledge allegiance to Mullah Omar, or the Haqqani network, which cast its lot with Omar but operated independently from the Taliban's Quetta command structure.

However, there is also a new player in Afghanistan's easternmost provinces: Pakistani Taliban factions that have been pushed across the border and have reorganized themselves under the ISIS banner. While the Pakistani government has long supported the Afghan Taliban, it is at war with their Pakistani namesakes, over which it has no control. With extreme violence, they have set up an insurgency within an insurgency. The Afghan Taliban and non-Taliban local militia work in parallel to protect their communities against the Pakistanis.

Diplomatic Hopes

When the national unity government took office in 2014, Ghani made a determined effort to cultivate closer ties with Pakistan. Since Islamabad was the main if unofficial sponsor of the Taliban, its cooperation wouldbe vital in negotiations to end the insurgency. In the past, Pakistan had prevented Taliban leaders from talking with the Afghan government, and until Islamabad changed its strategic policy of disrupting Afghanistan, the war would likely continue. (As in the 1990s, this policy assumed there was no reason to negotiate because the Taliban could win the war outright once the Americans left.) For many years, the cost of such disruption had been minimal. Afghanistan was at war but Pakistan was at peace. The United States needed Pakistan's cooperation to maintain a supply route to landlocked Afghanistan and so was unwilling to make it pay much of a price for its meddling.

That free ride began to end with the emergence of a domestic Pakistani Taliban whose focus was on Islamabad rather than Kabul. It was declared a terrorist group by the United States in 2010 and its leadership increasingly became the target of US drone strikes. In a series of ever more violent terrorist attacks on civilian targets, the Pakistani Taliban struck throughout the country, making many Pakistani cities more dangerous than Afghan ones.

The Pakistani army eventually responded by moving troops into the militants' northwestern strongholds, the Swat Valley and the so-called tribal areas of North and South Waziristan, pushing many of them across the border into Afghanistan. It was with no small sense of irony that Kabul responded coolly to complaints from Islamabad that insurgents were now using Afghan territory to stage attacks on Pakistan, which had long allowed militants to cross the border in the other direction.

Ghani was hopeful when Pakistan facilitated official talks with the Taliban in the summer resort of Murree in July 2015, a step that Pakistani Prime Minister Nawaz Sharif called "a major breakthrough." But the talks soon broke down; the Taliban were consumed with their succession struggle due to the death of Mullah Omar. Ghani's broader outreach to Pakistan, which included domestically unpopular security arrangements with the Pakistani military, fell apart after an enormous truck bomb leveled a poor neighborhood in Kabul and inflicted hundreds of casualties in early August. Blamed on the Pakistan-backed Haqqani network, the bombing forced Ghani into an admission that his policy of rapprochement had failed. Talks to strengthen ties with India, which had been put on hold in deference to Pakistan's concerns, now went forward.

China's new interest in Afghanistan constitutes the one remaining hope for peace talks between the government and the Taliban. Chinese President Xi Jinping made a state visit to Islamabad in April 2015 and announced a plan to build a $46 billion China–Pakistan Economic Corridor featuring a network of roads, railways, and pipelines linking Xinjiang in western China with Pakistan's Indian Ocean port of Gwadar in Baluchistan province. Such a huge investment could not be made if active insurgencies in both Afghanistan and the border regions of Pakistan persisted.

Also, China had long expressed its concerns about the danger Islamic radicals posed to its restive province of Xinjiang (which borders both Afghanistan and Pakistan), where the Muslim Uighur minority was increasingly viewed as potentially subversive. In May 2015, Beijing arranged for secret negotiations between the Taliban and the Kabul government in Urumqi, Xinjiang's capital, and in January 2016 it declared that as "a peaceful mediator of the Afghan issue, China supports the 'Afghan-led and Afghan-owned' reconciliation process."

Iran is another new player that might also help broker peace in Afghanistan. Released from an international sanctions and diplomatic isolation in January 2016, it has more in common with the United States than Pakistan on the question of stabilizing Afghanistan. Neither the United States nor Iran, which are fighting in parallel in Iraq and Syria, wants to see another state besieged by Sunni jihadists. That would bring an unwelcome wave of new Afghan refugees to Iran, which still hosts perhaps two million from previous wars.

Paying the Price

If international peace talks fail to gain traction, the more likely outcome is a continued stalemate in Afghanistan. But there is always the possibility of an internal deal among the Afghans themselves. What divides Afghans more than ethnicity or ideology is an unwillingness to share power and a winner-takes-all approach to politics. Historically this led to domestic stability after wars were settled because the losers either did not have the means to keep fighting or deemed it prudent to join the winners. Periods of intense violence were followed by long periods of peace. But this changed after 1978, when domestic factions began to draw on outside resources that encouraged continued fighting rather than compromise or surrender.

By playing on Cold War fears, the People's Democratic Party of Afghanistan gained Soviet support. The mujahedeen used the same Cold War fears to get money and weapons from the United States, and they played the Islamic brotherhood card to obtain funding from Saudi Arabia. With the fall of the Soviet Union, the Afghan state collapsed, but the subsequent civil war was funded by new flows of international support. Private donors in the Arab world such as Osama bin Laden backed the Taliban's Islamic emirate. Pakistan also gave money and arms to the Taliban in hopes of gaining "strategic depth" and an ally against India.

When Al-Qaeda, based in Afghanistan, attacked the United States in 2001, US forces and their Afghan allies toppled the Taliban in 10 weeks but failed to follow up on the victory. Washington did nothing to midwife a structure of peace at a time when all factions were open to reconciliation. Instead, Afghanistan drifted back into a vortex of conflict that fed on itself. Today, the United States and its allies fund the Kabul government while Pakistan funnels support to its opponents.

Throughout this period of international intervention, Afghan factional leaders (in the government or among the insurgents) have been more intransigent than their followers. Supported by outside money and arms, they have assumed there is no need to compromise since victory could be just over the horizon. Yet over the course of four decades, it never was, and it never will be. No people are more practical than the Afghans, but until they can agree among themselves that dying in conflicts over issues mostly irrelevant to their own lives is a price not worth paying, the war will continue.

Critical Thinking

1. How can China play a role in mediating the conflict in Afghanistan?
2. How has the Islamic State played a role in changing the conflict in Afghanistan?
3. Why was the relationship between President Karzai and President Obama strained?

Internet References

Afghanistan Research and Evaluation Unit
http://areu.org/af/default.aspx?

The Afghanistan Analysts Network
www.Afghanistan-analysts.org/

The Taliban in Afghanistan
http://www.cfr.org/afghanistan/taliban-afghanistan/p/0551

Thomas Barfield is a professor of anthropology at Boston University. His books include *Afghanistan: A Political and Cultural History* (Princeton University Press, 2010).

Article Prepared by: Robert Weiner, *University of Massachusetts, Boston*

ISIS's Next Move

DANIEL BYMAN

Learning Outcomes

After reading this article, you will be able to:

- Learn what the future strategy of ISIS is.
- Learn why ISIS will attack the West.

The Islamic State is on the ropes, yet the group may make a comeback. The U.S.-led coalition has driven it from much of its territory in Iraq and Syria, while most of its so-called "provinces" elsewhere in the Muslim world also have lost territory or stagnated. In July, U.S.-backed local forces took Mosul, the Islamic State's largest stronghold in Iraq, and then in October took the Syrian city of Raqqa, the Islamic State's capital. The caliphate may soon exist only as an idea. Once the most powerful jihadist group in modern history, the Islamic State is "now pathetic and a lost cause," claimed Brett McGurk, the U.S. envoy for the anti--Islamic State coalition.

Despite these impressive gains, the United States is not well prepared for the group's defeat. After losing control of key territory, the Islamic State may repeat the actions of its predecessor when the U.S.-led surge brought Al Qaeda in Iraq to the edge of defeat: go underground, disrupt politics and foster sectarianism; wage an insurgency; and then come roaring back. The United States cannot depend on its partners to counter this cycle, as local allies in Iraq and Syria are unprepared to govern and conduct effective counterinsurgency operations, while the very identity of long-term U.S. allies is unclear as Washington lacks a durable coalition in Iraq, let alone in Syria. Finally, the concepts the Islamic State promulgated are dangerous and may be exploited in the future by the Islamic State or successor organizations. As a result, the Islamic State's campaign of regional and international terrorism, already maintained at a high level despite the group's territorial setbacks, will likely continue and perhaps even grow in the near term.

President Donald Trump began or continued several positive counterterrorism policies—but also undertook initiatives that risk aggravating the danger the Islamic State poses. The administration improved relations with important allies like Saudi Arabia and continued the military campaign that began under former president Barack Obama to steadily drive the Islamic State from its strongholds in Iraq and Syria. However, the administration's anti-Muslim rhetoric and policies will likely alienate some American Muslims, increasing the risk of radicalization and discouraging cooperation between these communities and police and intelligence services. In addition, the administration's blanket embrace of the Saudi position in the Middle East will heighten sectarianism, which feeds the Islamic State and like-minded groups. Finally, a decline in foreign aid, the State Department budget and the number of national-security personnel diminishes U.S. diplomacy and the United States' ability to resolve conflicts—all necessary for fighting jihadist groups and preventing them from spreading to new areas. Although many positive changes seem unlikely under the Trump administration, efforts to fight the Islamic State more effectively would include continuing efforts to train allied military and security services (albeit with realistic expectations). The Trump administration and U.S. leaders in general should try to bolster American resilience, which current policies are undermining.

The Islamic State has steadily suffered a series of defeats in the last two years. Most important, its base in Iraq and Syria has shrunk dramatically. By fall of 2014, the Islamic State controlled much of eastern Syria and western Iraq, including Raqqa, Mosul and Tikrit, and by spring 2015, the group captured Ramadi in Iraq and Palmyra in Syria, while its so-called province in Libya seized Sirte and nearby territory. Since then, a mix of Iraqi government forces, Kurdish militias, local tribal groups and others have pushed the group from major cities in Iraq and Syria. The Islamic State is likely to lose almost all its territory in Iraq and Syria in the coming months.

The so-called Islamic State provinces have also suffered. However, in 2016, U.S.-supported militia groups drove the Islamic State province in Libya from its base around Sirte, dispersing it to southern Libya. Elsewhere, Islamic State provinces demonstrated little dynamism in recent years—a stark contrast to 2014 and 2015 when the group seemed to expand throughout the Middle East. Some Islamic State provinces, like the one in Sinai, are succeeding with a low-level insurgency that includes bloody terrorist attacks, but these have focused on their own societies and governments, not the United States and its allies. Although analysts fret that the Islamic State might relocate—and some fighters will inevitably find a new home—there is no credible substitute for Iraq and Syria as a base, as terrorism analyst Jason Burke contends.

Funding and recruitment also dried up. The Islamic State attracted more than forty thousand foreign fighters; in some months, more than a thousand foreign fighters would join its ranks. In the last year, the number of new foreign volunteers reduced to a trickle, and the organization's budget, which relies heavily on "taxing" local territory, also declined.

The Islamic State's decline perpetuates itself. The group appealed to foreign fighters and funders partly by marketing itself as a winner that successfully created an Islamic state with true Islamic governance. Its biggest boasts are now its biggest failures. Fewer foreigners want to join a group incapable of defending the caliphate and clearly losing to the enemies it vowed to vanquish. Additionally, local groups in Iraq and Syria allied with the Islamic State due to its perception of constant success and feared that they would end up vulnerable when the group inevitably triumphed. As the tables turn, even groups that embrace the Islamic State's ideology have a strong incentive to defect to its enemies.

The Islamic State recognizes its own pitiful position. As the group lost more of its strongholds, its rhetoric shifted to dismiss the importance of territorial control. Instead, the Islamic State emphasized the concept of a caliphate as the driving force behind the group's success. Abu Muhammad al-Adnani, the Islamic State's spokesman and senior operational figure, stated in a May 2016 recorded message,

> O America, would we be defeated and you be victorious if you were to take Mosul or Sirte or Raqqa? . . . Certainly not! We would be defeated and you victorious only if you were able to remove the Quran from Muslims' hearts.

Like so many other Islamic State leaders, al-Adnani is now dead.

As the Islamic State crumbles, its leaders will try to continue fighting father than surrender. They plan to regroup, maintain their relevance and eventually resurge, through a mix of international and regional terrorism and local insurgency, while keeping their cause alive.

The Islamic State is not a stranger to defeat: it emerged out of Al Qaeda in Iraq, which for several years was on death's door. In June 2010, Gen. Ray Odierno, then the commander of U.S. forces in Iraq, noted that in the last ninety days U.S. and Iraqi forces had "either picked up or killed 34 out of the top 42 Al Qaeda in Iraq leaders." The two top Al Qaeda in Iraq leaders died in a firefight that year. By 2011, CIA director Leon Panetta declared, "we're within reach of strategically defeating al-Qaeda."

In response, Al Qaeda in Iraq focused on terrorism—in Iraq, not abroad—to stay relevant and to intimidate its enemies. It waged a campaign of assassination against opposing tribal leaders and other Sunnis who cooperated with the Iraqi government, killing more than 1,300 Iraqi leaders from 2009 through 2013. Due to terrorism and local killings, many Iraqis distrusted their government to secure peace and feared openly defying the jihadists. Al Qaeda in Iraq's patience paid off: over time, the Iraqi government stepped up discrimination against Sunnis and the Syrian Civil War offered a sanctuary across the border. Taken together, these two factors allowed the group to rebuild.

The Islamic State will try to repeat this success. Al-Adnani referred to a "retreat into the desert" to rebuild forces in order to prepare for upcoming battles. This mirrors jihadis' relocation to remote areas along Iraq's borders after the 2007 U.S. surge.

Unfortunately, both Iraq and Syria offer promising areas for a rebirth. In Iraq, the government in Baghdad repeatedly implemented policies that discriminated against Sunnis while lacking the strength or support necessary to impose order on an unhappy population—in other words, it cannot coerce, and it seems unwilling to coopt. Shia militias are occupying many Sunni areas where the Islamic State once held sway. Already the Baghdad government is exchanging fire with Kurdish forces. Shia militias are committing abuses against local Sunnis in areas they conquered from the Islamic State. Revenge killings are common.

In Syria, the situation is even worse, with the regime of Bashar al-Assad committing unspeakable atrocities against Sunni Muslims. Tribes, Kurdish groups and other local actors that have worked with America against the Islamic State often regard each other as enemies, or at least have different interests, which will inhibit their ability to cooperate against Islamic State remnants. As their shared enemy declines and the competition for local power increases, these groups are more likely to war against each other. As such, the Islamic State will likely find many openings to exploit, allowing it to relieve pressure and ensure at least some sanctuary. In the scramble for power, many local groups may even shift from enemy to temporary ally.

To defeat the Islamic State as an underground insurgency, someone must develop good governance in its former territories in Iraq and Syria, convincing locals to help uproot the group—an unlikely feat for which there are no credible volunteers. Unlike its previous revival, the Islamic State now has two countries where it can exploit problems, as opposed to just Iraq in the past.

The Islamic State will probably further regionalize the conflict and seek opportunities beyond Iraq and Syria. Already the group regards Iraq and Syria as one theater, shifting assets between the two countries depending on its perceived dangers and opportunities. The group also maintains an extensive network in Turkey, and has attacked Jordan, Lebanon and Saudi Arabia, among other countries. The Islamic State will probe these areas for weaknesses, using its operatives and local supporters to conduct attacks and develop an enduring presence.

Most troubling for the long term, the Islamic State has nurtured the flame of jihad around the world. Even as the group declines, the ideas it champions—the necessity of a caliphate, the glory of brutality and the evil of Western states—have spread further, as the staggering volume of foreign fighters suggests. The Islamic State's propaganda is extensive and almost ubiquitous. It, or would-be successor organizations, will try to harvest the ideas that the early Islamic State leaders planted.

The Islamic State will likely continue, and may even focus on, terrorist attacks in the West. In 2015, Paris suffered the worst terrorist attack on French soil in history when Islamic State gunmen and suicide bombers killed 130 people. France and other European states saw smaller attacks that year and the next, and by October 2017, five attacks linked to or inspired by the Islamic State targeted the United Kingdom—its most lethal year in terrorism since 2005. These high-profile attacks allow the Islamic State to maintain relevance to would-be and current supporters, convincing them to fight for the group despite its setbacks in Iraq and Syria. The attacks also signify revenge for the group's tremendous losses. Although U.S. efforts to destroy the Islamic State's havens hinder the group's ability to carry out sophisticated attacks, some attacks involve an Islamic State facilitator who helps recruit or directs the attacker, but does not provide elaborate operational support. Many of these attacks are low-tech but quite bloody: the Bastille Day attacker in Nice in 2016, for example, killed eighty-six people by driving a truck through a crowd.

Internal dynamics make Europe a particularly likely target, and in the short term the terrorism threat may grow as the caliphate collapses. The Syrian conflict has attracted over six thousand European volunteers. Some of these European foreign fighters will die and some will stay in the war zone, but some will also likely return to their home countries. One EU official estimates that approximately 1,500 will return. A fraction of those who return home may commit terrorist attacks or recruit locals to join the cause. The potential size of that fraction is unclear, but even a small percentage out of 1,500 can frustrate local police and security services. Europe contains more radicalized Muslims relative to their overall population, as suggested by the dramatically higher number of foreign fighters from European states. In addition, many European Muslims integrate poorly into their broader communities, which discourages them from cooperating with local intelligence and law-enforcement services. Finally, European intelligence services vary in skill: some, including those of France and the United Kingdom, are highly skilled, while others, such as Belgium's, are under-resourced and less capable of responding to terrorism threats. Fortunately, with heavy U.S. prodding and support, European states have improved intelligence cooperation and otherwise tightened their defenses. But this will remain a long-term challenge.

In comparison with Europe, the Islamic State poses a more manageable threat to the U.S. homeland. Since the September 11 attacks, ninety-seven Americans have died in jihadist-related attacks in the United States (the figure was ninety-five until the October 2017 truck-ramming attack in New York City, which killed two Americans and six foreign visitors). The two deadliest attacks, in San Bernardino in 2015 and in Orlando in 2016, that together killed sixty-three Americans, involved individuals who claimed some allegiance to the Islamic State but acted independently of the group—often referred to as "lone wolves." Although any death from terrorism is deplorable, the number of American deaths in the U.S. homeland—ninety-seven—is far lower than many experts, both inside and outside of government, predicted. Multiple factors likely explain this relatively low level of violence. First, senior U.S. officials overestimated the number of radicals in the United States after 9/11 when they spoke of thousands of jihadist terrorists in the United States. Second, the American Muslim community regularly works with law enforcement, leading to many arrests. As former FBI director James Comey explained,

> They do not want people committing violence, either in their community or in the name of their faith, and so some of our most productive relationships are with people who see things and tell us things who happen to be Muslim.

"Lone wolf" attacks will likely continue. The trend towards "lone wolf" attacks has grown: although the absolute number of attacks remains low, the scholar Ramon Spaaij found that the number of "lone wolf" attacks since the 1970s grew by nearly 50 percent in the United States and by more than 400 percent in the other countries he surveyed. The Internet and social media explain part of this increase, as both aid the Islamic State in inspiring individuals to act in its name. The October 2017 attack in New York was lifted straight out of the Islamic State's propaganda organ Rumiyah, which called for using vehicles to mow down pedestrians and then for the attacker to exit and continue to attack. Would-be fighters who do not travel pose a danger as well: according to one 2015 study of the terrorist plots in the United States, 28 percent of returned foreign fighters participated in a plot, but a staggering 60 percent of those who considered but did not attempt to travel became involved in a terrorist plot. As travel to Iraq and Syria loses its luster or becomes infeasible, frustrated jihadists might attack at home. As one French jihadist told the scholar Amarnath Amarasingam, "We believe that even a small attack in dar ul-kufr [the land of disbelief] is better than a big attack in Syria. As the door of hijrah [going to the Islamic State] closes, the door of jihad opens." Over time this frustration will decline, as would-be fighters no longer have firsthand contact with friends or family who went to fight, but the short-term danger is quite real.

Although the Orlando attack suggests that "lone wolf" attacks can be bloody, most "lone wolves" are incompetent; they are unlikely to succeed compared to trained foreign fighters who return to their home countries. But "lone wolves" have a strategic impact by altering politics in the United States and Europe, thus shattering the relations between Muslim and non-Muslim communities that are so vital to counterterrorism and to democracy itself. "Lone wolf" attacks increase Islamophobia in the West. After the attacks in Paris and San Bernardino, concerns about terrorism spiked. In the weeks following the Paris attacks in November 2015, London's Metropolitan Police Service announced that attacks targeting Muslims had tripled. Meanwhile, in the United States, assaults against Muslims have increased to nearly 9/11-era levels.

This Islamophobia can also begin a dangerous spiral. As communities become suspect, they withdraw into themselves and become less trustful of law enforcement, which results in providing fewer tips. In contrast, if a community has good relations with the police and society, fewer grievances exist for terrorists to exploit and the community is more likely to point out malefactors in their midst. Even though he was never arrested, the attacker in Orlando came to the FBI's attention because a local Muslim was concerned by his behavior and reported him.

Such problems risk fundamental changes in politics and undermine liberal democracy. Far-right movements are growing stronger in several European countries. In the United States, Islamophobia and fears of terrorism—despite the less-than-anticipated number of attacks on U.S. soil since the 9/11 attacks—have fueled the rise of anti-immigrant politics.

The Trump administration continued the Obama administration's military campaign against Islamic State forces in Iraq and Syria and has loosened restrictions on military commanders and deployed additional forces to Syria—nearly doubling the number of previously deployed forces in the fight for Raqqa. Additionally, the administration has maintained the coalition of states and local actors that the previous administration cobbled together. Furthermore, the aggressive global intelligence campaign begun under President George W. Bush and continued under Obama remains robust. Taken together, such efforts have hindered Islamic State operations and steadily forced it underground.

In his first year in office, however, the president has taken several steps that may impede the struggle against jihadist terrorism. First, in his campaign rhetoric and through actions like Executive Order 13769 (the so-called "Muslim ban"), the Trump administration demonized American Muslims and damaged relations between religious communities—a traditional source of American strength, pride, and values. Such actions increase the allure of the Islamic State and other groups claiming that the West is at war with Islam, while also adding credibility and legitimacy to their ideas. In addition, these actions increase the likelihood that Muslim communities will fear the police, the FBI, and other government institutions, and thus be less likely to cooperate with them. This enables "lone wolves" to remain undetected and offers fodder for Islamic State virtual recruiters trying to convince Muslims that the West is the enemy.

Overseas, Trump embraced the Saudi perspective of the Middle East. Saudi Arabia is an important counterterrorism partner, and relations with the Saudis became strained under Obama. Though Trump's efforts to strengthen ties should be commended, the Saudi government continues to fund an array of preachers and institutions that promulgate an extreme version of Islam, enabling the Islamic State to recruit and otherwise gain support around the world. In addition, Saudi Arabia promotes an anti-Shia agenda that harms regional stability and fosters sectarianism, a key recruiting tool of the Islamic State.

Perhaps most troubling is how the president responded to the first significant jihadist attack on U.S. soil during his tenure—the October 2017 truck-ramming attack in New York City. At a time when a president should provide steady leadership, Trump (inevitably) began to tweet. He tried to turn the attack into a political issue, excoriating Sen. Chuck Schumer for the visa program that let the attacker into the country. He then stoked fears of immigration, called for the attacker to be sent to Guantanamo (and then apparently dropped that), and otherwise appeared erratic, partisan and lacking an understanding of the policy implications of his own words.

Instead of relaxing pressure as the Islamic State prepares to go underground, the United States must redouble its efforts. This will require crafting a sustainable coalition of local allies in Iraq and Syria that demands resources, skill millions on training programs in Iraq and high-level engagement.

Training allied forces remains vital, but this must be understood as a limited solution rather than a cure-all. In theory, training allies seems a Goldilocks answer to many policy questions: it is relatively low in cost, it minimizes direct risk to U.S. forces and it helps reduce terrorism in the long term when newly capable allies can police their own territory. Yet, especially in the Middle East, these efforts often fail. Despite spending hundreds of millions on training programs in Iraq, Syria and elsewhere, U.S.-trained forces have often crumbled in the face of the adversary. Regime corruption, divided societies, politicized militaries and other problems plague the region, and U.S. training can only move the needle slightly. Limited progress is better than no progress, but other policies must supplement training programs.

To prevent the Islamic State from reestablishing itself in Iraq and Syria or spreading elsewhere in the region, the United States also must adopt a broader conception of counterterrorism, recognizing the link between jihadist terrorist groups and civil wars. Groups like the Islamic State exploit civil wars and worsen them: if civil wars in the Muslim world are left to rage, we can expect jihadist groups in the region to remain strong actors. Resolving civil wars is a strategic as well as a humanitarian imperative. Programs for conflict resolution and sustained U.S.-led diplomacy are vital to ameliorate the effects of civil wars.

Furthermore, the United States must also support allies on the front line, like Jordan, that are vulnerable to jihadist meddling. The United States must also strengthen nascent democracies that have a significant jihadist problem, like Tunisia. As such, the administration should rescind the proposed dramatic cuts to the already-small foreignaid budget, and staff much of the Department of State, the civilian arm of the Department of Defense and other key agencies to strengthen the ability of the United States to use a whole-of-government approach to combating terrorism.

One significant problem is institutionalization. Since 9/11, the executive branch alone has executed counterterrorism policy, with some modifications by the courts. One branch of government, perhaps the most important in the long term, has been conspicuously absent under both parties' leadership: the U.S. Congress. Under both Bush and Obama, new and controversial counterterrorism instruments—targeted killings, increased domestic surveillance, aggressive FBI sting operations, detention without trial and so on—moved to the center of U.S. counterterrorism efforts without significant congressional input. In addition, the United States is bombing the Islamic State in both Iraq and Syria with only dubious legal justification.

Defeating the Islamic State could be the marquee foreign-policy accomplishment of the Trump administration. Doing so, however, will require more than just forcing the caliphate underground. Instead, the administration must maintain pressure on the group in the Middle East, work with allies around the world and shore up efforts at home. Failing to do so will result in at best a respite, not lasting victory.

Critical Thinking

1. What is Al-Qaeda's model that ISIS may follow?
2. Why can ISIS survive after losing its territorial Caliphate?
3. What problems in Syria and Iraq will facilitate the survival of ISIS?

Internet References

US Department of Homeland Security
www.dhs.gov

US Department of State, Bureau of Counter-Terrorism
www.state.gov/en

Daniel Byman is a professor and senior associate dean at Georgetown University's Edmund A. Walsh School of Foreign Service and a senior fellow at the Brooking Institution's Center for Middle East Policy.

Article

Prepared by: Robert Weiner, *University of Massachusetts, Boston*

Strategic Amnesia and ISIS

DAVID V. GIOE

Learning Outcomes

After reading this article, you will be able to:

- Understand the relevance of the U.S. military history to the war against Islamic State of Iraq and Syria (ISIS).
- Discuss the problems associated with expeditionary forces on foreign soil.

Mark Twain observed, "history doesn't repeat itself, but it does rhyme." The study of military history teaches us valuable lessons that are applicable to today's most intractable strategic problems; yet, these lessons are underappreciated in current American strategy formulation. Throughout the history of American armed conflict, the United States has discerned, at great cost, four critical lessons applicable to containing and combating the Islamic State.

First, as war theorist Carl von Clausewitz noted, war is a continuation of politics by other means; but resorting to war rarely yields the ideal political solution envisioned at the start of hostilities. Second, the use of proxy forces to pursue American geopolitical goals is rarely an investment worth making because proxies tend to have goals misaligned with those of their American sponsors. True control is an illusion. The corollary to this axiom is that supporting inept and corrupt leaders with American power only invites further dependency, does not solve political problems and usually prolongs an inevitable defeat. Third, conflating the security of a foreign power with that of America leads to disproportionate resource allocation and an apparent inability at the political level to pursue policies of peace and successful war termination. Fourth, alliance formation through lofty rhetorical positions imperils rational analysis of geopolitical and military realities. Publicly staking out inviable political end states invites a strategic mismatch between military capabilities and political wishes, endangering the current enterprise as well as future national credibility. America has paid for these lessons in blood; our leaders ought to heed them.

The Obama administration's effort to again increase the number of American military advisors in Iraq, coupled with the reconstruction of a new base at Al-Taqaddum in Anbar Province, has given rise to accusations among both Democrats and Republicans about either mission creep (from doves and non-interventionists) or weak incrementalism (from hawks and liberal interventionists). Former defense secretary Robert Gates observed in May 2015 that there simply was no American strategy in the Middle East. Congressional hawks have used Gates's observation to criticize the Obama administration's cautious efforts in any ground campaign against the Islamic State, and some have called for thousands of American boots back on the ground in Iraq. However, during a 2011 visit to the U.S. Military Academy at West Point, Gates also told the cadets that any adviser who counseled deploying large land forces to the Middle East should "have his head examined," suggesting that a larger military footprint should not be confused with a robust strategy.

Using rhetoric reminiscent of George W. Bush's "War on Terror," in 2014 President Obama pledged to "ultimately destroy" the Islamic State, but over a year later the Obama administration itself admitted that its strategy is not yet "complete." Indeed, even complete strategies often do not survive first contact with the enemy. As military personnel often quip, "the enemy gets a vote," and the Islamic State seems to be visiting the ballot box early and often. A candid comment about strategy formulation makes for an interesting sound bite (or cudgel). But discretion may be the better part of valor when facing the slippery slope of another open-ended commitment in Iraq. Many observers took President Bush to task for suggesting that an ideology could be defeated by applying military force, but their critiques could apply just as well to Obama's turn at the helm.

Pledging victory implies an end state that is ultimately acceptable to one's adversary—whether it's forced upon it (like the unconditional surrender of Japan in 1945) or a negotiated political solution (as in Korea eight years later). The Islamic State seems to show little taste for negotiation, and why should it? Most wars are prolonged in the hope that each side will come to the negotiating table with a better hand to play. The Pentagon service chiefs appear reticent to get further involved in Iraq absent a political solution. But the Obama administration has been bullied into increasing troop numbers by Congressional hawks who have conflated the security of Iraqis with that of Americans and the cohesiveness of the Iraqi state with core U.S. national-security interests. America has been here before.

Unlike at the height of its post-Cold War military power, America no longer has the ability to dictate events globally—to the overplayed extent that it ever did. This is particularly true in North Africa, the Middle East, and South Asia. Those who argue that an American military campaign could defeat ISIS in a durable way place too much confidence in the ability of any American administration to control events abroad, especially in deeply rooted internecine conflicts. In fact, although what is happening in the Islamic State's Iraqi strongholds is both primitive and shocking, the state of affairs in Baghdad is what should be cause for even greater concern in Washington. Iraq, under its Shia-dominated government, has marginalized Sunnis and alienated Kurds, perhaps to the point of no return. Indeed, the billions of dollars invested in training Iraqi forces are for naught if the controlling political entity is a house divided against itself.

Iraq did not slide into its current state of affairs without outside help. The United States cannot escape some culpability for what Iraq (and Syria) have become, but expensive U.S. efforts to encourage good governance and interreligious and intertribal dialogue and cooperation have fallen short. Calling for a strategy to defeat an ideology or repair Iraq is tantamount to demanding that a physician devise a strategy to treat a patient admitted to the emergency room with a shotgun blast to the head. Even with the best of intentions, unlimited resources and the best expertise available, there isn't much that can be done to reach the status quo ante helium.

Since the end of World War II, the American military has struggled to translate tactical military success on the battlefield into durable political gains

Since the end of World War II, the American military has struggled to translate tactical military success on the battlefield into durable political gains. Although America has no peer when it comes to accumulating post-9/11 tactical victories, the record is not enviable at the political level. Witness the bin Laden raid of May 2011 and the daring May 2015 Army Special Forces raid into Raqqa, Syria. These were spectacular tactical successes, and perhaps necessary from a moral perspective, but they achieved little at the strategic level. To be sure, the world is a better place without Osama bin Laden and ISIS financier Abu Sayyaf, and they richly deserved their fates. The problem is what comes next. The United States has been eliminating the leadership of Al-Qaeda since the end of 2001 and has transferred those lessons to effectively remove the leadership of many Al-Qaeda franchises in Yemen and North Africa as well. No doubt the United States will further apply its lethal craft to ISIS in the near term. However, military history reveals that accumulating tactical successes does not equal strategic victory. The German military learned these lessons the hard way in both twentieth-century world wars. The Germans, although well equipped and tactically sound, were unable to realize their broader political desires through violence.

Soviet leader Joseph Stalin is said to have suggested that quantity has a quality all its own. If this is true, we must recognize American tactical successes for what they have achieved, even absent a broader strategy. The American homeland is arguably safer because those that seek to do it harm are impeded by those tactical successes. American military and intelligence operations have made enemy communications more difficult and secure staging bases hard to come by. The U.S. military killed or captured the top leadership of Al-Qaeda and it's like, retarding their operational planning and derailing their efforts to undertake spectacular attacks. If American strategy is threat mitigation through sustained special-operations raids and intelligence-driven covert action, it is working. Still, it is an open question how long this is sustainable, especially on the back of a shrinking, all-volunteer military force. American military and political leaders have spoken of a "generational war," yet they also shrink from serious discussion about national service or a draft.

Assume for a moment that the Obama administration were to pour troops into Iraq and loosen their rules of engagement, permitting direct American participation in the fighting. Could a couple of U.S. divisions retake ISIS strongholds? Absolutely. It would come at a bloody cost to young American soldiers, as when Fallujah fell to American forces in 2004 with 560 American casualties (and thousands more with psychological wounds), but the U.S. military could surely retake the large cities of Anbar province. Could they hold them? Not indefinitely. With the forces and resources available to the Pentagon, the United States could hold it for a time while building Iraqi capacity. This is a key pillar of the Army's counterinsurgency doctrine, but recent experiences in Iraq and Afghanistan lay bare the failure of this approach, at least on a timeline not measured in decades. The only dramatic success in Anbar province was a political one: the Sunni tribes "awakened" to turn against Al-Qaeda of Iraq and the monstrous tactics of AQI leader Abu Musab al-Zarqawi. Political settlement manifested on the battlefield—a much more promising proposition than the other way around.

Most war theorists conceive of war as a contest of wills. Put another way, the party who wants it most—and will thus sacrifice the most—usually wins in the long run. Consider, for instance, a group of mujahideen repelling the Soviets after a decade of fruitless bloodletting in Afghanistan. Or, for that matter, consider some of the same fighters showing NATO the exit a generation later. War's fundamental character has not changed over time, and the contest of wills remains a bedrock principle. To apply the concept, consider what the average ISIS fighter would do to secure the success of the Islamic State against what the average American would do to roll it back. As things stand, the ISIS fighter is considerably more committed to his cause—particularly in the absence of convincing proof that ISIS poses an existential threat to the American way of life—and only 1 percent of Americans are actually involved in the so-called war on terror. In both of America's greatest military successes in the twentieth century—the world wars—America came late, but with a total mobilization that called on the resources of a significantly larger proportion of the population. Moreover, that population was considerably more unified in purpose.

Additionally, expeditionary wars on foreign soil represent challenges on several fronts. Deploying and supporting troops, heavy machinery and the routine supplies of war, especially over great distances into landlocked countries with rugged terrain, complicates logistics. This is also expensive and relies on the continued support of the citizenry back home. The expeditionary force is most often at a disadvantage in that it must secure victory while fighting far from home. Its opponents, comfortable on their home soil, do not have to win—they just have to wait out the invading force and not suffer catastrophic battlefield defeat.

During the American Revolution, George Washington employed a Fabian strategy against the expeditionary British force. Washington avoided large engagements on unfavorable terms, as his goal was to preserve a true fighting force and wear down his enemy until Westminster decided to stop throwing men and material at the Continental Army. The militarily superior British pulled the plug on their colonial undertaking after six years of active fighting in North America. They had other strategic considerations and made a difficult choice to concede defeat in the North American theater of a larger war. Unfortunately, today Washington has more in common with Westminster: it holds a losing hand against a determined enemy pursuing a Fabian strategy. The British experience in America suggests that a professional military in an expeditionary capacity may come up short.

To use a parallel from the American Revolution, those under ISIS rule may from their own Iraqi committees of correspondence, Iraqi Sons of Liberty and Iraqi minutemen.

Recent media coverage of daily life inside of the Islamic State suggests that U.S. officials should not be so condescending as to think that those living under harsh ISIS rule are mere sheep awaiting rescue. The millions of Iraqis and Syrians now living under ISIS domination far outnumber their new masters. Those under ISIS rule have few good options, and the costs of rash action are high. The Iraqi army is apparently unable to retake any of ISIS core territory, at least not without the help of American airpower, advisors and (most problematically) Iranian-backed Shia militias. It is perhaps not a foregone conclusion that Sunnis living in the Islamic State would prefer militias backed by Iran's Quds Force as their liberators from ISIS. They may well view this sort of "liberation" as out of the frying pan and into the fire.

Although patience in a 24-hour, crisis-to-crisis news cycle is notable for its absence, given time some promising developments in ISIS territory could come to pass that undermine the Islamic State from within. Any lasting governing entity relies on some level of support or at least consent of the governed. The actions of ISIS toward its subjects suggest that over the long term they might not achieve this. Fear and brutality only go so far. Parents fed up with their children being indoctrinated with fundamentalist hate at school, women who cannot leave the house with their faces uncovered or without male relatives, men who are being extorted for ISIS taxes, citizens disgusted by summary executions and floggings, fathers who dread their sons becoming brainwashed to be martyrs and mothers who want their daughters to enjoy equal rights will begin to find common cause against the Islamic State. They may decide to cautiously provide tips to the Iraqi army's special forces on the locations of ISIS leaders or their weapons caches. They may themselves begin to hide weapons and supplies for when the popular mood shifts.

To use a parallel from the American Revolution, those under ISIS rule may form their own Iraqi committees of correspondence, Iraqi Sons of Liberty and Iraqi minutemen. They may seek their own outside allies and develop their own internal intelligence networks. In short, they will eventually resist, as the early sparks of the ill-fated Arab Spring will attest. From the Orange Revolution to the Prague Spring, from the Polish Solidarity movement to the Warsaw Ghetto uprising, the oppressed eventually resist. The desire for life, liberty and the pursuit of happiness is not exclusively American, but Americans cannot be the sole guarantors either. History has shown that peoples who perceive themselves to be oppressed usually organize into a creditable rebellious force, although this usually takes years for suitable levels of organized frustration to congeal into a counterforce.

As in the American Revolution example, defeating a superior force often requires powerful allies. The French contribution to the American war effort was a key development, but the French refrained from becoming openly involved until after the American victory at Saratoga in the fall of 1777. It was only after the Americans proved to be a formidable fighting force and fully devoted to their cause that the French arrived. Indeed, foreign military assistance and training for "internal defense" can be critical to success, but first the will and ability to win must be convincingly demonstrated. An increasingly vocal minority of Americans, whom we now refer to as "Patriots" and "Founding Fathers," spent the greater part of the early 1770s organizing themselves and secretly preparing for violence.

Consider a familiar scenario, so familiar that it reads like many recent headlines from Iraq: Enemy forces are gaining momentum and seizing territory at an alarming rate. They have a stronghold in the shape of a triangle just over an hour's drive from the capital. With the political dysfunction in the capital city and, even after American training, local troops underperforming, it seems that only an American-led search-and-destroy mission could root out the enemy, protect the capital and shift the battlefield momentum. After a period of airstrikes, American armored and helicopter-borne infantry forces duly arrive in the insurgent triangle and for three weeks attempt to clear the area of enemy forces, but they are unable to discern civilian from insurgent—it's possible they are one and the same, but absent uniforms or a recognizable chain of command, they are uncertain. These American troops are killed by booby traps and snipers, but never identify the enemy. Eventually declaring the area "cleared," American soldiers destroy some enemy weapons caches, and American senior officers brand it a successful operation.

No, the above scenario isn't Baghdad, the nearby Sunni Triangle, the crumbling Iraqi National Army, and the advancing Islamic State. The year was 1967 and the operation, called Cedar Falls, was to be the largest of the Vietnam War. Its purpose was to clear the "Iron Triangle" of Viet Cong irregular forces that were threatening Saigon—a nearly failed state propped up by American power. Frustratingly, the enemy would not stand and fight in the face of overwhelming American tactical and material superiority. The Viet Cong forces moved across the porous border into Cambodia and simply returned when the American forces departed the area. After the operation had ended, and at the cost of more than four hundred American casualties, many senior American officers counted Cedar Falls as a success, pointing to the numbers of weapons stockpiles that were destroyed and the fleeing enemy. In retrospect, the failure of Cedar Falls was emblematic of American efforts in Vietnam. In reality, the residents of the Iron Triangle, much like their countrymen throughout the rest of South Vietnam, found aspects of the Viet Cong message appealing, and surely no worse than the repressive and corrupt government in Saigon.

Confusing a messy, localized civil war with an existential threat to American national security is a strategic mistake.

In hindsight, the theory that Vietnam's fall to the Communists would make the rest of Asia topple like dominos into the Soviet sphere proved to be alarmist and false. Further, what appeared in 1965 to be in America's core national-security interests was identified by 1970 as irrelevant, a significant diversion of American resources from more pressing concerns and a major source of political and social tension on the home front. American political leaders declared that the Republic of Vietnam should be responsible for its own security and promptly began a period of "Vietnamization," in which American forces trained and equipped the South Vietnamese military, paving the way for an American withdrawal. Saigon fell two years after the American withdrawal, but that would have happened no matter what year the Americans finally decided to pull the plug. This is worth remembering when hawks seek to blame the Obama administration for the current state of affairs in Iraq because it pulled out American troops in 2011. Even if the United States had kept thousands of troops in Vietnam until 1983, Saigon would have fallen by 1985. Political problems can be papered over with military force for a long time, but in the end the result is the same.

Such parallels from the Vietnam War are haunting, and should not be tossed aside in current strategy formulation. These are the lessons learned at the cost of 58,220 American soldiers who gave the last full measure of devotion in Southeast Asia. In Vietnam, America paid a heavy price in lives and treasure to prop up a corrupt and unrepresentative government, which it hoped could function as a regional ally and bulwark against the seemingly prevailing ideology in the region. American military personnel attempted to both provide population-centric security and bring massive firepower to bear on the enemy. With the benefit of hindsight, many military historians have declared the Vietnam War "unwinnable," yet America was not less secure because of the loss.

In the same way that Viet Cong forces took advantage of the porous border with Cambodia during operation Cedar Falls, ISIS fighters in Iraq would just as easily slip across the border (which they control) into Syria and wait for the Americans to leave. And just like the residents of the Iron Triangle outside Saigon, not all Sunni residents of Anbar province view being governed by ISIS as particularly worse than a corrupt Iranian proxy government in Baghdad. The politics of the region, particularly the animus between Shia and Sunni Islam, are a jumble of tribalism, mistrust, anarchy and greed. The government in Baghdad is helplessly divided and, as history consistently reveals, American military efforts cannot fix a political problem.

As in Iraq and Afghanistan, not all wars are worthy of continued American involvement, and hardly any wars in the U.S. history could be considered existential. During the Korean War, President Harry S. Truman correctly elected not to expand the war into China, despite the vociferous urging of General Douglas MacArthur. Likewise, President Johnson did not permit an invasion of North Vietnam, despite the fact that in both cases the enemy center of gravity lay beyond the local battlefields of Korea and South Vietnam. Neither president opted to unleash the supposed guarantor of continued American existence—the nuclear triad. While it is true that these conflicts were limited wars without existential risks, it is proper that they were conducted as such by the U.S. administrations that oversaw them. Escalation to total war, or an existential fight for national survival, is only appropriate in the direst circumstances, in which a loss on the battlefield might mean national calamity. Despite the repulsive and brutal conduct of ISIS, the stakes for the United States are not that high. Confusing a messy, localized civil war with an existential threat to American national security is a strategic mistake.

American power toppled the Taliban and Saddam rapidly with modestly sized forces, but the maelstrom and "surges" that followed pulled hundreds of thousands of American and allied troops into its wake. Like the process of Vietnamization, in both Iraq and Afghanistan U.S. forces sought to train, advise and equip allies in the hopes that they could stand on their own and American troops could leave with some political gains realized. In Iraq, the United States spent nearly a decade and approximately $20.2 billion on a dubious mission to train the Iraqi army to secure the country. This army had years to develop under the tutelage of the finest American instructors and was the beneficiary of millions of dollars of U.S. military hardware. Yet it is the black flag of ISIS that waves atop U.S.-made Humvees, armor and heavy weapons. This suggests that motivation, loyalty and esprit de corps matter more than the latest technology, hardware and training cadre.

Again, war is a contest of wills, and the U.S. policy at present is to stiffen the spine of the locals who are expected to do the fighting. The U.S. Marine Corps tried this in South Vietnam with Combined Action Platoons, a small group of Marines and a Navy Corpsman residing in a rural hamlet, strengthening the local militia forces. The cap program is often judged as successful because it denied sanctuary to enemy forces, but successes at such a low level had little impact. The political dysfunction in Saigon overshadowed stability in rural hamlets.

Stiffening the spine of local forces, sometimes referred to as Foreign Internal Defense, can work if there is a baseline level of common mission already in existence among the host nation forces. American advisors can provide expertise and technology, but vision and commitment need to be homegrown. In addition to the lack of discipline and esprit de corps that accompanies good militaries, a major failing in the Iraqi army is the lack of a shared vision of the end state for Iraq. It isn't obvious that a Shia soldier in the Iraqi Army, from Basra for instance, considers it a good idea to fight ISIS in Anbar province. He may not view Anbar as his home or even part of his conception of Iraq. It is understandable, then, that he may want victory there less than an ISIS fighter does. Anbar just doesn't mean as much to him.

The fear of "losing face" has led many commanders to attempt to turn straw into gold with new strategies.

In the early sixteenth century, Machiavelli observed that troops who are not fighting for their own homeland are not inclined toward bravery because their "trifle of a stipend" is acceptable until war comes and then they "run from the foe." This begs the question whether members of the Iraqi army can be said to be fighting for the U.S. conception of a single federated Iraq, or for their own religious sect or tribe. American military and political leaders hoped that a reliably paid and equipped Iraqi army would fight like those defending their homeland. In fleeing before the ISIS advance, they proved to resemble the "mercenaries" and "auxiliaries" of whose dubious dedication Machiavelli warned.

What is to be done? Washington continues to substitute tactical action for strategy, and thus continues to throw good money and American lives at the chimera of a pluralistic and tolerant Iraq (and Afghanistan) while at the same time breeding dependency on America. American decision-makers would be well served to avoid ideologically guided wishful thinking as this often tempts the strategist to ignore history's warnings. Americans aren't the only ones who brush aside historical lessons with wishful thinking. Why would Adolf Hitler open himself up to a two-front war and invade the Soviet Union in June 1941? His extreme ideology compelled him to brush aside Napoleon's harsh lessons about invading Russia.

Another step in the right direction is to stop speaking in euphemisms when discussing the performance of the Iraqi (or any) army. Defense Secretary Ashton Carter noted that the Iraqi army "showed no will to fight" in Ramadi, but a White House spokesman characterized the dismal performance as a "setback." Investing rhetorically in an ally is a slippery slope, and almost always comes at the cost of sober and dispassionate analysis of battlefield performance. If unchecked, when "their" performance turns into "ours" and "they" starts to be "us," two unfortunate things usually follow. First, it conflates the security of the Iraqi state with that of American national security. And, more insidiously, cutting losses becomes harder. The fear of "losing face" has led many commanders to attempt to turn straw into gold with new strategies. It gets harder to withdraw absent a plausible "mission accomplished" narrative because of the inevitable argument that cutting losses is tantamount to forfeiting American military credibility. As Clausewitz reminded his readers, once the blood and treasure expended exceeds the value of an objective, that objective must be given up. Giving up an advise-and-assist mission for Iraqi allies will be politically impossible when the effort transforms from a military analysis of "them" into face-saving political measures involving "us." Moreover, despite some marginal but real tactical differentiation, publicly referring to forward operating bases (fobs) or combat outposts (cops), as "lily pads"—implying fleeting presence—is another misleading battlefield euphemism. With the Obama administration being bullied into dripping the U.S. Army back into Iraq a few hundred soldiers at time, it would be unsurprising if these "lily pads" remain into the next decade.

Military history warns that observers with skin in the game are unable to see strategy unfolding as it actually is. Despite evidence that the government in Saigon was increasingly corrupt and repressive, President Johnson observed in 1967, "Certainly there is a positive movement toward constitutional government." In June 2005, the Bush administration claimed that there were 160,000 Iraqi security forces who were trained, equipped and on the verge of independent operations. The results of this training were on full display in the May 2015 ISIS victory in Ramadi. The list of misstatements goes on when we view our allies as we wish to see them, not as they are. Even if intelligence assessments in private offer more accurate assessments, their own skin in the game, coupled with a guiding ideological approach, will always color the vision of political leaders. Not succumbing to the temptation to offer a continual drumbeat of rosy analysis for public consumption is a critical first step to avoiding foreign-policy missteps, or at least reversing those errors already committed.

In fairness, a few commentators are advocating a full-scale return to Iraq, but that's not how long-term commitments are usually undertaken. Again, the case of Vietnam is instructive. In the early 1960s, the Kennedy administration sent advisors to South Vietnam on a rather modest advise-and-assist mission. Once it became clear that this would be insufficient to accomplish the desired objective, "they" became "us" and "their task" became "our task." The inevitable mission creep set in; by the end of the Vietnam War, more than 2.5 million American troops had rotated through a country roughly half the size of New Mexico—and still lost. Some U.S. officials would reject that the new Iraq mission is anything like Operation Iraqi Freedom (2003–11), but it may look increasingly similar with the passage of time.

Eventually, the American people will tire of nonstop war. After the horror of the Islamic State's barbarism and the resulting surge of patriotism has subsided, the public will question the costs borne by so many troops, especially those who have done many tours without seeing any real progress. It is up to the Iraqis, and perhaps the greater Middle East, to decide their own fate. The international system seeks balance, and this often occurs through violence. We're seeing that now in the Middle East. Given the cast of players, things may even get worse before they get better as regional competitors become more involved and the stakes get higher. Yet the American people cannot want a pluralistic and tolerant Iraq more than the Iraqis do. Clausewitz noted that, "One country may support another's cause, but will never take it so seriously as it takes its own." The study of military history reveals an abundance of material for defense strategists, commanders and policymakers. It is accessible and directly applicable to contemporary strategic dilemmas. Ignoring these lessons would be a disservice to those who made the ultimate sacrifice to reveal them.

Critical Thinking

1. Why can't the U.S. translate tactical victories against ISIS into political gains?
2. Has the U.S. counterinsurgency strategy worked? Why or why not?
3. What is the relevance of Machiavelli to U.S. strategy and tactics in Iraq?

Internet References

Department of Homeland Security
www.dhs.gov/

U.S. Department of State, Bureau of Counter-Terrorism
www.state.gov/j/ct/

David V. Gioe is an assistant professor of Military History at the U.S. Military Academy at West Point. He previously served as a CIA operations officer. The opinions expressed here are his own and do not necessarily reflect the U.S. Army, the Department of Defense or the U.S. government.

Article

Prepared by: Robert Weiner, *University of Massachusetts, Boston*

Arabia Infelix: The War Devouring Yemen

SHEILA CARAPICO

Learning Outcomes

After reading this article, you will be able to:

- Learn the details of how feuding presidents are contributing to the war in Yemen.
- Learn why the war is a humanitarian disaster.

For many centuries, European cartographers labeled the southwest corner of the otherwise mostly desert Arabian Peninsula as Arabia Felix, or "Happy Arabia." It was a place where towering mountains trapped clouds blown in from the Indian Ocean so that twice-annual monsoon rains blessed terraced slopes and lowland wadis with plentiful crops. Sadly, since fighting engulfed the country in late March 2015, Yemen has never been less felix.

The brutal war and humanitarian catastrophe in Yemen are the consequences of domestic and especially regional power struggles. Often cast as a sectarian battle between Sunni and Shia Muslims, it is more accurately understood on the one hand as a banal rivalry between two discredited presidents and on the other as an unprovoked counterrevolutionary war by wealthy Gulf monarchies against a poverty-stricken republic where citizens rose up against decades of inept, corrupt rule. As fighting raged and coalitions on both sides unraveled, the sectarian narrative also grew tattered.

Hundreds of thousands of Yemenis took to the streets in sustained, colorful, peaceful nationwide demonstrations for months on end in 2011, chanting the same slogan that toppled dictators in Tunisia and Egypt: "The people want the downfall of the regime." Armed tribesmen laid down their weapons to protest. Young people wrote manifestos and staged street performances. Northerners and Southerners, Zaydis and Shafa'is, peasants and urbanites, women and men turned out in veritable unison. The Nobel Peace Prize committee recognized a Yemeni woman, Tawakkol Karman, a gifted public speaker, as a symbolic leader of the Arab Spring.

The Gulf dynasties fretted over the North African uprisings, but were especially alarmed by the raucous clamor for revolutionary change on their proverbial doorstep. Eventually the carbon kingdoms persuaded a reluctant President Ali Abdullah Saleh to relinquish power to his long-serving vice president, Abdurabbu Mansur Hadi. Instead of regime change, the protesters achieved only a regime makeover.

Hadi, the only candidate, "won" a two-year term in an election in February 2012 hastily arranged under the auspices of the Gulf Cooperation Council (GCC), an association of the six oil monarchies of the Arabian Peninsula comprising Saudi Arabia and the smaller states of Kuwait, the United Arab Emirates, Bahrain, Oman, and Qatar. The so-called GCC Initiative also granted Saleh and his family immunity from prosecution and left him as the leader of the party with a parliamentary majority. Concurrently, it funded a National Dialogue Conference of over 650 Yemeni delegates from various constituencies that was intended to facilitate a post-Saleh, post-Hadi transition. The conference, meeting with international consultants in five-star hotels, dragged on inconclusively.

Eventually, the feuding presidents proved willing to destroy the country in order to restore dictatorship. Their respective alliances complicated and sectarianized the nasty rivalry between the two self-serving, nonideological secularists.

Strange Bedfellows

In a counterintuitive turn of events, in September 2014 Saleh joined forces with a group called Ansar Allah (Partisans of

God), also known as the Houthis—a ragtag militia from the far north, near the Saudi border, that had waged six separate regional rebellions against his regime. The Houthis are a self-styled Zaydi revivalist movement within the Shia school of Islam, opposed to Wahhabism, the ultraconservative Saudi brand of Sunni Islam. They marched on Sana'a, the capital, to protest policies announced by Hadi—especially a fuel price increase recommended by the International Monetary Fund and a scheme (devised by foreign consultants and announced after minimal domestic debate) to reorganize provinces into six "federal" regions that left the Houthi's homeland landlocked.

The national armed forces, in which Saleh loyalists retained essential commands, offered no resistance as the Houthis entered the capital. Hadi was placed under house arrest. Saleh's party, the General People's Congress, expelled him. In early 2015, he fled to his hometown of Aden, the onetime capital of the old People's Democratic Republic of Yemen (or South Yemen), which existed until 1990. In Aden and the south, an overwhelming majority had yearned for renewed separation from "unified" Yemen since the 1994 civil war. In late March, Hadi went into exile in Riyadh and appealed to the kingdom to reinstate him in the presidential palace in Sana'a. Frequently referred to as the "internationally recognized president," but lacking a domestic constituency or mandate, he nominally "governs" from exile.

On March 26, 2015, the new Saudi king, Salman, and his son Muhammad bin Salman, the defense minister, launched what they called Operation Decisive Storm on Hadi's behalf. The declared motivation for the Saudi-led intervention was and is to thwart Iranian influence in the peninsula. Because Zaydism is a sub-denomination of Shia Islam, the Saudis see Ansar Allah as an Iranian proxy. Despite the absence of any photographic or physical evidence, Riyadh continually insists that Iran is smuggling weapons to the Houthis—and US officials occasionally repeat the claim.

Thus the convoluted power struggle in Yemen is frequently characterized as a battle of Sunni against Shia. This simplified sectarian frame, widely reiterated in the international media, implicitly places the Saudi-led coalition alongside al-Qaeda in the fight against the Zaydi-Shia Ansar Allah. But in a separate conflict ongoing in Yemen, the Americans continue to fire drone missiles and occasionally other weapons at suspected al-Qaeda positions, while concurrently siding with the Saudis against the Houthis. The local branch, known as al-Qaeda in the Arabian Peninsula, has benefited strategically from the general mayhem, particularly in the southern regions where it has established a presence and seems to be fighting alongside Emirati forces against the Houthi and Saleh militias.

Wanton Devastation

The viciousness on all sides is extreme. The Saleh-Houthi alliance has committed atrocities on the ground, particularly in attempting to occupy, control, and subdue Aden and the third major city, Ta'iz, which they subjected to a bitter siege after encountering fierce local resistance. They also recruit and exploit child soldiers. When they manage to shoot a few missiles across the northern border into Saudi Arabia, they do so indiscriminately.

However, the most devastating humanitarian and economic repercussions are inflicted from above and beyond by the superior firepower of fighter-jets, helicopter gunships, and naval warships—all manufactured in the West and deployed by the Saudi-led coalition. The United States, the United Kingdom, and several other NATO powers explicitly or implicitly share responsibility for the onslaught. The Obama and Trump administrations arranged the sale of over $300 billion worth of Hellfire missiles, cluster munitions, and other advanced weaponry to Saudi Arabia alone, and have participated directly in operations with in-air refueling, surveillance, and training. American, British, and other arms industries profit from the wanton and largely pointless belligerence. Boeing, Lockheed Martin, Raytheon, and Honeywell Aerospace are the major US suppliers and contractors to the Saudis and the UAE.

Saudi Arabia, the UAE, and a few coalition partners such as Egypt have enforced a draconian air and naval embargo, ostensibly to block the importation of Iranian weapons but in fact, according to international humanitarian organizations, also obstructing shipments of food, fuel, medical supplies, and even the arrival of foreign journalists. The Sana'a airport's runways were bombarded. The main Red Sea port at al-Hudaydah was deliberately put out of commission; inspectors refused to allow replacement equipment to be imported. Air strikes in most parts of the country disabled water and power supplies, damaged or closed hospitals and clinics, left craters in roads, toppled bridges, and blasted markets, factories, gas stations, hotels, and cultural sites. Hadi and his foreign backers shut down the banking system, so that civil servants—including first responders, health-care providers, and sanitation workers—went without salaries indefinitely.

Two and a half years into the Saudi-led intervention, some 21 million people—three-quarters of Yemen's population—face food insecurity or severe malnutrition. Untreated sewage has contaminated drinking water, causing what the United Nations calls the world's largest cholera epidemic—nearly 900,000 cases as of November 2017. Cholera is treatable, but the crippled health care system was already overwhelmed by patients suffering from war wounds and starvation. Many of the afflicted can't reach hospitals because of damage to roads and bridges.

Meanwhile, as many as three million people have abandoned their homes. Prevented from traveling abroad by the air and naval blockades and strict border controls, and without systematic international assistance, they have sheltered with extended family, in makeshift tents, or in caves. It is worth underscoring that the victims of internal displacement, hunger, disease, and bombing are disproportionately the already underprivileged Afro-Yemenis of the Red Sea coastal region, who are Sunnis of the Shafa'i denomination, not Zaydi Shia. They are not supporters of either Saleh or the Houthis.

The internationally reported death toll from fighting alone reached about 10,000 by October 2016. After that, the Ministry of Health and other national and international agencies lost count, while the pace of airstrikes accelerated—from just under 4,000 in all of 2016 to 5,676 in the first half of 2017, according to tallies from the Office of the UN High Commissioner for Refugees.

Splintering Alliances

By the second half of 2017, events belied the simplistic story of a conflict driven by Sunni-Shia rivalries. Even the more convincing paradigm of geo-strategic rivalry between Iran and the dynastic autocracies of the Arab side of the Persian Gulf failed to pass the most rudimentary reality test when the Saudis and other GCC allies broke with Qatar. The UAE, a major party to the foreign intervention (unlike Saudi Arabia it put boots on the ground), has begun pursuing its own independent policy in South Yemen. The alliance of GCC ruling families no longer shares a common purpose; they are divided among themselves.

The temporary marriage of convenience between Saleh's supporters and the Houthis is also on the rocks. In September 2017, the two groups staged large, rival rallies in Sana'a (the capital has been under their joint control for three years). Unlike the joyful marches of 2011, when women and children were front and center, these were menacing, heavily armed, all-male power displays. As internecine rifts in both camps threaten to widen and deepen the violence, the overwhelming majority of victims will continue to be innocent civilians. The poorest country in the Middle East is paying a heavy price for reckless, convoluted struggles for supremacy among domestic, regional, and international powers.

Critical Thinking

1. Who are the Houthis?
2. Why has Saudi Arabia intervened in Yemen?
3. What is the US policy toward the war in Yemen?

Internet References

Embassy of Saudi Arabia in DC
https:////www.saudiembassy-net/

Special Envoy of the Secretary-General for Yemen
https://osesgy.unmissions.org/

The War in Yemen
https://www.cfr.org/interactive/global-conflict-tracker#1/conflict/war-in-yemen

SHEILA CARAPICO is a professor of political science and international studies at the University of Richmond.

Article

Prepared by: Robert Weiner, *University of Massachusetts, Boston*

A Perfect Storm: American Media, Russian Propaganda

Sarah Oates

Learning Outcomes

After reading this article, you will be able to:

- What factors have made it easier for Russia to engage in propaganda in the United States?
- Why foreign countries have an opportunity to influence Americans?

There has never been a better time for Russian information warfare. Long gone is the Soviet-style propaganda, with outdated Leninist clichés and improbable tales of Soviet superiority, that older generations of Americans may vaguely remember from the Cold War before the communist regime collapsed in 1991. Today's Russian propaganda is information designed to take advantage of the viral nature of new media, in the sense that it is both easy to spread and parasitical to the way in which news now circulates online in the West. The new Russian propaganda hides in plain sight, masquerading as just another point of view or an "alternative" source of information.

Whenever possible, the Kremlin is poised to exploit divisions among parties and even undermine trust in democracy itself through propaganda and misinformation. This is how a country that lacks significant political credibility, economic heft, or global popularity can still have influence in world politics. There is evidence of attempted Kremlin interference in several recent Western elections, including in France. But the most obvious such attempt played out in the 2016 US presidential campaign. There have been moments—such as the leaking of Hillary Clinton's campaign emails—that provide compelling evidence of Russian attempts to influence the US electorate.

What we don't really know is how much of the problem of "fake news" is an American phenomenon and how much has been fostered by Russian propaganda. The Kremlin would like us to think that American attitudes and vote choices are affected by Russian information warfare. So far, however, there is little direct evidence of persistent, effective Russian interference in US domestic politics. But the more the Western media writes about it and American politicians complain about it, the more powerful the reputation of Russian information warfare becomes.

Toxic Environment

What's certain is that the Russians have been helped enormously by three key factors: the weakened state of American journalism due to the lack of a robust business model, the Trump administration's attacks on the press, and outsized fears of Moscow's influence in an era of "fake news." While the Russians are making a considerable effort to spread propaganda, observers often attribute great influence to this factor while paying less attention to the US media environment, where much of the problem of misinformation lies.

American media, particularly print outlets, have been in financial and professional crisis for decades. The United States is the only powerful nation that does not have a large state-funded or public broadcaster to occupy the central information space. This means that corporations and even individual media outlets are left to uphold standards in training and professionalism. But newsrooms are shrinking fast. Newspapers almost everywhere were trimming staff and cutting costs even before the advent of the Internet. In just 11 years, US newspapers shed more than a third of their journalistic workforce, which fell to 41,400 newsroom workers by 2015, according to the Pew Research Center.

Meanwhile, the rise of the Internet and the shift of both attention and advertising to online venues caused even deeper economic woes, particularly for print publications, as people discovered they no longer needed to pay to read the news. Media outlets struggled to capitalize on the new media economy as search companies and social media platforms came to dominate the shifting information landscape. Television news, especially local news, remains relatively popular, but it is generally the printed press that pursues more detailed and analytical reporting.

The "objectivity" model of the US media also changed, at first most noticeably on cable television. Fox News created a successful business model by slanting news toward more conservative viewers both through story choice and bias. While liberals railed against Fox's favoritism, the network amassed a huge following and consolidated a new era in television news: no longer did all outlets share the same view of what it meant to be objective. The media became divided and divisive, and much of the audience followed suit. The way in which Americans consume news today, in particular via recommendations, shares, and likes on social media, means that crowd interest rather than careful editorial curation dictates much of the visibility of news stories.

During the 2016 presidential campaign, Pew found that 40 percent of Trump voters listed Fox as their main source of campaign news, more than all others they reported. On the other hand, barely any Clinton voters watched Fox—only 3 percent claimed it was their main campaign news source. More Clinton voters (18 percent) relied on CNN, but overall their media diet was more varied than that of Trump voters. Much in the way that Russian state television has supported President Vladimir Putin, ignoring or vilifying his competitors, Fox News was a key factor in Trump's victory. This blinkered pattern of news consumption by a sizable portion of the US electorate leaves the American media vulnerable to attacks by politicians (domestic or foreign) with little interest in a free press as a critical component of liberal democracy.

This is the polarized information environment in which Russian propaganda is operating today. There is an opportunity for foreign countries to influence Americans who are unmoored from traditional media outlets they no longer trust. Just as Fox acquired a large and loyal following by catering to the beliefs of one part of the American electorate, foreign propaganda can utilize the same tactics by producing selective and biased reports that mimic the form, but not the function, of news reporting.

Kremlin Narratives

The most visible face of Russian propaganda in the United States is the RT television network. RT (formerly Russia Today) is the expression of the Kremlin "line" in the countries in which it broadcasts, camouflaged as a television news channel. RT's core mission is not to report news; it is to support the Russian state. It uses the loosening of news norms and lack of balance on cable news outlets such as Fox as a cover. For example, it tries to hide Russian international crimes (such as the deployment of Russian troops in eastern Ukraine and the seizure of Crimea) by casting doubt on US media reports and sometimes falsifying information. It employs native English speakers, modern video techniques, and people with some claim to expertise who are willing to deliver a pro-Kremlin point of view. There are also shows and coverage that are not anti-American or pro-Kremlin in nature, providing camouflage for the propaganda. RT recently said it has been told by the Justice Department to register under the US Foreign Agent Registration Act, though this had not been confirmed by US officials at the time of writing.

It is important to note that all major countries seek to broadcast information to citizens in other nations. The United States funds a range of foreign news programming, including Radio Liberty's Russian service and the new Current Time (*Nastoyashchee Vremya*) television network targeted at Russian speakers worldwide. While these programs are designed for foreign citizens, they are obligated by US government rules to adhere to accuracy in their programming. In other words, unlike RT, the US foreign broadcasters must use news, rather than propaganda, to try to win hearts and minds.

Whereas the liberal American media environment has so far allowed RT to operate freely in the United States, Current Time is essentially unable to broadcast on Russian television. Authoritarian regimes such as Putin's have wide powers to block websites, punish journalists, and even arrest and imprison people who share unauthorized news stories online. In the United States, the commitment to media freedom allows foreign propagandists to operate relatively unhindered. This is not a level information battlefield.

Soviet-era international propaganda focused on extolling the superiority of the communist system, but recent Russian propaganda is mostly about attacking democracy. This is paired, to a degree, with portraying Russia as a strong nation that has deep historical and cultural roots. Unsurprisingly, the narratives promoted by the Kremlin have greatest resonance among the domestic audience: the resurgence of a great Russian nation, the longing of ethnic Russians in the near abroad to be reunited with the Motherland, as well as NATO's desire to destroy Russia.

Now the Kremlin would like to extend that narrative resonance to win over citizens in other countries. It has used communication warfare in tandem with military force in Crimea, eastern Ukraine, and Syria. There are fears about the Russian strategic narrative in other neighboring countries. This is especially true in the Baltic states, which have viewed with great trepidation efforts to mobilize their sizable Russian minorities, who could be used to justify a Russian invasion. Crimea and eastern Ukraine are living examples of that threat.

Yet it is not as though Kremlin officials can push a button to immediately win hearts and minds, even at home. The relationship between the media and rising authoritarianism in Russia remains very dynamic. While the Kremlin can control most media outlets through a range of financial and political tactics, the Russian audience is not a passive vessel. Rather, the Kremlin must craft a complex narrative that promotes Russian nationalism in the face of increasing international isolation and economic woes.

Much has been made of Russia's online "troll army" as well as the Russian use of automated "bots" to spread and respond to comments on social media. They are often unleashed against journalists and citizens who criticize the regime. Studies have found that it is relatively easy to detect paid commentary as well as bots, and the jury is still out on how much these tactics affect public opinion. Certainly, they discourage open discussion and investigative journalism. Yet the Internet is a reflection of real-world events and opinion, which limits the scope of deliberate manipulation.

There are still courageous and creative acts of journalism taking place within Russia that challenge the Kremlin narrative. It remains an open question whether the Kremlin can completely dominate the domestic information sphere as the Soviets once did. But the question that is being raised anxiously in Western countries, especially in the United States and Europe, is the extent to which Russian narratives are influencing Western citizens.

Fake News

It's time to consider different approaches to Russian communication warfare. First, it would be useful to disentangle it from the general morass of fake news and misinformation in the United States, which has been fostered in large part by the refusal of social-media platforms to take responsibility for information distribution. Although limited controls are now in place, companies such as Facebook and Google can do more to differentiate between professional reporting, advertising, promotion, and communication warfare. The relatively unregulated content of these platforms gives those seeking to spread foreign propaganda an excellent way to do so. This issue has been highlighted in recent weeks by *New York Times* articles detailing how Russians used fake profiles and targeted advertising on Facebook to promote anti-immigrant sentiment and other themes that resonated with Donald Trump's campaign message.

We should also more accurately measure the effect of Russian communication warfare and be careful not to give it more publicity than it warrants. Attributing Trump's election to Russian propaganda gives the Kremlin way too much credit. What percentage of news that Americans read comes from Russian sources? How much is circulated by Americans on a daily basis? Which stories find traction and which are ignored?

One tactic of Russian news outlets is to try out some more outrageous stories and quietly drop them if they get a skeptical response (such as a faked story on Russian television that Ukrainians had crucified a Russian toddler). Russian communication warfare is constantly beta-testing stories in this way, then running with those that are shared and liked more—and don't draw a large amount of pushback.

Finally, it is arguably not the Russians who are the problem, but American leaders such as Donald Trump who denigrate the press. It is puzzling that the same figures who were built up and propelled into office by the media now attack it. In Russia, denigration of a free press is designed to undermine any challenge to the oligarchic elite. When US leaders also attack the media as an institution, they are attacking democracy in a far more dangerous way than the Kremlin ever could.

Critical Thinking

1. What is fake news?
2. What is the core mission of RT news?
3. What did recent Russian propaganda focus on?

Internet References

The Kremlin Playbook

https://www.csis.org/analysis/kremlin-playbook

Russian Active Measures, Past, Present and Future

https://www.csis.org/analysis/russian-active-measures-past-present-and-future

---

SARAH OATES is a professor and senior scholar at the Philip Merrill College of Journalism, University of Maryland, College Park. Her latest book is *Revolution Stalled: The Political Limits of the Internet in the Post-Soviet Sphere* (Oxford University Press, 2013).

Article

Prepared by: Robert Weiner, *University of Massachusetts, Boston*

The New Russian Chill in the Baltic

MARK KRAMER

Learning Outcomes

After reading this article, you will be able to:

- Understand the effect of the conflict in Ukraine on the Baltic States.
- Understand the Russian military provocations in the Baltics.

In late February and March 2014, shortly after the violent overthrow of Ukrainian President Viktor Yanukovych, Russian President Vladimir Putin sent troops to occupy the Crimean Peninsula, which had long been part of Ukraine. Putin's subsequent annexation of Crimea sparked a bitter confrontation with Western governments and stoked deep anxiety in Central and Eastern Europe about the potential for Russian military encroachments elsewhere. Nowhere has this anxiety been more acute than in Poland and the three Baltic countries—Lithuania, Latvia, and Estonia—where fears have steadily mounted as Russia has helped to fuel a civil war in eastern Ukraine while undertaking a series of military provocations in the Baltic region.

Estonia, Latvia, and Lithuania were forcibly annexed by the Soviet Union in 1940 and remained an involuntary part of it until 1991, when they were finally able to regain their independence. Their relations with post-Soviet Russia have often been tense, even though Russian troops were withdrawn on schedule from Baltic territory by 1994. As a deterrent against possible threats from Russia, all three Baltic countries pressed hard to gain membership in NATO, a status that would entitle them to protection from the United States and other alliance members. Initially, the NATO governments were skeptical about bringing in the Baltic states, but in November 2002 the allied leaders invited Estonia, Latvia, and Lithuania to join. The three were formally admitted into the alliance in 2004.

Poland, for its part, had joined NATO several years earlier. After Communist rule came to an end in Poland in 1989, a broad consensus emerged among Polish elites and the public that membership in NATO would be crucial for the country's long-term security vis-à-vis Russia and other potential threats. In the early 1990s, leaders in Washington and other NATO capitals were wary of adding new members to an alliance that already included 16 countries. Over time, however, NATO shifted in favor of enlargement, and in 1997 the member states agreed to invite Poland and two other former Warsaw Pact countries (Hungary and the Czech Republic) to join. The three formally gained membership in 1999.

The subsequent entry of Estonia, Latvia, and Lithuania into NATO brought most of the Baltic region under the alliance's auspices. The only exceptions have been Finland and Sweden, both of which have chosen thus far to remain outside military alliances, as they have since 1945. However, one of the by-products of the Russia-Ukraine confrontation and the recent spate of Russian military provocations in northern Europe has been a surge of public discussion in both Finland and Sweden about the need for closer links with NATO and even possible membership in the alliance—a step that once would have been unthinkable.

Prior Misgivings

Well before the conflict between Russia and Ukraine erupted in 2014, concerns had arisen in the Baltic region about Russia's intentions. The August 2008 Russia-Georgia war, which saw the Russian Army quickly overwhelm and defeat the much smaller Georgian military and then carve off two sizable parts of Georgia's territory (the self-declared independent republics of South Ossetia and Abkhazia), stirred doubts in the Baltic countries and Poland about the willingness of the United States and other key NATO members to defend them against Russian military pressure or intervention. Even though Georgia was not a NATO member and thus had not received any guarantee of allied protection, the televised images of Russian forces sweeping across Georgian territory and overrunning Georgian military positions came as a jolt to both elites and the wider public in the Baltic countries. Their misgivings were reinforced by the conspicuous maneuvers undertaken by Russian ground forces along the Russian-Estonian border in late 2008 and by the provocative nature of Russia's "Zapad 2009" military exercises with Belarus in September 2009, which involved simulations of rapid offensive operations against NATO.

To allay misgivings in the Baltic countries, NATO leaders in December 2009 authorized the preparation of contingency plans for the reinforcement and defense of the whole Baltic region against an unspecified enemy. Contingency planning known as Eagle Guardian already existed for the defense of Poland, but until 2009 neither the United States nor Germany had wanted to produce additional blueprints to defend the Baltic states, for fear that such an effort would damage relations with Russia if it became publicly known. Polish officials initially expressed concern that the decision to include the Baltic countries would dilute Eagle Guardian, but they were eventually willing to embrace the expanded contingency plan, provided that Poland was treated separately and that US bilateral military support would increase.

Russian actions came perilously close to provoking a military confrontation or a collision with a passenger aircraft.

The new plan designated a minimum of nine NATO divisions—from the United States, Britain, Germany, and Poland—for combat operations to repulse an attack against Poland or the Baltic countries. US and German policy makers tried to keep the revised contingency planning secret, but some details began leaking to the press in early 2010. Soon thereafter, the unauthorized release of thousands of classified US State Department documents on the WikiLeaks website, including many items pertaining to Eagle Guardian and the concerns that led to its expansion, revealed the alliance's planning for all to see.

The public disclosure of NATO's behind-the-scenes deliberations and revised Eagle Guardian plans in late 2010 and early 2011 spawned hyperbolic commentary in the Russian press and drew a harsh reaction from the Russian Foreign Ministry. High-ranking officials claimed to be "bewildered" and "dismayed" that NATO, after issuing countless "proclamations of friendship," would be treating Russia as "the same old enemy in the Cold War." The Russian ambassador to NATO, Dmitri Rogozin—a notorious hard-liner—denounced the "sinister manipulations and intrigues" of the allied governments and accused them of engaging in "warmongering," "odious discrimination," "hateful anti-Russian propaganda," and "flagrant hypocrisy."

The ensuing tensions, coming at an early stage of Barack Obama's presidency, tarnished his administration's much-ballyhooed "reset" of relations with Moscow and eroded NATO's credibility in its dealings with Russia, including its repeated statements that "NATO does not view Russia as a threat." Perhaps if Dmitri Medvedev had stayed on as Russian president (as the Obama administration expected), the damage from the disclosures would have abated relatively quickly and would not have hindered closer ties via the NATO-Russia Council. But with the return of Putin as president in 2012 and the Russian government's growing predilection for flamboyant anti-Western rhetoric and policies, the adverse impact of the revelations persisted.

Among other things, the Russian Army stepped up its military exercises, including simulations of attacks against the Baltic countries and Poland. Russia's Zapad 2011 and Zapad 2013 exercises with Belarus, which were given wide publicity in the Russian and Belarusian media, featured simulated preventive nuclear strikes against Poland and large-scale thrusts toward the Baltic countries. In March 2012, Russian combat aircraft also began to conduct simulated attacks against military sites in Sweden. The Russian authorities' shift to a more belligerent posture throughout the Baltic region began when Medvedev was still president, and any prospect for a rapprochement between NATO and Russia disappeared after Putin returned to the presidency, lending an even sharper edge to the two sides' military rivalry vis-à-vis the Baltic countries and Poland.

NATO sought to offset Russian military activities in the region by carrying out major maneuvers of its own in 2012 and 2013, especially Exercise Steadfast Jazz in the Baltic countries and Poland in early November 2013, which included more than 1,000 mechanized infantry, 2,000 other troops, 3,000 command-and-control personnel, 40 combat aircraft, 2 submarines, and 15 surface vessels. All the NATO countries as well as Finland, Sweden, and Ukraine (then still headed by Yanukovych) took part in Steadfast Jazz, which was tied to plans for a NATO Response Force capable of deploying thousands of allied troops to "defend all member states" in the Baltic region against external attack at very short notice. Before the exercise began, Russian Deputy Defense Minister Anatoly Antonov complained that it would mark a return to the "chill of the Cold War." Although the results of Steadfast Jazz and other joint exercises helped to calm nerves among Baltic leaders, apprehension in the region about Russia's intentions was mounting long before Put in authorized the annexation of Crimea or began fueling a civil war in eastern Ukraine.

Grave Threat

No sooner had the Russian government embarked on the takeover of Crimea in late February and early March 2014 than senior officials in Estonia, Latvia, and Lithuania began warning about the "grave threat" to their own countries. At an emergency meeting of European Union leaders in Brussels in early March, Lithuanian President Dalia Grybauskaite warned that "Russia today is dangerous.. . . They are trying to rewrite the borders established after the Second World War in Europe." The vice speaker of Lithuania's parliament, Petras Auštrevičius, concurred: "Russia is presenting a clear threat, and, knowing the Russian leadership, there is a great risk they might not stop with Ukraine. There is a clear risk of an extension of [Russia's military] activities." Estonian President Toomas Hendrik Ilves emphasized that "no one in [the Baltic] countries can safely assume that Russia's predatory designs will end with the seizure of Crimea."

No country would see much point in belonging to an alliance that refused to protect its members against external aggression.

Baltic leaders' concerns about the prospect of Russian aggression intensified after Putin delivered a bellicose speech before the Russian parliament on March 18, 2014, announcing the annexation of Crimea and proclaiming a duty to protect ethnic Russian populations in other countries. Senior Baltic officials and military commanders urged the United States and other leading NATO countries to reaffirm and strengthen Article 5 of the North Atlantic Treaty, which stipulates that "an armed attack against one or more of them in Europe or North America shall be considered an attack against them all" and obliges every NATO country to take "such action as it deems necessary, including the use of armed force," to repulse an attack against another NATO member state. The commander-in-chief of the Estonian Defense Forces, Major General Riho Terras, declared that although Estonia faced no immediate military threat, the events in Crimea demonstrated that the Russian Army has "a very credible capability" of "doing various things" elsewhere in Europe, especially in countries like Estonia with large ethnic Russian minorities. "It is very important," Terras warned, "that we [the members of NATO] now seriously think about defense plans based on Article 5."

Terras's comments were echoed a week later by Estonian Prime Minister Taavi Rõivas, who said that, in light of the annexation of Crimea and Russia's conspicuous military activities in the Baltic region, it would be "extremely important for the alliance," especially the United States, to deploy "boots on the ground" in the Baltic countries in order to "increase the NATO presence and defend all allies" against any possible encroachments. Only through a robust and lasting troop presence, he implied, could NATO counter "external threats" and ensure the "security and well-being" of alliance members. Lithuanian Defense Minister Juozas Olekas expressed much the same view, arguing that the "very active" Russian troop movements in Kaliningrad Oblast (the Russian exclave on the Baltic Sea) along the border with Lithuania necessitated the deployment of "NATO ground forces in the [Baltic] region and visits from NATO navies" to deter "aggression from the East." Terras's predecessor as Estonia's commander-in-chief, General Ants Laaneots, addressed the issue even more bluntly in an interview with the Estonian press a few weeks later: "Putin has brought sense back to European minds regarding military dangers. I am happy that NATO and above all the EU members [of the alliance] have woken up after twenty years of self-delusion in the field of security."

These comments by Baltic officials were in line with broader public opinion in the Baltic countries and Poland. A survey in Poland in March 2014, as Russia's takeover of Crimea was unfolding and Russian military forces in and around the Baltic Sea were engaging in unscheduled large-scale maneuvers, indicated that 59 percent of Polish adults viewed Russia as a "threat to Poland's security." Surveys in Estonia and Lithuania in late March turned up even higher shares of respondents—74 percent of Estonians and 68 percent of Lithuanians—who saw Russia as the "greatest threat" to their countries and to the "whole of Europe."

After a civil war erupted in eastern Ukraine in the late spring of 2014 with Russia's active support of separatist rebels, anxiety in the Baltic countries and Poland steadily intensified. Polish Foreign Minister Radoslaw Sikorski warned in August 2014 that Putin "has moved beyond all civilized norms" and "thinks he's facing a bunch of degenerate weaklings [in NATO]. He thinks we wouldn't go to war to defend the Baltics. You know, maybe he's right." Sikorski was hardly alone in this view. Nearly every senior official in Poland and the Baltic countries expressed great unease. Political leaders and military commanders in the region increasingly warned that Putin was intent on undermining NATO's resolve to protect their countries. The Latvian and Estonian governments noted with consternation that Russian diplomats had stepped up efforts to give Russian passports and higher pensions to ethnic Russians in Latvia and Estonia. They urged the United States and other NATO countries to take full account of the "overriding danger" posed by Russia and the "evident desire by Moscow authorities to reestablish domination over their former empire in Europe."

Surging Provocations

Russia's annexation of Crimea and sponsorship of an armed insurgency in eastern Ukraine were accompanied by a surge of Russian military provocations in and around the Baltic region. Some of these actions were targeted at the Baltic countries and Poland, whereas others were aimed at Sweden, Finland, and

Norway. Although most of the incidents were little more than shows of strength and bravado, some proved highly dangerous. In a few cases, Russian actions came perilously close to provoking a military confrontation between Russia and one or more NATO countries or a collision with a passenger aircraft.

For more than a decade after the Soviet Union collapsed, Russian military forces engaged in relatively few exercises and kept a very low profile. Long training flights for Russian combat aircraft nearly ceased, and sea patrols by naval vessels and submarines were drastically curtailed. The situation began to change gradually in the early 2000s as Russia's economy started to recover from the steep output decline that followed the disintegration of the Soviet economy. By 2006, Russian military forces had returned to higher levels of readiness and resumed activities beyond Russia's borders, including lengthy training flights through international airspace. The extent of these activities was not quite as sizable as in the Soviet era, but on a gradually increasing number of occasions from 2006 through 2013 NATO fighter aircraft intercepted Russian military planes as they approached Polish and Baltic airspace. According to NATO's Combined Air Operations Center in Üdem, Germany, allied aircraft scrambled to intercept Russian combat planes roughly 45 times a year in the Baltic region from 2011 to 2013.

This pattern changed dramatically in 2014, as tensions mounted over Crimea and eastern Ukraine. Russian military activities of all sorts in northern Europe, especially harassing NATO countries and Sweden, precipitously increased. Russian warships intruded into Baltic countries' territorial waters, including one occasion when Russian vessels engaged in live-firing exercises that severely disrupted civilian shipping throughout the region. Russian fighter and bomber aircraft repeatedly buzzed warships and other naval vessels from NATO countries and Sweden in the Baltic and North Seas, carried out simulated attacks against NATO countries and Sweden (as well as a simulated volley of air-launched cruise missile targeting North America), intruded into NATO countries' airspace, and undertook armed missions against US and Swedish reconnaissance aircraft, forcing them to take evasive maneuvers.

NATO fighter aircraft in the region were scrambled to intercept Russian planes more than 130 times in 2014, roughly triple the number of interceptions in 2013. On many occasions, NATO fighters intercepted formations of Russian bombers and tanker aircraft as they approached or entered Baltic and Norwegian airspace. After the largest such incident, in late October 2014, NATO's Allied Operations Command reported that "the bomber and tanker aircraft from Russia did not file flight plans or maintain radio contact with civilian air traffic control authorities and they were not using onboard transponders. This poses a potential risk to civil aviation as civilian air traffic control cannot detect these aircraft or ensure there is no interference with civilian air traffic."

The quantity and provocative nature of Russian aerial incursions over northern Europe in 2014 marked a sharp departure from the pattern of earlier years. A report published in late 2014 by the European Leadership Network (a London-based think tank) highlighted the magnitude of the difference in the Baltic region from January to September 2014, when "the NATO Air Policing Mission conducted 68 'hot' identification and interdiction missions along the Lithuanian border alone, and Latvia recorded more than 150 incidents of Russian planes approaching its airspace." Also, "Estonia recorded 6 violations of its airspace in 2014, as compared to 7 violations overall for the entire period between 2006 and 2013." This pattern continued in late 2014 and 2015, far exceeding the number of incidents since the height of the Cold War.

Reckless Endangerment

Of particular concern to Polish and Baltic leaders were the seemingly deliberate efforts by Russian military forces to provoke armed clashes or to endanger civilian passenger aircraft. Provocations directed against NATO countries, Sweden, and Finland occurred throughout 2014 and early 2015 but were particularly frequent in the aftermath of the controversy surrounding the downing of Malaysia Airlines flight MH17 by Russian-backed insurgents in eastern Ukraine. In early September 2014, two days after Obama traveled to Estonia and pledged strong support to the three Baltic countries, Russian state security forces kidnapped at gunpoint an Estonian Internal Security Service officer, Eston Kohver, from a border post on Estonian territory and spirited him to Moscow. As of February 2015, Kohver was still being held without trial in Moscow's notorious Lefortovo Prison on charges of espionage.

Apprehension in the region about Russia's intentions was mounting long before Putin authorized the annexation of Crimea.

The same month Kohver was abducted, Russian strategic nuclear bombers carried out simulated cruise missile attacks against North America; Russian fighters buzzed a Canadian frigate in the Black Sea while Russian naval forces engaged in maneuvers nearby; Russian medium-range bombers intruded into Swedish airspace to test the reactions of air defense forces; and Russian warships seized a Lithuanian fishing vessel in international waters of the Barents Sea and brought it to Murmansk in defiance of the Lithuanian government's protests. The next month, Russian military forces not only kept up their aerial incursions in the Baltic region (including a mission against Swedish surveillance aircraft that was deemed "unusually provocative") but also dispatched submarines on a prolonged series of intrusions into Swedish territorial waters, causing sharp bilateral tensions and nearly provoking an armed confrontation at sea. The Swedish navy received authorization to use force if necessary to bring the submarines to the surface, but a 10-day search for the intruders proved unsuccessful.

Equally disturbing was the apparent willingness—indeed eagerness—of the Russian authorities to deploy military forces in ways that endangered civilian air traffic in northern Europe. The most egregious incident of this sort occurred in March 2014, when a Russian reconnaissance plane that was not transmitting its position nearly collided with an SAS passenger airliner carrying 132 people. A fatal collision was avoided only because of the alertness and skillful reaction of the SAS pilots. Other such incidents occurred in the spring and early summer. Even after the MH17 incident in July 2014 drew opprobrium from around the world, Russian military aircraft continued to pose dangers to civilian airliners. The frequency and audacity of the incidents left no doubt that they were deliberate.

By the end of 2014, it had become clear that, as the European Leadership Network report stated, the "Russian armed forces and security agencies seem to have been authorized and encouraged to act in a much more aggressive way toward NATO countries, Sweden, and Finland." Against a backdrop of large-scale Russian military exercises and force redeployments in the Baltic region and elsewhere in 2014 and early 2015, the long series of Russian military provocations has raised troubling questions about Moscow's intentions.

Boots on the Ground?

The surge of tensions over Ukraine and Russian military provocations in the Baltic region spurred officials from Poland and the three Baltic countries to push for a strong show of resolve by NATO and a concrete reaffirmation of Article 5. The US government moved relatively quickly to allay some of these concerns, announcing in March 2014 that it would send six F-15C fighters and two KC-135 tanker aircraft to the headquarters of NATO's Baltic Air Policing Mission at the Šiauliai air base in Lithuania, joining the four F-15Cs that had been on patrol since the mission was established in 2004 when the Baltic countries (which lack their own combat aircraft) entered the alliance. The United States also deployed twelve F-15s and F-16s to Poland to assist air defense operations there and augmented the US naval presence in the Baltic Sea. Subsequently, Denmark, France, and Britain sent additional fighter planes to Šiauliai to expand the air policing mission further and relieve some of the aircraft already on patrol. NATO also expanded its surveillance of the Baltic region with extra flights of allied Airborne Warning and Control Systems planes, which provided broad coverage around the clock.

These incremental increases of NATO's military presence in the Baltic region, and a decision by alliance foreign ministers in April 2014 to "suspend all practical civilian and military cooperation between NATO and Russia," were welcomed by the Baltic and Polish governments, but they urged the United States and other large NATO countries to go further with defense preparations. Officials in Warsaw, Tallinn, Riga, and Vilnius sought the permanent stationing of allied ground and air forces on Polish and Baltic territory. Estonian Prime Minister Rõivas's plea for "boots on the ground" was echoed by other leaders in the region, who hoped that their requests would be endorsed by the NATO governments at a summit meeting in Wales on September 4–5, 2014. For years, such a step had been precluded by the NATO-Russia Founding Act, signed by Russian and NATO leaders at a Paris summit in May 1997. To mitigate Moscow's aversion to the enlargement of NATO, the Founding Act established conditions for the deployment of allied troops on the territory of newly admitted member states:

> NATO reiterates that in the current and foreseeable security environment, the Alliance will carry out its collective defense and other missions by ensuring the necessary interoperability, integration, and capability for reinforcement rather than by additional permanent stationing of substantial combat forces. Accordingly, it will have to rely on adequate infrastructure commensurate with the above tasks. In this context, reinforcement may take place, when necessary, in the event of defense against a threat of aggression.

In accordance with this provision, the United States and other allied countries had always eschewed any prolonged deployment of "substantial" military forces on the territory of new NATO members in the Baltic region.

In the lead-up to the Wales summit, the Polish and Baltic governments argued that international circumstances had fundamentally changed since 1997 and that NATO should no longer be bound by anything in the Founding Act. Referring to the clause stating that NATO and Russia no longer regarded each other as enemies, General Terras of Estonia contended in May 2014 that the whole document had become obsolete: "Russia sees NATO as a threat, and therefore NATO should not view Russia as a friendly, cooperative country. That is very clear. The threat assessment of NATO needs to fit the current realistic circumstances." Terras returned to this theme a few months later, just before the Wales summit:

> No one now believes that friendly relations between NATO and Moscow can be reestablished. Russia today regards NATO as an enemy, and this must facilitate changes [in NATO's force posture]. Some changes have already taken place, and the NATO summit must give out a clear message to the allies and to Russia that NATO

is the world's most powerful military organization and is willing to do everything to protect its member states, including increasing its presence in areas bordering Russia.

Terras and other senior military and political officials in the Baltic region also argued that in light of Russia's actions in Ukraine and elsewhere, NATO should move ahead as expeditiously as possible with concrete military preparations for "defense against a threat of aggression," as stipulated in the Founding Act.

Limited Measures

Many officials in the United States and other NATO countries were sympathetic to the arguments of Polish and Baltic leaders, but the US government ultimately decided not to proceed with long-term deployments of "substantial" military forces in the Baltic region. US officials at the Wales summit did make an effort to address Baltic and Polish concerns, not least by joining with all the other NATO allies in vowing to uphold Article 5: "The Alliance poses no threat to any country. But should the security of any Ally be threatened we will act together and decisively, as set out in Article 5 of the Washington Treaty." The summit declaration made clear that this warning was meant for Russia.

In the military sphere, however, the summit mostly just endorsed and extended the relatively limited measures that had been adopted earlier in the year to expand the Baltic Air Policing Mission and to bolster Western naval forces in the Baltic Sea. Although the Wales summit participants welcomed the fact that more than 200 military exercises had been held in Europe in 2014, the reality was that few of these exercises were of any appreciable size. Moreover, although they endorsed the deployment of "ground troops in the eastern parts of the Alliance for training and exercises," they made clear that these troops were stationed there solely "on a rotational basis," not permanently. The allied leaders did adopt a Readiness Action Plan to enlarge the long-planned NATO Response Force (from 13,000 to 20,000 troops) and to put it on a higher state of readiness, with a "Spearhead Force" of up to several thousand troops and reinforcements that could be deployed to the Baltic region within a few days. Whether those projections will actually materialize in 2015 and 2016 remains to be seen, however. Even if the proposals are fully implemented, they fall well short of what the Baltic countries and Poland had been seeking.

In the months following the Wales summit, the NATO governments tried to fulfill several of the pledges they had adopted, most notably with the establishment of multinational command-and-control centers in the Baltic countries, Poland, Bulgaria, and Romania, consisting of "personnel from Allies on a rotational basis" who are to "focus on planning and exercising collective defense." At a February 2015 meeting in Brussels, NATO defense ministers pledged to increase the size of the Response Force to 30,000, including a Spearhead Force of 5,000. NATO military planners and individual governments took other concrete steps, including the upgrading of infrastructure and the prepositioning of weaponry and support equipment, to enhance the alliance's capacity to uphold Article 5 in the Baltic region.

Nevertheless, in the absence of large, permanent deployments of US and other NATO ground and air forces in the Baltic countries and Poland, doubts about the collective defense of the region are bound to persist. Terras highlighted this problem when he noted that although he himself did not doubt NATO's willingness to carry out its defense commitments, "the real question is whether Putin believes that Article 5 works." He warned, "We should not give any option of miscalculation for President Putin."

Gloomy Scenarios

Russia's actions in Crimea and eastern Ukraine, and the risky nature of Russian military operations and exercises in and around the Baltic region in 2014 and 2015, have sparked acute unease in Poland and the Baltic countries. Although the available evidence does not indicate that the Russian authorities will attack a NATO member state, Putin does seem intent on undermining NATO by raising doubts about the credibility of Article 5. Despite the strong pledges of support offered at the Wales summit, some uncertainty remains about what would happen if Russia undertook a limited military probe against one or more of the Baltic states. Certain European members of NATO might hinder a timely response, but if that were to happen the United States and some other NATO member states would likely act outside the alliance's command structure to defend the Baltic states, as envisaged in NATO's contingency defense planning. They would undoubtedly try to avoid escalation to all-out war against Russia, not least because that would require the NATO countries to fight in a region in which they would be at a serious geographic disadvantage.

However, if the United States and its allies failed to uphold Article 5 in the Baltic region and refrained from intervening against Russian military forces, this would gravely damage the credibility of all of NATO's defense commitments. No country would see much point in belonging to an alliance that refused to protect its members against external aggression. If Putin were fool-hardy enough to risk all-out war by embarking on military action in the Baltic region, NATO would have no fully reliable or attractive military and diplomatic options. But the worst option of all would be to do nothing and allow Russian military expansion to proceed unchecked.

These gloomy scenarios seem improbable for now, but the very fact that they are being discussed seriously in NATO circles as well as in Warsaw and the Baltic capitals is a sign of how gravely Russia's actions in Ukraine and elsewhere have affected the post—Cold War European security order. Peace and security in the Baltic region, which only a decade ago appeared more robust than ever, now seem all too precarious.

Critical Thinking

1. Why would Sweden and Finland consider joining NATO?
2. Should Poland and the Baltics be defended separately or together from external attacks?
3. What can the U.S. do to protect the Baltics from Russia?

Internet References

Latvian Foreign Ministry
mfa.gov.lv

Lithuanian Ministry of Foreign Affairs
urm.lt

Republic of Estonia, Ministry of Foreign Affairs
vm.el

Russian Ministry of Foreign Affairs
government.ru

Mark Kramer is director of the Cold War Studies program and a senior fellow of the Davis Center for Russian and Eurosian Studies at Harvard University.

Article

Prepared by: Robert Weiner, *University of Massachusetts, Boston*

Putin's Foreign Policy

The Quest to Restore Russia's Rightful Place

FYODOR LUKYANOV

Learning Outcomes

After reading this article, you will be able to:

- Discuss the basic elements of Putin's foreign policy.
- Identify the reasons for Putin's distrust of the West.

In February, Moscow and Washington issued a joint statement announcing the terms of a "cessation of hostilities" in Syria—a truce agreed to by major world powers, regional players, and most of the participants in the Syrian civil war. Given the fierce mutual recriminations that have become typical of U.S.–Russian relations in recent years, the tone of the statement suggested a surprising degree of common cause. "The United States of America and the Russian Federation . . . [are] seeking to achieve a peaceful settlement of the Syrian crisis with full respect for the fundamental role of the United Nations," the statement began. It went on to declare that the two countries are "fully determined to provide their strongest support to end the Syrian conflict."

What is even more surprising is that the truce has mostly held, according to the UN, even though many experts predicted its rapid failure. Indeed, when Russia declared in March that it would begin to pull out most of the forces it had deployed to Syria since last fall, the Kremlin intended to signal its belief that the truce will hold even without a significant Russian military presence.

The ceasefire represents the second time that the Russians and the Americans have unexpectedly and successful cooperated in Syria, where the civil war has pitted Moscow (which acts as the primary protector and patron of Syrian President Bashar al-Assad) against Washington (which has called for an end to Assad's rule). In 2013, Russia and the United States agreed on a plan to eliminate Syria's chemical weapons, with the Assad regime's assent. Few believed that arrangement would work either, but it did.

These moments of cooperation highlight the fact that, although the world order has changed beyond recognition during the past 25 years and is no longer defined by a rivalry between two competing superpowers, it remains the case that when an acute international crisis breaks out, Russia and the United States are often the only actors able to resolve it. Rising powers, international institutions, and regional organizations frequently cannot do anything—or don't want to. What is more, despite Moscow's and Washington's expressions of hostility and contempt for each other, when it comes to shared interests and common threats, the two powers are still able to work reasonably well together.

And yet, it's important to note that these types of constructive interactions on discrete issues have not changed the overall relationship, which remains troubled. Even as it worked with Russia on the truce, the United States continued to enforce the sanctions it had placed on Russia in response to the 2014 annexation of Crimea, and a high-level U.S. Treasury official recently accused Russian President Vladimir Putin of personal corruption.

The era of bipolar confrontation ended a long time ago. But the unipolar moment of U.S. dominance that began in 1991 is gone, too. A new, multipolar world has brought more uncertainty into international affairs. Both Russia and the United States are struggling to define their proper roles in the world. But one thing that each side feels certain about is that the other side has overstepped. The tension between them stems not merely from events in Syria and Ukraine but also from a continuing disagreement about what the collapse of the Soviet Union meant for the world order. For Americans and other Westerners, the legacy of the Soviet downfall is simple: the United States won the Cold War and has taken its rightful place as the world's sole superpower, whereas post-Soviet Russia has failed to integrate itself as a regional power in the Washington-led postwar liberal international order. Russians, of course, see things differently. In their view, Russia's subordinate position is the illegitimate result of a never-ending U.S. campaign to keep Russia down and prevent it from regaining its proper status.

In his annual address to the Russian legislature in 2005, Putin famously described the disappearance of the Soviet Union as a "major geopolitical disaster." That phrase accurately captures the sense of loss that many Russians associate with the post-Soviet era. But a less often noted line in that speech conveys the equally crucial belief that the West misinterpreted the end of the Cold War. "Many thought or seemed to think at the time that our young democracy was not a continuation of Russian statehood, but its ultimate collapse," Putin said. "They were mistaken." In other words: the West thought that Russia would forever going forward play a fundamentally diminished role in the world. Putin and many other Russians begged to differ.

In the wake of the 2014 Russian reclamation of Crimea and the launch of Russia's direct military intervention in Syria last year, Western analysts have frequently derided Russia as a "revisionist" power that seeks to alter the agreed-on post–Cold War consensus. But in Moscow's view, Russia has merely been responding to temporary revisions that the West itself has tried to make permanent. No genuine world order existed at the end of the twentieth century, and attempts to impose U.S. hegemony have slowly eroded the principles of the previous world order, which was based on the balance of power, respect for sovereignty, noninterference in other states' internal affairs, and the need to obtain the UN Security Council's approval before using military force.

By taking action in Ukraine and Syria, Russia has made clear its intention to restore its status as a major international player. What remains unclear is how long it will be able to maintain its recent gains.

No World Order

In January 1992, a month after the official dissolution of the Soviet Union, U.S. President George H. W. Bush announced in his State of the Union address: "By the grace of God, America won the Cold War." Bush put as fine a point as possible on it: "The Cold War didn't 'end'—it was won."

Russian officials have never made so clear a statement about what, exactly, happened from their point of view. Their assessments have ranged from "we won" (the Russian people overcame a repressive communist system) to "we lost" (the Russians allowed a great country to collapse). But Russian leaders have all agreed on one thing: the "new world order" that emerged after 1991 was nothing like the one envisioned by Mikhail Gorbachev and other reform-minded Soviet leaders as a way to prevent the worst possible outcomes of the Cold War. Throughout the late 1980s, Gorbachev and his cohort believed that the best way out of the Cold War would be to agree on new rules for global governance. The end of the arms race, the reunification of Germany, and the adoption of the Charter of Paris for a New Europe aimed to reduce confrontation and forge a partnership between the rival blocs in the East and the West.

But the disintegration of the Soviet Union rendered that paradigm obsolete. A "new world order" no longer meant an arrangement between equals; it meant the triumph of Western principles and influence. And so in the 1990s, the Western powers started an ambitious experiment to bring a considerable part of the world over to what they considered "the right side of history." The project began in Europe, where the transformations were mainly peaceful and led to the emergence and rapid expansion of the EU. But the United States led 1990–1991 Gulf War introduced a new dynamic: without the constraints of superpower rivalry, the Western powers seemed to feel emboldened to use direct military intervention to put pressure on states that resisted the new order, such as Saddam Hussein's Iraq.

Soon thereafter, NATO expanded eastward, mainly by absorbing countries that had previously formed a buffer zone around Russia. For centuries, Russian security strategy has been built on defense: expanding the space around the core to avoid being caught off guard. As a country of plains, Russia has experienced devastating invasions more than once; the Kremlin has long seen reinforcing "strategic depth" as the only way to guarantee its survival. But in the midst of economic collapse and political disorder in the immediate post-Soviet era, Russia could do little in response to EU consolidation and NATO expansion.

The West misinterpreted Russia's inaction. As Ivan Krastev and Mark Leonard observed last year in these pages, Western powers "mistook Moscow's failure to block the post–Cold War order as support for it." Beginning in 1994, long before Putin appeared on the national political stage, Russian President Boris Yeltsin repeatedly expressed deep dissatisfaction with what he and many Russians saw as Western arrogance. Washington, however, viewed such criticism from Russia as little more than a reflexive expression of an outmoded imperial mentality, mostly intended for domestic consumption.

From the Russian point of view, a critical turning point came when NATO intervened in the Kosovo war in 1999. Many Russians—even strong advocates of liberal reform—were appalled by NATO'S bombing raids against Serbia, a European country with close ties to Moscow, which were intended to force the Serbs to capitulate in their fight against Kosovar separatists. The success of that effort—which also led directly to the downfall of the Serbian leader Slobodan Milosevic the following year—seemed to set a new precedent and provide a new template. Since 2001, NATO or its leading member states have initiated military operations in Afghanistan, Iraq, and Libya. All three campaigns led to various forms of regime change and, in the case of Iraq and Libya, the deterioration of the state.

In this sense, it is not only NATO's expansion that has alarmed Russia but also NATO's transformation. Western arguments that NATO is a purely defensive alliance ring hollow: it is now a fighting group, which it was not during the Cold War.

Victors and Spoils

As the United States flexed its muscles and NATO became a more formidable organization, Russia found itself in a strange position. It was the successor to a superpower, with almost all of the Soviet Union's formal attributes, but at the same time, it had to overcome a systemic decline while depending on the mercy (and financial support) of its former foes. For the first dozen or so years of the post-Soviet era, Western leaders assumed that Russia would respond to its predicament by becoming part of what can be referred to as "wider Europe": a theoretical space that featured the EU and NATO at its core but that also incorporated countries that were not members of those organizations by encouraging them to voluntarily adopt the norms and regulations associated with membership. In other words, Russia was offered a limited niche inside Europe's expanding architecture. Unlike Gorbachev's concept of a common European home where the Soviet Union would be a co-designer of a new world order, Moscow instead had to give up its global aspirations and agree to obey rules it had played no part in devising. European Commission President Romano Prodi expressed this formula best in 2002: Russia would share with the EU "everything but institutions." In plain terms, this meant that Russia would adopt EU rules and regulations but would not be able to influence their development.

For quite a while, Moscow essentially accepted this proposition, making only minimal efforts to expand its global role. But neither Russian elites nor ordinary Russians ever accepted the image of their country as a mere regional power. And the early years of the Putin era saw the recovery of the Russian economy—driven to a great extent by rising energy prices but also by Putin's success in reestablishing a functioning state—with a consequent increase in Russia's international influence. Suddenly, Russia was no longer a supplicant; it was a critical emerging market and an engine of global growth.

Meanwhile, it became difficult to accept the Western project of building a liberal order as a benign phenomenon when a series of so-called color revolutions in the former Soviet space, cheered on (at the very least) by Washington, undermined governments that had roots in the Soviet era and reasonably good relations with Moscow. In Russia's opinion, the United States and its allies had convinced themselves that they had the right, as moral and political victors, to change not only the world order but also the internal orders of individual countries however they saw fit. The concepts of "democracy promotion" and "transformational diplomacy" pursued by the George W. Bush administration conditioned interstate relations on altering any system of government that did not match Washington's understanding of democracy.

The Iron Fist

In the immediate post-9/11 era, the United States was riding high. But in more recent years, the order designed by Washington and its allies in the 1990s has come under severe strain. The many U.S. failures in the Middle East, the 2008 global financial crisis and the subsequent recession, mounting economic and political crises in the EU, and the growing power of China made Russia even more reluctant to fit itself into the Western-led international system. What is more, although the West was experiencing growing difficulties steering its own course, it never lost its desire to expand—pressuring Ukraine, for example, to align itself more closely with the EU even as the union appeared to be on the brink of profound decay. The Russian leadership came to the conclusion that Western expansionism could be reversed only with an "iron fist," as the Russian political scientist Sergey Karaganov put it in 2011.

The February 2014 ouster of Ukrainian President Viktor Yanukovych by pro-Western forces was, in a sense, the final straw for Russia. Moscow's operation in Crimea was a response to the EU's and NATO's persistent eastward expansion during the post–Cold War period. Moscow rejected the further extension of Western influence into the former Soviet space in the most decisive way possible—with the use of military force. Russians had always viewed Crimea as the most humiliating loss of all the territories left outside of Russia after the disintegration of the Soviet Union. Crimea has long been a symbol of a post-Soviet unwillingness to fight for Russia's proper status. The return of the peninsula righted that perceived historical wrong, and Moscow's ongoing involvement in the crisis in Ukraine has made the already remote prospect of Ukrainian membership in NATO even more unlikely and has made it impossible to imagine Ukraine joining the EU anytime soon.

The Kremlin has clearly concluded that in order to defend its interests close to Russia's borders, it must play globally. So having drawn a line in Ukraine, Russia decided that the next place to put down the iron fist would be Syria. The Syrian intervention was aimed not only at strengthening Assad's position but also at forcing the United States to deal with Moscow on a more equal footing. Putin's decision to begin pulling Russian forces out of Syria in March did not represent a reversal; rather, it was a sign of the strategy's success. Moscow had demonstrated its military prowess and changed the dynamics of the conflict but had avoided being tied down in a Syrian quagmire.

Identity Crisis

There is no doubt that during the past few years, Moscow has achieved some successes in its quest to regain international stature. But it's difficult to say whether these gains will prove lasting. The Kremlin may have outmaneuvered its Western rivals in some ways during the crises in Ukraine and Syria, but it still faces the more difficult long-term challenge of finding a credible role in the new, multipolar environment. In recent years, Russia has shown considerable skill in exploiting the West's missteps, but Moscow's failure to develop a coherent economic strategy threatens the long-term sustainability of its newly restored status.

As Moscow has struggled to remedy what it considers to be the unfair outcome of the Cold War, the world has changed dramatically. Relations between Russia and the United States no longer top the international agenda, as they did 30 years ago. Russia's attitude toward the European project is not as important as it was in the past. The EU will likely go through painful transformations in the years to come, but mostly not on account of any actions Moscow does or does not take.

Russia has also seen its influence wane on its southern frontier. Historically, Moscow has viewed Central Asia as a chessboard and has seen itself as one of the players in the Great Game for influence. But in recent years, the game has changed. China has poured massive amounts of money into its Silk Road Economic Belt infrastructure project and is emerging as the biggest player in the region. This presents both a challenge and an opportunity for Moscow, but more than anything, it serves as a reminder that Russia has yet to find its place in what the Kremlin refers to as "wider Eurasia."

Simply put, when it comes to its role in the world, Russia is in the throes of an identity crisis. It has neither fully integrated into the liberal order nor built its own viable alternative. That explains why the Kremlin has in some ways adopted the Soviet model—eschewing the communist ideology, of course, but embracing a direct challenge to the West, not only in Russia's core security areas but far afield, as well. To accompany this shift, the Russian leadership has encouraged the idea that the Soviet disintegration was merely the first step in a long Western campaign to achieve total dominance, which went on to encompass the military interventions in Yugoslavia, Iraq, and Libya and the color revolutions in post-Soviet countries—and which will perhaps culminate in a future attempt to pursue regime change in Russia itself. This deep-rooted view is based on the conviction that the West not only seeks to continue geopolitical expansion in its classical form but also wants to make everyone do things its way, by persuasion and example when possible, but by force when necessary.

Even if one accepts that view of Western intentions, however, there is not much Moscow can do to counter the trend by military means only. Influence in the globalized world is increasingly determined by economic strength, of which Russia has little, especially now that energy prices are falling. Economic weakness can be cloaked by military power or skillful diplomacy, but only for a short time.

Angry or Focusing?

Putin and most of those who are running the country today believe that the Soviet collapse was hastened by perestroika, the political reform initiated by Gorbachev in the late 1980s. They dread a recurrence of the instability that accompanied that reform and perceive as a threat anything and anyone that might make it harder to govern. But the Kremlin would do well to recall one of the most important lessons of perestroika. Gorbachev had ambitious plans to create a profoundly different relationship with the West and the rest of the world. This agenda, which the Kremlin dubbed "new political thinking," was initially quite popular domestically and was well received abroad as well. But as Gorbachev struggled and ultimately failed to restart the Soviet economy, "new political thinking" came to be seen as an effort to compensate for—or distract attention from—rapid socioeconomic decline by concentrating on foreign policy. That strategy didn't work then, and it's not likely to work now.

It's doubtful that the Kremlin will make any significant moves on the Russian economy before 2018, when the next presidential election will take place, in order to avoid any problems that could complicate Putin's expected reelection. Russia's economy is struggling but hardly in free fall; the country should be able to muddle through for another two years. But the economic agenda will inevitably rise to the fore after the election, because at that point, the existing model will be close to exhausted.

Turbulence will almost certainly continue to roil the international system after the 2018 election, of course, so the Kremlin might still find opportunities to intensify Russia's activity on the world stage. But without a much stronger economic base, the gap between Russian ambitions and Russian capacities will grow. That could inspire a sharper focus on domestic needs—but it could also provoke even more risky gambling abroad.

"Russia is not angry; it is focusing." So goes a frequently repeated Russian aphorism, coined in 1856 by the foreign minister of the Russian empire, Alexander Gorchakov, after Russia had lowered its international profile in the wake of its defeat in the Crimean War. The situation today is in some ways the opposite: Russia has regained Crimea, has enhanced its international status, and feels confident when it comes to foreign affairs. But the need to focus is no less urgent—this time on economic development. Merely getting angry will accomplish little.

Critical Thinking

1. Should the next U.S. President try to "reset" relations with Russia? Why or why not?
2. Why did Russia intervene in the Syrian civil war?
3. Is there a new "Cold War" between Russia and the U.S.?

Internet References

Collective Security Treaty Organization
www.odkb.gov.ru/start/index_aengl.htm

Embassy of the Russian federation to the U.S.
www.RussianEmbassy.org/

Permanent Mission of the Russian federation to the United Nations
Russiaun.nu/en

The Eurasian Economic Union
https://en.wikipedia.org/wiki/Eurasian_Economic_Union

FYODOR LUKYANOV is an editor in chief of *Russia in Global Affairs*, Chair of the Presidium of the Council on Foreign and Defense Policy, and a research professor at the National Research University Higher School of Economics, Moscow.

Article

Prepared by: Robert Weiner, *University of Massachusetts, Boston*

How to Prevent an Iranian Bomb

The Case for Deterrence

MICHAEL MANDELBAUM

Learning Outcomes

After reading this article, you will be able to:

- Explain how the Cold War strategy of deterrence can be applied to Iran.
- Identify the basic elements of the nuclear deal with Iran.

The Joint Comprehensive Plan of Action (JCPOA), reached by Iran, six other countries, and the European Union in Vienna in July, has sparked a heated political debate in the United States. Under the terms of the agreement, Iran has agreed to accept some temporary limits on its nuclear program in return for the lifting of the economic sanctions the international community imposed in response to that program. The Obama administration, a chief negotiator of the accord, argues that the deal will freeze and in some ways set back Iran's march toward nuclear weapons while opening up the possibility of improving relations between the United States and the Islamic Republic, which have been bitterly hostile ever since the 1979 Iranian Revolution. The administration further contends that the agreement includes robust provisions for the international inspection of Iran's nuclear facilities that will discourage and, if necessary, detect any Iranian cheating, triggering stiff penalties in response.

Critics of the deal, by contrast, argue that it permits Iran to remain very close to obtaining a bomb, that its provisions for verifying Iranian compliance are weak, and that the lifting of the sanctions will give Iranian leaders a massive windfall that they will use to support threatening behavior by Tehran, such as sponsoring global terrorism, propping up the Syrian dictator Bashar al-Assad, and backing Hezbollah in its conflict with Israel (a country that the Iranian regime has repeatedly promised to destroy).

The American political conflict came to a head in September, when Congress had the chance to register its disapproval of the accord—although the president had enough support among Democrats to prevent a vote on such a resolution. Despite the conflict, however, both the deal's supporters and its critics agreed that the United States should prevent Iran from getting a bomb. This raises the question of how to do so—whether without the deal, after the deal expires, or if the Iranians decide to cheat. Stopping Iranian nuclear proliferation in all three situations will require Washington to update and adapt its Cold War policy of deterrence, making Tehran understand clearly in advance that the United States is determined to prevent, by force if necessary, Iranian nuclearization.

A Credible Threat

The English political philosopher Thomas Hobbes noted in *Leviathan* that "covenants, without the sword, are but words." Any agreement requires a mechanism for enforcing it, and the Iranian agreement does include such a mechanism: in theory, if Iran violates the agreement's terms, the economic sanctions that the accord removes will "snap back" into place. By itself, however, this provision is unlikely to prevent Iranian cheating. The procedures for reimposing the sanctions are complicated and unreliable; even if imposed, the renewed sanctions would not cancel contracts already signed; and even as the sanctions have been in place, Iran's progress toward a bomb has continued. To keep nuclear weapons out of Tehran's hands will thus require something stronger—namely, a credible threat by the United States to respond to significant cheating by using force to destroy Iran's nuclear infrastructure.

The term for an effort to prevent something by threatening forceful punishment in response is "deterrence." It is hardly a novel policy for Washington: deterring a Soviet attack on the

United States and its allies was central to the American conduct of the Cold War.

Deterring Iran's acquisition of nuclear weapons now and in the future will have some similarities to that earlier task, but one difference is obvious: Cold War deterrence was aimed at preventing the use of the adversary's arsenal, including nuclear weapons, while in the case of Iran, deterrence would be designed to prevent the acquisition of those weapons. With the arguable exception of Saddam Hussein's Iraq, the United States has not previously threatened war for this purpose and has in fact allowed a number of other countries to go nuclear, including the Soviet Union, China, Israel, India, Pakistan, and North Korea. Does the Iranian case differ from previous ones in ways that justify threatening force to keep Iran out of the nuclear club?

It does. An Iranian bomb would be more dangerous, and stopping it is more feasible. The Soviet Union and China were continent-sized countries that crossed the nuclear threshold before the U.S. military had the capacity for precision air strikes that could destroy nuclear infrastructure with minimal collateral damage. Israel and India, like the United Kingdom and France before them, were friendly democracies whose possession of nuclear armaments did not threaten American interests. Pakistan is occasionally friendly, is a putative democracy, and crossed the nuclear threshold in direct response to India's having done so. The United States is hardly comfortable with the Pakistani nuclear arsenal, but the greatest danger it poses is the possibility that after a domestic upheaval, it could fall into the hands of religious extremists—precisely the kind of people who control Iran now.

North Korea presents the closest parallel. In the early 1990s, the Clinton administration was ready to go to war to stop Pyongyang's nuclear weapons program, before signing an agreement that the administration said would guarantee that the communist regime would dismantle its nuclear program. North Korea continued its nuclear efforts, however, and eventually succeeded in testing a nuclear weapon during the presidency of George W. Bush. Since then, North Korea has continued to work on miniaturizing its bombs and improving its missiles, presumably with the ultimate aim of being able to threaten attacks on North America. It is worth noting that in 2006, two experienced national security officials wrote in *The Washington Post* that if Pyongyang were ever to achieve such a capability, Washington should launch a military strike to destroy it. One of the authors was William Perry, who served as secretary of defense in the Clinton administration; the other was Ashton Carter, who holds that position today.

Bad as the North Korean bomb is, an Iranian one would be even worse. For in the case of North Korea, a long-standing policy of deterrence was already in place before it acquired nuclear weapons, with the United States maintaining a strong peacetime military presence on the Korean Peninsula after the end of the Korean War in 1953. For this reason, in the years since Pyongyang got the bomb, its neighbors have not felt an urgent need to acquire nuclear armaments of their own—something that would be likely in the case of Iranian proliferation.

Nor would the Iranian case benefit from the conditions that helped stabilize the nuclear standoff between the United States and the Soviet Union. A Middle East with multiple nuclear-armed states, all having small and relatively insecure arsenals, would be dangerously unstable. In a crisis, each country would have a powerful incentive to launch a nuclear attack in order to avoid losing its nuclear arsenal to a first strike by one of its neighbors. Accordingly, the chances of a nuclear war in the region would skyrocket. Such a war would likely kill millions of people and could deal a devastating blow to the global economy by interrupting the flow of crucial supplies of oil from the region.

But if an Iranian bomb would be even worse than a North Korean bomb, preventing its emergence would be easier. A U.S. military strike against North Korea would probably trigger a devastating war on the Korean Peninsula, one in which the South would suffer greatly. (South Korea's capital, Seoul, is located within reach of North Korean artillery.) This is one of the reasons the South Korean government has strongly opposed any such strike, and the United States has felt compelled, so far, to honor South Korea's wishes. In the Middle East, by contrast, the countries that would most likely bear the brunt of Iranian retaliation for a U.S. counterproliferation strike—Saudi Arabia and Israel, in particular—have made it clear that, although they are hardly eager for war with Iran, they would not stand in the way of such a strike.

A Limited Aim

Deterring Iran's acquisition of nuclear weapons by promising to prevent it with military action, if necessary, is justified, feasible, and indeed crucial to protect vital U.S. interests. To be effective, a policy of deterrence will require clarity and credibility, with the Iranian regime knowing just what acts will trigger retaliation and having good reason to believe that Washington will follow through on its threats.

During the Cold War, the United States was successful in deterring a Soviet attack on its European allies but not in preventing a broader range of communist initiatives. In 1954, for example, the Eisenhower administration announced a policy of massive retaliation designed to deter communist provocations, including costly conventional wars like the recent one in Korea, by promising an overpowering response. But the doctrine lacked the credibility needed to be effective, and a decade later, the United States found itself embroiled in another, similar war in Vietnam.

In the case of Iran, the aim of deterrence would be specific and limited: preventing Iran's acquisition of nuclear weapons. Still, a policy of deterrence would have to cope with two difficulties. One is the likelihood of Iranian "salami tactics"—small violations of the JCPOA that gradually bring the Islamic Republic closer to a bomb without any single infraction seeming dangerous enough to trigger a severe response. The other is the potential difficulty of detecting such violations. The Soviet Union could hardly have concealed a cross-border attack on Western Europe, but Iran is all too likely to try to develop the technology needed for nuclear weapons clandestinely (the United States believes it has an extensive history of doing so), and the loopholes in the agreement's inspection provisions suggest that keeping track of all of Iran's bomb-related activities will be difficult.

As for credibility—that is, persuading the target that force really will be used in the event of a violation—this posed a major challenge to the United States during the Cold War. It was certainly credible that Washington would retaliate for a direct Soviet attack on North America, but the United States also sought to deter an attack on allies thousands of miles away, even though in that case, retaliation would have risked provoking a Soviet strike on the American homeland. Even some American allies, such as French President Charles de Gaulle, expressed skepticism that the United States would go to war to defend Europe. The American government therefore went to considerable lengths to ensure that North America and Western Europe were "coupled" in both Soviet and Western European eyes, repeatedly expressing its commitment to defend Europe and stationing both troops and nuclear weapons there to trigger the U.S. involvement in any European conflict.

In some ways, credibly threatening to carry out a strike against Iran now would be easier. Iran may have duplicated, dispersed, and hidden the various parts of its nuclear program, and Russia may sell Tehran advanced air defense systems, but the U.S. military has or can develop the tactics and munitions necessary to cause enough damage to lengthen the time Iran would need to build a bomb by years, even without the use of any ground troops. The Iranians might retaliate against Saudi Arabia or Israel (whether directly or through their Lebanese proxy, Hezbollah), or attack American military forces, or sponsor acts of anti-American terrorism. But such responses could do only limited damage and would risk further punishment.

The problems with deterring Iran's acquisition of nuclear weapons are not practical but rather political and psychological. Having watched American leaders tolerate steady progress toward an Iranian bomb over the years, and then observed the Obama administration's avid pursuit of a negotiated agreement on their nuclear program, Iran's ruling clerics may well doubt that Washington would actually follow through on a threat to punish Iranian cheating. U.S. President Barack Obama initially embraced the long-standing American position that Iran should not be permitted to have the capacity to enrich uranium on a large scale, then abandoned it. He backed away from his promise that the Syrian regime would suffer serious consequences if it used chemical weapons. He made it the core argument in favor of the JCPOA that the alternative to it is war, implying that American military action against Iran is a dreadful prospect that must be avoided at all costs. Moreover, neither he nor his predecessor responded to Iran's meddling in Iraq over the past decade, even though Tehran's support for Shiite militias there helped kill hundreds of U.S. troops. The mullahs in Tehran may well consider the United States, particularly during this presidency, to be a serial bluffer.

Doubt Not

All of this suggests that in order to keep Iran from going nuclear, the JCPOA needs to be supplemented by an explicit, credible threat of military action. To be credible, such a threat must be publicly articulated and resolutely communicated. The Obama administration should declare such a policy itself, as should future administrations, and Congress should enshrine such a policy in formal resolutions passed with robust bipartisan support. The administration should reinforce the credibility of its promise by increasing the deployment of U.S. naval and air forces in the Persian Gulf region and stepping up the scope and frequency of military exercises there in conjunction with its allies. As in Europe during the Cold War, the goal of U.S. policy should be to eliminate all doubts, on all sides, that the United States will uphold its commitments.

The debate about the Iran nuclear deal has become politically polarized, but a policy of deterrence should not be controversial, since all participants in the debate have endorsed the goal of preventing an Iranian bomb. In addition, a robust policy of deterrence would help address some of the shortcomings of the JCPOA without sacrificing or undermining its useful elements. And since the deterrence policy could and should be open ended, it would help ease worries about the provisions of the accord that expire after 10 or 15 years. As during the Cold War, the policy should end only when it becomes obsolete—that is, when Iran no longer poses a threat to the international community. Should the Islamic Republic evolve or fall, eliminating the need for vigilant concern about its capabilities and intentions, the United States could revisit the policy. Until then, deterrence is the policy to adopt.

Critical Thinking

1. Is the international verification scheme of the Iranian nuclear deal working?
2. Why did the U.S. negotiate a nuclear deal with Iran?
3. What is the effect of the nuclear deal on the balance of power in the Middle East?

Internet References

International Atomic Energy Agency
https://www.iaea.org/

The Iranian Nuclear Deal
http://apps.washington.post.com/g/documents/world/full-tex

Treaty on the Non-Proliferation of Nuclear Weapons
www.un.org/disarmament/wmd/nuclear/npt

MICHAEL MANDELBAUM is Christian A. Herter professor of American Foreign Policy at the Johns Hopkins School of Advanced International Studies and the author of the forthcoming book *Mission Failure: America and the World in the Post–Cold War Era.*

Article

Prepared by: Robert Weiner, *University of Massachusetts, Boston*

Containing North Korea

HARRY J. KAZIANIS

Learning Outcomes

After reading this article, you will be able to:

- Learn why a war between the United States and North Korea would be a catastrophe.
- Learn what the strategy of containment toward North Korea would consist of.

A specter is haunting Washington—the specter of nuclear war with North Korea. The idea that the Trump administration should endorse a military solution—and a full-blown war if necessary—to degrade or destroy North Korea's nuclear-weapons program is acquiring a new prominence. Advocates of war argue that the time to hit North Korea is now. They say that time is running out, and that Pyongyang will soon perfect its ability to attack America. Their contention is that America can knock out North Korea's nuclear program with some "shock and awe"—style bolt from the blue. Finally, they say that a war "over there" would be better than the death of innocent Americans "over here."

Such thinking is redolent of the Iraq War. Just as the war in Iraq evaded the prediction that it would be a "cakewalk," so a conflict over North Korea would likely issue in a calamity. There is no widespread public support, as a recent *Washington Post*–ABC News poll indicates, for a preemptive American strike on North Korea: 67 percent of Americans say Washington should act only if North Korea attacks it or our allies first. Before Washington experiences a fresh spasm of war fever on the Potomac, it's imperative to examine just why a conflict with North Korea is inimical to America's national interest. For the notion that America can "totally destroy" North Korea, as President Trump put it, with impunity is not quite persuasive. The result could even be a wider conflict, one that draws in great powers such as Russia and China that would seek to defend what they perceive as their own national interests.

Today, we live an age where America's greatest strategic advantage—two big oceans that protect us from the great geopolitical struggles in Europe and Asia, both past and present—is no longer the strategic safety blanket it once was, thanks to modern missile technology. Put simply, while the American military is the most destructive force ever devised in human history, such a force cannot guarantee that Washington will eliminate every single North Korean nuclear weapon. Nor can we ensure that if Pyongyang retaliates, potentially with whatever nuclear weapons we miss, that our missile defenses can keep us safe. Quite the contrary.

The truth is that a war with North Korea could be nothing like the First Gulf War, Yugoslavia, Kosovo, Afghanistan, the Second Gulf War or Libya. Such a conflict could be an epic struggle in which millions of people, on the Korean Peninsula, in Japan and even in the continental United States, could perish. The best path forward is to practice the foreign-policy doctrine that ended the Cold War peacefully: containment. Containment, rather than open conflict or outright appeasement, is not a panacea, but under the circumstances, it is the best of the options that Washington can pursue. As during the Cold War, a patient and vigilant strategy can wait out a hostile regime in the expectation that it will eventually crumble. A look at the possible outcomes of a nuclear conflict shows why it is more prudent to adopt this approach than to strike first.

When we consider the possibility of a military operation against North Korea, it is helpful to take a step back and consider from Pyongyang's perspective how it might respond. North Korean leader Kim Jong-un may not wish to counter such a strike by using the full range of his military options, fearing such an attack could unleash his worst fear: regime change.

This is where U.S. military planners begin to get nervous, as North Korea—even despite having an army that looks more like it was outfitted in the 1950s—has many ways to keep us guessing militarily. Kim could opt for less conventional means to instill fear and panic, attacking in an asymmetric manner that would be hard to counter.

For example, Kim could order an attack on South Korea's vast civilian nuclear infrastructure, unleashing deadly plumes of radioactive fallout. Seoul operates twenty-four nuclear power plants that could all come under various forms of North Korean attack, though they are relatively far from the North. With many of these facilities lumped together, Pyongyang could fire a salvo of missiles at these plants, creating an immediate humanitarian crisis.

The North Koreans could also use their special forces. They could infiltrate the South from existing tunnels to launch terror attacks against such facilities. If North Korea were to destroy just a few reactors, a disaster eclipsing Chernobyl could occur, killing tens of thousands and leaving millions of acres of South Korea an uninhabitable wasteland for generations.

The above example is just one of the most basic ways in which North Korea could respond asymmetrically. If we broaden our consideration of what a total war would look like, we find ourselves staring into the nuclear abyss.

Several years ago, I took part in a series of computer-based war games at a reputable think tank in Washington to examine what might happen if North Korea and America engaged in a nuclear conflict. Over the course of a few days, we simulated three scenarios to explore what a war with North Korea would look like, focusing on nuclear-weapons use, and conducting one full war per day in a fast-paced exercise.

The first of these exercises—a small nuclear war, if such a thing exists—imagined a conflict in coming decades, in which North Korea launches conventional weapons in a surprise attack on U.S. and allied forces on the Korean Peninsula, responding to reports that America is considering building up its military might in northeast Asia for a possible attack and regime-change operation.

Kim starts this Second Korean War with an artillery strike on Seoul while firing hundreds of small- and medium-range missiles on targets all over northeast Asia. U.S. and allied forces counterattack, focusing mostly on Kim's weapons of mass destruction, wiping out what they think is all of them.

Allied forces then move their way up the Korean Peninsula. But Kim, even with his forces battered and bloody, hid four nuclear weapons deep underground. He makes the ultimate decision: to launch a nuclear strike on Seoul and Tokyo. While the allies go on to win the war, Kim's nuclear attacks—with one warhead each making it past allied missile defenses, detonating in each city—result in over one million people dead and millions more wounded.

From here it gets even worse. As though the above wasn't bad enough, we pressed forward into a second scenario that begins with North Korea starting a conflict using nuclear weapons in a sort of atomic Pearl Harbor. In this exercise, we assume that U.S. forces are building up in the Asia-Pacific in preparation for a possible regime-change invasion of the North, responding to a demand from the international community for Pyongyang to give up its nuclear weapons. Sound familiar?

But Kim Jong-un is no fool. He knows that once U.S. forces are in place, they will try to take out his weapons of mass destruction, first from the air and then moving across the thirty-eighth parallel. Kim takes a gamble and decides that his only course of action is to strike first with his most powerful of weapons. Kim, in what thankfully was just a war game, launches nuclear attacks on Seoul, Pusan, Incheon, Tokyo, Sendai and Nagoya, with one nuclear weapon hitting each city. Over two million people perish in the atomic fire before Kim is defeated.

The last war game, however, was the most shocking of them all. We assumed a similar scenario, with allied forces preparing for a possible invasion, but this time Kim decides to launch a preemptive attack on the U.S. homeland—to take as many people to the grave with him as possible, a goal the North Koreans have declared in the past. In this last war game, North Korea attacks the cities in the second scenario with atomic weapons, but also launches successful nuclear strikes on Los Angeles, San Francisco, Seattle, and Portland. We were shocked to discover that the combined body count, across Asia and America, came to over three million people—before America's nuclear counterattack, which would add millions more. After North Korea retaliates with every weapon it has, launching more nuclear attacks along with chemical- and biological-weapons strikes, eight million people have lost their lives.

What is to be done? There are five potential pathways to mitigating the North Korea challenge. None involve a unilateral military strike; rather, they all embrace the idea that containing Pyongyang is our best and only option.

First, the Trump administration needs to be honest with itself and the problem it faces. There is no room for thinking we have more time to "stop" North Korea from getting nuclear weapons and long-range missiles, a line constantly repeated by experts across the political spectrum. The truth is that Pyongyang already has them.

Second, we need to embrace a financial-containment strategy—limiting the amount of illegal money going into Pyongyang's atomic and missile programs—as the only way to slow and potentially halt any further advances of North Korea's missile programs. We should insist that all nine UN Security Council resolutions are not only followed to the letter of international law, but strengthened every time North Korea tests another missile or nuclear weapon.

Third, U.S. and allied joint military capabilities in northeast Asia need to be significantly strengthened, considering the threat they face together. That means allied missile defenses need to be shored up in South Korea and Japan, as well as in the United States.

Fourth, whatever policies Washington pursues, it needs to take into consideration the interests of other great powers—especially China and Russia, two nations that have the power to block or negate any of America's strategies either at the United Nations or around the globe. The United States must be frank with Moscow and Beijing that its intent is not regime change, but the containment and management of a situation that also impacts their vital national interests. Washington should welcome any realistic diplomatic proposals they could put forward to mitigate the risks of a nuclear North Korea—especially from China, considering the potential influence it wields over North Korea and the prospect of Beijing advancing dialogue with Pyongyang. Unfortunately, the dual-freeze proposal first advanced by China would lock in North Korea's gains and require unreciprocated sacrifices from Washington and Seoul.

And lastly, while at the moment there is little possibility for direct U.S.-North Korea talks—Pyongyang currently holds three Americans captive and has little incentive to negotiate before it feels fully confident in its nuclear arsenal—some sort of direct negotiation, even if all that comes of it is the establishment of a communication channel, would be of great help. The United States and Soviet Union at the height of the Cold War enjoyed direct communication through their embassies. Washington and Pyongyang, in the event of a crisis, have no rapid means to communicate their unfiltered intentions.

One way to begin such talks could be a negotiation for the creation of interests sections in both nations, similar to what America and Cuba had during their decades-long estrangement. This would provide a way to communicate directly, but also begin to build some sort of trust between the two sides. While both sides may fail in their quest to extract concessions from one another, something of immense value would be achieved—a potential pathway out of a war that could be started by accident or through misperception. This is an outcome that the Trump administration should embrace, not dismiss.

Critical Thinking

1. What are the three scenarios for a war between North Korea and the United States?
2. How could a war between the United States and North Korea start?
3. What are the similarities between the containment of the Soviet Union during the Cold War and North Korea now?

Internet References

Arms Control Association
http://www.armscontrol.org

Special Envoy of the Secretary-General for Yemen
https://osesgy.unmissions.org/

38 North
38north.org

Harry J. Kazianis is a director of defense studies at the Center for the National Interest and an executive editor at the National Interest.

Article Prepared by: Robert Weiner, *University of Massachusetts, Boston*

Getting What We Need with North Korea

LEON V. SIGAL

Learning Outcomes

After reading this article, you will be able to:

- Understand what is involved in negotiating the denuclearization of North Korea.
- Discuss the relationship between China and North Korea.

While Washington's chattering classes were all atwitter about North Korean nuclear testing and rocket launching and China's backing for UN sanctions against Pyongyang in recent months, U.S. diplomats were tiptoeing to the negotiating table.

Any chance of a nuclear deal with North Korea depends on giving top priority to stopping the North's arming even if that means having Pyongyang keep the handful of weapons it has for the foreseeable future. Success will also require probing Kim Jong Un's seriousness about ending enmity, starting with a peace process on the Korean peninsula.

The revelation that Washington was willing to talk to Pyongyang without preconditions was a surprise to those who had not been tracking the evolution of U.S. policy closely. The Department of State confirmed that the United States held talks in New York last fall and rejected a proposal to begin negotiating a peace treaty. "To be clear, it was the North Koreans who proposed discussing a peace treaty," department spokesman John Kirby said on February 21. "We carefully considered their proposal, and made clear that denuclearization had to be part of any such discussion. The North rejected our response."[1]

Intriguingly, the revelation came on the eve of Chinese Foreign Minister Wang Yi's visit to Washington. Four days earlier, while signaling China's support for UN sanctions, Wang had made a more negotiable proposal of his own: "As chair country for the six-party talks, China proposes talks toward both achieving denuclearization and replacing the armistice agreement with a peace treaty." The proposal, Wang said, was intended to "find a way back to dialogue quickly."[2]

Wang's proposal was consistent with the September 19, 2005, six-party joint statement, which called for "the directly related parties" to "negotiate a permanent peace regime on the Korean Peninsula at an appropriate separate forum."[3] Those parties included the three countries with forces on the peninsula—North Korea, South Korea, and the United States—and China. They, along with Japan and Russia, agreed in six-party talks in September 2005 on the aim of "denuclearization of the Korean Peninsula" to be negotiated in parallel with a peace process in Korea and bilateral U.S.–North Korean and Japanese–North Korean talks on political and economic normalization.

Wang's initiative was also a way to bridge the gap between Washington and Pyongyang. North Korea has long sought a peace treaty. Its position hardened, however, after Washington, backed by Seoul and Tokyo, demanded preconditions—"pre-steps" in diplomatic parlance—to demonstrate its commitment to denuclearization before talks could begin. In response, Pyongyang began insisting that a peace treaty had to precede any denuclearization.

The Chinese proposal is a testament that sanctions are unlikely to curb North Korean nuclear and missile programs and that negotiation, however difficult, is the only realistic way forward. So is Washington's newfound openness to talks with Pyongyang.

Many in Washington and Seoul, however, still contend that negotiation is pointless if North Korea remains unwilling to give up the handful of crude nuclear weapons it has. That premise ignores the potential danger that an unbounded weapons program in North Korea poses to U.S. and allied security.

It also ignores the possibility that Pyongyang may be willing to suspend its nuclear and missile programs if its security concerns are satisfied. That was the gist of its January 9, 2015, offer of "temporarily suspending the nuclear test over which the U.S. is concerned" if the United States "temporarily suspends joint military exercises in South Korea and its vicinity this year."[4]

Like most opening bids, it was unacceptable. Instead of probing it further, however, Washington rejected it out of hand—within hours—and publicly denounced it as an "implicit threat."[5] That was a mistake Washington would not repeat in the fall.

Unofficial contacts later that January indicated that Pyongyang was prepared to suspend not just nuclear testing, but also missile and satellite launches and fissile material production. In return, the North was willing to accept a toning-down of the scale and scope of U.S.–South Korean exercises instead of the cancellation it had sought. This underscored the need for reciprocal steps to improve both sides' security.

Those contacts might have opened the way to talks at that time, but the initiative was squelched in Washington. Instead, U.S. officials continued to insist Pyongyang had to take unilateral steps to demonstrate its commitment to denuclearizing and ruled out reciprocity by Washington. Their stance was based on the flawed premise that the North alone had failed to live up to past agreements.[6] As Daniel Russel, assistant secretary of state for East Asian and Pacific affairs, put it on February 4, "North Korea does not have the right to bargain, to trade or ask for a pay-off in return for abiding by international law."[7] This crime-and-punishment approach, however warranted by North Korean flouting of international law, has never stopped North Korea from arming in the past, and it is unrealistic to think it would work now.

Tiptoeing Toward Talks

Last September 18, U.S. negotiator Sung Kim dropped Washington's preconditions for talks while still insisting that the agenda would be pre-steps North Korea would have to take to reassure Washington before formal negotiations could begin. "When we conveyed to Pyongyang that we are open to dialogue to discuss how we can resume credible and meaningful negotiations, of course we meant it. It was not an empty promise. We are willing to talk to them," Kim said. "And frankly for me, whether that discussion takes place in Pyongyang, or some other place, is not important. I think what's important is for us to be able to sit down with them and hear directly from them that they are committed to denuclearization and that if and when the six-party talks resume, they will work with us in meaningful and credible negotiations towards verifiable denuclearization."[8] In short, Washington would sit down with Pyongyang without preconditions in order to discuss U.S. preconditions for negotiations. That opened the door to contacts with the North Koreans in the New York channel in November.

In a November 3 interview, North Korean Foreign Ministry official Jong Tong Hak hinted at what the North might be proposing behind the scenes in New York. He said a permanent peace settlement on the Korean peninsula first required a North Korean–U.S. "peace agreement," perhaps a declaration committing the sides to negotiate peace. That was an advance. It was accompanied by a step backward from previous North Korean positions: "If the American government is serious about respecting the sovereignty of the DPRK [the Democratic People's Republic of Korea] and ending its ongoing hostile policy against the DPRK then it can be solved very easily between the two sides."[9] The apparent exclusion of South Korea made that proposal a nonstarter even if the North had been ready to suspend its nuclear and missile programs.

Sung Kim reiterated the U.S. position on November 10. "I think for us it's pretty straightforward: If [the North Koreans are] willing to talk about the nuclear issue and how we can move towards meaningful productive credible negotiations, [the United States would be] happy to meet with them anytime, anywhere," he said. He went on to respond to Jong obliquely: "It's not that we have no interest in seeking a permanent peace regime, peace mechanism or peace treaty. But I think they have the order wrong. Before we can get to a peace mechanism to replace the armistice, I think we need to make significant progress on the central issue of denuclearization."[10] The armistice agreement, signed in July 1953, established a cease-fire in the Korean War "until a final peaceful settlement is achieved."[11] That has yet to happen.

The Obama administration deserves praise for agreeing to meet in New York to explore what the North Koreans had in mind and not to reject a peace process out of hand. Disappointingly, North Korea proved unready to discuss denuclearization, which is stymieing talks for now.

The latest sanctions will squeeze Pyongyang but not enough to compel it to knuckle under and accept Washington's preconditions for negotiating.

The Limits of Sanctions

North Korea's January 6 nuclear test and February 7 satellite launch spurred more-stringent sanctions at the UN Security Council and in Washington. Even worse, the sanctions revived dreams of a North Korean collapse in Seoul, dreams that jeopardize a peace process.

Sanctions might have helped bring Iran around to negotiating, but North Korea is no Iran. It is far more autarkic and less dependent on trade with the rest of the world. It has no big-ticket items such as oil that require access to the global banking system to transact business.

An offer to ease sanctions may be of some utility in negotiations with Pyongyang, as it was with Tehran. The latest sanctions will squeeze Pyongyang but not enough to compel it to knuckle under and accept Washington's preconditions for negotiating. If anything, Pyongyang's nuclear advances have enhanced its leverage and given it greater confidence to proceed with negotiations on its own terms.

Wang and U.S. Secretary of State John Kerry acknowledged as much in their February 23 joint press conference announcing their agreement to move ahead on sanctions and negotiations. "China would like to emphasize that the Security Council resolution cannot provide a fundamental solution to the Korean nuclear issue. To really do that, we need to return to the track of dialogue and negotiation. And the secretary and I discussed this many times, and we agree on this," Wang said. Kerry echoed him, saying that the goal "is not to be in a series of cycling, repetitive punishments. That doesn't lead anywhere. The goal is to try to get Kim Jong Un and the DPRK to recognize that . . . it can rejoin the community of nations, it can actually ultimately have a peace agreement with the United States of America that resolves the unresolved issues of the Korean peninsula, if it will come to the table and negotiate the denuclearization."[12] Once again, news reports focused on China's willingness to endorse sanctions without paying attention to the U.S. commitment to negotiations.

Kirby, the State Department spokesman, improved that formulation on March 3: "We haven't ruled out the possibility that there could sort of be some sort of parallel process here. But—and this is not a small 'but'—there has to be denuclearization on the peninsula and work through the six-party process to get there."[13]

Many in Washington may question whether Beijing will enforce UN-mandated sanctions. By the same token, many in Beijing may wonder whether Washington will keep its commitment to negotiate.

Focus on the Urgent

North Korea's January 6 nuclear test, its fourth, was nothing to disparage. Even if it was neither a hydrogen bomb nor a boosted energy device, the test likely advanced Pyongyang's effort to develop a compact nuclear warhead that it can deliver by missile.

That is not all. The North has restarted its reactor at Yongbyon, which is working fitfully to generate more plutonium. It also is moving to complete a new reactor and has expanded its uranium-enrichment capacity. It has paraded two new longer-range missiles, the Musudan and KN-08, which it has yet to test-launch, and it is developing its first solid-fueled missile, the short-range Toksa.

That makes stopping the North's nuclear and missile programs a matter of urgency. Doing so should take priority over eliminating the handful of nuclear weapons Pyongyang already has, however desirable that may be. Such a negotiating approach is also more likely to bear fruit, given Kim Jong Un's goals.

What Is in it for Kim?

For nearly three decades, Pyongyang has sought to reconcile—end enmity—with Washington, Seoul, and Tokyo or, in the words of the 1994 Agreed Framework, "move toward full political and economic normalization."[14] To that end, it was prepared to suspend its weapons program or ramp it up if the other parties thwarted the reconciliation effort. Although North Korea's nuclear and missile brinkmanship is well understood, it is often forgotten that, from 1991 to 2003, North Korea reprocessed no fissile material and conducted very few test launches of medium-or long-range missiles. It suspended its weapons programs again from 2007 to early 2009.

U.S. negotiators need to probe whether an end to enmity remains Kim Jong Un's aim. He is not motivated by economic desperation, as many in Seoul and Washington believe. On the contrary, his economy has been growing over the past decade. Yet, he has publicly staked his rule on improving his people's standard of living, unlike his father and predecessor, Kim Jong Il. To deliver on his pledge, he needs to divert investment from military production to civilian goods.

That was the basis of his so-called *byungjin*, or "strategic line on carrying out economic construction and building nuclear armed forces simultaneously under the prevailing situation,"[15] meaning as long as U.S. "hostile policy" persists.

To curb military spending, Kim needs a calm international environment. Failing that, he will strengthen his deterrent, reducing the need for greater spending on conventional forces—a Korean version of U.S. President Dwight Eisenhower's bigger bang for the buck.

Pushback from the military on the budget may explain what prompted him to have his defense minister executed last spring.[16] It may also account for Kim Jong Un's exaggerated claims about testing an "H-bomb" in January. By crediting the party and the government, not the National Defense Commission, for the test, he was putting the military in its place.[17] The role that nuclear weapons play in putting a cap on defense spending was explicit in his March 9 claim of a "miniaturized" warhead deliverable by missile, which he called "a firm guarantee for making a breakthrough in the drive for economic construction and improving the people's standard of living on the basis of the powerful nuclear war deterrent."[18]

If Kim Jong Un still wants a fundamentally transformed relationship with his enemies or a calmer international climate in order to improve economic conditions in his country, a peace process is his way forward.

Testing whether North Korea means what it says about a peace process is also in the security interests of the United States and its allies.

Probing for Peace

Testing whether North Korea means what it says about a peace process is also in the security interests of the United States and its allies, especially now that North Korea has nuclear weapons.

North Korea's March 2010 sinking of a South Korean corvette, the *Cheonan*, in retaliation for the fatal November 2009 South Korean shooting up of a North Korean naval vessel in the contested waters of the West (Yellow) Sea showed that steps taken by each side to bolster deterrence can cause armed clashes. So did North Korea's November 2010 artillery barrage on Yeonpyeong Island in reprisal for South Korea's live-fire exercise. A peace process could reduce the risk of such clashes.

Negotiating a peace treaty is a formidable task. To be politically meaningful, it would require normalization of diplomatic, social, and economic relations and rectification of land and sea borders, whether those borders are temporary, pending unification, or permanent. To be militarily meaningful, it would require changes in force postures and war plans that pose excessive risks of unintended war on each side of the demilitarized zone (DMZ) separating the two Koreas. That would mean, above all, redeployment of the North's forward-deployed artillery and short-range missiles to the rear, putting Seoul out of range. Yet, to the extent Pyongyang would see that redeployment as weakening its deterrent against attack, it might be more determined to keep its nuclear arms.

A peace treaty is unlikely without a more amicable political environment. One way to nurture that environment is a peace process, using a series of interim peace agreements as stepping stones to a treaty. Such agreements, with South Korea and the United States as signatories, would constitute token acknowledgment of Pyongyang's sovereignty. In return, North Korea would have to take a reciprocal step by disabling and then dismantling its nuclear and missile production facilities.

A first step could be a "peace declaration." Signed by North Korea, South Korea, and the United States and perhaps China, Japan, and Russia, such a document would declare an end to enmity by reiterating the language of the October 12, 2000, U.S.–North Korean joint communiqué stating that "neither government would have hostile intent toward the other" and confirming "the commitment of both governments to make every effort in the future to build a new relationship free from past enmity." It could also commit the three parties to commence a peace process culminating in the signing of a peace treaty. The declaration could be issued at a meeting of the six foreign ministers.

A second step long sought by Pyongyang is the establishment of a "peace mechanism" to replace the Military Armistice Commission set up to monitor the ceasefire at the end of the Korean War. This peace mechanism could serve as a venue for resolving disputes such as the 1994 North Korea downing of a U.S. reconnaissance helicopter that strayed across the DMZ or the 1996 incursion by a North Korean spy submarine that ran aground in South Korean waters while dropping off agents. The peace mechanism would include the United States and the two Koreas.

The peace mechanism also could serve as the venue for negotiating a series of agreements on specific confidence-building measures, whether between the North and South, between the North and the United States, or among all three parties. A joint fishing area in the West Sea, as agreed in principle in the October 2007 North–South summit meeting, is one. Naval confidence-building measures, such as "rules of the road" and a navy-to-navy hotline, are also worth pursuing.

Lacking satellite reconnaissance, North Korea has conducted surveillance by infiltrating agents into the South. An "open skies" agreement allowing reconnaissance flights across the DMZ by both sides might reduce that risk. In October 2000, Kim Jong Il offered to end exports, production, and deployment of medium- and longer-range missiles. In return, he wanted the United States to launch North Korean satellites, along with other compensation. A more far-reaching arrangement might be to set up a joint North-South watch center that could download real-time data from U.S. or Japanese reconnaissance satellites. It is unclear how much such confidence-building measures will reduce the risk of inadvertent war, but they would provide political reassurance of an end to enmity.

A Starting Point

Before the sides can get to a peace process, they need to take steps to rebuild some trust. For Washington, that means verifiable suspension of all of North Korea's nuclear tests, missile and satellite launches, and fissile material production.

For Pyongyang, that means an easing of what it calls U.S. "hostile policy," starting with a toning-down of joint military exercises, partial relaxation of sanctions, and some commitment to initiating a peace process. Such reciprocal steps could lead to resumption of parallel negotiations among the six parties as envisioned in their September 2005 joint statement.

If the two sides can avoid deadly clashes triggered by the current joint military exercises, they may get back to exploring the only realistic off-ramp from the current impasse: reciprocal steps to open the way to negotiations that would address denuclearization and a peace process in Korea. That, however, would require a change of heart in Pyongyang, Seoul, and Washington. As the Rolling Stones put it, "You can't always get what you want/But if you try sometimes, well you just might find/You get what you need."

Endnotes

1. "U.S. Rejected Peace Talks before Last Nuclear Test," Reuters, February 21, 2016.
2. Lee Je-hun, "Could Wang's Two-Track Proposal Lead to a Breakthrough?" *Hankyoreh*, February 19, 2016.
3. U.S. Department of State, "Six-Party Talks, Beijing, China," n.d., http://www.state.gov/p/eap/regional/c15455.htm (text of the joint statement of the fourth round of six-party talks on September 19, 2005).
4. Korean Central News Agency (KCNA), KCNA Report, January 10, 2015, www.kcna.co.jp/item2015/201501/news10/20150110-12ee.htm.
5. Marie Harf, transcript of U.S. Department of State daily briefing, January 12, 2015, http://www.state.gov/r/pa/prs/dpb/2015/01/235866.htm.
6. On the history of reneging by various parties, see Leon V. Sigal, "How to Bring North Korea Back Into the NPT," in *Nuclear Proliferation and International Order*, ed. Olaf Njolstad (London: Routledge, 2010), pp. 65–82.
7. "US: No Sign Yet NKorea Serious on Nuke Talks," Associated Press, February 4, 2015.
8. Chang Jae-soon and Roh Hyo-dong, "U.S. Nuclear Envoy Willing to Hold Talks with N. Korea in Pyongyang," Yonhap, September 19, 2015.
9. "N. Korea Accuses U.S. of 'Nuclear Blackmail,'" Associated Press, November 4, 2015.
10. Chang Jae-soon, "Amb. Sung Kim: U.S. 'Happy to Meet' With N. Korean 'Anytime, Anywhere,'" Yonhap, November 11, 2015.
11. Armistice Agreement for the Restoration of the South Korean State, North Korea–U.S., July 27, 1953, 4 U.S.T. 234.
12. U.S. Department of State, remarks of Secretary of State John Kerry and Chinese Foreign Minister Wang Yi, Washington, DC, February 23, 2016, http://www.state.gov/secretary/remarks/2016/02/253164.htm.
13. John Kirby, transcript of U.S. Department of State daily briefing, March 3, 2016, http://www.state.gov/r/pa/prs/dpb/2016/03/253948.htm.
14. Bureau of Arms Control, U.S. Department of State, "Agreed Framework between the United States and the Democratic People's Republic of Korea," October 21, 1994, http://2001-2009.state.gov/t/ac/rls/or/2004/31009.htm.
15. KCNA, "Report on Plenary Meeting of WPK Central Committee," March 31, 2013, www.kcna.co.jp/item/201303/news31/20130331-24ee.htm.
16. "N. Korean Ex-Army Chief 'Locked Horns with Technocrats,'" *Chosun Ilbo*, May 15, 2015, http://english.chosun.com/site/data/html_dir/2015/05/15/2015051500971.html.
17. KCNA, "DPRK Proves Successful in H-Bomb Test," January 6, 2016, www.kcna.co.jp/2016/201601/news06/20160106-12ee.htm; KCNA, "WPK Central Committee Issues Order to Conduct First H-Bomb Test," January 6, 2016, www.kcna.co.jp/item/2016/201601/news06/20160106-11ee.htm.
18. KCNA, "Kim Jong-un Guides Work for Mounting Nuclear Warheads on Ballistic Rockets," March 9, 2016, http://www.kcna.kp/kcna.user.special.getArticlePage.kcmsf;jsessionid=823D154834DB0E5032E668C39EDE74B3.

Critical Thinking

1. Why does North Korea want the bomb?
2. How does North Korea's nuclear weapons capability affect the regional balance of power in East Asia?
3. What can the United States do to stop North Korea's nuclear weapons program?

Internet References

38 North
38north.org/

Arms Control Association
https://www.armscontrol.org.

International Atomic Energy Agency
https://www.iaea.org

U.S. Korea Institute at SAIS
uskoreainstitute.org/

LEON V. SIGAL is director of the Northeast Asia Cooperative Security Project at the Social Science Research Council in New York and author of *Disarming Strangers: Nuclear Diplomacy with North Korea* (1998). A portion of this article draws from a piece that appeared in the Nautilus Institute for Security and Sustainability's NAPSNet Policy Forum.

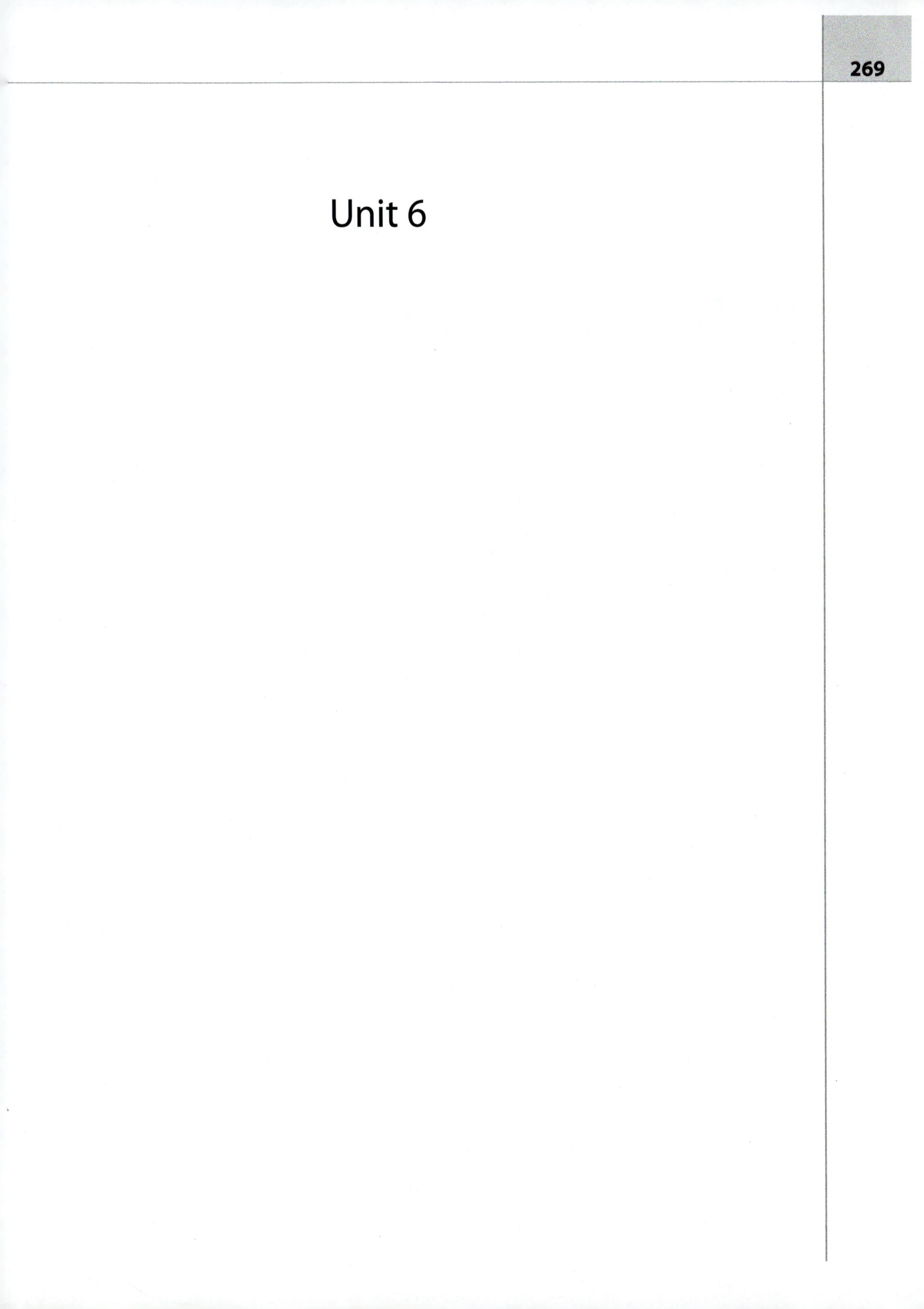

Unit 6

UNIT

Prepared by: Robert Weiner, *University of Massachusetts, Boston*

Ethics and Values

The international liberal order is also based on norms the foundation of which is the rule of law, both on the national and international level. Historically, a rule of law state can be traced back to the first Magna Carta or Great Charter that was signed in 1215. The Magna Carta not only established the rule of law in England, but also established a model for the rule of law around the world. A definition of a rule of law state includes a liberal democracy, which is characterized by free and fair elections, a competitive party system, areal rather than a nominal constitution, a free press, and an impartial judicial system, which is appointed and allowed to function by the government without intimidation and political interference. Liberal democracies also form an essential component of an international liberal order. However, as several articles in this unit indicate, the spread of liberal democracy around the world has regressed somewhat. A rule of law state is necessary to eliminate racism in such liberal democracies as the United States, where unarmed black males fall victim to a discriminatory system, in a society which is becoming more diverse. Rule of law on the international level means that gross and mass violations of human rights should be prevented and punished as long as states have the political will to do so. On the international level, although the international community hoped that genocide would never again take place, it has occurred again and again since the Second World War. For example, in Rwanda in 1994, genocide resulted in the deaths of at least 800,000 to 1 million Tutsis and moderate Hutus. In Srebrenica, Bosnia, in 1995, 7,000–8,000 Bosnian men and boys were massacred. Based on the genocide which occurred in Rwanda and Bosnia, a considerable amount of case law has been produced to provide a solid legal foundation. Most recently, a fact-finding commission of the United Nations has concluded that the military leadership of Myanmar had the intent of committing genocide against the Rohingya, a Muslim minority living in Myanmar. Genocide is a term which was invented by Raphael Lemkin, who almost single-handedly persuaded the international community in 1948 to adopt the Convention on the Prevention and Punishment of the Crime of Genocide. The term genocide is based on the Greek word "gens" which means people, and the Latin word "cide," which means killing. Until 1948, genocide was known as the "crime with no name." Genocide was finally recognized as a crime under international law after the Holocaust of World War II.

The Genocide Convention lists various acts of genocide which are designed to destroy in part or in whole a group of people based on race, ethnicity, religion, or nationality. The Convention was criticized for only focusing on these four groups and not including political, economic, or social groups. Moreover, the Convention did not contain a definition of group, and left it up to subsequent international criminal courts to come up with a definition of group. The national courts have wrestled for years with figuring out whether membership in a group can be defined by objective or subjective factors, or both. Furthermore, the Convention did not explain what was meant by "in part." The courts have tried to determine whether "in part" refers to a substantial or significant part of a group, sometimes arriving at contradictory opinions. Finally, an international judicial body as mentioned in the Convention did not exist at the time the Convention was adopted, and took 50 years to appear as the International Criminal Court in 1998. Although human rights advocates have been disappointed by the narrow legal basis of the Genocide Convention, the extension of the definition of crimes against humanity has filled in the gap that has been left by the Convention.

An objective judicial system is critical in rooting out corruption in states like China. There have been a series of show trials involving select high-level elites in China, in an effort to maintain the legitimacy of the ruling party. Kleptocracy is related to corruption, and not only exists in China, but is a widespread phenomenon in a number of states, ranging from Ukraine to Nigeria. This unit contains an interesting discussion of corruption in Latin America, especially in Brazil.

As some of the readings in this unit also point out, the globalization of democracy has suffered a regression recently. Moreover, for example, freedom of the press is also an important part of a liberal democracy. The World Press Freedom Index for 2017, published by Reporters Without Borders, indicates that press freedom around the world continues to be under attack. Attacks against the media are not only occurring in authoritarian states such as Russia and China, but also in democratic states like the United States. In the United States, the media has been attacked by the Trump administration, as the "enemy of the people" and for disseminating "fake news." Attacks on freedom of the press undermine one of the basic pillars of liberal democracy. The United States in 2017 was ranked as 45th on the World Press Freedom Index, just behind Romania. Norway stands at the top of the list as the most free, while North Korea is at the bottom of the list as number 180. Freedom House, a US Non-Governmental Organization, has also ranked states around the world for "Freedom on the Net." Governments have increasingly engaged in repressive policies to stifle dissent and censor criticism on the internet. Freedom House issued a report for 2017, which covers 65 countries and 87 percent of the users of the Internet. Governments have become more technically sophisticated in controlling virtual private networks which Internet users employ to get around government censorship. Governments also block live video streaming, have increased the use of BOTS (automated robotic systems), shut down mobile devices, and created more fake news sites to spread disinformation and manipulate news on the social media. As the Russian hacking into the US Presidential elections indicates, the Internet and social media have become weaponized by governments which has the effect of undermining the international liberal order.

Article

Prepared by: Robert Weiner, *University of Massachusetts, Boston*

Xi's Corruption Crackdown: How Bribery and Graft Threaten the Chinese Dream

James Leung

Learning Outcomes

After reading this article, you will be able to:

- Explain the scope of corruption in China.
- Understand the relationship between the Party and corruption.

In a series of speeches he delivered shortly after taking office in 2012, Chinese President Xi Jinping cast corruption as not merely a significant problem for his country but an existential threat. Endemic corruption, he warned, could lead to "the collapse of the [Chinese Communist] Party and the downfall of the state." For the past two years, Xi has carried out a sweeping, highly publicized anticorruption campaign. In terms of sheer volume, the results have been impressive: according to official statistics, the party has punished some 270,000 of its cadres for corrupt activities, reaching into almost every part of the government and every level of China's vast bureaucracy. The most serious offenders have been prosecuted and imprisoned; some have even been sentenced to death.

The majority of the people caught up in Xi's crackdown have been low- or midlevel party members and functionaries. But corruption investigations have also led to the removal of a number of senior party officials, including some members of the Politburo, the group of 25 officials who run the party, and, in an unprecedented move, to the expulsion from the party and arrest of a former member of the Politburo's elite Standing Committee.

Xi's campaign has proved enormously popular, adding a populist edge to Xi's image and contributing to a nascent cult of personality the Chinese leader has begun to build around himself. And it has the quiet support of the aristocratic stratum of "princelings," the children and grandchildren of revolutionary leaders from the Mao era. They identify their interests with those of the country and consider Xi to be one of their own. But there has been pushback from other elites within the system, some of whom believe the campaign is little more than a politically motivated purge designed to help Xi solidify his own grip on power. Media organizations in Hong Kong have reported that Xi's two immediate predecessors, Jiang Zemin and Hu Jintao, have asked him to dial back the campaign. And some observers have questioned the campaign's efficacy: in 2014, despite Xi's efforts, China scored worse on Transparency International's Corruption Perceptions Index than it had in 2013. Even Xi himself has expressed frustration, lamenting a "stalemate" in his fight to clean up the system while pledging, in grandiose terms, not to give up: "In my struggle against corruption, I don't care about life or death, or ruining my reputation," he reportedly declared at a closed-door Politburo meeting last year.

There is no doubt that Xi's campaign is in part politically motivated. Xi's inner circle has remained immune, the investigations are far from transparent, and Xi has tightly controlled the process, especially at senior levels. Chinese authorities have placed restrictions on foreign media outlets that have dared to launch their own investigations into corruption, and the government has detained critics who have called for more aggressive enforcement efforts.

But that doesn't mean the campaign will fail. The anticorruption fight is only one part of Xi's larger push to consolidate his authority by establishing himself as "the paramount leader within a tightly centralized political system," as the China expert Elizabeth Economy has written. So far, Xi seems capable of pulling off that feat. Although this power grab poses other risks, it puts Xi in a good position to reduce corruption significantly—if not necessarily in a wholly consistent, apolitical manner.

This might seem paradoxical: after all, too much central power has been a major factor in creating the corruption epidemic. That is why, in the long term, the fate of Xi's anticorruption fight will depend on how well Xi manages to integrate it into a broader economic, legal, and political reform program. His vision of reform, however, is not one that will free the courts, media, or civil society, or allow an opposition party that could check the ruling party's power. Indeed, Xi believes that Western-style democracy is at least as prone to corruption as one-party rule. Rather, Xi's vision of institutional reform involves maintaining a powerful investigative force that is loyal to an honest, centralized leadership. He seems to believe that, over the course of several years, consistent surveillance and regular investigations will change the psychology of bureaucrats, from viewing corruption as routine, as many now do, to viewing it as risky—and, finally, to not even daring to consider it.

Stamping out graft, bribery, and influence peddling could very well help China's leaders maintain the political stability they fear might slip away as economic growth slows and geopolitical tensions flare in Asia. But if Xi's fight against corruption becomes disconnected from systemic reforms, or devolves into a mere purge of political rivals, it could backfire, inflaming the grievances that stand in the way of the "harmonious society" the party seeks to create.

I'll Scratch Your Back . . .

One school of thought holds that corruption is a deeply rooted cultural phenomenon in China. Some political scientists and sociologists argue that when it comes to governance and business, the traditional Chinese reliance on guanxi—usually translated as "connections" or "relationships"—is the most important factor in explaining the persistence and scope of the problem. The comfort level that many Chinese citizens have with the guanxi system might help explain why it took so long for public outrage to build up to the point where the leadership was forced to respond. But all cultures and societies produce a form of guanxi, and China's version is not distinct enough to explain the depth and severity of the corruption that inflicts the Chinese system today. The main culprits are more obvious and banal: one-party rule and state control of the economy. The lack of firm checks and balances in a one-party state fuels the spread of graft and bribery; today, no Chinese institution is free of them. And state control of resources, land, and businesses creates plenty of opportunities for corruption. In the past three decades, the Chinese economy has become increasingly mixed. According to Chinese government statistics, the private sector now accounts for around two-thirds of China's GDP and employs more than 70 percent of the labor force. And the Chinese economy is no longer isolated; it has been integrated into the global market. Nevertheless, the private sector is still highly dependent on the government, which not only possesses tremendous resources but also uses its regulatory and executive power to influence and even control private businesses.

When it comes to government purchasing and contracting and the sale of Chinese state assets (including land), bidding and auctioning processes are extremely opaque. Officials, bureaucrats, and party cadres exploit that lack of transparency to personally enrich themselves and to create opportunities for their more senior colleagues to profit in exchange for promotions. Midlevel officials who oversee economic resources offer their superiors access to cheap land, loans with favorable terms from state-owned banks, government subsides, tax breaks, and government contracts; in return, they ask to rise up the ranks. Such arrangements allow corruption to distort not just markets but also the workings of the party and the state.

Similar problems also exist in government organizations that do not directly control economic resources, such as China's military. To win promotion, junior military officers routinely bribe higher-ranking ones with gifts of cash or luxury goods. Last year, the authorities arrested Xu Caihou, a retired general who had served as a member of the Politburo and had been the vice chair of the Central Military Commission. In his house, they discovered enormous quantities of gold, cash, jewels, and valuable paintings—gifts, the party alleged, from junior officers who sought to advance up the chain of command. After the party expelled him, Xu confessed, according to Chinese state media; a few months later, he died, reportedly of cancer.

Direct state ownership, however, is hardly a prerequisite for self-dealing. The immense regulatory power that Chinese authorities hold over the private sector also helps them line their own pockets. In highly regulated industries, such as finance, telecommunications, and pharmaceuticals, relatives of senior government officials often act as "consultants" to private businesspeople seeking to obtain the licenses and approvals they need to operate. Zheng Xiaoyu, the former head of the State Food and Drug Administration, accepted around $850,000 in bribes from pharmaceutical companies seeking approval for new products. In 2007, after more than 100 people in Panama died after taking contaminated cough syrup that Zheng had approved, he was tried on corruption charges; he was found guilty and executed a few months later.

Corruption has also infected law enforcement and the legal system. Organized criminal groups pay police officers to protect their drug and prostitution rings. Criminal suspects and their relatives often bribe police officers to win release from jail or to avoid prosecution. If that fails, they can try their luck with prosecutors and judges. And of course, since China's judiciary is not independent, there are always party and government officials who might be able and willing to intervene in a case—for the right price. Authorities allege that Zhou Yongkang, a former member of the party's Standing Committee who oversaw legal and internal security affairs, personally intervened in many court cases after accepting bribes. Zhou was arrested, charged, and expelled from the party last year and is currently awaiting trial—the first time in decades that the state has pursued a criminal case against a former member of the Standing Committee.

As China's domestic markets have grown, multinational companies and banks have learned that getting access means knowing whose palms to grease. Many firms have taken to hiring the children of senior government officials, sometimes even paying their tuition at Western universities. Others have opted for a more direct route, paying hefty "consulting" fees to middlemen in order to participate in stock offerings or to win preferential treatment in bidding for government contracts. This environment has discouraged some multinational companies from investing and conducting business in China, especially those constrained by U.S. anticorruption laws.

Meanwhile, officials have taken advantage of loose financial controls and a lack of transparency to safeguard their illicit profits. Many officials hold a number of Chinese passports, often under different names but with valid visas, and use them to travel abroad and stash their money in foreign bank accounts.

But corruption is hardly limited to official circles and big business; every aspect of society feels its effects. Consider education. To give their child a shot at getting into one of the relatively small number of high-quality Chinese primary and secondary schools and universities, parents often have to bribe admissions officers or headmasters. Similarly, the scarcity of good hospitals and well-trained medical personnel has led to the practice of supplying doctors or medical administrators with a hongbao—a "red packet" of cash—to secure decent treatment.

Keep It Clean

Faced with this far-reaching problem, Xi has promised more than a mere Band-Aid, envisioning a long-term process of systemic reform. The first phase has been the heavily stage-managed crackdown of the past two years. So far, the campaign has contained an element of populism: it has targeted only officials, bureaucrats, and major business figures whom the party suspects of corrupt dealings; no ordinary Chinese people have felt the sting.

The campaign seeks not only to punish corruption but to prevent it as well. In late 2012, the party published a set of guidelines known as the "eight rules and six prohibitions," banning bureaucrats from taking gifts and bribes; attending expensive restaurants, hotels, or private clubs; playing golf; using government funds for personal travel; using government vehicles for private purposes; and so on.

The government has also required all officials and their immediate family members to disclose their assets and income, to make it harder to hide ill-gotten gains. At the same time, the party has sought to reduce incentives for graft by narrowing the income gaps within the system. In the last year, it raised the salaries and retirement benefits of military officers, law enforcement personnel, and other direct government employees, while sharply cutting the higher salaries enjoyed by top managers of state owned enterprises.

Still, to date, Xi's campaign has been chiefly an enforcement effort. Investigations are led by the party's Central Commission for Discipline Inspection (CCDI), which sends inspection teams to examine every ministry and agency and every large state-owned enterprise. The teams enjoy the unlimited power to investigate, detain, and interrogate almost anyone, but mainly government officials, the vast majority of whom are party members. Once the teams believe they have gathered sufficient evidence of wrongdoing, the CCDI expels suspects from the party and then hands them over to the legal system for prosecution.

Xi has declared that no corrupt official will be spared, no matter how high his position. In practice, however, the CCDI has chosen its targets very carefully, especially at senior levels. The decision to go after Zhou was heralded as setting a new precedent—since the late 1980s, the party has followed an unspoken rule against purging a member or former member of the Standing Committee. And yet Zhou's removal and prosecution remain unique; they appear to have been less a signal of things to come than a shot across the bow, intended to scare off any potential opposition to Xi within the leadership. Zhou was vulnerable because he was retired and no longer had direct control or power. Also, Zhou had backed a group of senior party officials who had challenged Xi's power and authority early in his tenure; among them was Bo Xilai, the influential party chief of Chongqing, who in 2013 was brought down by a scandal involving corruption and a murder plot in which his wife participated. Finally, Zhou and his immediate family members were particularly flagrant in their corrupt pursuits, which made him an easy target. Some media reports have indicated that authorities are investigating the family members of other retired Standing Committee members. But so far, no ranking member of the "red aristocracy" has yet been targeted, and all the highest-level targets, including Zhou and

Xu, have been part of a single loose political network. Apparently, there are still lines Xi is not willing to cross.

It is also worth noting that although Xi has allowed investigations of the country's key military institutions, he has yet to make any major personnel changes within the Commission for Discipline Inspection of the Central Military Commission, the armed forces' equivalent of the CCDI. Xi still needs more time to consolidate his control over the military and its institutions.

A number of other elements of Xi's campaign are also problematic, because they present opportunities for abuse and run contrary to the spirit of the legal reforms that Xi is pursuing. Xi claims that he wants to improve due process and reduce abusive police and judicial practices. But the CCDI itself does not always follow standard legal procedures. For example, Chinese law allows police to detain a suspect for only seven days without formally charging him, unless the police obtain express permission from legal authorities to extend the detention. The CCDI, on the other hand, has kept suspects in custody for far longer periods without seeking any approval and without issuing any formal charges, giving the appearance of a separate standard.

Meanwhile, with its newfound authority, the CCDI is gradually becoming the most powerful institution within the party system. Unless the party balances and limits the agency's power and influence, the CCDI could grow unaccountable and become a source of the very kinds of conduct it is supposed to combat.

Perhaps the biggest potential obstacle to the success of the campaign is strong resistance to it within the bureaucratic system. Xi has launched a direct attack on the interests of many entrenched bureaucrats and officials; even those who have escaped prosecution have watched their prosperity and privilege shrink. Many officials might also resent the idea that there is something fundamentally wrong with the way they are accustomed to conducting themselves. They may feel that they deserve the benefits they get through graft; without their work, after all, nothing would get done—the system wouldn't function.

Early in Xi's tenure, some officials seemed to believe that although the days of flagrant self-dealing were over, it would still be possible to exploit their positions for profit; they would just need to be a bit more subtle about it. In 2013, *The New York Times,* citing Chinese state media, reported that a new slogan had become popular among government officials: "Eat quietly, take gently, and play secretly." But that sense of confidence has evaporated as it has become clear that Xi is serious about cracking down. During the past two years, party members and state bureaucrats have become extremely cautious about running afoul of the new ethos, although many are quietly seething about the situation. This has interfered with the traditional wheel-greasing function of corruption and contributed to China's economic slowdown. If corruption no longer assists entrepreneurs in slipping past bureaucratic barriers, it will put additional pressure on Xi to institute economic reforms that genuinely reduce those obstacles.

The Politics of Anticorruption

Since the anticorruption campaign is just one of a number of major changes taking place in the Xi era, it's difficult to forecast what path it might take. In a pessimistic scenario, the campaign would end in failure after strong resistance within the top party leadership and the bureaucratic system forces Xi to back down. That outcome would be a catastrophe. Corruption would likely rise to pre-2012 levels (at the very least), destabilizing the economy, reducing investor confidence, and seriously eroding Xi's authority, making it difficult for him to lead.

In a more optimistic scenario, Xi would manage to overcome internal resistance and move on to broader economic, legal, and political reforms. Ideally, the campaign will strengthen Xi's power base enough and win him the support necessary to reduce the party's tight grip on policy and regulatory and administrative power, creating a favorable environment for the growth of a more independent private sector. Xi has no interest in creating a Western-style democratic system, but he does think that China could produce a cleaner and more effective form of authoritarianism. To better serve that goal, Xi should consider adding a number of more ambitious elements to the anti-corruption crusade, including a step that both Transparency International and the G-20 have called for: improving public registers to clarify who owns and controls which companies and land, which would make it harder for corrupt officials and businesspeople to hide their illicit profits.

At the moment, there is more reason for optimism than pessimism. Xi has already consolidated a great deal of control over the state's power structures and is determined and able to remove anyone who might resist or challenge his authority or policies. So far, within the senior leadership and the wider bureaucratic system, resistance to the anticorruption campaign has been passive rather than active: some bureaucrats have reportedly slowed down their work in a rather limited form of silent protest. Meanwhile, the anticorruption campaign continues to enjoy strong public support, especially from low- and middle-income Chinese who resent the way that corruption makes the Chinese system even more unfair than it already is. Anti-corruption thus represents a way for the party to ease the social tensions and polarization that might otherwise emerge as the economy slows, even as dramatic economic inequalities persist. To maintain this public support, the trick for Xi will be calibrating the scope and intensity of the campaign: not so narrow or moderate as to seem halfhearted, but not so broad or severe as to seem like a form of abuse itself.

Critical Thinking

1. What lines is president Xi willing to cross in the battle against corruption?
2. Why is corruption so endemic in China?
3. What path will the anti-corruption campaign take?

Internet References

Embassy of the People's Republic of China in the United States
china-embassy.org

Freedom House
https://freedom house.org

Kleptocracy Initiative
kleptocracyinitiative.org

James Leung is a pseudonym for an economist with extensive experience in China, Europe, and the United States.

Article Prepared by: Robert Weiner, *University of Massachusetts, Boston*

Latin Americans Stand Up to Corruption

The Silver Lining in a Spate of Scandals

Jorge G. Castañeda

Learning Outcomes

After reading this article, you will be able to:

- Define the problem of kleptocracy in Latin America.
- Define about the problem of kleptocracy in Brazil.

Just a few years ago, Latin America was on a roll. Its economies, riding on the back of the Chinese juggernaut, were flourishing. A boom in commodity prices and huge volumes of foreign direct investment in agriculture and natural resources generated a golden decade. Ambitious government programs began to reduce inequality. And relations with the United States, long a source of friction, were improving—even as they became less important to the region's success.

Today, the picture looks very different. Latin America's economies are grinding to a halt: in 2015, average GDP growth slipped below 1 percent. Inequality is still declining, but more slowly. And according to the annual Latinobarómetro poll, satisfaction with democracy in Latin America is lower than it is in any other region and is at its lowest point in almost a decade, at 37 percent. In Brazil and Mexico, it has descended to just 21 percent and 19 percent, respectively.

Yet on one count at least, the lands south of the Rio Grande are faring better than ever: Latin Americans are denouncing corruption as never before. In decades past, residents of the region seemed resigned to the problem, treating it as an ordinary, if lamentable, part of everyday life. In 1973, for example, Argentines elected Juan Perón to a third term as president despite his infamous criminality; as a popular saying put it, "Mujeriego y ladrón, pero queremos a Perón" (Philanderer and thief, we still want Perón).

Such tolerance is now a thing of the past. According to the same Latinobarómetro poll, the region's inhabitants identify corruption as their third most important problem, behind crime and unemployment but above inflation and poverty. Latin Americans have also started judging their politicians based on their perceived trustworthiness. Of the five most unpopular chief executives in Latin America today—Brazil's Dilma Rousseff, Mexico's Enrique Peña Nieto, Paraguay's Horacio Cartes, Peru's Ollanta Humala, and Venezuela's Nicolás Maduro—three come from the countries with the worst ratings for government transparency (Brazil, Mexico, and Peru).

Several factors explain this change in attitude. First, the economic growth of the last 15 years has created a large middle class (now estimated at almost a third of the region's population, according to the World Bank, up from around 20 percent a decade ago, although higher in Brazil, Chile, Mexico, and Uruguay) with high expectations. Second, the region has grown more democratic. As the recent economic downturn has highlighted the damage corruption causes, this newly enlarged middle class has used its new freedoms to vent its frustration with those in charge.

Of all the region's recent uprisings against corruption, the most dramatic have been in Brazil.

Foreigners have also played a role. As Latin America has become more integrated into the world economy, international media and civil society organizations have begun to direct intense opprobrium at corrupt leaders and to lavish praise on reformers. Outside forces have also helped the region's more independent judiciaries and media outlets expose official malfeasance.

Together, all these forces have created a combustible mix, and when cases of graft have come to light in recent years, they have sparked major scandals in one country after another. High-level Latin American officials and business leaders have found themselves denounced by the media. Prosecutors and courts have issued indictments, and protesters have taken to the streets. Although few of the governments implicated in the scandals have actually fallen—and few others are likely to—the sheer scale of the social and political protest has been astonishing and represents an important positive trend in a part of the world with an otherwise gloomy forecast.

Brazilian Bribes

Of all the region's recent uprisings against corruption, the most dramatic has unfolded in Brazil. The problems began in late 2013, a time when popular discontent with the government was already running high. The previous president, Luis Inácio Lula da Silva (known as Lula), had been tarnished by a corruption scandal years earlier. Now the economy was stagnating, and protests had begun to erupt over Brazil's lavish spending on the coming World Cup. Then, in late 2014, shortly after Rousseff narrowly won reelection, the so-called *petrolâo* scandal hit.

The scale of the revelations was unprecedented. In November 2014, federal police arrested 18 people, including senior executives of Petrobras, Brazil's state oil company, for corruption in the first raid of the investigation. Numerous firms had paid high-ranking government officials, including members of Rousseff's Workers' Party, enormous sums of money to obtain contracts from Petrobras. The bribes were thought to have totaled around $3 billion. Prosecutors charged executives from more than a dozen of the country's largest construction companies with corruption and money laundering. As several Petrobras executives turned state's witness, the police investigation, known as Operation Car Wash, continued to expand. Before long, many Brazilians concluded that Rousseff, who served as chair of Petrobras' board from 2003 to 2010, must be guilty herself. Although she has not been charged, Brazil's Supreme Court has ruled that her predecessor, Lula, can be called in for questioning, and on September 1, his former chief of staff was charged with racketeering, receiving bribes, fraud, and money laundering.

Lula was already under intense official scrutiny at the time: just a few months earlier, prosecutors had concluded that they had enough evidence to launch a full investigation into allegations that Brazil's biggest building company had paid Lula to lobby overseas on the firm's behalf. In yet another case, Rousseff has been accused by Brazil's Controller General's Office of illegally using funds earmarked for social programs and development to cover up budget deficits. Taken together, all these charges have helped push Rousseff's approval ratings down into the single digits; talk of her impeachment is now in the air. As demonstrations continue and the economy languishes—Brazil is now in its worst recession in decades—it's looking increasingly likely that Rousseff will not manage to serve out her term, which ends in 2018.

Bad as all these revelations have been for Brazil, the public reckoning that has followed can also be read as a sign of progress. The fact that the police, prosecutors, and judges have been willing to investigate the country's most powerful politicians and business leaders has highlighted the independence of Brazil's judicial system. And the unprecedented level of anger suggests that business as usual will no longer satisfy a Brazilian public increasingly intolerant of high-level corruption.

Another country north of Brazil has also recently turned a corner. In early 2015, after it emerged that officials had siphoned off millions of dollars in customs revenue, thousands of Guatemalans took to the streets, gathering every Saturday for weeks in the central square of the capital. In May, they forced the resignation of the country's vice president and several cabinet ministers. But the protests continued, and on September 1, legislators from President Otto Pérez Molina's own party stripped him of his immunity. On September 2, he resigned and within hours was jailed on corruption charges—an extraordinary event in a country where politicians have long enjoyed great impunity.

The fight against corruption in Guatemala has benefited from outside support. That help has come in the form of the International Commission against Impunity in Guatemala (CICIG), which was created in 2006 as part of a larger agreement between the UN and Guatemala. The body, which was initially intended to investigate crimes committed during the civil war, is financed by the European Union, supported by the U.S. embassy, and led by a Colombian; it now numbers more than 200 foreign officials. It has become one of the most powerful instruments in the campaign against corruption. As a high government official told me in August, "It hurts to admit that we are unable to clean up our own house, but it is better that someone else does it than that nobody does."

Guatemala's example has sparked similar protests across Central America, in neighboring Honduras and to a lesser extent in El Salvador. In Tegucigalpa, the capital of Honduras, thousands of protesters gather every Friday in a *marcha de las antorchas*, or "march of the torches," to demand an investigation into the defrauding of the Honduran Institute of Social Security by the governing party. President Juan Orlando Hernández has attempted to placate the demonstrators by creating a commission similar to the CICIG, although without prosecutorial powers, but so far, these attempts have failed. As long as it lacks the teeth of its Guatemalan counterpart, such a commission is unlikely to satisfy the protesters, and if more scandals come to light, calls for Hernández's resignation will mount. And in El Salvador, there are calls for an external investigation into Alba Petróleos, the subsidized energy venture set up by the then Venezuelan President Hugo Chávez several years ago, which is suspected of making financial contributions to El Salvador's ruling party.

No Más

Foreign influence has played a crucial role in Mexico as well. The American press is responsible for unveiling three of the most important corruption cases in recent history: Walmart's bribing of Mexican municipal officials (which *The New York Times* reported on in 2012); the revelation that Luis Videgaray, the country's finance minister, had acquired property under suspicious circumstances (a story *The Wall Street Journal* broke in 2014); and the concealed purchase of multimillion-dollar condos in Manhattan and elsewhere by a former governor (another story broken by the *Times*, this one in 2015).

Yet Mexico's domestic media have also done their part. The radio reporter Carmen Aristegui, the newsweekly *Proceso*, and the daily *Reforma* have helped expose numerous scandals. Aristegui revealed that the $7 million modernist mansion in Mexico City built for the Peña Nieto family was in fact owned by a company to which the president had awarded hundreds of millions of dollars in public contracts and that was headed by a personal friend of his. A government investigation has since exonerated Peña Nieto, but many in Mexico have dismissed the inquiry as a cover-up. Other accusations have been leveled at the interior minister and several governors. These scandals have all generated a great deal of anger and unhappiness in Mexican society. But so far, not much more has come of them.

Latin Americans are denouncing corruption as never before.

Yet, public opinion has come to matter more and more in today's democratic Mexico. Online social networks now provide the new middle class with an outlet halfway between public protest and private complaint: Mexicans use Facebook and Twitter to vent their anger and share information (not all of it accurate) about high-level corruption. This allows for a measure of catharsis but has yet to produce actual change: although many think that Peña Nieto's government is Mexico's most corrupt since the late 1980s, so far calls for the president's resignation have foundered.

A major corruption scandal is causing a similar reaction in Chile. It began in February 2015 with accusations of influence peddling against the son and daughter-in-law of President Michelle Bachelet. Other scandals soon emerged, involving tax fraud and campaign finance crimes on the part of opposition leaders and members of the governing coalition, several of whom were jailed. As of May, most had been released, but some were under house arrest. Even Marco Enríquez-Ominami, a former independent candidate for president and one of Chile's most popular politicians, has been caught up in a controversy regarding campaign finance, according to Chilean media reports. Large financial and mining conglomerates—one of them led by the ex-son-in-law of Chile's former dictator, Augusto Pinochet—have been accused of fraudulently contributing to electoral campaigns.

Yet as in Mexico, the scandals have prompted only muted protests so far. That may be a consequence of their relatively small scale. Historically, Chile's has ranked as one of the more honest governments in Latin America, and the amounts at stake in the country's recent scandals pale in comparison to those in Brazil and Mexico: the most serious charge against a Chilean official involves a loan of $10 million.

Yet, the allegations still represent the most serious challenge Bachelet has faced in her two terms in power. Although she tried to show that she takes the issue seriously by asking for the resignation of her entire cabinet in May and by calling for a new constitution, by August, her approval rating had dropped by 30 percentage points in one year. Popular protests are likely to become louder unless Chile's elites take genuine steps to reduce corruption, especially if the country's copper-dependent economy doesn't pick up soon. But once again, this represents good news as well as bad. Chile's independent and honest judiciary and its free press were central to uncovering the corruption scandals—an important sign of the growing effectiveness of Chile's democratic institutions.

Even in Venezuela, where flagrant corruption is still the norm, there are signs that the public's patience is running out. According to Latinobarómetro, Venezuela is the region's second least transparent country, and at the end of 2015, Maduro was its third most unpopular president. The United States recently leaked accusations that many of the country's leaders—including Diosdado Cabello, the head of the National Assembly and Maduro's closest aide—have used illegal means to enrich themselves immensely, partly through links with Colombian drug cartels. In June, the deterioration of Venezuela's economy and the increase in violence and human rights violations forced Maduro to call elections for December. Although candidates and voters mostly focused on the economy, violence, and repression, more and more of them raised corruption as well.

A Glass Half Full

Yet, there are exceptions to this trend. In Argentina, there are a few positive developments in the fight against corruption. The outgoing vice president, Amado Boudou, is awaiting trial for corruption, but many suspect the charges will be dismissed. Allegations also surround outgoing President Cristina Fernández de Kirchner, whose net wealth surged to a reported $6 million over the 13 years she and her late husband ruled the country, but the chances of a prosecution are slim. Outsiders have less influence in Argentina than in many of its neighbors—the country's tradition of Peronist nationalism makes it hostile to perceived meddling. And the public seems resigned to the status quo: although hundreds of thousands of Argentines joined demonstrations when the prosecutor Alberto Nisman died under mysterious circumstances—as he was investigating Kirchner in connection with the 1994 bombing of a Buenos Aires Jewish community center—they have remained stubbornly passive when it comes to corruption.

The people of Nicaragua, meanwhile, seem even more complacent. President Daniel Ortega is currently focused on an enormous undertaking: an attempt to build a second interoceanic canal just north of the existing Panama Canal. A Chinese businessman has agreed to underwrite the cost, which could reach up to $100 billion. Some Nicaraguans think that the businessman is working for the Chinese government, but given China's economic problems, it is in no position to foot the bill, and more than two years after the project was announced, excavation has yet to begin. Many Nicaraguans believe that the whole venture is nothing more than an elaborate scheme designed to enrich the Ortega family and that no canal will ever be built. Yet, Nicaraguans have done little to register their displeasure: there have been no massive protests, for example.

As all these stories suggest, corruption remains deeply embedded in Latin American political and social life. Some countries have seen little improvement from the bad old days decades ago. Yet, the outraged reactions to the wave of scandals currently sweeping the continent may be the first sign that Latin American publics are no longer prepared to tolerate systemic dishonesty in their governments. The region's new middle classes, aided by pressure from abroad and by increasingly confident and independent domestic institutions, have begun demanding better governance.

The outcome of all these movements is still uncertain. Some may generate new institutions: autonomous controller's offices, more powerful and independent judiciaries, greater transparency, and more active and conscious civil societies. Others may take a populist turn, as candidates for office run on antielite platforms. And in some countries, the movements will subside. But in all cases, something will have changed in Latin America, and much for the better.

Critical Thinking

1. What are the causes of kleptocracy in Latin America?
2. Why should the U.S. be concerned about corruption in Brazil?
3. What are the factors that have stimulated opposition to corruption in Latin America?

Internet References

Kletocracy Initiative
kleptocracy/initiative.org

Natural Resources Governance Initiative
www.resourcegovernance.org/

Transparency International
www.transparency.org/

Jorge G. Castañeda is a global distinguished professor of Politics and Latin American and Caribbean Studies at New York University. He served as Mexico's Foreign Minister from 2000 to 2003. Follow him on Twitter @JorgeGCastaneda. This article draws on columns he wrote for *Project Syndicate* last year.

Article

Prepared by: Robert Weiner, *University of Massachusetts, Boston*

Race in the Modern World: The Problem of the Color Line

KWAME ANTHONY APPIAH

Learning Outcomes

After reading this article, you will be able to:

- Explain the difficult efforts of scholars to come up with a definition of race.
- Discuss the importance of Pan-Africanism.

In 1900, in his "Address to the Nations of the World" at the first Pan-African Conference, in London, W. E. B. Du Bois proclaimed that the "problem of the twentieth century" was "the problem of the color-line, the question as to how far differences of race—which show themselves chiefly in the color of the skin and the texture of the hair—will hereafter be made the basis of denying to over half the world the right of sharing to their utmost ability the opportunities and privileges of modern civilization."

Du Bois had in mind not just race relations in the United States but also the role race played in the European colonial schemes that were then still reshaping Africa and Asia. The final British conquest of Kumasi, Ashanti's capital (and the town in Ghana where I grew up), had occurred just a week before the London conference began. The British did not defeat the Sokoto caliphate in northern Nigeria until 1903. Morocco did not become a French protectorate until 1912, Egypt did not become a British one until 1914, and Ethiopia did not lose its independence until 1936. Notions of race played a crucial role in all these events, and following the Congress of Berlin in 1878, during which the great powers began to devise a world order for the modern era, the status of the subject peoples in the Belgian, British, French, German, Spanish, and Portuguese colonies of Africa—as well as in independent South Africa—was defined explicitly in racial terms.

Du Bois was the beneficiary of the best education that North Atlantic civilization had to offer: he had studied at Fisk, one of the United States' finest black colleges; at Harvard; and at the University of Berlin. The year before his address, he had published *The Philadelphia Negro,* the first detailed sociological study of an American community. And like practically everybody else in his era, he had absorbed the notion, spread by a wide range of European and American intellectuals over the course of the nineteenth century, that race—the division of the world into distinct groups, identifiable by the new biological sciences—was central to social, cultural, and political life.

Even though he accepted the concept of race, however, Du Bois was a passionate critic of racism. He included anti-Semitism under that rubric, and after a visit to Nazi Germany in 1936, he wrote frankly in *The Pittsburgh Courier,* a leading black newspaper, that the Nazis' "campaign of race prejudice . . . surpasses in vindictive cruelty and public insult anything I have ever seen; and I have seen much." The European homeland had not been in his mind when he gave his speech on the color line, but the Holocaust certainly fit his thesis—as would many of the centuries' genocides, from the German campaign against the Hereros in Namibia in 1904 to the Hutu massacre of the Tutsis in Rwanda in 1994. Race might not necessarily have been *the* problem of the century—there were other contenders for the title—but its centrality would be hard to deny.

Violence and murder were not, of course, the only problems that Du Bois associated with the color line. Civic and economic inequality between races—whether produced by government policy, private discrimination, or complex interactions between the two—were pervasive when he spoke and remained so long after the conference was forgotten.

All around the world, people know about the civil rights movement in the United States and the antiapartheid struggle in

South Africa, but similar campaigns have been waged over the years in Australia, New Zealand, and most of the countries of the Americas, seeking justice for native peoples, or the descendants of African slaves, or East Asian or South Asian indentured laborers. As non-Europeans, including many former imperial citizens, have immigrated to Europe in increasing numbers in recent decades, questions of racial inequality there have come to the fore, too—in civic rights, education, employment, housing, and income. For Du Bois, Chinese, Japanese, and Koreans were on the same side of the color line as he was. But Japanese brutality toward Chinese and Koreans up through World War II was often racially motivated, as are the attitudes of many Chinese toward Africans and African Americans today. Racial discrimination and insult are a global phenomenon.

Of course, ethnoracial inequality is not the only social inequality that matters. In 2013, the nearly 20 million white people below the poverty line in the United States made up slightly more than 40 percent of the country's poor. Nor is racial prejudice the only significant motive for discrimination: ask Christians in Indonesia or Pakistan, Muslims in Europe, or LGBT people in Uganda. Ask women everywhere. But more than a century after his London address, Du Bois would find that when it comes to racial inequality, even as much has changed, much remains the same.

Us and Them

Du Bois' speech was an invitation to a global politics of race, one in which people of African descent could join with other people of color to end white supremacy, both in their various homelands and in the global system at large. That politics would ultimately shape the process of decolonization in Africa and the Caribbean and inform the creation of what became the African Union. It was politics that led Du Bois himself to become, by the end of his life, a citizen of a newly independent Ghana, led by Kwame Nkrumah.

But Du Bois was not simply an activist; he was even more a scholar and an intellectual, and his thinking reflected much of his age's obsession with race as a concept. In the decades preceding Du Bois' speech, thinkers throughout the academy—in classics, history, artistic and literary criticism, philology, and philosophy, as well as all the new life sciences and social sciences—had become convinced that biologists could identify, using scientific criteria, a small number of primary human races. Most would have begun the list with the black, white, and yellow races, and many would have included a Semitic race (including Jews and Arabs), an American Indian race, and more. People would have often spoken of various subgroups within these categories as races, too. Thus, the English poet Matthew Arnold considered the Anglo-Saxon and Celtic races to be the main components of the population of the United Kingdom; the French historian Hippolyte Taine thought the Gauls were the race at the core of French history and identity; and the U.S. politician John C. Calhoun discussed conflicts not only between whites and blacks but also between Anglo-Canadians and "the French race of Lower Canada."

People thought race was important not just because it allowed one to define human groups scientifically but also because they believed that racial groups shared inherited moral and psychological tendencies that helped explain their different histories and cultures. Of course, there were always skeptics. Charles Darwin, for example, believed that his evolutionary theory demonstrated that human beings were a single stock, with local varieties produced by differences in environment, through a process that was bound to result in groups with blurred edges. But many late-nineteenth-century European and American thinkers believed deeply in the biological reality of race and thought that the natural affinity among the members of each group made races the appropriate units for social and political organization.

Essentialism—the idea that human groups have core properties in common that explain not just their shared superficial appearances but also the deep tendencies of their moral and cultural lives—was not new. In fact, it is nearly universal, because the inclination to suppose that people who look alike have deep properties in common is built into human cognition, appearing early in life without much prompting. The psychologist Susan Gelman, for example, argues that "our essentializing bias is not directly taught," although it is shaped by language and cultural cues. It can be found as far back as Herodotus' *Histories* or the Hebrew Bible, which portrayed Ethiopians, Persians, and scores of other peoples as fundamentally other. "We" have always seen "our own" as more than superficially different from "them."

What was new in the nineteenth century was the combination of two logically unrelated propositions: that races were biological and so could be identified through the scientific study of the shared properties of the bodies of their members and that they were also political, having a central place in the lives of states. In the eighteenth century, the historian David Hume had written of "national character"; by the nineteenth century, using the new scientific language, Arnold was arguing that the "Germanic genius" of his own "Saxon" race had "steadiness as its main basis, with commonness and humdrum for its defect, fidelity to nature for its excellence."

If nationalism was the view that natural social groups should come together to form states, then the ideal form of nationalism would bring together people of a single race. The eighteenth-century French American writer J. Hector St. John de Crèvecoeur's notion that in the New World, all races could be "melted into a new race of man"—so that it was the nation that made the race, not the race the nation—belonged to an older way of thinking, which racial science eclipsed.

The Other Dismal Science

In the decade after Du Bois' address, however, a second stage of modern argumentation about human groups emerged, one that placed a much greater emphasis on culture. Many things contributed to this change, but a driving force was the development of the new social science of anthropology, whose German-born leader in the United States, Franz Boas, argued vigorously (and with copious evidence from studies in the field) that the key to understanding the significant differences between peoples lay not in biology—or, at least, not in biology alone—but in culture. Indeed, this tradition of thought, which Du Bois himself soon took up vigorously, argued not only that culture was the central issue but also that the races that mattered for social life were not, in fact, biological at all.

In the United States, for example, the belief that anyone with one black grandparent or, in some states, even one black great-grandparent was also black meant that a person could be socially black but have skin that was white, hair that was straight, and eyes that were blue. As Walter White, the midcentury leader of the National Association for the Advancement of Colored People, whose name was one of his many ironic inheritances, wrote in his autobiography, "I am a Negro. My skin is white, my eyes are blue, my hair is blond. The traits of my race are nowhere visible upon me."

Strict adherence to thinking of race as biological yielded anomalies in the colonial context as well. Treating all Africans in Nigeria as "Negroes," say, would combine together people with very different biological traits. If there were interesting traits of national character, they belonged not to races but to ethnic groups. And the people of one ethnic group—Arabs from Morocco to Oman, Jews in the Diaspora—could come in a wide range of colors and hair types.

In the second phase of discussion, therefore, both of the distinctive claims of the first phase came under attack. Natural scientists denied that the races observed in social life were natural biological groupings, and social scientists proposed that the human units of moral and political significance were those based on shared culture rather than shared biology. It helped that Darwin's point had been strengthened by the development of Mendelian population genetics, which showed that the differences found between the geographic populations of the human species were statistical differences in gene frequencies rather than differences in some putative racial essence.

In the aftermath of the Holocaust, moreover, it seemed particularly important to reject the central ideas of Nazi racial "science," and so, in 1950, in the first of a series of statements on race, unesco (whose founding director was the leading biologist Sir Julian Huxley) declared that national, religious, geographic, linguistic and cultural groups do not necessarily coincide with racial groups: and the cultural traits of such groups have no demonstrated genetic connection with racial traits. . . . The scientific material available to us at present does not justify the conclusion that inherited genetic differences are a major factor in producing the differences between the cultures and cultural achievements of different peoples or groups.

Race was still taken seriously, but it was regarded as an outgrowth of sociocultural groups that had been created by historical processes in which the biological differences between human beings mattered only when human beings decided that they did. Biological traits such as skin color, facial shape, and hair color and texture could define racial boundaries if people chose to use them for that purpose. But there was no scientific reason for doing so. As the unesco statement said in its final paragraph, "Racial prejudice and discrimination in the world today arise from historical and social phenomena and falsely claim the sanction of science."

Construction Work

In the 1960s, a third stage of discussion began, with the rise of "genetic geography." Natural scientists such as the geneticist Luigi Luca Cavalli-Sforza argued that the concept of race had no place in human biology, and social scientists increasingly considered the social groups previously called "races" to be social constructions. Since the word "race" risked misleading people on this point, they began to speak more often of "ethnic" or "ethnoracial" groups, in order to stress the point that they were not aiming to use a biological system of classification.

In recent years, some philosophers and biologists have sought to reintroduce the concept of race as biological using the techniques of cladistics, a method of classification that combines genetics with broader genealogical criteria in order to identify groups of people with shared biological heritages. But this work does not undermine the basic claim that the boundaries of the social groups called "races" have been drawn based on social, rather than biological, criteria; regardless, biology does not generate its own political or moral significance. Socially constructed groups can differ statistically in biological characteristics from one another (as rural whites in the United States differ in some health measures from urban whites), but that is not a reason to suppose that these differences are caused by different group biologies. And even if statistical differences between groups exist, that does not necessarily provide a rationale for treating individuals within those groups differently. So, as Du Bois was one of the first to argue, when questions arise about the salience of race in political life, it is usually not a good idea to bring biology into the discussion.

It was plausible to think that racial inequality would be easier to eliminate once it was recognized to be a product of sociology and politics rather than biology. But it turns out that all sorts of status differences between ethnoracial groups can persist long after governments stop trying to impose them. Recognizing that institutions and social processes are at work rather than innate qualities of the populations in question has not made it any less difficult to solve the problems.

Imagined Communities

One might have hoped to see signs that racial thinking and racial hostility were vanishing—hoped, that is, that the color line would not continue to be a major problem in the twenty-first century, as it was in the twentieth. But a belief in essential differences between "us" and "them" persists widely, and many continue to think of such differences as natural and inherited. And of course, differences between groups defined by common descent can be the basis of social identity, whether or not they are believed to be based in biology. As a result, ethnoracial categories continue to be politically significant, and racial identities still shape many people's political affiliations.

Once groups have been mobilized along ethnoracial lines, inequalities between them, whatever their causes, provide bases for further mobilization. Many people now know that we are all, in fact, one species, and think that biological differences along racial lines are either illusory or meaningless. But that has not made such perceived differences irrelevant.

Around the world, people have sought and won affirmative action for their ethnoracial groups. In the United States, in part because of affirmative action, public opinion polls consistently show wide divergences on many questions along racial lines. On American university campuses, where the claim that "race is a social construct" echoes like a mantra, black, white, and Asian identities continue to shape social experience. And many people around the world simply find the concept of socially constructed races hard to accept, because it seems so alien to their psychological instincts and life experiences.

Race also continues to play a central role in international politics, in part because the politics of racial solidarity that Du Bois helped inaugurate, in co-founding the tradition of pan-Africanism, has been so successful. African Americans are particularly interested in U.S. foreign policy in Africa, and Africans take note of racial unrest in the United States: as far away as Port Harcourt, Nigeria, people protested against the killing of Michael Brown, the unarmed black teenager shot to death by a police officer last year in Missouri. Meanwhile, many black Americans have special access to Ghanaian passports, Rastafarianism in the Caribbean celebrates Africa as the home of black people, and heritage tourism from North and South America and the Caribbean to West Africa has boomed.

Pan-Africanism is not the only movement in which a group defined by a common ancestry displays transnational solidarity. Jews around the world show an interest in Israeli politics. People in China follow the fate of the Chinese diaspora, the world's largest. Japanese follow goings-on in São Paulo, Brazil, which is home to more than 600,000 people of Japanese descent—as well as to a million people of Arab descent, who themselves follow events in the Middle East. And Russian President Vladimir Putin has put his supposed concern for ethnic Russians in neighboring countries at the center of his foreign policy.

Identities rooted in the reality or the fantasy of shared ancestry, in short, remain central in politics, both within and between nations. In this new century, as in the last, the color line and its cousins are still going strong.

Wouldn't It Be Nice?

The pan-Africanism that Du Bois helped invent created, as it was meant to, a new kind of transnational solidarity. That solidarity was put to good use in the process of decolonization, and it was one of the forces that helped bring an end to Jim Crow in the United States and apartheid in South Africa. So racial solidarity has been used not just for pernicious purposes but for righteous ones as well. A world without race consciousness, or without ethnoracial identity more broadly, would lack such positive mobilizations, as well as the negative ones. It was in this spirit, I think, that Du Bois wrote, back in 1897, that it was "the duty of the Americans of Negro descent, as a body, to maintain their race identity until . . . the ideal of human brotherhood has become a practical possibility."

But at this point, the price of trying to move beyond ethnoracial identities is worth paying, not only for moral reasons but also for the sake of intellectual hygiene. It would allow us to live and work together more harmoniously and productively, in offices, neighborhoods, towns, states, and nations. Why, after all, should we tie our fates to groups whose existence seems always to involve misunderstandings about the facts of human difference? Why rely on imaginary natural commonalities rather than build cohesion through intentional communities? Wouldn't it be better to organize our solidarities around citizenship and the shared commitments that bind political society?

Still, given the psychological difficulty of avoiding essentialism and the evident continuing power of ethnoracial identities, it would take a massive and focused effort of education, in schools and in public culture, to move into a postracial world. The dream of a world beyond race, unfortunately, is likely to be long deferred.

Critical Thinking

1. What are W. E. B. Dubois' major ideas about race?
2. What is meant by an imagined community?

Internet References

Brennan Center for Justice
www.brennancenter.org

W. E. B. Dubois papers
credo.library.umass.edu

Kwame Anthony Appiah is Professor of Philosophy and Law at New York University. His most recent book is *Lines of Descent: W. E. B. Du Bois and the Emergence of Identity.* Born to a British mother and Ghanaian father, Kwame Anthony Appiah grew up traveling between his two homelands, an experience that has shaped his wideranging writing on ethnicity, identity, and culture. He is the author of numerous books, including *Lines of Descent,* and has won scores of prizes, among them the National Humanities Medal and the Arthur Ross Book Award. Now a professor at New York University, Appiah explores the past, present, and future of thinking about race in "Race in the Modern World" (page 1).

Article

Prepared by: Robert Weiner, *University of Massachusetts, Boston*

The End of Human Rights?

Learning from the failure of the Responsibility to Protect and the International Criminal Court.

DAVID RIEFF

Learning Outcomes

After reading this article, you will be able to:

- Learn why the human rights movement has failed to live up to expectations.
- Learn why the human rights movement is not able to progress.

THERE IS NO DOUBT that the human rights movement is facing the greatest test it has confronted since its emergence in the 1970s as a major participant in the international order.

A bellwether of this crisis has been the essays that Kenneth Roth, the executive director of Human Rights Watch, has written introducing his organization's annual reports. One has to go back to 2014 to find Roth writing in a relatively sanguine way about the future of human rights across the globe. That year's report is couched in the positive terms of its title: "Stopping Mass Atrocities, Majority Bullying, and Abusive Counterterrorism." By 2016, he was musing on "how the politics of fear and the crushing of civil society imperil global rights." And the following year, Roth warned Human Rights Watch's supporters that the rise of populism "threatens to reverse the accomplishments of the modern human rights movement."

And though in the 2018 report Roth claims that things may not be as bad as they have been for the previous three years, he leaves no doubt that they remain very bad indeed. Roth concludes that a "fair assessment of global prospects for human rights should induce concern rather than surrender—a call to action rather than a cry of despair."

Strip away the activist language and what emerges is a human rights movement forced to refight and relitigate battles it once thought won. Human Rights Watch is not alone in calling for an all-hands-on-deck response from its supporters. In its own 2017-2018 report, Amnesty International states: "Over the past year, leaders have pushed hate, fought against rights, ignored crimes against humanity, and blithely let inequality and suffering spin out of control." But, like Roth, the authors of the Amnesty report conclude that "while our challenges may never be greater, the will to fight back is just as strong."

The question remains as to why the most prominent international human rights organizations seemed to have missed the gathering storm until, with the rise of populism in Europe, it reached them.

Of course, outside critics and scholars of the human rights movement—such as Stephen Hopgood, Samuel Moyn, and Eric Posner—had already predicted that the legalism of the human rights movement no longer sufficed. Implicit in the liberal human rights narrative is the idea that once binding legal norms are set, realities on the ground will eventually conform to them. It is a legal approach that simply has no place for German scholar Carl Schmitt's idea of the law as inseparable from politics, rather than above it. As far as the human rights movement has been concerned, once what the writer Michael Ignatieff called the post-World War II "revolution of moral concern" got fully underway, it was a matter of when—not if — an international system based on human rights would prevail throughout the world.

But for the moment, at least, Brexit, Donald Trump's presidency, and the steady rise of China have shattered the human rights movement's narrative that progress is inevitable.

Nothing is inevitable in history—except of course, sooner or later, the mortality of every civilization and system—and both Human Rights Watch and Amnesty are quite right to refuse to concede defeat. It is possible, though not likely, that the human rights movement will be more effective with its collective back against the wall: an underground dissident church as it was during its beginnings, rather than the secular church of liberal globalism that it was at its apogee. What is clear, however, is that the global balance of power has tilted away from governments committed to human rights norms and toward those indifferent or actively hostile to them. Into the latter camp fall, most obviously, China, Russia, Turkey, the Philippines, and Venezuela. Roth all but admits as much when, in the 2018 report, he speaks of powers that "have withdrawn" from the struggle for human rights, even if he holds out some hope that small and middle-sized nations will fill the void.

What the human rights movement has been unwilling to do is accept some of the blame for the greatly weakened position in which it finds itself. This is predictable. If your expectations are millenarian—if you believe there is a right side of history, yours, and a wrong side of history that is doomed to defeat—skepticism about the human rights project, let alone voices of opposition, is unlikely to sway your position. Given that perspective, why consider any change in your approach that goes beyond tactical adjustments?

This is what Ignatieff, though one of the human rights community's most important advocates, warned of in his prescient 2001 book, *Human Rights as Politics and Idolatry*. "In the next fifty years," he wrote, "we can expect to see the moral consensus that sustained the Universal Declaration [of Human Rights] in 1948 splintering still further. . . . There is no reason to believe that economic globalization entails moral globalization."

But this seems to have been exactly one of the main drivers of what a sympathizer with the human rights movement would call its moral serenity and a skeptic would call its hubris. Nowhere has this hubris been more evident than in the fate of institutional structures and frameworks meant to allow internationally sanctioned, state-sponsored intervention to prevent genocide, crimes against humanity, and war crimes or to bring to account those guilty of such horrors.

The first of these frameworks is the International Criminal Court (ICC), established in 2002. The second is the so-called Responsibility to Protect (R2P) doctrine, which the United Nations adopted at its World Summit in 2005 and reconfirmed in 2009.

R2P sets up an elaborate series of nonviolent measures that need to be tried before resorting to international military intervention on human rights or humanitarian grounds. Force, according to proponents of R2P, should only be used if both a reasonable chance of success and the proportionality of the response are possible. But as an internationally binding norm, it nonetheless obliges outside powers, albeit only if sanctioned by the U.N. Security Council, to intervene to halt a genocide or crimes of mass atrocity in countries where either the government of the country in question is committing the crimes at hand or is otherwise unable to prevent these horrors from continuing.

The claims made for both R2P and the ICC were sweeping. One of R2P's principal architects, former Australian Foreign Minister Gareth Evans, wrote that its emergence brought us much closer to "ending mass atrocity crimes once and for all." The promise contained in the vow "Never Again"—first coined by the prisoners of the Buchenwald concentration camp just after their liberation and repeated ad infinitum, if hollowly, since that moment in 1945—was at last to become a reality.

The promises regarding what the ICC was going to accomplish were only slightly less extravagant. When the Rome Statute, the treaty that paved the way for the court, was signed, then-U.N. Secretary-General Kofi Annan hailed it as a "gift of hope to future generations and a giant step forward in the march toward universal human rights and the rule of law." Indeed, Annan concluded, "It is an achievement [that], only a few years ago, nobody would have thought possible."

But only a few years later, both R2P and the ICC look like just that: doctrines that are not possible in the world as it actually exists. Some of these wounds were self-inflicted. Politically, it was a huge mistake on the part of the ICC's first chief prosecutor, Luis Moreno Ocampo, to appear to focus his investigations almost exclusively on Africa—even if he was right on legal grounds, since a disproportionate number of the early referrals to the court were from African governments themselves. The result has been a widespread perception that Africa is being unfairly targeted. In 2017, a number of African countries even attempted to organize a massive withdrawal of African Union members from the ICC. The fact that this effort was beaten back should not be taken as evidence that the ICC's crisis of legitimacy in Africa is over and done with.

And the global obligation, articulated in R2P, to act militarily *in extremis* to stop mass atrocity crimes has taken place only once: in Libya in 2011. But the intervention in Libya to protect the civilian population soon morphed into regime change, as a minority of supporters of R2P have since conceded. The widely held view among R2P champions is that the Libyan intervention was right—it's just that the implementation was faulty.

Regardless of whether it was right or wrong, there is very little likelihood of another R2P intervention in the foreseeable future. Syria, Yemen, and the ethnic cleansing of the Rohingya in Myanmar have demonstrated that all too painfully.

Deeper moral and political reasons explain why the ICC and R2P have failed to live up to what, in retrospect, seem like completely outlandish expectations for what they could achieve. In the case of the ICC, the court was created without a police force to carry out its instructions. Moreover, several of the world's most powerful states—China, the United States, India, and

Russia—haven't ratified or joined the Rome Statute. A legal institution that is only in a position to target war criminals who don't enjoy the protection of powerful states is likely to be intermittently effective at most. It will also be of questionable legitimacy no matter how many other nations officially recognize it. Legitimacy and legality, of course, do not necessarily go together. The intervention in Libya was legal; the intervention in Kosovo in 1999 was not. It is anything but an outlandish view to believe that the former was morally illegitimate and the latter undertaken on a far sounder moral basis.

An institution that is based on a double standard, as the ICC seems likely to remain for the foreseeable future, cannot be seriously considered to be an important step toward universal justice.

In the case of R2P, the nonmilitary features of the doctrine have been successful in a number of instances. Annan invoked R2P in his back-channel negotiations with the Kenyan government after deadly riots broke out during the country's national elections at the end of 2007, which almost led to Civil War. But useful as it was as a negotiating tool to Annan in Kenya, R2P has not transformed classical diplomacy. Instead, its moral force came from its claim to be able to halt genocide and mass atrocity crimes.

Defenders of R2P and the ICC might argue that the world is better off in the long run, as the court and the doctrine will eventually lead to the desired transformations of reality on the ground. But this is precisely the same mistaken assumption that has thrown the human rights movement into crisis as democracy is rolled back across the globe. Both the ICC and R2P were, from the beginning, unworkable ideas for the world we live in, one in which authoritarianism is growing stronger.

Calls to action by human rights activists, therefore, are not enough, given that the move away from democracy and toward authoritarianism may be resisted but is highly unlikely to be reversed in the foreseeable future. If the human rights movement has a future at all, it should consist of defending what remains of Ignatieff's revolution of moral concern, not pretending that—for now at least—it can be expanded.

Critical Thinking

1. Has the Responsibility to Protect (R2P) failed or succeeded?
2. Has the International Criminal Court (ICC) failed or succeeded? Why or why not?
3. Are Human Rights Watch and Amnesty International optimistic or pessimistic about the future of the human rights movement? Why or why not?

Internet References

Human Rights Watch
https://www.hrw.org/

Office of the High Commissioner for Human Rights
https://www.ohchr

Article

Prepared by: Robert Weiner, *University of Massachusetts, Boston*

Where Myanmar Went Wrong: From Democratic Awakening to Ethnic Cleansing

ZOLTAN BARANY

Learning Outcomes

After reading this article, you will be able to:

- Learn why Suu Kyi's government cannot stop the ethnic cleansing of the Rohingya.
- Learn about the history of the Rohingya in Myanmar.

Late last year, when news broke that Myanmar's military had been systematically killing members of the country's Muslim Rohingya minority, much of the world was shocked. In recent years, Myanmar (also known as Burma) had been mostly a good news story. After decades of brutal dominance by the military, the country had seen the main opposition party, the National League for Democracy, score an all-too-rare democratic triumph, winning the 2015 national elections in a landslide. The NLD's leader, Aung San Suu Kyi, an internationally celebrated dissident who had received the 1991 Nobel Peace Prize for her efforts to democratize Myanmar, became Myanmar' s de facto head of state. Many analysts and officials concluded that the county was finally on the path to democratic rule. Support poured in from Western democracies, including the United States. Myanmar had long been isolated, relying almost exclusively on China, which was content to turn a blind eye to human rights abuses. Now, many hoped, Suu Kyi would lead the country into the Western-backed international order.

But such hopes overlooked a fundamental reality, one that was brought into stark relief by the slaughter of the Rohingya: Myanmar's generals continue to control much of the country's political and economic life. Suu Kyi must strike a delicate balance, advancing democratic rule without stepping on the generals' toes. Her government has no power over the army and can do little to end the military's brutal campaign against the Rohingya—which, in any event, enjoys massive popular support. Yet Suu Kyi has taken the bad hand she was dealt and made it worse. She has adopted an autocratic style. She has failed to make progress in the areas where she does have influence. And she has alienated erstwhile allies in the West.

Citizens of Nowhere

Myanmar, which has a population of 54 million, officially recognizes 135 ethnic groups—but not the Rohingya. In fact, Myanmar authorities, including Suu Kyi, refuse to even use the term "Rohingya." But the Rohingya are indisputably a distinct group with a long history in Myanmar. They are the descendants of people whom British colonial authorities, searching for cheap labor, encouraged to emigrate from eastern Bengal (contemporary Bangladesh) to the sparsely populated western regions of Burma in the nineteenth and early twentieth centuries. Today, there are around 2.5 million Rohingya, who constitute the world's largest stateless population. But fewer than half a million currently reside in Myanmar; the rest have fled decades of official repression and exclusion, often crossing the border into Bangladesh, where they inhabit sprawling, squalid refugee camps. Those who have remained in Myanmar are a subset of the country's Muslim community. The majority of Myanmar's Muslims live in urban areas, speak Burmese, have

Burmese names, and are Myanmar citizens. The Rohingya are different: most speak a dialect of Bengali, have traditionally Muslim names, and have never received citizenship. The Rohingya in both Bangladesh and Myanmar have led unusually difficult lives even by the region's humble standards, marked by poverty, the absence of legal status, and multifaceted discrimination. Owing to their lack of resources and extreme vulnerability, the Rohingya have largely failed in their attempts at political mobilization, which have generated further resentment against them. For instance, the 1950–54 Rohingya resistance movement, which demanded citizenship and an end to discriminatory policies, was eventually crushed by the army.

Perhaps not surprisingly, a militant Rohingya faction also emerged: the Arakan Rohingya Salvation Army, which formed in 2013. Most of the ARSA's leaders are from Bangladesh or Pakistan, and some of them have received training from jihadist veterans of the wars in Afghanistan. (The group's chief leader was born in Pakistan and later became an imam in Saudi Arabia.) Arsa likely has fewer than 600 active members. But Myanmar officials consider it a dangerous organization. In the early morning hours of August 25, 2017, for example, about 150 ARSA militants staged coordinated attacks on police posts and an army base in Rakhine State. The confrontation ended with the deaths of 77 ARSA fighters and 12 police officers and touched off a crackdown by Myanmar's army, which burned down scores of Rohingya villages, murdered dozens of civilians, and launched a campaign of rape against Rohingya women and girls, according to Human Rights Watch. The un labeled the operation ethnic cleansing, and others, including French President Emmanuel Macron and eight Nobel Peace Prize laureates, have described it as an act of genocide.

By the end of 2017, 650,000 Rohingya had fled to neighboring Bangladesh, joining approximately 200,000 more who had escaped earlier waves of discrimination and violence in recent years. The large-scale forced migration seemed to have stopped by the end of last year. And last November, bowing to international pressure, Myanmar signed a Chinese-brokered agreement with Bangladesh for the tentative repatriation of the refugees to newly constructed villages. The fulfillment of this plan is at best questionable, however: it calls for Myanmar authorities to verify that each refugee did, in fact, reside in Myanmar before he or she can return. But most Rohingya have no documents to prove their prior residency. More important, few of them wish to return to a country that has persecuted them for generations.

Daughter of the Revolution

Myanmar's government, and especially its army, known as the Tatmadaw, has earned worldwide condemnation for the campaign against the Rohingya. Last September, the UN's top human rights official, Zeid Ra'ad al-Hussein, denounced the army's "brutal security operation" as a "textbook case of ethnic cleansing." Critics singled out Suu Kyi for at best inaction and at worst providing political cover for the army's atrocities. Regardless of how one interpreted her motives, it was hard to square her actions with her status as a human rights icon. Suu Kyi is the daughter of Aung San, the revered Burmese revolutionary who shepherded his country to independence from the United Kingdom in the 1940s. In the 1990s, "the Lady," as she is referred to in Myanmar, led the NLD to victory in national elections. But the military nullified the results and placed her under house arrest for 15 of the next 21 years, before releasing her in 2010 as a gesture meant to highlight the government's nascent liberalization program. The disappointment in her lack of action to stop the bloodshed—or, worse, her complicity in it—has been profound. Archbishop Desmond Tutu of South Africa, one of numerous Nobel Peace Prize laureates who have expressed their disillusionment, lamented the silence of his "dearly beloved sister" and said that it was "incongruous for a symbol of righteousness to lead such a country."

But Suu Kyi's response should not have come as a great surprise. She has a long record of downplaying the Rohingya's plight. In March 2017, Suu Kyi's office dismissed detailed descriptions of Rohingya women suffering sexual violence at the hands of Myanmar's armed forces as "fake rape." Once a defender of press freedom, Suu Kyi has remained mum about the case of two Reuters journalists who were arrested by the military last December after investigating the military's involvement in the killing of ten Rohingya civilians. Suu Kyi's government did create a commission, headed by former UN Secretary-General Kofi Annan, to study the Rohingya issue, and has promised to implement its recommendations. But last January, Bill Richardson, a former U.S. ambassador to the UN and a longtime Suu Kyi supporter, quit a separate ten-member international advisory board on the Rohingya crisis that the Myanmar government had set up, calling it "a whitewash" and "a cheerleading squad for the government." As for Suu Kyi, Richardson said, "I like her enormously and respect her. But she has not shown moral leadership on the [Rohingya] issue."

Suu Kyi deserves a great deal of criticism. But in faulting her for not publicly confronting the military, let alone restraining the generals, some critics have ignored two fundamental realities of contemporary Myanmar. First is the intensity of anti-Rohingya sentiment in the country. Hatred of the Rohingya is widespread and deep-seated, stirred up by influential extremist Buddhist monks who are the military's political allies and who have incited violence against Rohingya. The ugly truth is that the vast majority of Burmese, including most of the NLD's supporters, approve of the anti-Rohingya campaign. Making pro-Rohingya statements and gestures would be tantamount to political suicide for Suu Kyi and her government and would only strengthen the army's public support.

Second, the civilian-led government has no control over the armed forces nor any means of reining them in. Even if Suu Kyi wanted to limit the military's campaign against the Rohingya, it would be almost impossible to do so. Myanmar's constitution, crafted by the military in 2008, ensures that the military remains far and away the country's strongest political institution. Amending the constitution requires more than 75 percent of the votes in the legislature—and 25 percent of parliamentary seats are set aside for armed forces personnel, which ensures that no changes can be made without the military's cooperation. In addition, the constitution reserves three key ministries for the armed forces: Defense, Border Affairs, and Home Affairs. The last of these oversees the General Administration Department, the administrative heart of the state, which is responsible for the day-to-day running of every regional and state-level government and the management of thousands of districts and townships. The constitution further safeguards the army's interests by allowing its commander in chief to name six of the 11 members of the National Defense and Security Council, a top executive body.

The army also sets its own budget and spends it without any civilian oversight: in 2017, the budget amounted to $2.14 billion, representing 13.9 percent of government expenditures—around three percent of the national GDP and more than the combined total allotted to long neglected health care and education. Perhaps just as consequential as the military's political dominance is its economic clout. By some estimates, active and retired military officers and their associates control over 80 percent of the economy.

Drafting the constitution and then holding a referendum to gain the public's endorsement represented two important steps in the military's long-term plan to manage and control a cautious move toward a "disciplined democracy," in its words, and to transfer responsibilities over day-to-day politics to a civilian government. Having shed the burden of governance, military elites focused on their own interests: modernizing the army and tending to their business empires. They gave up little that was dear to them, and the changes they have permitted remain easily reversible. No further democratization will occur unless the generals relinquish their constitutionally granted privileges.

Iron Lady

Suu Kyi has been unable to alter this basic dynamic. Following the 2015 elections, she failed to persuade the military brass to amend the constitution by removing its prohibition against anyone who has family members who hold foreign passports from serving as president. This clause directly targets Suu Kyi, whose late husband, Michael Aris, was British and whose two children are British citizens. In March 2016, the NLD-controlled legislature elected a confidant of Suu Kyi's, Htin Kyaw, as president; he has served a mostly ceremonial role. Suu Kyi created and took the position of "state counselor," giving herself a role akin to that of a prime minister—a fully defensible workaround to the military's move to block her from becoming president.

Less justifiable are the autocratic inclinations Suu Kyi has demonstrated since taking office and the extraordinary degree to which she has centralized power in her own hands. In addition to serving as state counselor, she also heads the Ministry of Foreign Affairs and retains the presidency of the NLD. As party chief, she has personally chosen every member of the party's Central Executive Committee—a violation of party rules. She is a micromanager who finds it difficult to delegate; most consequential decisions require her approval, which has led to bottlenecks. She sits on at least 16 governmental committees, all of which seldom produce concrete decisions. In November 2017, the government established a new ministry, dubbed the Office of the Union Government, just to help Suu Kyi cope with her workload.

Suu Kyi has also decided to act as her own spokesperson, but she has done a poor job of communicating her administration's policies. She prefers limited transparency: according to several NLD members of parliament with whom I have spoken, she has instructed them to not ask tough questions during parliamentary sessions and to avoid speaking to journalists. Her preference for personal loyalty over competence was illustrated by her appointment of several cabinet members with scant qualifications.

Suu Kyi is in her early 70s yet has no apparent successor, and her party is dominated by other septuagenarians who enjoy her trust but lack the energy, imagination, and skills necessary to carry out the comprehensive renewal the country needs. Although Suu Kyi has been exceedingly critical of the constitution, she has used its antidemocratic provisions when they have suited her purposes. For instance, she appointed two NLD members as chief ministers in Rakhine and Shan States, both of which are home to large minority ethnic communities—even though in both places, a candidate from a local party that represents those groups had won the popular vote.

The complex political situation in which Suu Kyi operates requires a leader with a firm hand and a clear sense of purpose. She remains very popular among ordinary Burmese, who admire her tenacity, respect her authority, and consider her the one indispensable leader. Her autocratic style and silence on the Rohingya crisis might be less troubling if her government had made significant progress on economic reform or on reconciliation with other ethnic minority groups. But it has not.

Words and Deeds

Although the NLD has been Myanmar's main opposition group since 1988, it has never formulated a policy program beyond vague promises of democracy, the rule of law, and economic reform. One might argue that it did not need detailed proposals to succeed as an opposition party: it had an iconic leader, and it stood against the army. But even after two years in power, major questions remain about the government's economic policies, positions on ethnic and religious issues, and plans for persuading the military to leave politics. Notwithstanding its limited room to maneuver, the government should have accomplished much more since taking office. Decades of military control of the economy have turned Myanmar into a desperately poor country. In 2017, its per capita GDP of $1,300 was the lowest in Southeast Asia, about half of that of Laos and one fifth of Thailand's. GDP grew by more than six percent in 2016 and 2017, but that was a slower rate of growth than the country enjoyed in the early years of the decade. Inflation has been nearing double digits, commodity prices have increased, and the job market's expansion has been anemic. Millions of Burmese have been forced to find employment abroad, mostly in so-called 3D jobs: tasks that are dirty, dangerous, and demeaning.

Reforming the economy should be the NLD government's most critical task, but it waited until July 2016 to present its first major statement on the issue. The document turned out to be little more than a wish list, a general outline that identified neither policy instruments nor specific objectives to achieve within a given time frame. So far, the government's main economic achievement has been the partial modernization of the legal framework governing investment. In January 2016, the legislature passed an arbitration law intended to boost investor confidence. Last year, it passed new rules on investment that are designed to simplify and harmonize existing regulations and that specify the privileges that will be granted to domestic and foreign investors. But the NLD has offered scant details on how the new rules will be implemented. The arrival of the NLD government had fueled hopes of increased foreign direct investment, but partly as a result of its lack of action, such investment has actually tapered off since 2015.

Suu Kyi's record on other pressing economic issues has been even less impressive. Agriculture represents 37 percent of Myanmar's GDP and employs, directly or indirectly, about 70 percent of the country's labor force. But farmers tend to be extremely poor, and farming profits are among the lowest in Asia. The government must find a way to provide farmers with what they most need to increase their earnings: high-quality seeds and fertilizers, improved water control and irrigation facilities, and access to affordable credit.

Farmers also suffer from a lack of land rights. For several decades, the military expropriated hundreds of thousands of acres from helpless peasants, offering little or no compensation. In 2016, groups of farmers sent letters to the army's commander in chief, Senior General Min Aung Hlaing, requesting the return of their land. The military's response was to threaten the farmers and their lawyers with defamation lawsuits. Suu Kyi's government has said that dealing with the landownership issue is a priority, but she has done little.

The military regime also grossly neglected the country's infrastructure. Roads, railways, and public transportation systems all lie in a pitiful state of disrepair. Even more serious is the shortage of electricity: only one-third of the population has access to it, and blackouts are frequent, even in Yangon's luxury hotels. Economic growth will put even more pressure on the electricity supply, and shortages will likely get worse. These weaknesses affect every economic sector and scare off potential investors. But Suu Kyi's government seems to have realized the importance of infrastructure only recently. A number of plans have been drawn up, and the government has held some summits on the issue. But on this, too, there has been little action to match the government's rhetoric.

A Double Bind

Despite the lack of progress, relations between the civilian government and the military have settled into what Suu Kyi has described as a "normal" routine. The most charitable interpretation of Suu Kyi's accommodation of the military is that she hopes that, over time, the generals will conclude that their interests would be best served by leaving politics. The army appears to be taking its time: Min Aung Hlaing has said that the Tatmadaw intends to reduce its presence in parliament, but he has refused to set a timetable.

Part of the problem is that the military has two conflicting goals. The generals want to transform the army—which is plagued by obsolete equipment, archaic training methods, and poor morale—into a professional force comparable to its counterparts in other countries in the region. In order to do that, the military needs help from more developed, powerful countries—help that, in the case of Western governments, is conditioned on the military leaving politics. But Western governments also insist that the government must do more to resolve its many conflicts with ethnic minority groups. In September 2017, the United Kingdom's Ministry of Defence announced that it would suspend educational courses it provided for the Tatmadaw, citing ongoing violence and human rights abuses. Similarly, the following month, the Trump administration announced the withdrawal of U.S. military assistance from officers and units participating in the operations in Rakhine State and rescinded invitations to senior members of Myanmar's security forces to U.S.-sponsored events. Then, in November, a bipartisan group of U.S. representatives introduced the Burma Act of 2017, which, among other things, would reinstate sanctions against the Tatmadaw that were lifted the year before to reward the country for its putative progress and to incentivize more steps in the direction of democracy. That legislation has yet to be put to a vote. In December, however, the U.S. House of Representatives passed a resolution condemning "the ethnic cleansing of the Rohingya," and the Trump administration imposed new sanctions on Major General Maung Maung Soe, who has overseen the brutal campaign against the Rohingya.

But the military's troubling treatment of minorities extends far beyond the Rohingya. Several ethnic communities have been at war with the government for long periods—in some cases, ever since Burma proclaimed its independence in 1948. Together, these conflicts form something like a low-level, multifaceted civil war. Some ethnic groups bear long-held grudges against others, sometimes related to overlapping land claims. Individual ethnic communities themselves are often divided by sectarian differences. Aside from causing thousands of deaths and displacing millions, ethnic violence has prevented the consolidation of central authority over the country, as well as the formation of a shared national identity.

For decades, the military has prolonged ethnic conflicts in a bid to justify its continued rule. The fighting has also given cover to generals who profit from the drug trade (Myanmar is a major source of opium) and from the illegal export of gems, gold, and timber. But in recent years, the Tatmadaw has appeared more determined to end the civil war. In October 2015, prior to Suu Kyi's electoral victory, eight ethnic armed organizations and the government signed the National Ceasefire Agreement, brokered by the military, although some of the largest and most influential ethnic groups stayed away.

During her campaign, Suu Kyi repeatedly identified achieving ethnic peace as her number one priority. After the realization of that objective, she was to pursue the creation of a federal system of the sort first promised by her father in the 1940s. There is no agreement on what precise shape that system would take, except that it would grant more autonomy to ethnic groups but stop short of giving them the right to secede. Suu Kyi's promise to pursue ethnic peace was a tactical mistake, however, since she has little influence over how the military wages its wars against ethnic armed organizations, and the idea of a federal system is anathema to the generals.

Nevertheless, her administration organized conferences in August 2016 and May 2017 with the aim of persuading more ethnic armed organizations to sign the National Ceasefire Agreement; the talks brought together armed groups, the military, and the government. Predictably, the meetings achieved little besides providing a forum for grand speeches and gestures, and in the aftermath of the conferences, the fighting actually intensified in several regions. In February 2018, two additional rebel groups signed on to the ceasefire amid much fanfare. But the groups that represent four-fifths of all the ethnic armed personnel in the country remain as opposed to signing as ever. Meanwhile, the generals adamantly refuse to create a federal army that would represent the country's ethnic groups and regions, which is one of the ethnic armed organizations' key demands; the military falsely contends that the armed forces are already inclusive and fair. At the same time, the armed groups have refused to disavow secession—a position that the military insists they must take as part of any final agreement.

Optimists believe that the ethnic armed organizations' chief objective is to maximize their gains on the ground in preparation for eventual peace negotiations. In reality, their ultimate goal is the establishment of a federal system. Such a system represents a redline for the Tatmadaw: although military elites have adopted an increasingly pragmatic approach toward negotiation with the ethnic armed organizations, they continue to see federalism as the first step toward the country's disintegration. The word "federalism" is no longer taboo in public discourse, as it had been for decades, but the top brass are unlikely to relax their long-standing opposition to a federal system anytime soon.

Beijing Beckons

In the face of Western opprobrium over Myanmar's treatment of the Rohingya, there are signs that the military might abandon its relatively recent quest to placate Western governments and instead return to a strategy of reliance on its traditional patron, China. China has always been Myanmar's top trading partner and biggest investor, and for decades, Beijing was the main sponsor of Myanmar's military junta. Suu Kyi's first major trip abroad as state counselor, in August 2016, took her to Beijing. Her discussions there centered on business and trade issues, especially a few large infrastructure projects, such as a $7.2 billion deep-sea port in Rakhine State that China plans to build to give Chinese ships access to the Indian Ocean. Since then, the relationship between Myanmar and China has improved; in November 2017, Chinese President Xi Jinping described this moment in Chinese-Myanmar military relations as being the "best ever." It helps that Chinese officials, unlike Western ones, do not admonish Suu Kyi and her government for their human rights violations.

But the Chinese are pressing for progress on the civil war. Chinese leaders have endorsed negotiations between the Myanmar government and the country's ethnic armed organizations, and Beijing facilitated the participation of some recalcitrant groups in the May 2017 conference. China has played a complex role in the civil war for decades, backing the government but also providing shelter, weapons, and training to some of the belligerent groups; such contacts have allowed China to extract natural resources (mostly illegally), such as jade, gold, and timber, from regions where militants operate. But it now seems that the Chinese want the violence to end because the rebels' objectives have shifted from merely resisting government forces to improving their status within Myanmar, an aim that is more conducive to internal stability, and because the fighting has impeded economic development and trade. What's more, the Chinese want to be seen as peacemakers in a region where they have long been regarded as a destabilizing presence. Compared with Beijing, Washington today has little sway over Myanmar. That is a recent development: the Obama administration, in one of its undisputed foreign policy successes, managed to convince the country to take steps toward democracy. The United States was a steadfast supporter of Suu Kyi for years before she took office, and both President Barack Obama and Secretary of State Hillary Clinton made historic visits to Yangon. When Suu Kyi visited Washington in September 2016, she asked Obama to lift most of the remaining U.S. sanctions on Myanmar in order to help her government grow the country's economy. Obama obliged her; in hindsight, that was likely a mistake. Obama wanted to reward progress. But lifting the sanctions robbed Washington of precisely the kind of leverage it now needs. Indeed, democratic activists in Myanmar and elsewhere had hoped that the sanctions would stay in place until the antidemocratic features of the 2008 constitution were abolished.

Since the Rohingya crisis erupted last year, there have been few official interactions between the United States and Myanmar. Under the Trump administration, Myanmar has lost the special place it enjoyed on Washington's foreign policy agenda during the Obama years. Then U.S. Secretary of State Rex Tillerson made a five-hour visit to Myanmar in November 2017. In meetings with Suu Kyi and the army chief, Min Aung Hlaing, he raised concerns about ethnic violence. At a news conference, Tillerson said that there had been "crimes against humanity," but he did not back the idea of new economic sanctions against Myanmar. Pope Francis visited the country a few days later and called for peace and mutual respect. But neither Tillerson nor the usually outspoken pope used the term "Rohingya" during his discussions with Myanmar officials or in his public statements, likely out of a fear that doing so would aggravate an already highly charged situation and, in the case of the pope, out of a fear that it could endanger Myanmar's small and vulnerable Catholic community.

Admittedly, the United States has few appealing policy options for stopping the ethnic cleansing. Restoring the sanctions or placing more new ones on the generals would likely just drive the military further into the welcoming arms of the Chinese, who are keen to fill the vacuum left by Washington's flagging interest. Denunciations from Washington and other foreign capitals have failed to affect the government's position on the Rohingya and have actually increased domestic support for the Tatmadaw, as evidenced by a number of major promilitary rallies held throughout Myanmar last fall.

Still, there are ways for the United States to push for progress. For starters, it should suspend all military-to-military engagement with Myanmar and expand assistance to a number of sophisticated but underfunded nongovernmental organizations, such as Mosaic Myanmar, a civil society group that promotes tolerance between the majority Buddhist community and Christian and Muslim minorities.

Furthermore, the United States should establish or sponsor programs in Myanmar focused on health care, educational opportunities, and cultural exchanges. In a country that tends to be at best cautious of foreigners' intentions, the United States is generally held in high regard, according to Asian Barometer surveys of public opinion—a sharp contrast to the suspicious attitudes toward China and India that prevail throughout Myanmar. There are few societies where prudent U.S. democracy promotion could find more fertile ground or where it would be more gratefully accepted. Finally, in lieu of public expressions of indignation, Washington should communicate its displeasure privately—especially through politicians with long histories of supporting Suu Kyi and Myanmar's democratization, such as Clinton and Republican Senator Mitch McConnell of Kentucky.

As for Suu Kyi, her reduced stature abroad might further reduce her already limited leverage with the generals. She is boxed in to a degree that many critics fail to appreciate. But she has made her own situation worse through poor management and a lack of focus on issues that are under her administration's control: improving the economy, shoring up infrastructure, and revamping the health-care and educational systems. The government should adopt a personnel policy that emphasizes merit and accomplishment instead of personal loyalty to Suu Kyi. Instead of alienating ethnic minorities and their political parties and ignoring civil society organizations, Suu Kyi ought to open a meaningful dialogue with them with a view to forming a big-tent political and social coalition that might, in time, challenge the military's political supremacy. Suu Kyi and her administration should reverse their attacks on media freedoms. And even though the government cannot control the military, it must stop denying and defending the Tatmadaw's atrocities and start actively protecting those who have suffered so terribly from the army's repression.

Most important, Suu Kyi must shift gears quickly. International patience with her is almost extinguished. If she does not change course soon, she will lose what little goodwill remains.

Critical Thinking

1. Are the Rohingya victims of ethnic cleansing or genocide? Why or why not?
2. Why hasn't Suu Kyi shown moral leadership on the Rohingyan issue?
3. Discuss the state of Myanmar's economy.

Internet References

Special Envoy of the Secretary-General for Yemen
https://osesgy.unmissions.org/

Report of the Independent Fact-Finding Commission
https://www.ohchr.org/

The Rohingya Refugee Crisis
https://www.doctorswithoutborders.org

ZOLTAN BARANY is Frank C. Erwin, Jr., Centennial Professor of Government at the University of Texas and the author of *How Armies Respond to Revolutions and Why.*

Article

Prepared by: Robert Weiner, *University of Massachusetts, Boston*

The Pyrrhic Victories of Venezuela's President

Alejandro Velasco

Learning Outcomes

After reading this article, you will be able to:

- Learn the causes of Venezuela's economic crisis.
- Learn why the Maduro regime has lost support in Venezuela.

Christmas Eve should have been a time for making savory *hallacas* and dancing *gaitas* in Ciudad Bolívar, some 350 miles southeast of Caracas, as in the rest of Venezuela. But as evening fell, reports of protests and looting in some of the city's working-class neighborhoods began to circulate on social and conventional media. By the time police restored order, several grocery and liquor stores had been ransacked and dozens of people arrested. Over the following week, similar protests spread to working-class areas of other major cities, eventually turning lethal: in Caracas on New Year's Eve, a National Guard sergeant shot dead 18-year-old Alexandra Colopoyn, 25 weeks pregnant, when tensions flared in a crowd waiting for food long promised by the government of President Nicolás Maduro.

A deadly protest in the waning hours of 2017 seemed a fitting end to a year that saw no shortage of either deaths or demonstrations, while everything from food to medicine to cash grew scarcer amid possibly the worst economic crisis in Venezuela's modern history. Between April and July, massive daily rallies against Maduro roiled Caracas and other cities. Over 120 people died as security forces suppressed the increasingly violent demonstrations. Meanwhile, dwindling oil revenues led to drastically slashed imports. In a country ever more dependent on foreign goods as its domestic productive sector collapses, this aggravated already alarming scarcities of consumer staples. Coupled with a currency worth less and less as the government printed more and more cash to finance a vast state bureaucracy, the result by year's end was a hyperinflationary spiral growing worse by the day.

Even against this backdrop, the turmoil at the end of December, spreading fast and now extending into the new year, was remarkable. Unlike the months-long demonstrations earlier in the year, led and organized primarily by anti-government activists seeking Maduro's ouster, these protests stem from popular sectors of society whose support the government has relied on to survive. Food-related riots have become increasingly common across Venezuela as the economy free-falls. But the latest protests have struck at a moment when Maduro appeared, at last, to have consolidated his political power over both the fractious pro-government movement and a once-ascendant opposition. Taken together, these developments signal a major shift in Venezuela's political dynamic, portending an unprecedented wave of social unrest that may prove the government's undoing just when it seemed to have outmaneuvered its enemies.

Winning the Battle

Before the outbreak of the year-end protests, Maduro had reason to feel bullish. For the first time in the nearly five years since he won the presidency by a razor's-edge margin following Hugo Chávez's death in March 2013, Maduro seemed fully in control after vanquishing his opponents outside and within *chavismo*, the late president's popular movement. In August, a Constituent Assembly he convened to sidestep the opposition-controlled National Assembly and rewrite the constitution was seated despite worldwide condemnation. That followed four months of some of the most intense protests Venezuela had seen in over a decade, sparked by a decision by the Maduro-appointed Supreme Court that effectively nullified the National Assembly.

As repression increased and the death toll mounted, so too did speculation that the governing elite would abandon Maduro. But only one major figure—Attorney General Luisa Ortega Díaz—broke ranks, eventually fleeing the country. Instead, Maduro emerged in a stronger position once the protests ended after the Constituent Assembly's installation. In October, pro-government candidates took 18 of 23 states in long-delayed gubernatorial elections, and in December even more sweeping victories in municipal elections gave the government control over almost all cities and towns in Venezuela.

It was a dramatic turn of events, particularly given the economy's exponential collapse over the same period and the impact on everyday life. Venezuela began 2017 with a world-leading 500-percent inflation rate, and ended it in hyperinflation of over 2,000 percent. The black market for dollars began the year at an astonishing 15,000 bolívars to one dollar—compared with the official rate of 10 to one—and closed 2017 at over 130,000, effectively dollarizing much of the economy. Meanwhile, imports contracted 35 percent as the government prioritized paying bondholders instead of declaring default and risking the loss of oil industry assets abroad.

The combined effect was that fewer products were available, and those mainly at black market prices. Most basic necessities, from rice and meat to toilet paper and aspirin, were hard to obtain. And infrastructure crumbled due to lack of maintenance and spare parts, subjecting Venezuelans to roads in disrepair, deficient public transit, irregular trash collection, and water and power outages. More and more people flocked to neighboring Colombia and Brazil, part of an exodus no longer limited to the middle classes seeking opportunities. Increasingly they were from the popular sectors, seeking basic sustenance.

In part, Maduro's improbable political turnaround despite a brutal crisis reflected an outright turn to authoritarianism over the previous year. To be sure, abuses of power are long-standing in *chavismo*—opponents have been targeted via violent repression, legal artifice, and unfair electoral conditions. Maduro drew on each of those methods in order to remain in power at all costs. But 2017 also marked a decided departure from *chavismo* in two significant ways that brought political survival for the president at a price that might ultimately prove his victories pyrrhic.

The first was sacrificing electoral legitimacy. Elections constituted a key pillar not just of Hugo Chávez's plebiscitary style of politics but also of his movement's larger appeal to supporters. Holding and winning elections offered *chavistas*, many of whom had felt disenfranchised before Chávez came to power, a sense of identity and of belonging to a popular movement that could legitimately claim a majority. The opposition was comprised largely of middle and upper classes who stubbornly, at times antidemocratically, refused to acknowledge the government's democratic underpinnings. Cries of electoral fraud had been raised since 2004, when Chávez overwhelmingly won a referendum to defeat an attempt to recall him from office. But absent credible evidence, such claims rather confirmed the suspicions of even weakly identified *chavistas* that the opposition sought power, not democracy. And while *chavismo* often resorted to skewing electoral processes in its favor—gerrymandering, blacklisting opposition candidates, misusing public funds for campaigns, and more—elections nonetheless were widely understood to reflect the will of the electorate.

But the mystique of electoral legitimacy began to collapse in 2016. After a long-splintered opposition united to win control of the National Assembly in a landslide in late-2015 elections, it seized the momentum to demand a referendum on recalling Maduro, as the constitution permitted. But with its support plummeting in polls as the economy nosedived, the government delayed and eventually nixed the referendum effort on the thinnest of technicalities. That move sparked an international outcry and massive, peaceful, cross-class marches in Venezuela. However, it was the government's suspension in late 2016 of regularly scheduled gubernatorial elections, with even less of an explanation, that punctured for government supporters the sense that *chavismo* did not fear the ballot box.

Maduro began 2017 with some of the lowest approval ratings of his presidency, losing support even among strongly identified *chavistas*. By mid-July, following months of often violent street protests, the opposition staged a plebiscite in which it claimed that 7.5 million people rejected Constituent Assembly elections scheduled for later that month. Although impossible to confirm, the figure was symbolically powerful, putting pressure on the government to surpass it in two weeks' time. But when Maduro announced that just over 8 million votes had been cast in the elections, which lacked many of the oversight protocols of previous contests, representatives of the voting-machine company the government had used since 2004 refused to certify the results. The process made it clear that Maduro would sooner steal than lose an election if his political survival was at stake.

That was a message aimed not only at anti-*chavistas* but also at critics within *chavismo* itself. Beyond sacrificing electoral legitimacy, the Constituent Assembly blew open a long gestating schism: between those loyal to Chávez's 1999 constitution and its calls for participatory democracy, and those more in line with Chávez's later vision of a socialist Venezuela. In principle, Maduro's convocation of a Constituent Assembly addressed that second, more ideologically committed group of grassroots *chavistas* who had long advocated for just such a move. In practice, however, it left them out.

Despite protests by grassroots *chavistas* against vote rigging, most seats went to Maduro loyalists. Since the Constituent Assembly opened in August, they have offered no proposals for a new constitution but rather acted at Maduro's behest to circumvent the opposition-controlled National Assembly. Meanwhile, the government has moved against some of Chávez's former ministers who had begun to criticize Maduro, as well as grassroots *chavista* activists who opposed pro-Maduro candidates in municipal elections.

The sacrifice of electoral legitimacy for the Constituent Assembly and the attendant consolidation of Maduro's inner circle dramatically conveyed, to both *chavistas* and opponents alike, that Maduro would stay in power not because but in spite of the ballot box. It was at this critical moment that Maduro departed in a second, immensely consequential way from *chavismo*. He pressed ahead to become the only game in town.

Losing the War

If legitimate elections were a key pillar of *chavismo*, an opposition it could credibly present as a threat to its power was another. Throughout his presidency Chávez both stoked and benefited from an opposition singularly focused on ousting him from office. Although its strategies for doing so changed over time, from coups to strikes to elections, anti-*chavismo* long proved skillful at "snatching defeat from the jaws of victory," as the sociologist and Venezuela analyst David Smilde has observed. Again and again, whether due to internal divisions, overreach, or miscalculation, the opposition squandered political gains just when the government was weakest. Chávez, and Maduro earlier in his term, deftly exploited this dynamic, riding the ebbs and flows of the opposition from one crisis to the next but always providing it just enough room to maneuver in order to keep alive the claim that economic elites might return to power and roll back social and political gains made under Chávez.

Immediately following the Constituent Assembly elections, major clashes that had roiled Caracas for almost four months stopped cold. The failure of protests, a plebiscite, and growing international condemnation to derail Maduro's plans dealt a crushing blow to the opposition, leaving it adrift and demoralized. The end of daily protests and of their disruptions to the everyday lives of most people already struggling to get by lifted Maduro's popularity, less in support than in relief.

With the opposition in disarray and his poll numbers rising, Maduro made his next move. This time, he was the one who overreached. In September, Constituent Assembly officials announced that gubernatorial elections, inexplicably postponed a year before and rescheduled for December 2017, would in fact take place in October. The move further split a reeling opposition between those advocating a boycott and those choosing to field candidates. The opposition won five states, but when Maduro required incoming governors to swear loyalty to the Constituent Assembly, all but one did so, effectively neutralizing them. Pro-government candidates swept municipal elections held in December in a landslide, as erstwhile *chavistas* returned to the polls, drawn by Maduro's promise that he alone could deliver relief for the dire and mounting needs of Venezuelans.

But relief is precisely what Maduro cannot deliver—despite the enormous expectations he generated around Local Food Production and Provision Committees (called CLAPs, their Spanish acronym). Formed in April 2016 as an emergency response to already grave food shortages, CLAPs use existing get-out-the-vote mechanisms at the neighborhood level to distribute food staples at highly subsidized prices on a biweekly basis. But massive corruption plagues the process. Direct delivery of food was meant to bypass so-called *bachaqueo*, whereby products imported at the official exchange rate are sold at black market rates locally or smuggled abroad. Maduro tapped military commanders to manage food distribution; increasingly they are most responsible for diverting subsidized products to the black market or abroad, reaping enormous profits in the process.

The problem is only in part about corrupt distribution chains. As Venezuela runs out of money to maintain even minimum import levels, the key issue is supply. Massive debt contracted during years of unchecked spending at the height of the oil boom between 2004 and 2014 has left the government strapped for cash to purchase goods. It has stubbornly avoided even basic reforms—like ending the preferential exchange of dollars that fuels massive corruption—and has prioritized paying bondholders in order to avoid default, which would leave oil industry assets such as oil tankers and refineries vulnerable to seizure abroad. Still, 2017 ended with a spate of late bond payments, targeted defaults, and public pleas by Maduro to private creditors for restructuring talks.

Sanctions imposed by the United States on trading in Venezuelan bonds have sapped the country's ability to issue more debt in order to raise revenues. In the past, bailouts from Russia and China have provided temporary relief. But China, to which Venezuela already owes over $50 billion, has signaled it will offer no more loans, leaving only Russia as a source for credit. Even rising oil prices offer little hope for reprieve, since lack of investment in the oil industry has decimated production, now at its lowest level in decades.

As the crisis worsens exponentially, more and more Venezuelans, especially among popular sectors with no access to dollars, have come to depend on biweekly CLAPs for subsistence, while Maduro relies on the CLAPs to generate support. Ahead of the municipal elections in December, the government promised to add *pernil*—roast pork leg, a traditional Venezuelan Christmas dish—to the list of subsidized items for CLAPs to distribute, with payments made up front. But after the electoral sweep, and as Christmas neared, *perniles* failed to materialize. Lacking a domestic opposition to credibly blame, Maduro accused Portugal and Colombia of holding up the promised pork, allegations promptly denied by both governments.

The protests and looting that broke out in Ciudad Bolívar on Christmas Eve have spread across the country, with few signs of winding down. And unlike protests they can dismiss, and attack, as opposition-led, these have sprouted among popular sectors whose support Maduro and his government need to stay in power.

At no time since it first came to power in 1999 have conditions been more dire for *chavismo*—and for reasons almost entirely of its own doing. Whereas a year ago it was possible to envision the government losing electorally in order to survive politically, 2017 proved that Maduro and those around him are willing to sacrifice the movement in order to stay in power. At a moment when he controls the political landscape, Maduro alone stands to be blamed for a crisis that worsens by the day. The protests—and what they portend for the year ahead—are an indication of the government's success in consolidating political power at a price that seems pyrrhic, at a time when it can least afford it. The end of 2017 may have been just the beginning of Maduro's undoing.

Critical Thinking

1. How did President Maduro consolidate his power?
2. How did Maduro depart from Chavism?
3. What are the main features of the economic collapse of Venezuela?

Internet References

National Endowment for Democracy
https://www.ned.org

Venezuela, The World Factbook, CIA
https://www.cia.gov/library/publications/the-world-factbook

Alejandro Velasco is an associate professor of Latin American history at New York University and the author of *Barrio Rising: Urban Popular Politics and the Making of Modern Venezuela* (University of California Press, 2015).

Article

Prepared by: Robert Weiner, *University of Massachusetts, Boston*

The Age of Insecurity: Can Democracy Save Itself?

RONALD INGLEHART

Learning Outcomes

After reading this article, you will be able to:

- Learn what are the differences between materialists and postmaterialists.
- Learn about the relationship between cultural wars and the rise of populism and authoritarianism in advanced democracies.

Over the past decade, many marginally democratic countries have become increasingly authoritarian. And authoritarian, xenophobic populist movements have grown strong enough to threaten democracy's long-term health in several rich, established democracies, including France, Germany, the Netherlands, Sweden, the United Kingdom, and the United States. How worried should we be about the outlook for democracy?

The good news is that ever since representative democracy first emerged, it has been spreading, pushed forward by the forces of modernization. The pattern has been one of advances followed by setbacks, but the net result has been an increasing number of democracies, from a bare handful in the nineteenth century to about 90 today. The bad news is that the world is experiencing the most severe democratic setback since the rise of fascism in the 1930s.

The immediate cause of rising support for authoritarian, xenophobic populist movements is a reaction against immigration (and, in the United States, rising racial equality). That reaction has been intensified by the rapid cultural change and declining job security experienced by many in the developed world. Cultural and demographic shifts are making older voters feel as though they no longer live in the country where they were born. And high-income countries are adopting job-replacing technology, such as artificial intelligence, that has the potential to make people richer and healthier but also tends to result in a winner-take-all economy.

But there is nothing inevitable about democratic decline. Rising prosperity continues to move most developing countries toward democracy—although, as always, the trajectory is not a linear one. And in the developed world, the current wave of authoritarianism will persist only if societies and governments fail to address the underlying drivers. If new political coalitions emerge to reverse the trend toward inequality and ensure that the benefits of automation are widely shared, they can put democracy back on track. But if the developed world continues on its current course, democracy could wither away. If there is nothing inevitable about democratic decline, there is also nothing inevitable about democratic resurgence.

By Popular Demand

Over the past two centuries, the spread of democracy has been driven by the forces of modernization. As countries urbanized and industrialized, people who were once scattered over the countryside moved into towns and cities and began working together in factories. That allowed them to communicate and organize, and the economic growth driven by industrialization made them healthier and wealthier. Greater economic and physical security led successive generations to place less emphasis on survival and more on intangible values, such as freedom of expression, making them more likely to want democracy. Economic growth also went hand in hand with more education, which made people better informed, more articulate, more skilled at organizing, and therefore more effective at pushing for democracy. Finally, as industrial societies matured, jobs shifted from manufacturing to knowledge sectors. Those new occupations involved less routine and more independence. Workers had to think for themselves, and that spilled over into their political behavior.

Moreover, democracy has a major advantage over other political systems: it provides a nonviolent way to replace a country's leaders. Democratic institutions do not guarantee that the people will elect wise and benevolent rulers, but they do provide a regular and nonviolent way to replace unwise and malevolent ones. Nondemocratic leadership successions can be costly and bloody. And since democracy enables people to choose their leaders, it reduces the need for repressive rule. Both these advantages have helped democracy survive and spread.

For the past few decades, the most striking alternative to the democratic path has come in China. Since its disastrous experience under Mao Zedong, the country has been governed by an exceptionally competent authoritarian elite. That reflects the political genius of Mao's successor, Deng Xiaoping. In addition to guiding China toward a market economy, Deng established norms that limited top leaders to two five-year terms in office and mandated retirement at age 70. He then selected some of the country's most competent 60-year-olds to run the government and installed a carefully chosen group of 50-year-olds below them. For roughly two decades after Deng's retirement, China was governed by the people he had selected. In 2012, that group chose a new generation of leaders. Despite growing cronyism and corruption, this group also seems competent, but its leader, Xi Jinping, is maneuvering to establish himself as dictator for life, abandoning Deng's system of predictable, nonviolent successions. If Xi succeeds, China's government is likely to become less effective.

Most authoritarian countries, however, are not governed nearly as effectively as contemporary China (nor was China under Mao). During the early stages of industrialization, authoritarian states can attain high rates of economic growth, but knowledge economies flourish best in open societies. In the long run, democracy seems to be the best way to govern developed countries.

Fits and Starts

The long-term trend toward democracy has always had ups and downs. At the start of the twentieth century, only a few democracies existed, and even they were not full democracies by today's standards. The number increased sharply after World War I, with another surge following World War II and a third at the end of the Cold War. Sooner or later, however, each surge was followed by a decline.

Democracy's most dramatic setback, which came in the 1930s, when fascism spread over much of Europe, was partially driven by economic decline. Under relatively secure conditions in 1928, the German electorate viewed the Nazi Party as a lunatic fringe, giving it less than three percent of the vote in national elections that year. But in July 1932, with the onset of the Great Depression, the Nazis won 37 percent of the vote, becoming the largest party in the Reichstag, before taking over the government the next year. Each period of democratic decline brought a widespread belief that democracy's spread had ended and that some other system—fascism, communism, bureaucratic authoritarianism—would be the wave of the future. But the number of democracies never fell back to its original level, and each decline was eventually followed by a resurgence.

The defeat of the Axis powers in World War II largely discredited authoritarian parties in the developed world: from 1945 to 1959, they drew an average of about seven percent of the vote across the 32 Western democracies that contained at least one such party. Then, in the 1960s, as the unprecedented prosperity of the postwar era took hold, their support fell even further, to about five percent, and it remained low during the 1970s.

After 1980, however, support for authoritarian parties surged. By 2015, they were drawing an average of more than 12 percent of the vote across those 32 democracies. In Denmark, the Netherlands, and Switzerland, authoritarian parties became the largest or second-largest political bloc. In Hungary and Poland, they won control of government. Since then, they have grown even stronger in some countries. In the 2016 U.S. presidential election, the Republican candidate Donald Trump campaigned on a platform of xenophobia and sympathy toward authoritarianism, yet he won 46 percent of the vote (and the Electoral College). In Austria's 2016 presidential election, Norbert Hofer, the far-right Freedom Party candidate, narrowly lost with 46 percent of the vote. In France, Marine Le Pen, the leader of the National Front, won 34 percent of the vote in last year's presidential election, almost double her party's previous high. Ever since World War II, Germans have had a strong aversion to authoritarian, xenophobic parties, which for decades never surpassed the five percent threshold required for representation in the Bundestag. But in 2017, the authoritarian, xenophobic Alternative for Germany won 13 percent of the vote, becoming Germany's third-largest party.

The Era of Not-so-good Feelings

To a large degree, the shifts between democracy and authoritarianism can be explained by the extent to which people feel that their existence is secure. For most of history, survival was precarious. When food supplies rose, population levels rose with them. When food grew scarce, populations shrank. In both lean and fat times, most people lived just above the starvation level. During extreme scarcity, xenophobia was a realistic strategy: when a tribe's territory produced just enough food to sustain it, another tribe moving in could spell death for the original inhabitants. Under these conditions, people tend to close ranks behind strong leaders, a reflex that in modern times leads to support for authoritarian, xenophobic parties.

In rich countries, many people after World War II grew up taking their survival for granted. They could do so thanks to unprecedented economic growth, strong welfare states, and peace between the world's major powers. That security led to an intergenerational shift in values, as many people no longer gave top priority to economic and physical security and no longer felt as great a need to conform to group norms. Instead, they emphasized individual free choice. That sparked radical cultural changes: the rise of antiwar movements, advances in racial and gender equality, and greater tolerance of the LGBTQ community and other traditional out-groups.

Those shifts provoked a reaction among older people and those holding less secure positions in society (the less educated, the less well off) who felt threatened by the erosion of familiar values. During the past three decades, that sense of alienation has been compounded by an influx of immigrants and refugees. From 1970 to 2015, the Hispanic population of the United States rose from five percent to 18 percent. Sweden, which in 1970 was inhabited almost entirely by ethnic Swedes, now has a foreign-born population of 19 percent. Germany's is 23 percent. And in Switzerland, it is 25 percent.

All this dislocation has polarized modern societies. Since the 1970s, surveys in the United States and other countries have revealed a split between "materialists," who stress the need for economic and physical security, and "postmaterialists," who take that security for granted and emphasize less tangible values.

In the U.S. component of the 2017 World Values Survey, respondents were asked a list of six questions, each of which required choosing which of two goals was most important for their country. Those who chose things such as spurring economic growth, fighting rising prices, maintaining order, and cracking down on crime were defined as materialists. By contrast, those who gave top priority to things such as protecting freedom of speech, giving people more say in important government decisions, and having greater autonomy in their own jobs were designated postmaterialists.

In recent U.S. presidential elections, this split has had a major influence on voting patterns, dwarfing the effects of other demographic traits, such as social class. Consider the 2012 election: those who gave priority to materialist values in all six of their choices were 2.2 times as likely to have voted for the Republican candidate, Mitt Romney, as they were for the Democratic candidate, Barack Obama, and those who gave priority to postmaterialist values in all six choices were 8.6 times as likely to have voted for Obama as they were for Romney. This relationship grew even stronger in 2016, when Trump, an openly racist, sexist, authoritarian, and xenophobic candidate, ran against Hillary Clinton, a liberal and cosmopolitan one, who was also the first woman nominated by a major party. Pure materialists were now 3.8 times as likely to vote for Trump as they were for Clinton, and pure postmaterialists were a stunning 14.3 times as likely to vote for Clinton as they were for Trump.

Economic insecurity can exacerbate these cultural pressures toward authoritarianism. In 2006, the Danish public was remarkably tolerant when protesters burned Danish embassies in several Muslim-majority countries in response to a cartoon of the Prophet Muhammad published by a Danish newspaper. At the height of the crisis, there was no Islamophobic backlash in Denmark. The next year, the anti-Muslim Danish People's Party won 14 percent of the vote. But in 2015, in the wake of the Great Recession, it won 21 percent, becoming Denmark's second-largest party. A backlash against the European migrant crisis was the immediate cause of the party's support, but rising economic insecurity strengthened the reaction.

For Richer, For Poorer

Economic insecurity need not take the form of absolute hardship to undermine democracy. In the vast literature on democratization, researchers disagree on many issues, but one point draws almost unanimous acceptance: extreme inequality is incompatible with democracy. Indeed, it is not surprising that the rise in support for authoritarian parties over the last three decades roughly parallels the rise in inequality over the same period.

According to data compiled by the economist Thomas Piketty, in 1900, in France, Germany, Sweden, and the United Kingdom, the wealthiest ten percent of the population took home 40 to 47 percent of total income before taxes and transfers. In the United States, the figure was 41 percent. By around 1970, things had gotten better, and the share going to the top ten percent in all five countries fell to levels ranging from 25 percent to 35 percent. Since 1980, however, income inequality has risen in all five countries. In the United States, the top ten percent now takes home almost half of the national income. In all but one of the countries in the Organization for Economic Cooperation and Development for which data are available, income inequality rose from 1980 to 2009.

Although inequality in almost all developed countries has followed a U-shaped pattern, there are striking differences between them that reflect the effects of varying political systems. Sweden stands out: although it had substantially higher levels of inequality than the United States in the early twentieth century, by the 1920s, it had lower income inequality than the other four countries in Piketty's study, and it has maintained that to this day. The advanced welfare state introduced by Sweden's long-dominant Social Democrats is largely responsible for the country's low inequality. Conversely, the conservative policies implemented by U.S. President Ronald Reagan and British Prime Minister Margaret Thatcher in the 1980s weakened labor unions and sharply cut back state regulation, leading to higher levels of income inequality in the United States and the United Kingdom than in most developed countries.

As long as everyone was getting richer, rising inequality did not seem to matter much. Some people might have been rising faster than others, but everyone was going in the right direction. Today, however, everyone isn't getting richer. For decades, the real income of the developed world's working classes has been declining. Fifty years ago, the largest employer in the United States was General Motors, where workers earned an average of around $30 an hour in 2016 dollars. Today, the country's largest employer is Walmart, which in 2016 paid around $8 an hour. Less educated people now have precarious job prospects and are shut out from the benefits of growth, which have overwhelmingly gone to those above them.

Rising inequality and a stagnant working class are not the inevitable results of capitalism, as Piketty claims. Instead, they reflect a society's stage of development. The transition from an agrarian to an industrial economy creates a demand for large numbers of workers, increasing their bargaining power. Moving to a service economy has the opposite effect, undermining the power of organized labor as automation replaces humans. This first reduces the bargaining power of industrial workers and then, with the transition to a society dominated by artificial intelligence, that of highly educated professionals.

The Machine Age

The problems of cultural change and inequality in rich democracies are being compounded by the rise of automation, which threatens to create an economy in which almost all the gains go to the very top. Because most goods in a knowledge economy, such as software, cost almost nothing to replicate and distribute, high-quality products can sell for the same price as lower-quality ones. As a result, there is no need to buy anything but the top product, which can take over the entire market, producing enormous rewards for those making the top product but nothing for anyone else.

It is often assumed that the most important part of the knowledge economy, the high-tech sector, will create large numbers of well-paid jobs. But that sector's share of all jobs in the United States has remained flat since statistics first became available about three decades ago. Canada, France, Germany, Sweden, and the United Kingdom show the same pattern in their high-tech sectors. Unlike the transitions from an agrarian economy to an industrial economy and then to a knowledge economy, the move toward artificial intelligence is not generating large numbers of secure, well-paid jobs.

That is because computers are fast reaching the point where they can replace even highly educated professionals. Artificial intelligence has already made huge strides toward replacing human labor in analyzing legal documents, diagnosing patients, and even writing computer programs. As a result, although U.S. politicians and voters often blame global trade and offshoring for their country's economic difficulties, between 2000 and 2010, over 85 percent of U.S. manufacturing jobs were eliminated by technological advances, whereas only 13 percent were lost to trade.

Although artificial intelligence is rapidly replacing large numbers of jobs, its effects are not immediately visible: the global economy is growing, and unemployment is low. But these reassuring statistics conceal the fact that in the United States, 94 percent of the job growth from 2005 to 2015 was among low-paid security guards, housekeepers, janitors, and others who report to subcontractors. Moreover, the top-line unemployment figure hides the large numbers of people who have been driven by dismal job prospects to drop out of the workforce altogether. The U.S. unemployment rate is 4.1 percent. But the percentage of adults either working or actively seeking a job is near its lowest level in more than 30 years. In 2017, for every unemployed American man between 25 and 55 years old, another three were neither working nor looking for work. Work rates for women rose steadily until 2000; since then, those have also declined.

Life as a labor-force dropout is not easy. Working-age men who are out of the labor force report low levels of emotional well-being, and a 2016 study by the National Bureau of Economic Research found that nearly half of all working-age male labor-force dropouts—roughly 3.5 million men—took pain medication on a daily basis. Not surprisingly, they tend to die early. From 1999 to 2013, death rates rose sharply for non-Hispanic white American men with high school degrees or less, the group most likely to have left the labor force recently. So-called deaths of despair—suicides, liver cirrhosis, and drug overdoses—accounted for most of the increase. From 1900 to 2012, U.S. life expectancy at birth rose from 47 to 79 years but then leveled off, and in both 2015 and 2016, life expectancy at birth for all Americans declined slightly.

Getting Democracy Right

Whether this latest democratic setback proves permanent will depend on whether societies address these problems, which will require government intervention. Unless new political coalitions emerge in developed countries that represent the 99 percent, their economies will continue to hollow out and most people's economic security will carry on declining. The political stability and economic health of high-income societies require greater emphasis on the redistributive policies that dominated much of the twentieth century. The social base of the New Deal coalition and its European counterparts is gone, but the reappearance of extreme wealth concentrated in the top one percent has created the potential for new coalitions.

In the United States, taking a punitive approach to the top one percent would be counterproductive, as it includes many of the country's most valuable people. But moving toward a more progressive income tax would be perfectly reasonable. In the 1950s and 1960s, the top 1 percent of Americans paid a much higher share of their income in taxes than they do today. That did not strangle growth, which was stronger then than now. Two of today's wealthiest Americans, Warren Buffett and Bill Gates, advocate higher taxes for the very rich. They also argue that the inheritance tax is a relatively painless way to raise badly needed funds for education, health care, research and development, and infrastructure. But powerful conservative interests are moving the United States in the opposite direction, sharply reducing taxes on the rich and cutting government spending.

From 1989 to 2014, as part of the World Values Survey, pollsters asked respondents around the world which statement better reflected their views: "Incomes should be made more equal" or "Income differences should be larger to provide incentives for individual effort." In the earliest surveys, majorities in 52 of the 65 countries polled at least twice supported greater incentives for individual effort. But over the next 25 years, the situation reversed itself. In the most recent available survey, majorities in 51 of the 65 countries, including the United States, favored making incomes more equal.

The rise of automation is making societies richer, but governments must intervene and reallocate some of the new resources to create meaningful jobs that require a human touch in health care, education, infrastructure, environmental protection, research and development, and the arts and humanities. Governments' top priority should be improving the quality of life for society as a whole, rather than maximizing corporate profits. Finding effective ways to achieve this will be one of the central challenges of the coming years.

Democracy has retreated before, only to recover. But today's retreat will be reversed only if rich countries address the growing inequality of recent decades and manage the transition to the automated economy. If citizens can build political coalitions to reverse the trend toward inequality and preserve the possibility of widespread, meaningful employment, there is every reason to expect that democracy will resume its onward march.

Critical Thinking

1. What are the differences between materialists and postmaterialists?
2. How does economic insecurity undermine democracy?
3. What has been the cause of growing income insecurity in the United States?

Internet References

Foundation for Defense of Democracies
www.defenddemocray.org/

National Endowment for Democracy
https://www.ned.org

World Values Survey Database
www.worldvaluessurvey.org

Article

Prepared by: Robert Weiner, *University of Massachusetts, Boston*

Democracy and Its Discontents

JOHN SHATTUCK

Learning Outcomes

After reading this article, you will be able to:

- Explain why liberal democracy has been replaced by illiberal democracy in Eastern Europe.
- Define what is meant by illiberal democracy.

What's happening to democracy in Eastern Europe? A new authoritarianism, "illiberal governance," has taken over in Hungary and Poland. It's been boosted by the Paris and Brussels attacks and the fear of terrorism. Hungary and Poland are not isolated cases. A trend away from democratic pluralism is also sweeping through Western Europe. Where will it lead?

In trying to answer this question, I'll follow the advice of Václav Havel, to "keep the company of those who seek the truth, but run from those who claim to have found it." I promise to make no such claim, but I will seek the truth about "illiberal governance"—the new threat to democracy that is prominent in the headlines these days. My tentative answer is that this modern form of "soft authoritarianism" may not prove sustainable.

Competing forces were unleashed by the fall of the Berlin Wall. Forces of integration broke down barriers, promoted democratic development, created economic interdependence, and facilitated the digital revolution. Forces of disintegration tore apart failed states, stimulated ethnic and religious violence, and spurred nationalist leaders to challenge transnational entities like the European Union. There are conflicting scenarios about how these forces will play out. An optimistic view envisions slow and steady progress toward the universal realization of democracy. A negative, almost dystopian, perspective sees the increasing clash of cultures and civilizations—a steady regress toward ongoing conflict among cultures, religions, and societies. These two visions have been caricatured as alternative realities of the post–Cold War world, but they provide a useful starting point for understanding what's happening today to democracy in Europe and the United States.

I

Discontent with democracy is widespread. In 2014, a European Commission poll revealed that 68 percent of Europeans distrusted their national governments, and 82 percent distrusted the political parties that had produced these governments.[1] In the United States, a Gallup poll in the same year found that 65 percent of Americans were dissatisfied with their system of government and how it works—a striking increase from only 23 percent in 2002.[2]

One reason for this discontent may be a growing sense that the world is spinning out of control, and that democracy is only making matters worse. A deeper reason may be that people today are confused about the meaning of democracy—demanding both greater participation in their own governance and greater efficiency in the way government operates. The very idea of democracy may be at war with itself; people look to democratic governments to solve their problems but are unwilling to recognize their own responsibility for keeping democracy healthy.

Digging deeper, the roots of discontent can be found in four democratic revolutions of the last 50 years. As my colleague Ivan Krastev has written, these four upheavals have simultaneously strengthened and weakened democracy in Europe and the United States.

The Cultural Revolution of the 1960s gave birth to a modern world of individual rights and freedoms. At the same time, the rights revolution reduced the sense of collective purpose essential to democratic governance. It transformed democratic society, but a counterrevolution pushed back, turning the struggle for human rights and civil liberties into an endlessly divisive political battleground.

The Market Revolution of the 1980s released the power of the market economy to produce economic growth. It also cut way back on the role of government in regulating the economy, destroying the Keynesian consensus about the social benefits of a mixed economy and a welfare state. It paved the way for the rise of new economic elites, globalization, and inequality, while breeding political resentment among the overwhelming majority left behind.

The Political Revolution of 1989 marked the end of Communism and the Cold War, the opening of borders, and the beginning of a transition to democracy and market freedom in Eastern Europe. But it also marked the collapse of longstanding social support systems in the East, and in the West an end to the informal social contract between economic elites and the people.

The most recent democratic upheaval, the Internet Revolution, opened the floodgates of information, creating unlimited opportunities for peer-to-peer communication and horizontal grassroots pressure for change. At the same time, it spawned vast echo chambers and ghettoes of communication, reducing discourse across political divides and increasing the polarization of democratic societies.

II

Democratic discontent is especially acute in Eastern Europe, where the roots of democratic governance are shallow. Eastern Europeans were ruled for centuries by successive empires of Ottoman, Russian, Hapsburg, fascist, and communist authoritarian regimes. A long-suppressed hunger for national identity and honor among the peoples of the region constantly fueled their anger against outside oppressors—the Hapsburgs, who executed the first elected Hungarian prime minister in 1849; the Russians, who dominated Poland throughout the 19th century; and the Turks, who defeated the Serbs in the Battle of Kosovo Polje at the end of the 14th century. The collective memory of this ancient defeat in Serbia was so powerful that Slobodan Milosevic was able to invoke it 600 years later, when he launched his notorious ethnic cleansing campaign against the Kosovar Muslims.

In the 20th century, communism destroyed civil society in Eastern Europe by limiting civic engagement to activities relating to or mandated by the state. It also destroyed the sense of personal responsibility to the community that is essential for the growth of democracy. In Prague in the 1990s "volunteering" still meant collaborating with the regime. In Budapest today common spaces in apartment buildings are still rarely cared for by the residents. Communism's alternative to civil society was state employment and social security, but of course these were dismantled after the fall of the Berlin Wall.

After 1989, hopes in Eastern Europe that democracy would bring immediate economic benefits went unfulfilled. Standards of living failed to keep pace with popular expectations, especially after the financial crisis hit the region in 2009. In this neuralgic environment, Eastern Europeans found themselves attracted to political leaders who claimed they could defend the people against outsiders, like the foreign banks that had called in their mortgages when the financial markets collapsed.

These festering resentments were the building blocks of a new nationalism. Two basic elements went into its construction.

First was the politics of national identity. The longing for national identity had been largely ignored by the proponents of post–Cold War European integration, but it was taken up with a vengeance by nationalist leaders who developed new narratives to appeal to a resentful and confused populace.

In Hungary, which had been on the losing side of both world wars as an ally of Germany, the new nationalist narrative depicted Hungarians as victims, stripped of ⅔ of their lands and separated from their compatriots by the Treaty of Trianon after the First World War, then occupied by Germany and allegedly forced to participate in the Holocaust at the end of the Second World War. A particularly dangerous charge in this twisted national narrative was that "Brussels is the new Moscow."[3] After decades of being dictated to by a distant Soviet regime, Hungarians were susceptible to this claim. Casting the European Union as a hostile foreign power served the interests of nationalist politicians like Viktor Orban whose popularity was bolstered whenever EU authorities questioned the quality of Hungarian democracy.

A second building block of nationalism is the politics of fear. Today, leaders are linking the threat of terrorism in their countries to the refugees fleeing the violence in the Middle East. In Hungary, Slovakia, and Poland, the governing parties have called Muslim refugees "a threat to Christian civilization." Not to be outdone, the Hungarian government has warned that refugees in Europe are all potential terrorists, and is now preparing to enact an antiterror law to give the government emergency powers to declare "a state of terror threat" and suspend the constitution for 60 days, subject to continuous extension.

III

Once an Eastern European nationalist state was fully constructed, its form of government was given a new name—illiberal democracy. The term was coined in July 2014 by Prime Minister Viktor Orban of Hungary for the Hungarian government. He asserted that Hungary and its Eastern European neighbors had rejected the liberal values of individual rights and were returning to the traditional collective values of their nation-states. To emphasize his point, he asserted that "the Hungarian nation is not a pile of individuals" like people in the West after the rights revolution of the 1960s.[4] Orban claimed that liberal democracy was a failure, pointing to political division and economic inequality in the United States, and dysfunction in the EU on issues of financial policy and migration. In his view, countries that are "capable of making us competitive" in the global economy "are not Western, not liberal democracies, maybe not even democracies," citing as models the governments of Russia, China, Turkey, and Singapore.[5]

What are the elements of an "illiberal democracy?" The entry point is an election, to establish its claim—however tenuous—to be a democracy. Beyond that, the critical feature is majoritarian rule, implemented by a parliamentary supermajority that guarantees total control by the ruling party. In Hungary, this supermajority has opened the door to constitutional changes abolishing checks and balances and other key distinguishing features of a pluralist democracy.

The central claims of the new illiberal system are its promises of efficiency, collective purpose, and national pride. The trade-offs to achieve these goals are the centralization of power and the curtailment of individual rights. A question mark hanging over the system is whether it is sustainable, especially when it is inside a larger transnational system like the European Union. In his 2014 speech, Viktor Orban challenged the EU, claiming, "I don't think our EU membership precludes building an illiberal new state based on a national foundation."[6]

The Hungarian government has rejected the values and structures of a liberal democratic order. These values and institutional structures are intended to maximize accountability and liberty within a framework of democratic governance—checks and balances; freedoms of expression and assembly; due process of law; independence of the judiciary and the media; the protection of minorities; a pluralist civil society; and the rule of law.

The European Union was built on these values. They are at the heart of the political culture that has promoted the integration of Europe, but the new illiberal regimes of Eastern Europe are alien to this culture, and their neo-authoritarian leaders are rejecting it. Forces of disintegration unleashed by the refugee crisis and the Eurocrisis, combined with Viktor Orban's challenge to European values, are threatening the very concept of European integration.

Last fall, this new model of illiberal democracy galvanized nationalists across Europe when Hungary constructed razor wire fences on its borders and stationed its army and police to keep out refugees. The result was a huge boost to the governing party's flagging popularity at home, and the Hungarian Prime Minister's emergence on the European stage as a challenger to German Chancellor Angela Merkel, whose response to the refugee crisis was based on the liberal values of the EU. The refugee crisis provided a golden opportunity for Viktor Orban to burnish his illiberal credentials without having to make the kinds of sober compromises that a liberal leader like Merkel has had to do to support both European values and European security. To paraphrase the Polish sociologist Zygmunt Bauman, illiberal leaders use chaos to create the opportunity for imposing order.

The new Polish government is now emulating the Hungarian model. It made the refugee issue a central feature of its election campaign last fall, promising that religious and ethnic nationalism would protect Poles from an invasion of Muslims into Poland's homogeneous Catholic society. The government took a page out of the Hungarian playbook by attacking the Polish Constitutional Court and the independence of the Polish judiciary.

A pitched battle is now shaping up in Europe between liberal and illiberal democracy. At stake are the values that safeguard Europe against a repeat of its catastrophic experience with 20th century fascism and communism. These values are challenged not only by the proponents of illiberal democracy but also from within liberal democracies in Europe and the United States. Disturbing signs are everywhere about the health of Western democracies—their steady decline in voter participation, their broad distrust of political leaders, their alienation from distant decision-makers, their susceptibility to the influence of money in politics, their inability to make decisions on urgent issues like the Eurocrisis, refugees and immigration, and their increasing polarization and gridlock. Out of this discontent, new nationalists and demagogues on both sides of the Atlantic like Marine Le Pen and Donald Trump are gaining popularity.

IV

This is why the winter of our discontent will not be ending soon. But if we step back and ask some questions, I think some surprising answers may indicate the state of democracy may not be as bleak in the long run as it may seem today.

Can the EU Survive the Challenge Posed to it from within by Illiberal Governance?

The EU is clearly vulnerable. Without major structural reforms, EU institutions make easy targets for nationalist movements. The Brussels bureaucracy is remote and voters have no real connection to it. Only the member states participate directly in EU governance, and so far, their leaders have shown little inclination to discipline a member state like Hungary or Poland that defies EU rules and principles—probably, because they may want to do so themselves one day, as many British leaders are doing now by promoting "Brexit," or British exit from the EU.

Paradoxically, Eastern European illiberal states may not be as big a threat to the EU as they appear because the benefits they receive are far greater than their costs of staying in. Two basic factors tie Hungary and its Eastern European neighbors to the EU—money and politics.

The money is plentiful, and flows freely in the form of structural funds with few strings attached. Over the next five years, Hungary is guaranteed to receive EUR 22 billion from the EU. Many of the country's major capital projects, public investment opportunities, and employment strategies are connected to this beneficent and benign funding source.

The second factor is politics. The EU provides an attractive political target for Eastern European politicians who benefit from biting the hand that feeds them with their rallying cry that "Brussels is the new Moscow." And despite their assault on the

EU's liberal values, Eastern European countries benefit substantially from the Schengen rules on freedom of movement within the EU that guarantee employment mobility for their citizens. Without the EU, Hungary and its neighbors would be cast adrift in a chaotic environment. They have no natural resources, and would become economic vassals of the two big illiberal states to the East—Russia and Turkey—whose economic and security situation is far more uncertain even than that of the EU. This is why Viktor Orban is trying to prevent the EU from detaching Eastern Europe from the Schengen zone, and also why he is seeking to maintain social benefits for Hungarian workers in the UK. These may be losing battles for him, especially if he continues to resist the EU quota rules on accepting refugees, but they show how much he and his neighbors need the EU.

Are the New Illiberal Democracies in Eastern Europe Sustainable?

If an illiberal government can be changed by democratic means, then the system may be sustainable. But if the centralization of power is so successful that the government can fend off any democratic challenge, then, paradoxically, an illiberal system may not be sustainable in the long run. There are four key weaknesses in the system.

First, the legacy of state control over the economy and its eventual collapse under communism show that it may be difficult for centralized illiberal regimes to deliver economically to their citizens without liberalizing their political institutions. This is particularly true for countries like Hungary and Poland that have been incorporated into a much larger interconnected market economy like the EU. Russia and China, the two main countries cited by Viktor Orban as models of illiberal governance, are both faltering economically because of the way they are governed politically.

Second, illiberal governance tends to lead to systemic corruption, which is a drag on economic growth and a source of instability, as the situation in Russia shows. Eastern European countries have unfavorable ratings compared to other EU member states on Transparency International's European Corruption Index.

Third, illiberal governance is vulnerable to the digital revolution, which allows increased peer-to-peer flows of information and creates horizontal pressures for change. Traditional media may have fallen under the control of illiberal regimes, but digital media have not. In Hungary, over 100,000 people took to the streets in 2014 when the government threatened to tax the use of the internet, and the government had to back down.

Fourth, as the Internet tax controversy shows, illiberal regimes have few institutional safety valves for citizen discontent. When popular pressures build, the regime must either back down or resort to coercion. The Euromaidan protests in Ukraine demonstrated that the use of violence by an illiberal regime can lead to greater public discontent and pressure for more radical change.

A far greater challenge to the EU than illiberal governance in Eastern Europe is coming from one of the world's oldest democracies in the West—the United Kingdom. Now that the EU has given Prime Minister David Cameron what he has been asking for, it would be devastating for both sides of the Channel if the Brexit referendum were to pass.

Is Liberal Democracy in Recession, or a State of Permanent Decline?

This question can be answered in different ways. If one looks at the increasing popular demands for participation in governance and engagement in decision-making—as demonstrated by democracy movements around the world from Euromaidan to Taksim Gezi Park, to Tahrir Square, to Hong Kong, to Black Lives Matter in the United States—the ideas of democracy have greater appeal today than ever, even as the supply of healthy democratic governance may be diminishing.

On the other hand, if one looks at the popular appeal of the politics of national identity and security, and the demand for stability and efficiency in governance, as the opinion polls in Europe and the United States seem to show, then liberal democracy with its aging pluralist institutions and short-term election perspectives may be in decline.

In the end, it will depend on democracy's capacity to reform itself—to use the tools of the digital revolution to stimulate participation while leveling the playing field and curtailing the economic power of the top 1 percent to exercise disproportionate influence over decision-making. It will also depend on liberal democracy giving more recognition to national identity and security, and creating new channels for national participation in supranational structures like the EU.

What about the United States—Will They Elect a Nationalist, Populist, Unilateralist, Illiberal President?

There are certainly threats to liberal values in the United States from the far right—on immigration, racial issues, and women's rights, to name a few. But there's also plenty of energy, especially on the left, for economic and political reforms to strengthen liberal democracy. On foreign policy, no one should mistake populist discontent for support for foreign intervention. Military deployment is deeply unpopular in the wake of the disastrous 2003 intervention in Iraq. If anything, I'm concerned that the United States is being swept up in a wave of neo-isolationism that may keep it from engaging as a leader in the world, and particularly from working with Europe and Russia to address the crises in Ukraine and Syria, and manage the global refugee crisis.

My prediction is that the United States will not elect a nationalist, populist, unilateralist, illiberal president, but that gridlock and polarization will continue to plague American politics unless one party wins both the presidency and the Congress, especially now that the Supreme Court is up for grabs. This is a sorry commentary on the state of democracy in America. Democratic politics are about compromise and negotiation between opposing viewpoints, not about zero-sum scorched-earth attacks on anyone who does not follow the orthodoxy of one political group. The Tea Party movement was the harbinger of contemporary anticompromise, antidemocracy politics in America, and Donald Trump is its apotheosis. Trump may not succeed in capturing the presidency, but what he represents is a more dangerous American version of the nationalist illiberal democracy movements in Europe.

V

The rise of illiberal governance in Eastern Europe is rooted in a long legacy of authoritarianism. Democratic solutions must come from within and will take time to develop. These regimes do not pose an existential threat to the European Union—in fact, the benefits the EU provides them may make them stronger EU supporters than liberal democracies in the West like the UK. Illiberal democracies stimulate and feed on popular fears and anxieties, but without an institutional safety valve for popular discontent, they may not be sustainable in the long run.

The popular demand for democratic participation is growing, but it needs new language and new structures beyond those of traditional liberal democracy.

Democracy always sparks discontent, but discontent can also spark change. While democracy offers a path for change, illiberal governance is a dead end: its proponents are determined to control all the levers of power, and block all the avenues for change. In the end, democracy, as Winston Churchill famously pointed out, is the worst form of government, apart from all the others.

To return to Václav Havel, his words sum up very well the challenge of democracy and its discontents: "I'm not an optimist because I don't believe all ends well. I'm not a pessimist because I don't believe all ends badly. Instead, I'm a realist who carries hope, and hope is the belief that democracy has meaning, and is worth the struggle."

Legal Topics

For related research and practice materials, see the following legal topics:

Communications Law, Related Legal Issues, Taxation, Immigration Law, Refugees Eligibility, International Trade Law, Trade Agreements General Overview.

Footnotes

1. European Commission, Directorate-General for Communication. *Standard Eurobarometer 82 "Public Opinion in the European Union, First Results.*" December 2014, http://ec.europa.eu/public_opinion/archives/eb/eb82/eb82_first_en.pdf (accessed April 11, 2016).
2. McCarthy, Justin. "In U.S., 65% Dissatisfied with How Gov't System Works." Gallup. January 22, 2014, http://www.gallup.com/poll/166985/dissatisfied-gov-system-works.aspx (accessed April 19, 2016).
3. "Rechtsruck in Ungarn," DW, January 1, 2013, http://www.dw.com/de/rechtsruck-in-ungarn/a-16561346 (accessed April 19, 2016).
4. "Viktor Orban's illiberal world,"*Financial Times*, July 30, 2014, http://blogs.ft.com/the-world/2014/07/viktor-orbans-illiberal-world/ (accessed April 19, 2016).
5. "Orban Wants to Build Illiberal State,"*EU Observer*, July 28, 2014, https://euob-server.com/political/125128 (accessed April 19, 2016).
6. "Orban Says He Seeks to End Liberal Democracy in Hungary,"*Bloomberg*, July 28, 2014, http://www.bloomberg.com/news/articles/2014-07-28/orban-says-he-seeks-to-end-liberal-democracy-in-hungary (accessed April 19, 2016).

Critical Thinking

1. Why is illiberal democracy threatening European integration?
2. Why has illiberal democracy developed in Poland and Hungary?
3. Why can illiberal democracy lead to either fascism or communism?

Internet References

Democracy and Rule of Law Program, Carnegie Endowment for International Peace
www.carnegieendowment.org

Freedom House
www.freedomhouse.org/

Organization for Security and Cooperation in Europe, Office for Democratic Institutions and Human Rights
www.osce.org/odihr

UN Electoral Assistance Unit
www.un.org/undp/en/elections

Article

Prepared by: Robert Weiner, *University of Massachusetts, Boston*

The Global Challenge of the Refugee Exodus

GALLYA LAHAV

The horrific terrorist attacks in Paris on November 13, 2015, and those on American soil on September 11, 2001, have much more in common than the involvement of radicalized foreigners and international networks. Both of these events came immediately on the heels—within three days, in fact—of major international agreements to facilitate human mobility between sending and receiving countries. Just as then-President Vicente Fox of Mexico secured a deal for his citizens in the North American labor market days before the attack on the World Trade Center, embattled ministers and heads of states from the European Union met with African leaders in Valetta, Malta, on November 11 to work out a practical redistribution plan for dealing with the mass exodus of refugees trying to reach Europe.

Both of these initiatives to forge international policy cooperation on migration and refugee movements were quickly dashed by seemingly knee–jerk reactions across Western liberal states to temporarily close borders, lock down civil society, suspend rights and privileges, and contravene their own treaties and laws. Labor, trade, and humanitarian considerations were quashed by national security and "public order" exigencies. Fueled by public outrage and political protests, the resurgence of nationalist and populist sentiment following the Paris attacks all but shelved urgent relocation plans for the massive influx of refugees from protracted wars in the Middle East.

Not only did these events cement the link between human mobility and security; they catapulted refugee and migration politics onto the foreign and security policy agenda, prompting a proliferation of intergovernmental and international meetings. The salience of migration in the security agenda was best summarized by German Chancellor Angela Merkel's proclamation that "immigration is the largest problem facing Europe in this decade." The "new security" paradigm has put a spotlight on emerging threats like human mobility, fundamentalism, environmental degradation, smuggling, and terrorism—global issues that cross boundaries.

The movement of impoverished masses making their way to safer shores from regions including the Middle East, sub-Saharan Africa, and Central America, has grown over the past five years, and rose to unprecedented levels in 2014, according to the United Nations High Commissioner for Refugees (UNHCR). The numbers arriving in Europe reached crisis levels in the sweltering political and summer heat of 2015. Personified by the piercing image of a young Syrian boy's corpse retrieved from the surf by a Turkish soldier, the human toll was inescapable.

Yet the humanitarian narrative was quickly overshadowed by the spectacle of Paris, a symbol of liberty, under assault by Islamist radicalism and terrorism inspired from abroad. This sequence of tragic events showed how swiftly politicization of such issues could upset a fragile balance and move the debate from a humanitarian to a security framework. The sudden shift in the discourse surrounding refugee movements underscored the fluid interests and competing trade-offs of refugee politics in the post–Cold War era.

The terrorist suspects who surfaced in Paris embodied the range of threats facing liberal democracies as they deal with refugees and migration. Among the suspected perpetrators, Western officials identified a disguised or bogus asylum seeker of Syrian origin; a Belgian-born and Western-educated middle-class Muslim of Tunisian-Moroccan origin; and a number of other radicalized EU nationals from ethnic minority backgrounds in France and Belgium. These profiles encapsulated the multiple internal and external dimensions of the threats that inform policies on border control, minority integration, and identity politics in the 21st century.

The refugee crises of this decade amount to an emerging global challenge facing almost all industrialized liberal democracies, pitting their humanitarian norms against materialist values of survival and well-being. These crises have acutely tested the delicate immigration and asylum policy consensus that largely prevailed across Western countries throughout the Cold War period. Until the political earthquake of 9/11, this equilibrium was founded on the premise that each dimension of immigration policy could be addressed in relative isolation, and that decisions concerning one dimension did not significantly circumscribe the options for others. Since 2001, subsequent terrorist attacks have suggested to some that open economic borders, humanitarian passage, and immigrant integration policies now conflict with the core responsibility of liberal states to safeguard the physical safety of their citizens.

The new security context of the post–Cold War world poses what I have elsewhere called a political "trilemma" when it comes to balancing markets, rights, and security interests in dealing with human mobility. Liberal democracies have struggled with contradictory goals of maintaining open markets for trade and allowing freedoms for ethnically diverse populations while protecting their borders from the security threats associated with global mobility. How can liberal states in an international system reconcile the need to open borders—for the sake of human mobility, demographic balance, sustainable development, global markets, tourism, and human rights norms—with political, societal, and security pressures to effectively protect their citizens and control their borders?

Cooperation among states on human mobility has been largely based on restrictive policies.

Europe's Challenge

This difficulty has been most evident at the EU regional level, where democratic member states are forced to balance national impulses favoring protectionism with communitarian demands for more cooperation. On what basis might states with different historical exigencies and approaches to migration find their interests merging? Collective action among 28 diverse member states has proved intractable (in instructive ways).

The ongoing European refugee emergency is emblematic of the challenges generated by forced migration. It constitutes an enormous crisis for the vision of European integration. Moreover, assuming that EU integration is representative of the larger globalizing goal of free movement for all four economic factors of production (goods, capital, services, people), its struggles are revealing of the challenges faced by all advanced liberal democracies.

To the extent that achieving the aim of a single market, enshrined in the 1957 Treaty of Rome, rests on the success of freedom of movement, a common immigration and asylum policy is essential to founder Jean Monnet's concept of a frontier-free Europe. The functional rationale for this goal was that the pursuit of economic and social well-being, within a framework of human rights and democratic norms, would create a rational incentive for states to cooperate and further integrate with each other.

We have seen an incremental development of European instruments on the supranational level, such as the Common European Asylum System, the Schengen open borders system, the Dublin system for handling asylum claims, and the cross-border enforcement agency Frontex. This trend has been reinforced by a notable shift since the 2009 Lisbon Treaty toward deferring to the EU on policy decisions. Despite these encouraging signs of cooperation, the current refugee crisis has reopened serious rifts between member states over national borders. As demonstrated by the disputes between French and British authorities over migrants converging on Calais to seek passage to Britain through the Channel Tunnel, and by the abrupt border closures by some eastern and southern European countries, member states have diverged dramatically in fulfilling their obligations.

The disparate perspectives of member states toward humanitarian movements are reflected in the uneven reception of refugees and burden-sharing proclivities among the EU countries. Attitudes have varied widely, from generous Sweden and Germany, Europe's main economic powerhouses in northern Europe, to the economically embattled countries in the south and east (Hungary, Croatia, Greece, Austria) that have been at the geographic forefront of the crisis. Some countries, like Slovakia and Poland, have taken to specifying the types of people in need that they are willing to help—namely, Christian refugees. Other EU nations such as Greece and Italy have been struggling to avoid saddling their citizens, already exasperated by economic hardship, with the potentially catastrophic burden of massive refugee influxes.

The challenge is faced not only by individual member states (especially transit countries in the south and east) with weak migration infrastructures and beaten-down economies. The core challenge confronts the entire European Union, founded on the principles of free mobility and solidarity. The lack of a comprehensive, common asylum policy (including refugee quotas, reception centers, and a common list of safe third countries) that also recognizes member states' capacities and the public mood is a danger to the entire EU enterprise. At the moment, Brussels risks reneging on previous steps it has taken regarding human mobility (including the Dublin Regulations and Schengen Agreements). It also risks losing the support of national publics and even some member states, including key ones such as Britain, which are threatening to exit the union altogether because of migration concerns.

Policy responses that include detention, deportation, or refoulement represent a slow erosion of liberal norms.

Imperiled Principles

The current refugee crisis casts doubts over three major principles inscribed in the conscience of Europeans: free movement of persons, human rights protections, and social harmonization or solidarity. First, a Schengen breakdown, symbolized by the temporary shutdown of national borders (currently allowed for up to 90 days), compromises the principle of a free human mobility zone within the single market. Second, the buck-passing by safe "first-arrival" countries that are sending refugees on to their neighbors represents a serious breach of the Dublin Convention, and thereby jeopardizes the principle of human rights protections. Finally, the nature of a "peoples' Europe," as set out in the Maastricht Treaty of 1992, stressing the universality of human rights throughout the Union, is in flux. Even if the first two challenges are surmountable, the looming question is: What kind of Europe does the EU want to be? What is its identity, and where do Muslims fit within the rapidly redrawn lines between insiders and outsiders?

Lurking behind much of modern European identity building has been the ghost of Christendom; and as in the United States, relations with Muslim minorities have become strained. Amid rising anxiety over security, concern about the cultural impact of migrants and refugees has extended beyond perceived threats to language and customs. Fears of radical, anti-Western political culture in Muslim communities are prevalent. These tensions present a serious crisis for democracies. The religious cleavage of secularism or pluralism versus fundamentalism complicates refugee politics and lends some unwelcome credence to the late political scientist Samuel Huntington's contentious and gloomy prophesy about rivalry between "civilizations" becoming the fault line of future conflicts in world politics.

European integration, like globalization, compounds the challenges that refugee and migration issues have long posed to the exercise of sovereignty by states seeking to control territory, identity, and citizenship. The reinvention of borders has compelled Europeans to rethink fundamental questions of identity—of "us" and "them." The growing tendency toward restrictive and protectionist migration policies across Europe stems less from demographic changes than from the reactions of policymakers and ordinary citizens to migration in the context of changing borders.

Indeed, the rush to control migration seems initially puzzling in a Europe built on the principle of free movement, dependent on global mobility, committed to maintaining a robust welfare system, and facing a serious demographic crisis of aging populations and falling birthrates. It also runs counter to rising public expectations in the EU, especially among the young generations. Eurobarometer polls of European youths between 15 and 24 years old from 2005 to 2015 found that "free movement" ranked higher in importance than any other motivations for regional integration, including the euro, social protection, and peace. The hardening of migration and refugee controls despite the liberalization of borders for other global economic reasons is one of the contradictions of incomplete integration.

Asylum Redefined

Refugees have a sacred and separate space in migration politics. And yet, as was noted by a pair of eminent scholars, the late Aristide Zolberg and Astri Suhrke, their forced movements may be defined in at least three ways: legally (as stipulated in national or international law), politically (as interpreted to meet political exigencies), and sociologically (as reflected in empirical reality). Legally speaking, the modern right to asylum has its roots in the aftermath of World War I and the Russian Revolution. Forced to flee by the Bolsheviks and famine, an unprecedented wave of 1.5 million Russians who had been stripped of their citizenship were resettled by the League of Nations.

The humanitarian system broke down under Nazi aggression and its aftermath, until the United Nations took up the task of rebuilding it at the international level with the establishment of the UNHCR in 1950. Since then, the main pillar of refugee protections has been firmly institutionalized in the Geneva Convention of 1951 and its 1967 protocol. The narrow definition of refugees (broadened only slightly in 1967 to extend beyond the original European refugees of World War II) has remained the standard and template for all other international and regional instruments dealing with forced migrations.

Despite the tenacity of these legal standards, political definitions have shifted dramatically from the Cold War ideological competition to the post–Cold War geopolitical preoccupation with religious and ethnic conflict. Nation-states interpret their legal and humanitarian obligations in the context of shifting political and foreign policy concerns. Cuban and Soviet refugees, for example, are no longer guaranteed asylum in the West.

The extent to which accepted asylum applications to Europe show overrepresentation from countries such as Eritrea, Afghanistan, and Iraq relative to other countries of origin, such as Serbia, Kosovo, Pakistan, and Albania, is striking. The numbers reflect neither legal nor sociological considerations but political affinities. So, too, changes in the types of refugees (which now include, for example, those facing female genital mutilation, environmental calamities, and gang violence) mean that the numbers of people being pushed out involuntarily have increased greatly, belying limited legal definitions and institutional capacities. Clearly, the contemporary refugee crisis stems from the growing incongruence between narrow and anachronistic legal definitions and evolving political and sociological realities.

While the relative size of these flows is not unprecedented, their compositional breakdown is revealing. In 1945, 20 million European refugees were resettled; today, there reportedly are 19.5 million refugees in the world. In contrast to those earlier, mainly European flows, most of today's asylum applicants are fleeing violence and conflict outside Europe. In 2014, the world's largest source of refugees was the Middle East and North Africa. According to UNHCR statistics, one in every five displaced people worldwide came from Syria. More tellingly, the vast majority of refugees in 2014 were from countries in the developing world, such as Ethiopia, Kenya, and Pakistan: nearly 9 out of 10 refugees lived in such countries, compared with 7 out of 10 a decade earlier.

This period has seen the breakdown of countries in the former Soviet Bloc such as Ukraine or Kyrgyzstan, the collapse of states such as Afghanistan, Iraq, Libya, and Yemen, and the reconfiguration of others, such as Sudan, Eritrea, and Somalia. The range of potential candidates for refugee status has been vastly expanded by ethnic and religious conflicts in the Middle East and the wider region (particularly in Syria, Iraq, Afghanistan, and Libya) and in South and Southeast Asia (in Pakistan, Myanmar, and Bangladesh, locus of the Rohingya refugee crisis); war, poverty, and repression in Africa (for instance in Eritrea, Somalia, Nigeria, Ivory Coast, Mali, Burundi, Central African Republic, and the Democratic Republic of Congo); and the fraying of national boundaries elsewhere (prompting migrations of Roma, Kurds, and other ethnic groups).

Absorbing the Flow

The overwhelming displacement of Syrians and Libyans since 2011 has pushed refugees next door, to Jordan, Lebanon, Iraq, Turkey, Egypt, and Tunisia. The absorption of refugees in those neighboring countries, while keeping them as far from Western borders as possible, has also led to further destabilization of already fragile states. However, as these countries have become oversaturated, they have closed their borders, forcing the West to deal with the inevitable diversion of the refugee flow to other regions.

Although in 2015 only 10 percent of Syrians moved to Europe out of the roughly 4 million who have left their homeland, the political challenge now faced by Europe—which has agreed to resettle and distribute 160,000 refugees, according to the last EU relocation plan—is rather minimal by comparative standards. However, while past refugees were settled "temporarily" in close proximity to their country of origin, the current reality is based on long-term projections of permanent absorption. As UNHCR statistics suggest, only 126,800 refugees were resettled in their home countries in 2014—the lowest number in 31 years. This means cost–benefit assessments extend well beyond migration admissions and quotas and must include a consideration of permanent settlement.

Whereas vulnerable people tried to get to the Balkans from Germany during World War II, or fled from Serbian ethnic cleansing in the 1990s to countries such as Hungary, the route today is reversed, as refugees from the Middle East and Africa try to get to Serbia on their way to Austria and Germany. Asylum-seekers headed for Europe often start in Greece, which they can reach via a short boat trip from Turkey. Then, they move on through Macedonia and Serbia and into Hungary, where thousands have been crossing the border every day, crawling over or under a razor-wire fence meant to keep them out. Most go from there to other countries in the EU, sometimes paying smugglers to drive them. The danger of drowning has led migrants to increasingly seek land routes to Europe, especially through the Western Balkans.

The unprecedented number of deaths among people trying to reach Europe, which exceeded 2,500 in 2015 alone, reveals that all routes, by land or sea, have been closing. The reintroduction of archaic border fences by countries such as Spain (in its North African territories of Ceuta and Melilla), Bulgaria (on its border with Turkey), or Hungary (on its border with Serbia) has been designed to keep unwanted flows out, and as far away as possible. As a result, refugees seek alternative, dangerous routes through the Arctic Circle via Russia to the Nordic countries, or through harsh deserts, the Gulf of Aden, or the Red Sea to other unlikely countries that are culturally distant, such as Israel, Ethiopia, and Iran; or to others like Jordan, Malaysia, and Pakistan, which are not signatories of the Geneva Convention. And of course, despite the lukewarm reception in some countries, many still come to Europe. If they are lucky, some make it even farther, to Australia, the United States, and Canada.

Gray Areas

As the empirical and political redefinition of refugees outpaces legal definitions, scholars and policymakers alike are forced to reconsider the old distinctions between voluntary (mostly economically driven) and involuntary (humanitarian) migration. It is increasingly apparent that the refugee crisis is also a migration crisis. The elusiveness of policy categories not only deflects institutional responsibility, it neglects the gray areas which include unaccompanied minors and victims of natural catastrophe, trafficking, female genital mutilation, and other forms of discrimination. An untold number of those people fall through the terminological cracks in definitions of protected status.

How long can legal definitions maintain the differences between those who flee persecution on the basis of race, nationality, religion, or belonging to certain political or social groups, and those who flee other life-threatening events such as food insecurity, gang wars (which have driven unaccompanied children from Central America), or economic displacement? The link between climate change and massive human mobility goes beyond boundaries, as do civil strife, sustainable development issues, and other "new security" threats. Population movements are driven by compounding factors. Among the initial sparks for the imploding ethnic and sectarian conflicts in Syria and in Sudan's Darfur region were severe droughts and other ecological shocks, which aggravated fierce economic competition for scarce resources.

The definition of refugees has expanded in scope and complexity, and so have the potential solutions. Yet legal formulations have not been keeping pace. According to some estimates, there are now approximately 60 million uprooted, forcibly displaced, or stateless persons around the globe (equal in population to some of the larger European countries), most of whom are precluded from seeking the protections of existing legal rights. The scale of this problem obliges states to address the changing notions of refugee status and to align them with empirical realities.

The task for the international community is to uphold and adjust legal standards to meet the times. This involves bridging the enormous gap between generalized threats such as gang warfare, climate displacement, and food insecurity (which are not covered by the Geneva Convention), on the one hand, and narrowly defined forms of persecution, on the other. It also requires attention to the failure to uphold legal principles ratified by the world community. Policy responses that include detention, deportation, or *refoulement* (the return of refugees to a country where they face persecution) represent a slow erosion of liberal norms set out by international and supranational instruments such as the UN Refugee Convention, the Schengen Agreement, the Dublin Convention, and the Convention Against Torture. They also prevent any meaningful policy fixes.

Long-Term Solutions

Most scholars and observers of the post–World War II period have concluded that liberal principles are embedded in the evolution of the contemporary Western world. The principle of free movement of all factors of production has dominated the prevailing discourse. Globalization and regional economic entities such as the European Union have ensured efficient flows across borders. Liberal markets presupposed Adam Smith's "invisible hand," assuming that the international system would neutralize inequalities and find equilibrium if the poor regions could send their impoverished risk-takers to faraway capital-rich ones. Liberal norms for international mobility were institutionalized in the Bretton Woods system, facilitating efficient flows of foreign or guest labor.

Globalization is both a boon and a hindrance to international migration.

In the same spirit of postwar thinking, human rights norms sought to ensure compassionate migration flows. While the Geneva regime instituted refugee protections of non-*refoulement* and nondiscrimination, international human rights instruments guaranteed basic protections to all individuals regardless of citizenship.

But the breakdown of the Cold War system has unleashed new dilemmas for the world, testing the liberal paradigm on which the current migration–asylum equilibrium has rested. The increasing inclination of national governments to view refugee and migration questions through the prism of national security has both compelled and repelled greater bilateral and multilateral cooperation. The security paradigm has disclosed a series of paradoxes and unintended consequences looming in the background of refugee politics.

Before celebrating international cooperation and further integration, we need to recall that unlike other areas of globalization (such as trade), cooperation among states on human mobility has been largely based on restrictive policies. Indeed, with specific exceptions, such as the US Bracero program for Mexican guest workers from the 1940s to the 1960s, cooperation on migration has predominantly existed in the form of prevention. This is also true of refugee policies. In the EU, these have been less about establishing a common European asylum system and more concerned with reducing migration pressures.

Long-term perspectives should factor in lessons learned; international cooperation may be more compatible with national interests than is often presumed. Contrary to conventional theories of globalization and regional integration, cooperation may bolster, not compromise, state sovereignty. International and transnational organizations can serve as an opportunity for increasing, rather than constraining, the regulatory power of nation-states. States may deal more effectively with migration challenges by joining international or supranational institutions like the EU.

The tendency to outsource refugees to other countries that are already crumbling in the Middle East or elsewhere is shortsighted. The presence of 4–6 million homeless and stateless people undermines the goal of helping to stabilize those compromised countries, to which rejected asylum seekers reluctantly return or where they are stranded on the edge of society. It is rather duplicitous to offer development aid and humanitarian assistance, as the European Neighborhood Policy has done, to strategic partners like Ukraine, Libya, Morocco, Tunisia, Egypt, and Turkey—countries hardly known for their civil rights records—to help them monitor migrants and asylum-seekers. Beyond the human toll, the prospective costs, in terms of the further regional destabilization that comes with growing numbers of stateless and displaced persons, are immeasurable.

Finally, the piecemeal attempts to tackle the refugee crisis have belied the externalities of migration policy. The growing interdependence of migration with other policy domains means that outcomes are contingent on developments in other areas, from foreign affairs to welfare policy. Long-term solutions require holistic and comprehensive approaches that include diplomatic and military engagement, social and cultural integration, labor and demographic considerations, development aid, and environmental protections. They also require extending burden sharing (beyond financial assistance) to include more affluent countries in the area such as Saudi Arabia, the United Arab Emirates, and Qatar, and outside the region, such as Japan, Singapore, Russia, and the United States.

An alternative to disengagement in unstable regions of the world is population movement outside of them. In today's world, power is no longer commensurate with military might. Non-state actors such as the Islamic State (ISIS), Al-Qaeda, and Boko Haram, as well as states with poor military infrastructure, can deploy what the political scientist Kelly Greenhill calls "weapons of mass migration." Global strategies therefore need to attend to the insidious psychological trauma among uprooted, suffering, and marginalized peoples. The antidote to jihadist ideology is to prevent extremism and alienation at home.

Short-term fixes to current crises undermine long-term solutions in an increasingly interdependent world. The double-edged sword of globalization is both a boon and a hindrance to international migration. The expanded regulatory apparatus for migration includes global high-tech surveillance, cross-border intelligence, and real-time databases and information systems that are equally available to sophisticated smuggling networks. Desperate refugees may also rely on smartphones and social networks, a striking feature of the current exodus.

The massive flows of Syrians that dominate today's headlines are fleeing ISIS and a brutally oppressive regime at the same time. Ultimately, amid lagging rates of minority integration by the multicultural societies in the West, they may be abandoned to the alienation and hopelessness that feed radicalization and help terrorist organizations recruit. A responsible and holistic approach to integration needs to address threats including growing populist parties, the radicalization of alienated youths, and domestic violence, along with rising economic disparities. When globalization's own weapons are turned against itself, they threaten to undermine its core liberal values.

Critical Thinking

1. Why are some members of the European Union preventing refugees from crossing their borders?
2. What steps should the European Union take to deal with the refugee crisis?

Internet References

Doctors Without Borders
www.doctorswithoutborders.org/

The International Organization for Migration
www.iom.int/

The International Rescue Committee
www.rescue.org

The United Nations High Commissioner for Refugees
www.unhcr.org/en-us

GALLYA LAHAV is an associate professor of political science at Stony Brook University, the State University of New York.

Article

Prepared by: Robert Weiner, *University of Massachusetts, Boston*

Just and Unjust Leaks: When to Spill Secrets

MICHAEL WALZER

Learning Outcomes

After reading this article, you will be able to:

- Learn the difference between leaking and whistle-blowing.
- Learn when it is appropriate to whistle-blow on government secrets.

All governments, all political parties, and all politicians keep secrets and tell lies. Some lie more than others, and those differences are important, but the practice is general. And some lies and secrets may be justified, whereas others may not. Citizens, therefore, need to know the difference between just and unjust secrets and between just and unjust deception before they can decide when it may be justifiable for someone to reveal the secrets or expose the lies—when leaking confidential information, releasing classified documents, or blowing the whistle on misconduct may be in the public interest or, better, in the interest of democratic government.

Revealing official secrets and lies involves a form of moral risk-taking: whistle-blowers may act out of a sense of duty or conscience, but the morality of their actions can be judged only by their fellow citizens, and only after the fact. This is often a difficult judgment to make—and has probably become more difficult in the Trump era.

Lies and Damned Lies

A quick word about language: "leaker" and "whistle-blower" are overlapping terms, but they aren't synonyms. A leaker, in this context, anonymously reveals information that might embarrass officials or open up the government's internal workings to unwanted public scrutiny. In Washington, good reporters cultivate sources inside every presidential administration and every Congress and hope for leaks. A whistleblower reveals what she believes to be immoral or illegal official conduct to her bureaucratic superiors or to the public. Certain sorts of whistle-blowing, relating chiefly to mismanagement and corruption, are protected by law; leakers are not protected, nor are whistle-blowers who reveal state secrets.

Before considering the sorts of official deception where the stakes are high and the whistle-blower's decisions and the public's judgment of them are especially difficult, it's important to look at the way secrets and lies affect everyday politics, where the dilemmas are simple—and, most of the time, not much is at stake. Consider the many politically engaged men and women who insist that they are not running for office even while they are secretly raising money and recruiting help for a campaign. They don't want assaults on their records to begin before they have developed the resources they will need to counterattack. Citizens expect deception of this sort and commonly see through it: the practice is tolerable even if it is not fully justifiable.

But what about a candidate who tries to conceal political positions she has held in the past or who lies about her policy commitments for the future? Someone inside the candidate's campaign who exposes such lies is disloyal, but the disclosure is certainly not unjust. The leaker is a good citizen even though she may not be a desirable colleague in a conventional political enterprise.

Now imagine a politician who is particularly ruthless: she wins the election and then uses the power of the government to destroy records of her previous actions, removing documents from archives and threatening people who know too much.

Anyone breaking the silence or leaking the documents would be a public hero—and a welcome colleague to the vast majority of citizens who are sure that they would never destroy records or threaten anyone. Self-aggrandizing deception and ruthless attempts to cover it up invite moral exposure.

But now consider a politician who shouts lies at election rallies and solicits money from unsavory characters in order to defeat a particularly awful opponent—a neo-Nazi, for example, who threatens to dismantle the institutions of democratic government. Here is a politician with dirty hands. She has gotten her hands dirty for a good cause—but the good cause doesn't wash them clean. She is a lying and possibly corrupt politician. Still, I wouldn't defend someone inside her campaign who exposed the lies or revealed the source of the campaign funds and claimed something like a Kantian categorical imperative. "I had to do it," the leaker might say. "No, you didn't," I would respond. Lying to one's fellow citizens and seeking funds that the candidate doesn't dare talk about are certainly practices that should not be generalized. If all candidates acted in that way (and far too many do), democracy itself would be at risk. But if democratic institutions were already at risk, most citizens would want to make an exception for a politician they were sure would defend those institutions—even if she did not adhere to democratic norms while seeking office.

The Secret Sharer

Government secrets and deceptions are equally common but often harder to judge than the secrets and deceptions of individual candidates or elected officials. A relatively easy case can help establish some of the contours. It was militarily necessary and therefore justified for the U.S. government to keep the date of the 1944 D-Day invasion secret from the Germans and, in order to ensure secrecy, to withhold the information from almost everyone else, too. Governments justifiably conceal such information from anyone who does not need to know it. Similarly, Washington's and London's efforts to deceive the Germans about the location of the invasion were also justified, as were all the lies that officials told as part of those efforts. Providing that information to the press would not have been a good thing to do; in fact, someone who revealed it would probably have been charged with treason.

But contemporary U.S. military operations often do invite whistle-blowing—as in cases in which the people being kept in the dark are not U.S. enemies who know a good deal about what's going on since their operatives or soldiers are already engaged with American ones. Rather, it's the American people who don't know. Think of drone attacks or special operations that the public has never been told about, in places that most Americans have never heard of; recent U.S. military activities in Niger offer a good example. Soldiers die, and officials struggle to explain the mission—and, with even greater difficulty, the reasons for concealing it in the first place. In the wake of such incidents, it's plausible to argue that the truth should have been revealed earlier on by someone with inside knowledge. The whistle-blower in this case would be a good citizen, one might argue because the use of force abroad should always be the subject of democratic debate. Still, such a disclosure might not be justified if the operation was defensible—necessary for national security, for example, or intended to help people in desperate trouble—and if blowing the whistle would shut down any prospect of success. A disclosure might also be unjustified if it put the lives of U.S. operatives or armed forces at risk. Government officials usually claim that both the operation and U.S. personnel have been endangered. The case at hand, they regularly insist, is just like D-Day.

But U.S. leaders often choose secrecy for a very different reason: they fear that an operation would not survive public scrutiny or a democratic decision-making process. Or an operation has been debated and democratically approved but has taken on a different character in the field. Mission creep is common and often results in an entirely new mission, different from the one that citizens debated and Congress voted on. The new mission may be strategically and morally justifiable, but the democratic process has been cut short or avoided altogether. If the operation is kept secret, however, Americans don't know that it hasn't been democratically authorized; they don't know that it is going on at all. And obviously, they can't weigh official justifications, since they have never heard a government official justify the operation.

By contrast, a potential whistle-blower knows that the operation is going on and that it hasn't been democratically authorized. But who is she to judge its strategic or moral value? In recent years, many government whistleblowers have been very young people—members, perhaps, of a generation of "digital natives," who believe that everything should be revealed. But government employees and contractors take oaths or sign agreements that commit them to obey secrecy rules; their superiors and fellow workers trust them to protect the confidentiality of their common enterprise, whatever it is.

If the enterprise is clearly illegal or monstrously immoral, a government employee or contractor should certainly break that promise, violate the trust of her coworkers, and blow the whistle. Officials or operatives engaged in illegal or immoral activities don't deserve her protection. This argument is similar to one often made in the case of humanitarian intervention: if a massacre is going on, anyone who can stop it should stop it, regardless of the costs imposed on the killers. If the U.S.

government is engaged in an illegal and immoral operation, anyone who can stop it should.

Consider a rough analogy. U.S. soldiers are required by international law and by the Uniform Code of Military Justice to refuse to obey illegal commands—and they should assume that monstrously immoral commands are always illegal. Discipline and obedience are more crucial to a military than they are to a civilian bureaucracy, and yet soldiers are commanded to disobey illegal orders even on the battlefield. Citizens might excuse a soldier who obeyed an illegal order under coercion or who evaded rather than defied the order—as did the U.S. soldiers at the 1968 My Lai massacre in Vietnam who shot into the air, deliberately missing the civilians they had been ordered to kill. There are civilian equivalents of this kind of evasion, such as slowing down the work required to prepare for an operation or doing the work so badly that the operation has to be postponed or canceled. Whistle-blowing, by contrast, is closer to deliberate disobedience on the battlefield.

There is a difference between the two contexts, however: a soldier often has to decide whether to obey in an instant; a whistle-blower has more time. Bureaucracies move slowly, so a whistle-blower, thinking about a clearly illegal or immoral operation, can appeal to her superiors to stop the operation. She can deliberate at length about the costs of what she is preparing to do. She can talk to coworkers whom she trusts (although there probably won't be any). Publicly blowing the whistle may mean losing her job and perhaps going to prison. Yet assuming she has exhausted the options for internal dissent, this is her obligation. And if she blows the whistle, her fellow citizens should recognize the value of what she has done, after the fact.

But what if the operation isn't clearly illegal or morally monstrous? What if there are arguments to support it, and the would-be whistleblower has heard them, even though her fellow citizens haven't? How can she claim the right to judge the official account of what's going on and the justifications of her coworkers and superiors, many of whom have more experience than she has? Such a situation is very different from the case of a soldier on the battlefield who can see pretty clearly the meaning of what she is being ordered to do—who might even look into the eyes of the innocent civilians she has been told to kill.

Whistle-blowing generally involves decision-making under conditions of uncertainty. Americans elect officials and ask them (and their appointees) to make decisions under those conditions. These officials may not be any more qualified than ordinary citizens, but they have been given and they have accepted a charge and the responsibilities that go with it—which include, crucially, the obligation to worry about the consequences of their decisions. Officials have at their disposal a multitude of researchers, analysts, and advisers who presumably reduce the uncertainty and help with the worrying. By contrast, a whistle-blower is usually alone; her uncertainties are private, and the public cannot know how much she worries. Indeed, one of the things the public should be concerned about is how well a whistle-blower understands the uncertainties. Is she a good worrier? It can be dangerous when whistle-blowers make their decisions on the basis of some ideological fixation or long-standing prejudice. That's a danger for officials, too—but they are being watched by coworkers (and, to an extent, by Congress and the media), whereas whistleblowers act in the shadows.

A Whistle in the Dark

Does it make a difference if whistleblowers are (or claim to be) conscientious? "Conscience" originally meant what the word suggests: "co-knowledge," shared, as the early Protestants said, between a man and "his God." But in the case of a whistleblower, the knowledge is uncertain and limited to the individual: good enough, perhaps, to justify someone's refusal to serve in the military, but not good enough to justify decisions that affect large numbers of other people. I am sure that many whistle-blowers have consciences, but they have to defend their actions in other terms.

If American citizens are good democrats, they will always be suspicious of government officials, and that will make them receptive to the information that whistleblowers provide. But they ought to be suspicious of whistleblowers, too. Citizens may not need to know the information that a whistleblower provides—indeed, the whistleblower might be acting for profit or publicity and not out of a desire for more democratic decision-making or a concern for law and morality. Sometimes, however, whistle-blowing opens a debate that should have started long before and exposes government activities that many citizens strongly oppose.

Imagine a military or intelligence operation that originally made a lot of sense and that the government has successfully defended to the public but that has expanded in ways that U.S. citizens didn't anticipate and haven't been told about. The operation now requires a degree of force far greater than officials had originally planned for, and its geographic range has expanded. The potential whistleblower knows what is going on, and she knows that there hasn't been anything resembling a democratic decision. Is that enough knowledge to justify revealing details about the operation to the media? Probably not: she has to make some judgment about the character of the expanded operation, and she has to consider the possible consequences of her revelations—and she is, remember, no better a judge than anyone else.

Arguably, the goal of empowering citizens by supplying them with crucial but secret information justifies whistle-blowing—as long as there are good reasons to believe that secrecy isn't a legitimate requirement of the mission and as long as the revelation results in no negative consequences for U.S. personnel in the field. Those two qualifications, however, will probably mean that whistle-blowing cannot be justified in many cases. But now imagine that the expanded operation involves terrible brutality or potential danger to civilians abroad or in the United States. And the whistleblower believes that ordinary Americans would recognize the brutality or the danger, and so she isn't merely acting on her own judgment: she is assuming that most of her fellow citizens would judge the situation in the same way—and giving them the chance to do so.

This is the best way to think about whistle-blowing: it involves a kind of moral risk-taking, and it can be justified only after the fact, if other citizens recognize its morality. Of course, its morality will always be contested, with government officials arguing that an important mission has been undercut and that agents in the field have been endangered. This might be true, or it might be a lie, which would justify further whistle-blowing. The whistleblower herself is counting on her fellow citizens to defend her judgment—to affirm it, in fact, and say, "Yes, this is an operation that we should have been told about, and it is one that we would have rejected." If most of her fellow citizens agree—or, rather, most of those who are paying attention, since majority rule would not work here—then exposing the operation was likely justified.

The case is the same if U.S. citizens are both the objects of the operation and the ones from whom it is being concealed. The best-known contemporary American whistleblower, the former National Security Agency contractor Edward Snowden, revealed the large-scale surveillance of Americans by their own government. He bet that most of his fellow citizens would not think that the danger they faced was great enough to warrant such a massive invasion of their privacy. With some difficulty, I can imagine circumstances in which large-scale secret surveillance by an otherwise democratic state might be justifiable or at least defensible. But what Snowden revealed was an operation that could not be justified by any actually existing danger; this was something that American citizens needed to know about. Unfortunately, however, Snowden revealed much more than what Americans needed to know—and not only to his fellow citizens: in addition to sharing secrets about the surveillance of American citizens with journalists from *The Washington Post* and *The Guardian*, he provided the *South China Morning Post* with information about U.S. intelligence operations against non-American targets in mainland China. That disclosure put Americans at risk, and Snowden had no reason to believe that what the United States was doing in China was either illegal or immoral—or anything other than routine.

Judgments in cases like this one will obviously be shaped by political views but not, one hopes, by partisan loyalties. Many liberals and Democrats, along with some conservatives and a few Republicans, condemned the domestic surveillance that Snowden revealed and defended his decision to do so. The first year of the Trump administration, however, has seen many leaks that have derived from and invited partisanship. Consider the leaked details of the president's May 2017 conversation with Russian officials in the Oval Office, after he had fired FBI Director James Comey, who had been investigating whether Donald Trump's election campaign had coordinated with the Kremlin. "I just fired the head of the FBI. He was crazy, a real nut job," Trump said, according to a source quoted by *The New York Times*. "I faced great pressure because of Russia. That's taken off." *The Washington Post* reported that during the same meeting, Trump shared highly classified information with the Russians that "jeopardized a critical source of intelligence on the Islamic State." The leakers to the *Times* and the *Post* certainly meant to raise questions about the president's competence on foreign policy. Americans who already doubted Trump's abilities welcomed the leak. The president's supporters obviously did not.

There is no way to make an objective judgment here—not, at least, about the leakers. But the journalists who reported this and many other leaks, and who worked hard to make sure of their accuracy, were doing their job and ought to be commended. They did not confront a moral dilemma. Leaks of this sort are grist for the mill of a free press.

Bureaucratic Outlaws

As for whistle-blowing, as opposed to leaking, a truly detached and fully informed observer would probably be able to make an objective judgment about any particular revelation. But that sort of judgment isn't likely in the fraught world of politics and government—although a consensus might take shape, slowly, over time, as in the case of the Pentagon Papers: it seems likely that most Americans have come to believe that the military analyst Daniel Ellsberg did the right thing in sharing the documents with the press. Whistle-blowers such as Ellsberg appeal to their fellow citizens, and there really isn't any further appeal to make. If the citizens don't agree among themselves about the justifiability of the disclosure, there can be no definitive verdict.

But suppose that most Americans recognize the brutality or the danger that has driven the whistle-blower to act. Her action was justified, but she has violated the commitments she made when she took her job, and she may have broken the law. When soldiers disobey an illegal order, they are in fact obeying the official army code. But there is no official code that orders civil servants to refuse to keep secrets about an illegal or immoral operation. Soldiers are obligated to disobey; civil servants are not obligated to blow the whistle. They are, however, protected from official retaliation and punishment by the Whistleblower Protection Act of 1989 if they reveal a range of illegal government actions: gross mismanagement, the waste of public funds, or policies that pose a substantial and specific danger to public health and safety.

If whistleblowers are fired or demoted for revelations such as those, they can file an appeal to the U.S. Merit Systems Protection Board. These appeals are most often denied—but not always. In 2003, Robert MacLean, an employee of the Transportation Security Administration, told an MSNBC reporter that in an effort to reduce spending on hotels, the TSA would be removing air marshals from many long-distance flights. He was subsequently fired. After appealing the decision—first to the MSPB, then to a federal appeals court—he was finally reinstated in 2013. The Supreme Court upheld that decision in 2015. It was a rare judicial victory for whistle-blowing.

But blowing the whistle on government action abroad or on security-related surveillance at home isn't protected by the Whistleblower Protection Act. And revealing classified information is not legal even if public health and safety are at issue. If a whistleblower reveals secrets that the government doesn't believe should be revealed, she has broken the law, regardless of her intentions or public sentiment about her actions. She is a disobedient civil servant, a bureaucratic outlaw.

Citizens might well consider her action a form of civil disobedience. But an act must meet certain conditions for that term to apply. First, the whistle-blower must have tried to convince a superior that the government's operation was illegal or immoral. Before going outside the government, she must have done the best she could inside, among her coworkers. Second, she must act in person and in public, without any attempt to hide who she is—even though this means that she won't see any more secrets. Many leaks can come from a single concealed leaker, but whistle-blowing is almost certainly a one-time act. If internal dissent doesn't work, then going public is a kind of principled resignation. Third, the whistle-blower must take responsibility for the revelation she has made; she must not hand secret documents to agents about whose subsequent behavior she can't be reasonably confident. She has a purpose for blowing the whistle, and she has to do her best to make sure that her purpose, and no other, is served. Snowden initially chose *The Guardian*, *The Washington Post*, and *The New York Times* (among other media outlets) as venues for his leaked secrets, and this seems the right kind of choice since these are newspapers whose publishers have had, along with a desire to sell papers, a long-standing commitment to democratic government. But Snowden showed less careful judgment in choosing to share information with the *South China Morning Post*, an organization that he had no reason to believe was committed to democratic decision-making in the United States.

A similarly flawed judgment also affected the case of another well-known American whistle-blower, Chelsea Manning, who in 2010 provided a massive trove of classified diplomatic cables to WikiLeaks. In contrast to newspapers with long records of public service, WikiLeaks is the wrong kind of intermediary between a whistleblower and the American people. Its directors may or may not have democratic commitments, but they also have narrowly partisan and personal aims, about which the public has learned a great deal in recent years.

Tough Calls

A civil whistleblower is making the same appeal to her fellow citizens that civil rights activists in the 1960s made—in similar defiance of the law and with a similar willingness to accept legal punishment. Whistle-blowers can and probably should be punished for revealing state secrets, even if the secrecy is unjust. Judges and juries should try to make the whistle-blower's punishment fit her crime, and her crime must be weighed against the government's subversion of the democratic process and the illegality and immorality of the revealed operation: the more significant the subversion and the greater the brutality or danger, the milder the sentence should be.

There must be some punishment for people who break secrecy laws, to serve justice when someone blows the whistle recklessly and to deter others from doing so. The fear of punishment focuses the mind and forces a potential whistle-blower to think hard about what she is doing. Citizens should respect a whistle-blower's willingness to pay the price of her disobedience, and at the same time, they should make their own judgments about whether what she did was right or wrong. Her action may require a complicated verdict: for example, perhaps she was right to open the democratic debate but wrong in her assumption of what the outcome of the debate should be. In any case, the public owes her a reflective response—not knee-jerk hostility or knee-jerk support.

Democracies live uneasily with secrecy, and governments keep too many secrets. Greater transparency in government decision-making would certainly be a good thing, but it has to be fought for democratically, through the conventional politics of parties and movements. Whistle-blowing probably does not lead to greater transparency; in the long run, it may only ensure that governments bury their secrets more deeply and watch their employees more closely. Still, so long as there are secrets, whistle-blowing will remain a necessary activity. Whistleblowers have a role to play in a democratic political universe.

But it is an unofficial role, and one must recognize both its possible value and its possible dangers.

Critical Thinking

1. What is the relationship between whistle-blowing and democracy?
2. What is the difference between refusing illegal orders and whistle-blowing and leaking government secrets?
3. Under what circumstances is whistle-blowing suspicious?

Internet References

Edward Snowden
https://en.wikipedia.org/wiki/edward_snowden

National Endowment for Democracy
https://www.ned.org

WikiLeaks
Wikileaks.org

Michael Walzer is a professor emeritus at the Institute for Advanced Study.